Recent Trends in Chaotic, Nonlinear and Complex Dynamics

WORLD SCIENTIFIC SERIES ON NONLINEAR SCIENCE

Editor: Leon O. Chua
University of California, Berkeley

Series B. **SPECIAL THEME ISSUES AND PROCEEDINGS**

Volume 1: Chua's Circuit: A Paradigm for Chaos
Edited by *R. N. Madan*

Volume 2: Complexity and Chaos
Edited by *N. B. Abraham, A. M. Albano, A. Passamante, P. E. Rapp and R. Gilmore*

Volume 3: New Trends in Pattern Formation in Active Nonlinear Media
Edited by *V. Perez-Villar, V. Perez-Munuzuri, C. Perez Garcia and V. I. Krinsky*

Volume 4: Chaos and Nonlinear Mechanics
Edited by *T. Kapitaniak and J. Brindley*

Volume 5: Fluid Physics — Lecture Notes of Summer Schools
Edited by *M. G. Velarde and C. I. Christov*

Volume 6: Dynamics of Nonlinear and Disordered Systems
Edited by *G. Martínez-Mekler and T. H. Seligman*

Volume 7: Chaos in Mesoscopic Systems
Edited by *H. A. Cerdeira and G. Casati*

Volume 8: Thirty Years After Sharkovskiĭ's Theorem: New Perspectives
Edited by *L. Alsedà, F. Balibrea, J. Llibre and M. Misiurewicz*

Volume 9: Discretely-Coupled Dynamical Systems
Edited by *V. Pérez-Muñuzuri, V. Pérez-Villar, L. O. Chua and M. Markus*

Volume 10: Nonlinear Dynamics & Chaos
Edited by *S. Kim, R. P. Behringer, H.-T. Moon and Y. Kuramoto*

Volume 11: Chaos in Circuits and Systems
Edited by *G. Chen and T. Ueta*

Volume 12: Dynamics and Bifurcation of Patterns in Dissipative Systems
Edited by *G. Dangelmayr and I. Oprea*

Volume 13: Modeling and Computations in Dynamical Systems: In Commemoration of the 100th Anniversary of the Birth of John von Neumann
Edited by *E. J. Doedel, G. Domokos and I. G. Kevrekidis*

Volume 14: Dynamics and Control of Hybrid Mechanical Systems
Edited by *G. Leonov, H. Nijmeijer, A. Pogromsky and A. Fradkov*

Volume 15: From Physics to Control Through an Emergent View
Edited by *L. Fortuna, A. Fradkov and M. Frasca*

Volume 16: Frontiers in the Study of Chaotic Dynamical Systems with Open Problems
Edited by *Z. Elhadj and J. C. Sprott*

Volume 17: Advances on Nonlinear Dynamics of Electronic Systems
Edited by *A. Buscarino, L. Fortuna and R. Stoop*

Volume 18: Generalized Synchronization and Generalized Consensus of System Arrays
Edited by *L. Min and G. Chen*

Volume 19: Recent Trends in Chaotic, Nonlinear and Complex Dynamics
Edited by *J. Awrejcewicz, R. Shanmuganathan and M. Ragulskis*

WORLD SCIENTIFIC SERIES ON NONLINEAR SCIENCE Series B Vol. 19
Series Editor: Leon O. Chua

Recent Trends in Chaotic, Nonlinear and Complex Dynamics

edited by

Jan Awrejcewicz
Lodz University of Technology, Poland
Rajasekar Shanmuganathan
Bharathidasan University, India
Minvydas Ragulskis
Kaunas University of Technology, Lithuania

World Scientific
NEW JERSEY • LONDON • SINGAPORE • BEIJING • SHANGHAI • HONG KONG • TAIPEI • CHENNAI • TOKYO

Published by

World Scientific Publishing Co. Pte. Ltd.
5 Toh Tuck Link, Singapore 596224
USA office: 27 Warren Street, Suite 401-402, Hackensack, NJ 07601
UK office: 57 Shelton Street, Covent Garden, London WC2H 9HE

British Library Cataloguing-in-Publication Data
A catalogue record for this book is available from the British Library.

World Scientific Series on Nonlinear Science Series B — Vol. 19
RECENT TRENDS IN CHAOTIC, NONLINEAR AND COMPLEX DYNAMICS

ISBN 978-981-122-189-7 (hardcover)
ISBN 978-981-122-190-3 (ebook for institutions)
ISBN 978-981-122-190-3 (ebook for individuals)

For any available supplementary material, please visit
https://www.worldscientific.com/worldscibooks/10.1142/11872#t=suppl

Typeset by Stallion Press
Email: enquiries@stallionpress.com

https://doi.org/10.1142/9789811221903_fmatter

Preface

The research monograph *Recent Trends in Chaotic, Nonlinear and Complex Dynamics* is dedicated to the 60th anniversary of Professor Miguel A. F. Sanjuán. All chapters are authored by prominent researchers and most of them did give their talks at the special session on the occasion of the celebration of Miguel A. F. Sanjuán's 60th anniversary organized during the *15th International Conference "Dynamical Systems — Theory and Applications"* (*DSTA 2019*). The session gathered several dozen scientists to honor Prof. Miguel A. F. Sanjuán by presenting research related to the fields of his scientific interests.

DSTA has an over 25-year-long history. It takes place every two years in Lodz, the third biggest city of Poland and is organized by the Department of Automation, Biomechanics and Mechatronics of the Lodz University of Technology under the patronage of Mechanics Committee of the Polish Academy of Sciences. Its aim is to provide a common platform for exchange of new ideas and results of recent research in the field of scientific and technological advances in modern nonlinear dynamical systems. The latest edition, *DSTA 2019*, that took place in Lodz, Poland from the 2nd to 5th of December of 2019 hosted over 210 participants representing 40 countries from all over the world. It resulted in the program of the conference covering both theoretical and experimental approaches to different topics in nonlinear dynamics, including bifurcations and chaos, control in dynamical systems, asymptotic methods in nonlinear dynamics, stability of dynamical systems, lumped mass and continuous systems vibrations, original numerical methods of vibration analysis, nonsmooth systems, dynamics in life sciences and bioengineering, engineering systems and differential equations.

The significance and impact of Sanjuán's scientific work is evidenced by the large number of awards and honors he has received during his career. To commemorate Sanjuán's 60th birthday, this book is a collection of chapters celebrating Prof. Miguel A. F. Sanjuán's achievements in nonlinear dynamics, chaos theory, and complex systems, written by some of his international colleagues, collaborators, and long-term friends.

We would like to express our sincere appreciation to the World Scientific Publishing Company for their assistance in producing this book.

Lodz, Tiruchirappalli, Kaunas
Winter 2021

Jan Awrejcewicz
Shanmuganathan Rajasekar
Minvydas Ragulskis

https://doi.org/10.1142/9789811221903_fmatter

Contents

https://doi.org/10.1142/9789811221903_0001

Chapter 1

New Trends and Recent Advances — An Introduction

Jan Awrejcewicz[*,§], Shanmuganathan Rajasekar[†,¶] and Minvydas Ragulskis[‡,||]

[*] *Department of Automation, Biomechanics and Mechatronics, The Lodz University of Technology, Stefanowskiego 1/15, Lodz 90-924, Poland*
[†] *School of Physics, Bharathidasan University, Tiruchirappalli 620024, India*
[‡] *Department of Mathematical Modelling, Kaunas University of Technology, Studentu 50-146, Kaunas 51368, Lithuania*
[§] *jan.awrejcewicz@p.lodz.pl*
[¶] *srj.bdu@gmail.com*
[||] *minvydas.ragulskis@ktu.lt*

Enormous progress has been made in the area of nonlinear dynamics, particularly in chaos and complex phenomena during the recent years. This introductory chapter presents an overview of the contributions to this book across a wide range of topics covering chaotic, nonlinear and complex dynamics. This unique book presents advances made in theory, analysis, numerical simulation, experimental realization and promising novel practical applications on various topics of current level interest on chaos and related fields in nonlinear dynamics.

1.1. Introduction

The research monograph *Recent Trends in Chaotic, Nonlinear and Complex Dynamics* is in honor of Professor Miguel A. F. Sanjuán on the occasion of his 60th anniversary. It contains a collection of original research papers presented in the special session devoted to the celebration of the 60th birthday of Miguel A. F. Sanjuán in the *15th International Conference on Dynamical Systems — Theory and Applications* (*DSTA 2019*).

The present book constitutes an invaluable reference source of recent trends on chaotic and complex dynamics for researchers and newcomers to the field of nonlinear dynamics. Particularly, it covers such areas as the complex behavior of the dipole-segment with equal masses and arbitrary rotation, the dynamics of the retina system, stability of anti-bunched buses and local unidirectional Kuramoto oscillators, the role of the noise in the chaotic intermittency, control of quasi-passive dynamic walker based on entrainment effect, numerical errors and exact orbits of quadratic maps, feedback delay control of the complex Ginzburg–Landau equation with local and nonlocal delayed perturbation, the features of the physics-inspired particle swarm optimization algorithm, stochastic sensitivity analysis of noise-induced phenomena in discrete systems, electromechanical models of micro- and nanoresonators, enhanced vibrational resonance induced by an amplitude modulated force, generation of self-excited and hidden attractors in multistable systems, modal asynchronicity in pre-stressed continuous beams, collective sustained oscillations in complex systems, weak decay of autocorrelations in periodic barrier billiards, the interaction of vortices in stratified incompressible flows and fractal basin of attraction in planar switched systems.

Apart from the present chapter the book contains 18 chapters. These chapters broadly cover the following three aspects:

Part I. Fundamental Results in Nonlinear Dynamics

Chapter 2. Enhanced vibrational resonance by an amplitude-modulated force by Chinnathambi *et al.* [1].

Chapter 3. Generation of self-excited and hidden multiscroll attractors in multistable system by Escalante-González and Campos [2].

Chapter 4. Dynamics of the dipole-segment with equal masses and arbitrary rotation by Elipe *et al.* [3].

Chapter 5. The interaction of two vortices near a boundary in rotating stratified incompressible flows by Carton *et al.* [4].

Chapter 6. Modal asynchronicity in pre-stressed continuous simply-supported beams with transversal and rotational oscillators by Ribeiro *et al.* [5].

Chapter 7. On the weak decay of autocorrelations in periodic barrier billiards by Courbage [6].

Part II. Stochastic Dynamics, Fractal Structure Analysis and Numerical Errors

Chapter 8. Stochastic sensitivity analysis of noise-induced phenomena in discrete systems by Bashkirtseva and Ryashko [7].

Chapter 9. The role of noise in chaotic intermittency by del Rio and Elaskar [8].

Chapter 10. Fractal structures in a binary Schwarzschild black hole system by de Souza Filho *et al.* [9].

Chapter 11. Characterizing fractal basin of attraction in planar switched systems by Zhang [10].

Chapter 12. Numerical errors and exact orbits of quadratic maps by Gallas [11].

Part III. Recent Trends in Applications of Chaotic and Complex Systems

Chapter 13. Electromechanical models of micro- and nanoresonators by Lukin *et al.* [12].

Chapter 14. The retina as a dynamical system by Cessac [13].

Chapter 15. Collective sustained oscillations in complex systems by Zheng [14].

Chapter 16. Stability of anti-bunched buses and local unidirectional Kuramoto oscillators by Yue Chew *et al.* [15].

Chapter 17. Feedback delay as a control tool: the complex Ginzburg–Landau equation with local and nonlocal delayed perturbations by Casal and Díaz [16].

Chapter 18. Nonlinear analysis and control for quasi-passive dynamic walker based on entrainment effect by Li *et al.* [17].

Chapter 19. Physics-inspired swarm optimization: the general algorithmic search by Hernández *et al.* [18].

1.2. Fundamental Results in Nonlinear Dynamics

Resonance refers to a realization of a maximum response of a system induced by means of one or more external forces. Nonlinear systems are capable of exhibiting a variety of resonances depending on the nature of the external forces. Vibrational resonance (VR) induced by a biharmonic force is a fundamental phenomenon with a variety of practical applications in physics, engineering, and biology. Part I begins with the chapter on the VR by an amplitude modulated force by Chinnathambi, Rajasekar and Sanjuan [1]. First, the occurrence of VR is analyzed in the paradigmatic Duffing oscillator with a single-well potential and the external force consisting of one low-frequency and one high-frequency component. Analytical expressions and theoretical predictions for the response amplitude are verified through numerical simulation. Next, the authors show that a higher number of resonance peaks, hysteresis and a jump phenomenon are due to the presence of the second high-frequency component in the external forcing. Finally, the system with a double-well potential is investigated. The mechanism of the enhanced VR is explained by using the phase portrait and the width of the orbit.

Escalante-González and Campos present recent approaches to generate self-exited and hidden multiscroll attractors in multistable systems in Chapter 3 [2]. A vector field is defined by affine linear systems such that each equilibrium is a saddle point. The space is partitioned in hyperbolic set and the generated self-excited multiscroll attractors are based on heteroclinic orbits. The authors report an interesting phenomenon when the equilibria are separated by pairs and the coexistence of different self-excited double-scroll attractors arise. A class of systems with 1D, 2D and 3D scrolls attractors are generated around multiple self-excited attractors. This class of systems presents the coexistence of self-excited double-scroll chaotic attractors and a hidden multiscroll attractor.

Many bodies in the solar system have irregular shape. Planned missions to orbit around and land on such bodies require to model the gravity fields of such bodies. In Chapter 4, Elipe, Abad and Ferreira investigate the dynamics of the dipole-segment system (DSS) (two spherical masses placed at the endpoints of a finite massive straight element) [3]. The gravity function of the DSS is expressed in closed form, without a series expansion. The authors compute the equilibria, their stability, and find families of periodic orbits. It appears that the dynamics of the DSS is very rich, and it opens the way to new studies from the dynamical systems point of view.

Chapter 5 by Carton, Tchuenkam and Vic analyze the interaction of two vortices near a boundary in rotating stratified incompressible flows [4]. Point vortex dynamics is applied to the study of oceanic vortices. The evolution of the North Brazil current rings, idealized as point vortices, in a stratified rotating fluid, along a straight or angular wall, representing the northern coast of South America is analyzed. It is demonstrated that the point vortex model with initial two-vortex configuration, though highly idealized, can exhibit time evolution qualitatively comparable with reality. Point vortex models and mathematical models of vortex dynamics are thus an efficient tool for the exploration of the behavior of oceanic vortices.

Analysis of modal asynchronicity in pre-stressed continuous simply-supported beams with transversal and rotational oscillators is presented in Chapter 6 by Ribeiro, Lenci and Mazzilli [5]. They show that even very simple models, with few degrees of freedom can exhibit the effect of modal localization. However, the situation becomes much more complicated in continuous systems. This chapter demonstrates that modal localization occurs in a pre-stressed continuous simply-supported beam after transversal and rotational oscillators are added, and then parameter fine-tuning is performed. These results pave the way towards the construction of novel vibration controllers and efficient energy harvesters.

Chapter 7 by Courbage is concerned with the periodic barrier billiards problem [6]. The author shows that the process of reversing the direction of velocity in barrier billiards can have a singularly continuous spectrum. These long memory effects can coexist with other strange aspects, as an exponential decay of correlations in zero entropy maps. Further, the analysis is focused on the investigation of such dynamical systems in the gap between chaos and order.

1.3. Stochastic Dynamics, Fractal Structure Analysis and Numerical Errors

Investigation of features of noise-induced phenomenon are of great applications in various contexts. The first chapter of Part II, Chapter 8 by Bashkirtseva and Ryashko reports noise-induced phenomena in map-based stochastic systems [7]. It is demonstrated that the interrelation of nonlinearity and stochasticity often leads to new phenomena that are not observable in the original deterministic models. The authors discuss the novel approach for the study of noise-induced phenomena based on the stochastic sensitivity method and the confidence domain method. The stochastic sensitivity analysis of equilibria, cycles, closed invariant curves, and chaotic attractors is presented in this chapter. The suggested analytical approach is illustrated for diverse stochastic phenomena in discrete systems with quadratic map and coupling.

Further, on the effect of noise, in Chapter 9 the role of the noise on the intermittency is presented by del Rio and Elsakar [8]. Intermittency can be characterized by spontaneous transitions between laminar and chaotic dynamics. This chapter presents a thorough description of the effects of the additive noise to systems that have intermittency. The theory of intermittency is extended in order to reproduce statistic behavior far from classical predictions. The authors show that the effects of the noise on the chaotic region also affects the statistical dynamics in the laminar region. A good agreement is demonstrated between the proposed theory and the numerical simulations. Finally, the new intermittency effects found in continuous systems are discussed.

The dynamics of open Hamiltonian systems with chaotic scattering can be characterized by fractal structures. For example, a light ray approaching a pair of super massive black holes can diverge to infinity, fall down into one of the black holes or can orbit around the black hole pair as periodic orbits. In Chapter 10, Filho, Mathias and Viana investigate the escape basins and their fractal boundaries [9]. The fractality of the escape basin is investigated by using a scattering map. The fractal nature of the structure of escape basin is quantitatively investigated by computing the basin entropy and the basin boundary entropy. The authors show that the distance between the black holes is an important parameter impacting the complexity of the basin structure.

Furthermore, on fractal structures, Zhang devoted Chapter 11 for the fractal and Wada basin dynamics of the switched systems [10]. These systems comprise an important class of dynamical systems that consist of a finite number of subsystems and a switching rule between these subsystems. This chapter investigates the existence of fractal and Wada basin boundaries to the generally discrete-time switched systems with time-dependent switching. Analytical tools for the characterization of the properties of fractal basin boundaries for planar switched systems (with time-dependent switching) are constructed. It is demonstrated that the basin boundaries of an auxiliary dynamical system (related to the switched system) provide an evidence of the Wada basin boundaries for switched systems. Wada basin boundaries are used to provide the insight into the qualitative unpredictability and uncertainty of the switched systems.

The accumulation of errors and the numerical noise poses serious problems when performing large number of iterates in computational simulations of nonlinear systems. Gallas considers two problems related to computational analysis of dynamical systems in Chapter 12 [11]. The first problem is focused on the impact of small numerical errors in calculations of orbital points of the two-dimensional Hénon map. Numerical errors are determined by measuring the distance between the points of departure and return in the original and the inverse maps. Secondly, exact analytical results are given for the one-dimensional quadratic map in the fully chaotic partition generating limit. The presented algorithm allows to extract precise coordinates of points belonging to periodic orbits. Such techniques provide valuable tools which help to gauge the propagation of errors in computational analysis of chaotic nonlinear maps.

1.4. Recent Trends in Applications of Chaotic and Complex Systems

Part III begins with Chapter 13 by Lukin, Privalova, Popov, Skubov and Shytukin on the study of the periodical oscillations of mono- and multilayer nanocapacitors located in the electric field [12]. The authors observe that the time of overcharge of nanocapacitor is much less than the period of excited oscillations. One of possible applications of graphene nanoresonators is using of these devices as detectors of the mass of nanoparticles adherent on the graphene layer. The adherence of the particle (biological cell or even molecules) to the elastic surface changes its resonance frequency. This effect provides means for the determination of the mass of the nanoparticle.

Chapter 14 deals with the retina as a high-dimensional system [13]. The specific and hierarchical structure of the eye retina allows it to pre-process visual information, at different scales, in order to reduce redundancy and increase the speed, efficiency and reliability of visual response. In this chapter, Cessac demonstrated that the theory of dynamical systems, the theory of bifurcations, and the ergodic theory can provide useful insights on retinal behavior and dynamics. The retina is investigated as a high-dimensional, non-autonomous dynamical system, layered and structured, with non-stationary and spatially inhomogeneous entries. Finally, the dynamics of the system is generalized to more general perspectives.

Collective dynamics of spatiotemporally coupled elements can be considered as the emergence of the order from microscopic self-organization at the macroscopic scale. Zheng in Chapter 15 explores the mechanism of collective self-sustained oscillations in networks of coupled non-oscillatory units in [14]. It is demonstrated that the network topology plays a significant role in the dynamics of excitable networks and gene regulated networks. This chapter shows that the topology reduction is the key procedure in accomplishing the dimension-reduction description of a complex system.

Urban planning plays a crucial role in empowering development and sustainability of major cities around the world. In Chapter 16, Chew, Saw and Pang introduce a no-boarding policy as a form of kicking force to achieve a stable staggered state in the bus loop system [15]. The bus bunching phenomenon is related to the effect of synchronization of Kuramoto phase oscillators with local unidirectional coupling. It is shown that stable anti-bunched states can exist (without any additional kicking force) if the number of oscillators is at least five. This similarities between the bus loop system and the local unidirectional Kuramoto oscillators provide the insight on how the bus loop system can remain staggered.

In Chapter 17, the role of feedback delay as a control tool in the complex Ginzburg–Landau equation with local and nonlocal delayed perturbations is discussed by Casal and Díaz [16]. The authors consider the global delayed problem in which two real parameters, namely, the feedback intensity and the delay time play a fundamental role. The basic questions of the existence and uniqueness of solutions to the associate initial-boundary value problem with initial past history are rigorously discussed. Novel results on the stabilization of the uniform oscillations for the complex Ginzburg–Landau equation by means of global delayed feedback are presented. Further, several bifurcation effects produced by the delay time in the behavior of the complex Ginzburg–Landau equation are analyzed.

In classical passive walking, the locomotion power comes solely from the gravity. The walking cycles are sustained on a gentle downhill and the system converges to a limit cycle, in which gravitational potential energy and energy loss due to foot-landing cancel each other. In Chapter 18, Li, Tokuda, Asano and Yan present an overview on the passive dynamical walking as a typical example of nonlinear dynamics in the field of robotics [17]. A control mechanism which positively utilizes the intrinsic walking property, while preserving its efficiency is introduced. The authors demonstrate both numerically and experimentally that the well–known Arnold tongues are observed as a typical entrainment property of the limit cycle walker in two-dimensional parameter space spanned by the wobbling forcing and frequency.

In Chapter 19, Hernández, Duran and Amigó present a physics-inspired particle swarm algorithm for global optimization [18]. A stochastic, single-objective method that evolves a swarm of agents in search of global extremum is designed and discussed. The governing principles of the presented optimization algorithm are based on physical processes observable in dynamical systems. Numerical simulations are performed with 31 test functions with dimensions ranging between 2 and 45. It is shown that the general algorithmic search outperforms such global optimization algorithms as basin hopping, Cuckoo search, and differential evolution algorithms.

1.5. Conclusions

This research monograph covers new and novel emerging topics and expounds the state-of-the-art advancements in the field, including theory, analysis, numerical simulations, novel applications and future directions of research on chaos and complex system dynamics. It presents a comprehensive picture of recent developments in chaos and complex dynamics that is useful to a wide audience including newcomers, lecturers and researchers in physics, engineering, and applied mathematics.

References

[1] V. Chinnathambi, S. Rajasekar and M.A.F. Sanjuan, Enhanced vibrational resonance by an amplitude modulated force, in J. Awrejcewicz, S. Rajasekar and M. Ragulskis, eds. *Recent Trends in Chaotic, Nonlinear and Complex Dynamics* (World Scientific, Singapore, 2021).

[2] R. J. Escalante-González and E. Campos, Generation of self-excited and hidden multiscroll attractors in multistable systems, in J. Awrejcewicz, S. Rajasekar, and M. Ragulskis, eds. *Recent Trends in Chaotic, Nonlinear and Complex Dynamics* (World Scientific, Singapore, 2021).

[3] A. Elipe, A. Abad and A. F. S. Ferreira, On the dynamics of the dipole-segment with equal masses and arbitrary rotation, in J. Awrejcewicz, S. Rajasekar and M. Ragulskis, eds. *Recent Trends in Chaotic, Nonlinear and Complex Dynamics* (World Scientific, Singapore, 2021).

[4] X. Carton, F. C. K. Tchuenkam and A. Vic, The interaction of two vortices near a boundary in rotating stratified incompressible flows, in J. Awrejcewicz, S. Rajasekar and M. Ragulskis, eds. *Recent Trends in Chaotic, Nonlinear and Complex Dynamics* (World Scientific, Singapore, 2021).

[5] E. A. R. Ribeiro, S. Lenci and C. E. N. Mazzilli, Modal asynchronicity in pre-stressed continuous simply-supported beams with transversal and rotational oscillators, in J. Awrejcewicz, S. Rajasekar, and M. Ragulskis, eds. *Recent Trends in Chaotic, Nonlinear and Complex Dynamics* (World Scientific, Singapore, 2021).

[6] M. Courbage, On the weak decay of autocorrelations in periodic barrier billiards, in J. Awrejcewicz, S. Rajasekar and M. Ragulskis, eds.

Recent Trends in Chaotic, Nonlinear and Complex Dynamics (World Scientific, Singapore, 2021).

[7] I. Bashkirtseva and L. Ryashko, Stochastic sensitivity analysis of noise-induced phenomena in discrete systems, in J. Awrejcewicz, S. Rajasekar and M. Ragulskis, eds. *Recent Trends in Chaotic, Nonlinear and Complex Dynamics* (World Scientific, Singapore, 2021).

[8] E. del Rio and S. Elaskar, The role of noise in chaotic intermittency, in J. Awrejcewicz, S. Rajasekar and M. Ragulskis, eds. *Recent Trends in Chaotic, Nonlinear and Complex Dynamics* (World Scientific, Singapore, 2021).

[9] E. E. de Souza Filho, A. C. Mathias and R. L. Viana, Fractal structures in a binary Schwarzschild black hole system, in J. Awrejcewicz, S. Rajasekar and M. Ragulskis, eds. *Recent Trends in Chaotic, Nonlinear and Complex Dynamics* (World Scientific, Singapore, 2021).

[10] Y. Zhang, Characterizing fractal basin of attraction in planar switched systems, in J. Awrejcewicz, S. Rajasekar, and M. Ragulskis, eds. *Recent Trends in Chaotic, Nonlinear and Complex Dynamics* (World Scientific, Singapore, 2021).

[11] J. A. C. Gallas, Numerical errors and exact orbits of quadratic maps, in J. Awrejcewicz, S. Rajasekar and M. Ragulskis, eds. *Recent Trends in Chaotic, Nonlinear and Complex Dynamics* (World Scientific, Singapore, 2021).

[12] A. V. Lukin, O. V. Privalova, I. A. Popov, D. Yu. Skubov and L. V. Shyukin, Electromechanical models of micro- and nanoresonators, in J. Awrejcewicz, S. Rajasekar and M. Ragulskis, eds. *Recent Trends in Chaotic, Nonlinear and Complex Dynamics* (World Scientific, Singapore, 2021).

[13] B. Cessac. The retina as a dynamical system, in J. Awrejcewicz, S. Rajasekar, and M. Ragulskis, eds. *Recent Trends in Chaotic, Nonlinear and Complex Dynamics*, (World Scientific, Singapore, 2021).

[14] Z. Zheng, Collective sustained oscillations in complex systems, in J. Awrejcewicz, S. Rajasekar, and M. Ragulskis, eds. *Recent Trends in Chaotic, Nonlinear and Complex Dynamics* (World Scientific, Singapore, 2021).

[15] L. Y. Chew, V.-L. Saw and Y. E. I. Pang, Stability of anti-bunched buses and local unidirectional Kuramoto oscillators, in J. Awrejcewicz, S. Rajasekar, and M. Ragulskis, eds. *Recent Trends in Chaotic, Nonlinear and Complex Dynamics* (World Scientific, Singapore, 2021).

[16] A. C. Casal and J. I. Díaz. Feedback delay as a control tool: the complex Ginzburg–Landau equation with local and nonlocal delayed perturbations, in J. Awrejcewicz, S. Rajasekar, and M. Ragulskis, eds.

Recent Trends in Chaotic, Nonlinear and Complex Dynamics (World Scientific, Singapore, 2021).

[17] L. Li, I. T. Tokuda, F. Asano and C. Yan, Nonlinear analysis and control for quasi-passive dynamic walker based on entrainment effect, in J. Awrejcewicz, S. Rajasekar, and M. Ragulskis, eds. *Recent Trends in Chaotic, Nonlinear and Complex Dynamics* (World Scientific, Singapore, 2021).

[18] S. Hernández, G. Duran and J.M. Amigó, Physics-inspired swarm optimization: The general algorithmic search (GAS), in J. Awrejcewicz, S. Rajasekar, and M. Ragulskis, eds. *Recent Trends in Chaotic, Nonlinear and Complex Dynamics* (World Scientific, Singapore, 2021).

Part I

Fundamental Results in Nonlinear Dynamics

https://doi.org/10.1142/9789811221903_0002

Chapter 2

Enhanced Vibrational Resonance by an Amplitude-Modulated Force

V. Chinnathambi[*,§], S. Rajasekar[†,¶] and M. A. F. Sanjuan[‡,||]

[*] *Department of Physics, Sadakathullah Appa College, Tirunelveli 627 011, Tamilnadu, India*
[†] *School of Physics, Bharathidasan University, Tiruchirapalli 620 024, Tamilnadu, India*
[‡] *Nonlinear Dynamics, Chaos and Complex Systems Group, Departamento de Física, Universidad Rey Juan Carlos, Tulipán s/n, 28933 Móstoles, Madrid, Spain*
[§] *drchinnathambi@sadakath.ac.in*
[¶] *srj@bdu.ac.in*
[||] *miguel.sanjuan@urjc.es*

Vibrational resonance (VR) in the presence of an amplitude-modulated external force consisting of a low-frequency (ω) component and two high-frequencies ($\Omega + \omega$) and ($\Omega - \omega$) with $\Omega \gg \omega$ is investigated in the paradigmatic Duffing oscillator. There are certain valuable VR features induced by the two high-frequency components compared to the case driven by a high-frequency force in both single-well and double-well potential systems. A detailed comparison of VR in both cases is presented here. We show the enhanced response amplitude at the low-frequency ω, showing more number of resonance peaks, a nondecay of the response amplitude and hysteresis and a jump phenomenon on the response amplitude curve due to the amplitude-modulated force. By exploring the phase portrait and the width of the orbit, we explain the resonance mechanism and account for the other features of the resonance curve.

2.1. Introduction

Nonlinear systems are capable of displaying a variety of simple and complicated dynamics. Among the various nonlinear phenomena, certainly resonance is a fundamental one with practical applications in physics, engineering and biology. The phenomenon of resonance refers to a realization of a maximum response of a dynamical system, which can be induced by means of external forces. Different kinds of external forces may induce distinct types of resonances [1]. Here, we are interested in vibrational resonance (VR). To realize this resonance, a nonlinear system is driven by both a low-frequency (ω) and a high-frequency (Ω) force. When the amplitude or frequency of the high-frequency force is varied from a small value, then the response amplitude at the low-frequency ω denoted as $Q(\omega)$ of a system becomes maximum at one or more values of the control parameter.

Landa and McClintock [2] first reported the VR in a bistable system. Later on, a theoretical treatment for analyzing VR has been proposed by Gitterman [3] and Blekhman and Landa [4]. After these seminal works, the features of this resonance have been studied theoretically, numerically and experimentally in a variety of systems. For example, its occurrence has been analyzed in a noise-induced structure [5], the FitzHugh–Nagumo equation [6–8], a quintic oscillator [9, 10], a deformable potential system [11], a multistable system [12], a driven plasma [13], overdamped bistable systems [14–16], coupled oscillators [17–19], fraction-order damped systems [20–25], time-delayed systems [26–30] and networks [31–35]. Besides, much experimental work on VR has been carried out in a bistable optical cavity [36, 37], an analog simulation of the overdamped Duffing equation [14], an electronic circuit with a Chua's diode [6], a Chua's circuit [38], a twin-well oscillator [39] and an array of hard limiters [40]. Furthermore, applications of VR on energy harvesting [41] and bearing fault diagnosis [42] have been also explored.

Generally, in the VR studies, a nonlinear system driven by a weak periodic force, say, $f \cos \omega t$, is further subjected to a high-frequency force $g \cos \Omega t$ with $\Omega \gg \omega$ [2], that is, the total external force is $F_2(t) = f \cos \omega t + g \cos \Omega t$. For a certain type of nonlinear systems, the theoretical approach developed by Blekhman and Landa [4] can be successfully employed to determine an analytical expression for the response amplitude at the low-frequency denoted as $Q(\omega)$, as well as the values of the control parameter g at which $Q(\omega)$

becomes maximum. Exploring the features of VR in systems with different types of setup of external force is of great significance. Occurrence of VR by nonsinusoidal and arbitrary binary shape periodic input signals has been studied [1, 43]. The goal of the present chapter is to analyze VR by an amplitude-modulated force $F_{\mathrm{AMF}}(t) = (f + 2g\cos\Omega t)\cos\omega t$ which can also be written as $F_3(t) = f\cos\omega t + g\cos(\Omega+\omega)t + g\cos(\Omega-\omega)t$. When $\Omega \gg \omega$, the amplitude-modulated force F_{AMF} can also be treated as consisting of a low-frequency force $f\cos\omega t$ and two high-frequency forces with frequencies $\Omega+\omega$ and $\Omega-\omega$. There are certain notable earlier studies on nonlinear systems subjected to an amplitude-modulated force without the restriction $\Omega \gg \omega$ [44–49]. There are some interesting VR features in the Duffing oscillator driven by an amplitude-modulated force $F_{\mathrm{AMF}}(t)$ compared to the case when it is driven by the force $F_2(t)$. When the oscillator is subjected to the force $F_2(t)$, at most only two resonance peaks can occur, the value at resonance of the response amplitude $Q(\omega) = 1/(d\omega)$ where d is the damping coefficient, and after the last resonance $Q(\omega)$ decays to zero with the control parameter g, which represents the amplitude of the high-frequency force. Interestingly, and in contrast to this result, when the oscillator is subjected with the external force $F_{\mathrm{AMF}}(t)$, more than two resonance peaks occur, where $Q(\omega) \gg 1/(d\omega)$ at some resonance peaks and $Q(\omega)$ does not decays to zero even for very large values of g. Further, the response amplitude curve displays hysteresis and a jump phenomenon. These are the main results of this research work.

The plan of the present chapter is as follows. To start with, we introduce first the Duffing oscillator in Sec. 2.2. In Sec. 2.3, we consider the Duffing oscillator with a single-well potential and with the external force $F_2(t)$ consisting of one low-frequency and one high-frequency components. We review the VR realized with this force through a theoretical approach. We obtain analytical expressions for the response amplitude $Q(\omega)$ and the value of the control parameter g at which VR occurs and verify these theoretical predictions through numerical simulation. We takeup the system with the force $F_{\mathrm{AMF}}(t) = F_3(t)$ in Sec. 2.4. First, we illustrate that the theoretical procedure used to analyze VR in the case of $F(t) = F_2(t)$ is not applicable for $F(t) = F_3(t)$. Therefore, we perform a numerical investigation. We show the occurrence of an enhanced $Q(\omega)$, a higher number of resonance peaks, hysteresis and a jump phenomenon due to the presence of the second high-frequency component with frequency $\Omega - \omega$ in $F_3(t)$, and we explain them. We demonstrate that the slow motion of the system with frequency ω is modulated by the slow–fast motion. We compare the VR features in the presence of $F_3(t)$ with that of $F_2(t)$. Section 2.5 is devoted to the system with a double-well potential. Here, first we briefly summarize the results for the case of $F(t) = F_2(t)$. Next, we report our investigation on the oscillator with $F(t) = F_3(t)$. Later, we explain the mechanism of resonance by using the phase portrait and the width of the orbit. And finally, in Sec. 2.6 we briefly summarize the results of our present study.

2.2. The Duffing Oscillator

The Duffing oscillator equation is an ubiquitous nonlinear evolution equation and is realized in many physical, engineering and biological systems. Originally, the Duffing equation was introduced by the German engineer Georg Duffing in 1918 [50] to describe the hardening spring effect observed in many mechanical problems. The equation of motion of the linearly damped and driven by an amplitude-modulated periodic force $F_{\mathrm{AMF}}(t) = F_3(t)$ is given by

$$\ddot{x} + d\dot{x} + \omega_0^2 x + \beta x^3 = (f + 2g\cos\Omega t)\cos\omega t, \quad \Omega \gg \omega, \tag{2.1}$$

where the amplitude f of the low-frequency (ω) periodic force is modulated by the high-frequency (Ω) periodic force with amplitude g. With the use of the formula $2\cos\Omega t\cos\omega t = \cos(\Omega + \omega)t + \cos(\Omega - \omega)t$, Eq. (2.1) takes the form

$$\ddot{x} + d\dot{x} + \omega_0^2 x + \beta x^3 = f\cos\omega t + g\cos(\Omega + \omega)t + g\cos(\Omega - \omega)t, \quad \Omega \gg \omega. \tag{2.2}$$

For $g = 0$ and small values of β and f, the system (2.1) exhibits resonance with hysteresis and a jump phenomenon when the control parameter ω is varied from a small value [1, 51]. Equation (2.2) can also be thought of as the equation of motion for a unit mass particle in the potential well

$$V(x) = \frac{1}{2}\omega_0^2 x^2 + \frac{1}{4}\beta x^4 \tag{2.3}$$

subjected to a viscous drag force of strength d and driven by the force $F_3(t)$. $V(x)$ becomes a single-well potential for ω_0^2, $\beta > 0$; double-well potential for $\omega_0^2 < 0$, $\beta > 0$; single-well with a double-hump for $\omega_0^2 > 0$, $\beta < 0$; inverted single-well form for ω_0^2, $\beta < 0$. The inverted single-well potential is not of physical relevance. In the double-hump potential system, bounded motion exists as long as $x(t)$ is confined between the humps at $x = \pm\sqrt{\omega_0^2/|\beta|}$. In our present investigation of VR in Eq. (2.2), we consider the single-well and double-well potentials of the system.

2.3. Single-Well Duffing Oscillator with One High-Frequency Force

Consider the system (2.2) with $F(t) = F_2(t) = f\cos\omega t + g\cos(\Omega+\omega)t$. For $\Omega \gg \omega$, one can seek a solution of Eq. (2.2) as $x = X + \psi$ where $X(t)$ is a slow motion with period $2\pi/\omega$ and $\psi(t)$ is a fast motion with period $2\pi/(\Omega+\omega)$. The occurrence of VR in this system was first reported by Landa and McClintock [2] and theoretically analyzed in Refs. [3, 4].

2.3.1. *Theoretical description of VR*

Assume that the mean value of ψ, denoted as $\langle\psi\rangle$, with respect to time $\tau = (\Omega + \omega)t$ is $\langle\psi\rangle = (1/(2\pi))\int_0^{2\pi}\psi \mathrm{d}\tau = 0$. Substitution of $x = X + \psi$ in Eq. (2.2) with $\langle\psi\rangle = 0$ and adding and subtracting $3\beta\langle\psi^2\rangle X + \beta\langle\psi^3\rangle$ give the following equations for X and ψ:

$$\ddot{X} + d\dot{X} + \left(\omega_0^2 + 3\beta\langle\psi^2\rangle\right)X + \beta X^3 + \beta\langle\psi^3\rangle = f\cos\omega t, \tag{2.4a}$$

$$\ddot{\psi} + d\dot{\psi} + \left(\omega_0^2 + 3\beta X^2\right)\psi + 3\beta X\left(\psi^2 - \langle\psi^2\rangle\right) + \beta\left(\psi^3 - \langle\psi^3\rangle\right) = g\cos(\Omega+\omega)t, \tag{2.4b}$$

where $\langle\psi^m\rangle = (1/(2\pi))\int_0^{2\pi}\psi^m \mathrm{d}\tau$.

In view of ψ being rapidly varying, Eq. (2.4b) can be approximated as $\ddot{\psi} = g\cos(\Omega+\omega)t$ with the solution

$$\psi = -\frac{g}{(\Omega+\omega)^2}\cos(\Omega+\omega)t \ \text{ and } \ \langle\psi^2\rangle = \frac{g^2}{2(\Omega+\omega)^4}, \ \ \langle\psi^3\rangle = 0\,. \tag{2.5}$$

Then, Eq. (2.4a) becomes

$$\ddot{X} + d\dot{X} + C_1 X + \beta X^3 = f\cos\omega t, \ \ C_1 = \omega_0^2 + \frac{3\beta g^2}{2(\Omega+\omega)^4}\,. \tag{2.6}$$

The effective potential of X in the absence of damping and the low-frequency periodic force is

$$V_{\text{eff}}(X) = \frac{1}{2}C_1X^2 + \frac{1}{4}\beta X^4. \tag{2.7}$$

In general, by varying g or Ω, the shape of V_{eff} and the number of equilibrium points can be changed. Slow motions occur about the equilibrium point $X = X^*$. For $\omega_0^2 > 0$ and $\beta > 0$, the parameter $C_1 > 0$ and hence V_{eff} is always of a single-well form and hence slow motion occurs about $X^* = 0$.

Next, introduction of the change of variable $Y = X - X^*$ (deviation of X from X^*) in Eq. (2.6) gives

$$\ddot{Y} + d\dot{Y} + \omega_{\text{r}}^2 Y + \beta Y^3 + 3\beta X^* Y^2 = f\cos\omega t, \tag{2.8}$$

where the resonant frequency ω_{r} is given by

$$\omega_{\text{r}}^2 = \omega_0^2 + \frac{3\beta g^2}{2(\Omega+\omega)^4} + 3\beta X^{*2}. \tag{2.9}$$

In the limit of $t\to\infty$ for $f \ll 1$, we assume that $|X| \ll 1$ and neglect the nonlinear term in Eq. (2.8). This makes Eq. (2.8) linear and its solution in the limit of $t\to\infty$ is $Y = A_{\text{L}}\cos(\omega t+\phi)$, where

$$A_{\text{L}} = \frac{f}{\sqrt{(\omega_{\text{r}}^2-\omega^2)^2 + d^2\omega^2}}, \quad \phi = \tan^{-1}\left(\frac{d\omega}{\omega^2-\omega_{\text{r}}^2}\right). \tag{2.10}$$

VR is characterized using the response amplitude $Q(\omega)$, where

$$Q(\omega) = \frac{A_{\text{L}}}{f} = \frac{1}{\sqrt{S}}, \quad S = \left(\omega_{\text{r}}^2-\omega^2\right)^2 + d^2\omega^2. \tag{2.11}$$

From Eq. (2.11), we note that each local minimum of S represents a resonance. When g is varied, then a resonance occurs at $g = g_{\mathrm{VR}}$ where g_{VR} is a root of the equation $S_g = \mathrm{d}S/\mathrm{d}g = 0$ and $S_{gg}|_{g=g_{\mathrm{VR}}} > 0$. The resonance condition obtained from $S_g = 0$ is $\omega_{\mathrm{r}}^2 = \omega^2$. Substitution of $\omega_{\mathrm{r}}^2 = \omega^2$ and $X^* = 0$ in Eq. (2.9) gives

$$g_{\mathrm{VR}} = \sqrt{2(\Omega + \omega)^4 \left(\omega^2 - \omega_0^2\right) / (3\beta)}\,. \tag{2.12}$$

From Eq. (2.12), we infer that the system will exhibit only one resonance for $\omega^2 > \omega_0^2$ and no resonance for $\omega^2 \leq \omega_0^2$.

2.3.2. *Analysis of VR*

To compare with the theoretical Q given by Eq. (2.11), we compute Q from the numerical solution of Eq. (2.2). Numerical $Q(\omega)$ is given by

$$Q(\omega) = \frac{\sqrt{Q_{\mathrm{s}}^2 + Q_{\mathrm{c}}^2}}{f}, \tag{2.13a}$$

where

$$Q_{\mathrm{s}} = \frac{2}{nT}\int_0^{nT} x(t)\sin\omega t\mathrm{d}t, \quad Q_{\mathrm{c}} = \frac{2}{nT}\int_0^{nT} x(t)\cos\omega t\mathrm{d}t \tag{2.13b}$$

with $T = 2\pi/\omega$ and $n = 200$.

We fix $\omega_0^2 = 1$, $\beta = 1$, $d = 0.5$, $f = 0.1$, $\omega = 1.5$ and $\Omega = 15$. Figure 2.1(a) presents both a theoretically and numerically calculated Q as a function of g. We clearly observe a very good agreement of theoretical Q with numerically computed Q. The theoretical and numerical values of g_{VR} are 205.40 and 202.85, respectively. Figure 2.1(b) shows the plot of ω_{r}^2 vs. g. At $g = g_{\mathrm{VR}}$, we find $\omega_{\mathrm{r}}^2 = \omega^2$. At resonance $Q_{\max} = 1/(d\omega) = 4/3 \approx 1.33$.

2.4. Single-Well System with a Two High-Frequency Force

2.4.1. *Equations for slow and fast variables*

We consider the Duffing oscillator subjected to the amplitude-modulated periodic force, that is, for the system governed by

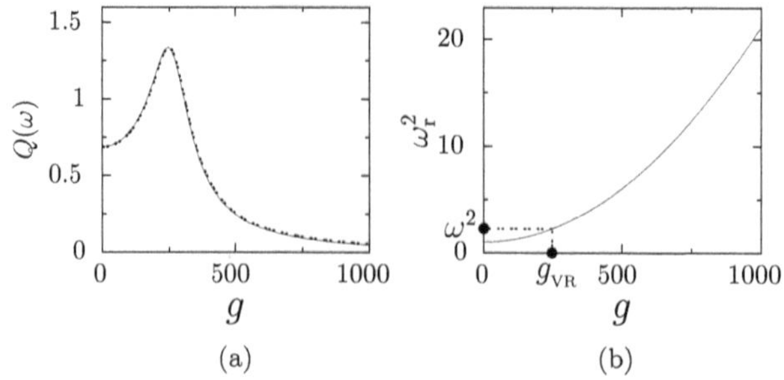

Fig. 2.1. (a) Variation of the theoretical $Q(\omega)$ (represented by a dashed line) and the numerically computed $Q(\omega)$ (represented by a continuous line) with g for the Duffing oscillator with $\omega_0^2 = 1$, $\beta = 1$, $d = 0.5$, $f = 0.1$, $\omega = 1.5$ and $\Omega = 15$. (b) Plot of the theoretical value of ω_r^2 vs. g.

Eq. (2.1) or (2.2), then the equations for X and ψ are given by Eq. (2.4) with the additional force $g\cos(\Omega-\omega)t$ in Eq. (2.4b). In this case

$$\psi = -\frac{g}{(\Omega+\omega)^2}\cos(\Omega+\omega)t - \frac{g}{(\Omega-\omega)^2}\cos(\Omega-\omega)t \tag{2.14}$$

and

$$\omega_r^2 = \omega_0^2 + 3\beta\langle\psi^2\rangle + 3\beta X^{*2}. \tag{2.15}$$

Q is given by Eq. (2.11). The theoretical value of Q is found to be highly different from the numerically computed Q. This implies that Eq. (2.4a) for the slow variable X with the replacement of ψ^m by $\langle\psi^m\rangle$, which yielded a very good approximate expression for Q for the case of a single high-frequency force, is now not applicable for the case of the system driven by two high-frequency forces. Actually, having $x = X + \psi$ in Eq. (2.2) gives

$$\ddot{X} + d\dot{X} + \left(\omega_0^2 + 3\beta\psi^2\right)X + \beta X^3 = f\cos\omega t, \tag{2.16a}$$

$$\ddot{\psi} + d\dot{\psi} + \left(\omega_0^2 + 3\beta X^2\right)\psi + \beta\psi^3$$
$$= g\cos(\Omega+\omega)t + g\cos(\Omega-\omega)t. \tag{2.16b}$$

We have numerically integrated Eq. (2.16) and computed $Q(\omega)$ $(=A_L(\omega)/f)$ of X and $A_L(\Omega+\omega)$ and $A_L(\Omega-\omega)$ of ψ. According to the assumed approximate solution ψ given by Eq. (2.14),

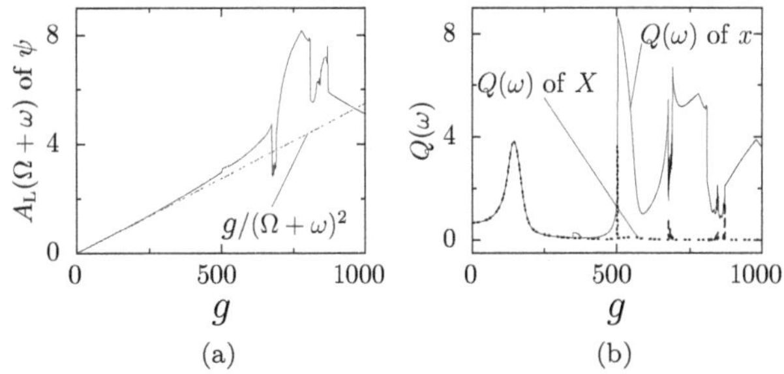

Fig. 2.2. (a) Plot of $A_{\rm L}(\Omega+\omega)$ of ψ as a function of g for the single-well Duffing oscillator driven by the amplitude-modulated force. (b) Variation of the numerically computed $Q(\omega)$ of x and $Q(\omega)$ of X with the control parameter g. The parameter values of the system are as in Fig. 2.1.

the theoretical approximations are as follows: $A_{\rm L}(\Omega+\omega) = g/(\Omega+\omega)^2$ and $A_{\rm L}(\Omega-\omega) = g/(\Omega-\omega)^2$. In Fig. 2.2(a), a continuous line is the numerically computed $A_{\rm L}(\Omega+\omega)$, while the theoretical approximation is represented by a dashed line. A theoretical approximation $A_{\rm L}(\Omega+\omega)$ is close to the numerical approximation $A_{\rm L}(\Omega+\omega)$ only for, say, $g < 400$. A similar result is found for $A_{\rm L}(\Omega-\omega)$. Figure 2.2(b) shows numerically computed $Q(\omega)$ of x (obtained from Eq. (2.2)) and $Q(\omega)$ of X (obtained from Eq. (2.16)). Q of X is in close agreement with Q of x for $g < 400$. The observation is that writing $x = X + \psi$ and using the equations of motion of X and ψ to compute $Q(\omega)$ is only valid for a certain range of values of g. In this range of g though

$$\ddot{\psi} = g\cos(\Omega+\omega)t + g\cos(\Omega-\omega)t \qquad (2.17)$$

is a valid approximation (as evident from Fig. 2.2(a)) replacing ψ^m by $\langle\psi^m\rangle$ and proceeding further to obtain an analytical expression for $Q(\omega)$ is not found to yield a good approximate $Q(\omega)$. Therefore, we compute Q by numerically solving Eq. (2.2).

2.4.2. *Enhanced $Q(\omega)$ by the amplitude-modulated force*

There is a large number of interesting and important results on VR due to the amplitude-modulated force compared to the single high-frequency force. To illustrate this, we can compare Fig. 2.2(b) with

Fig. 2.1(a). When only one high-frequency force is presented, $Q(\omega)$ decays to zero as g increases beyond $g_{\rm VR}$ (at which resonance occurs). Here, only one resonance peak is possible. The maximum value of $Q(\omega)$ is $1/(d\omega)$ and is ≈ 1.33 for $d = 0.5$ and $\omega = 1.5$. In Fig. 2.2(b) corresponding to the system subjected to two high-frequency forces, $Q(\omega)$ does not decrease continuously beyond the first resonance peak. $Q(\omega)$ is maximum at more than one value of g. Further, $Q(\omega)$ is $\gg 1/(d\omega)$ not only at the first resonance but also for a wide range of values of g. That is, an enhanced response amplitude at the low-frequency ω can be achieved by applying two high-frequency forces.

The solution $x(t)$ can be decomposed into a slow motion $X(t)$ and a fast motion $\psi(t)$. In Fig. 2.2(b), $Q(\omega)$ is computed from Eq. (2.16), and closely matches with $Q(\omega)$ of x for $g < 400$. Therefore, we report the influence of g on the slow motion for $g < 400$. When g is varied from a small value, the first resonance occurs at $g = g_{\rm VR} = 143.5$. Figure 2.3 shows the phase portrait of the slow oscillation for five values of g chosen far before, at and after the first resonance peak. For $g \neq 0$ in the limit of $t \to \infty$, X is not strictly described as $A_{\rm L} \cos(\omega t + \phi)$. In Fig. 2.3, we can clearly observe modulations in the solution X which are induced by the high-frequency forces $g \cos(\Omega + \omega)t + g \cos(\Omega - \omega)t$. Because of this, the analytical expression for $A_{\rm L}(\omega)$ or $Q(\omega)$ of a slow motion of the system given by Eq. (2.2) with two high-frequency forces obtained through the

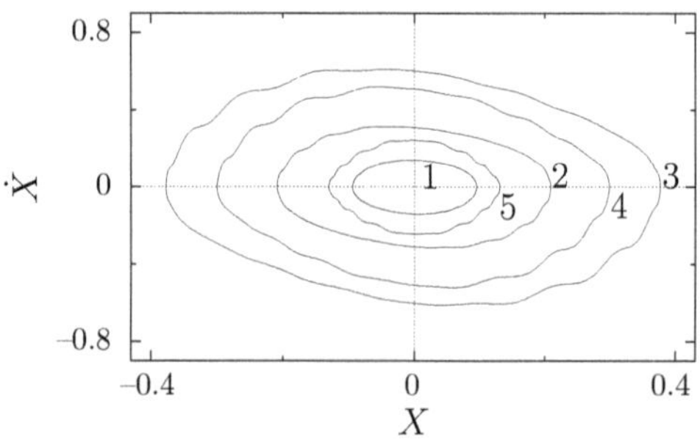

Fig. 2.3. Phase portrait of a slow oscillation for five values of g for the single-well Duffing oscillator driven by the amplitude-modulated force. The values of g for the orbits 1–5 are 70, 115, 143.5($= g_{\rm VR}$), 160 and 180, respectively. Other parameter values of the system are as in Fig. 2.1.

theoretical approach used for the case of only one high-frequency force is not in agreement with the numerically computed $A_{\mathrm{L}}(\omega)$ or $Q(\omega)$. As noted earlier for $g < 400$, the approximate equation of motion for ψ given by Eq. (2.17) is reasonable and its solution given by Eq. (2.14) is a valid approximation. With this ψ, the equation of motion for X given by Eq. (2.16a) can be regarded as the Duffing oscillator driven by both additive and parametric forces. The parametric force ψ^2 has periodic components with frequencies 2ω, 2Ω, $2(\Omega+\omega)$ and $2(\Omega-\omega)$. Analysis of such a complicated system to get an analytical expression for $Q(\omega)$ is tedious. However, Fig. 2.4 suggests that we can write the solution X of Eq. (2.16a) as

$$X(t) = z(t) + z(t)\phi(t), \tag{2.18}$$

where $z(t)$ is a slow motion (with period $2\pi/\omega$) and the last term is a slow–fast motion. By appropriately choosing the periodic form of $\phi(t)$, the equation for $z(t)$ becomes

$$\ddot{z} + (d + u_1(t))\,\dot{z} + \left(\omega_0^2 + u_2(t)\right) z + u_3(t)z^2 + u_4(t)z^3 = f\cos\omega t. \tag{2.19}$$

For small $z(t)$, we can neglect the nonlinear terms and consider only the linear system. The important observation is that not only the natural frequency ω_0^2 is modified, but also the damping coefficient d is modified into, say, d'. Since it is possible that $d' < d$, it can happen that the response amplitude $Q(\omega) > 1/(d\omega)$ for a certain range of values of the control parameter g. This is the reason for

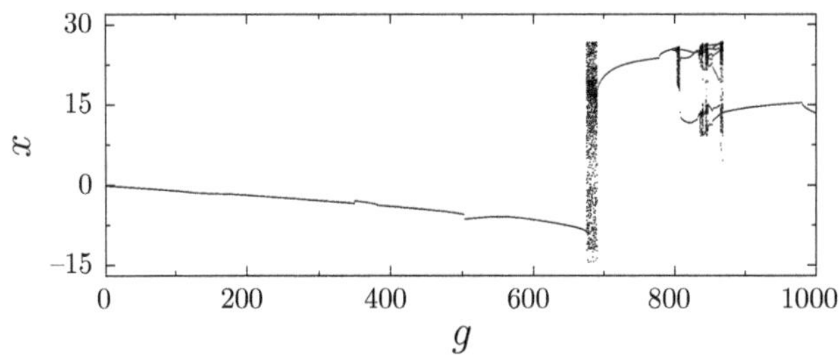

Fig. 2.4. Bifurcation diagram of the single-well Duffing oscillator driven by the amplitude-modulated force. The parameter values of the system are as in Fig. 2.1.

an enhanced $Q(\omega)$ to be realized in the system subjected to an amplitude-modulated force compared to the system subjected to one high-frequency force alone.

2.4.3. *Nonsmooth variation of $Q(\omega)$*

In Fig. 2.2(b), a nonsmooth and irregular variation of $Q(\omega)$ is clearly seen for a certain range of values of g. To find the reason for this fact, we consider the bifurcation diagram Fig. 2.3 where the values of x are collected at $t = n(2\pi/\omega)$ $n = 1, 2, \ldots, 100$ after leaving sufficient transient motion. In the interval of g, where x is periodic with period $T = 2\pi/\omega$, the response amplitude Q varies smoothly. For $g \in [672.5, 690]$ and $[796, 869.5]$ either higher periodic or chaotic motion occurs, and Q is found to vary irregularly. However, the value of Q is nonnegligible.

2.4.4. *Hysteresis and jump phenomenon*

In Fig. 2.2(b), Q of x is numerically calculated by varying g from 0 with increment $\Delta g = 0.25$ is plotted. For $g = 0$, the initial conditions are $x(0) = 0.1$ and $\dot{x}(0) = 0$. For other values of g, the values of $x(0)$ and $\dot{x}(0)$ are chosen as the last values of $x(t)$ and $\dot{x}(t)$, respectively, corresponding to $g - \Delta g$. A different Q is found in certain intervals of g when it is decreased from, say, 1000.

Figure 2.5(a) presents Q obtained by varying g in the forward and reverse directions. We can clearly observe hysteresis and a jump phenomenon. Figure 2.5(b) shows the magnification of the Q curve in the interval $g \in [440, 540]$. Q is found to follow different paths as indicated by an arrow when g is varied in the forward and reverse directions. Such a behavior of Q is not observed in Fig. 2.1(a) corresponding to the system driven by only one high-frequency force.

We describe the response curves in Fig. 2.5(b). For g values just below 471.5, there exists a periodic orbit with period $T(= 2\pi/\omega)$. This orbit has Fourier components with frequencies ω, $\Omega + \omega$ and $\Omega - \omega$. The value of $Q(\omega)$ is small. For $g \in [471.75, 503.5]$, two orbits coexist, one with a small value of $Q(\omega)$ and another with a large value of $Q(\omega)$, but both are periodic with the same period T.

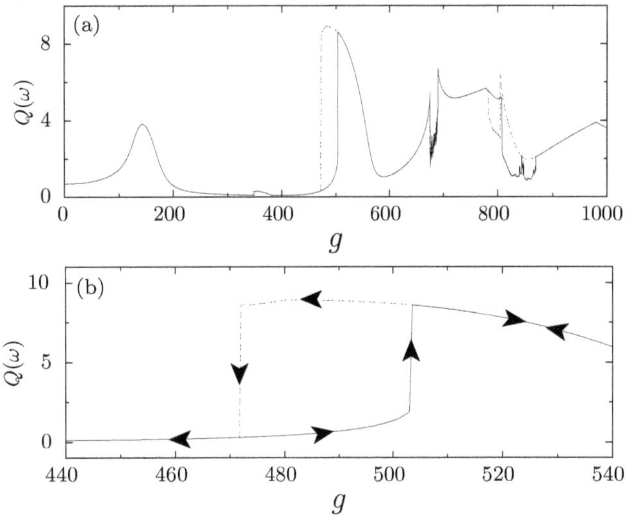

Fig. 2.5. (a) Response amplitude curves obtained by varying the control parameter g from 0 to 1000 (continuous curve) and from 1000 to 0 (dashed curve) for the single-well Duffing oscillator driven by the amplitude-modulated force. (b) Magnification of the $Q(\omega)$ curve in the interval $g \in [440, 540]$ indicating hysteresis and jumps in Q. The parameter values of the system are as in Fig. 2.1.

An example of two coexisting periodic orbits is shown in Fig. 2.6 for $g = 474$. The orbits shown in Figs. 2.6(a) and 2.6(b) have the response amplitude values 0.33 and 8.63, respectively. At $g = 503.5$, the orbit with small Q suddenly disappears and only the orbit with large Q persists. Consequently, there is a sudden jump in the Q curve at $g = 503.5$ when g is increased in the forward direction. Similarly, a jump in Q occurs at $g = 471.75$ when g is varied in the reverse direction.

We wish to point out here that in a recent experimental work in a nonlinear nano-electromechanical resonator subjected to an ac forcing voltage $V(t) = V_0 \cos(2\pi\nu_{\mathrm{f}} t)$, hysteresis and a jump phenomenon are found in the response amplitude curve when ν_{f} is varied [52]. Analysis of switching between the coexisting states in the hysteresis region and the associated VR with

$$V(t) = V_0 \left[1 + \nu \cos(2\pi\nu_{\mathrm{m}} t) + \delta \cos(2\pi\nu_{\mathrm{HF}} t)\right] \cos(2\pi\nu_{\mathrm{f}} t), \quad (2.20)$$

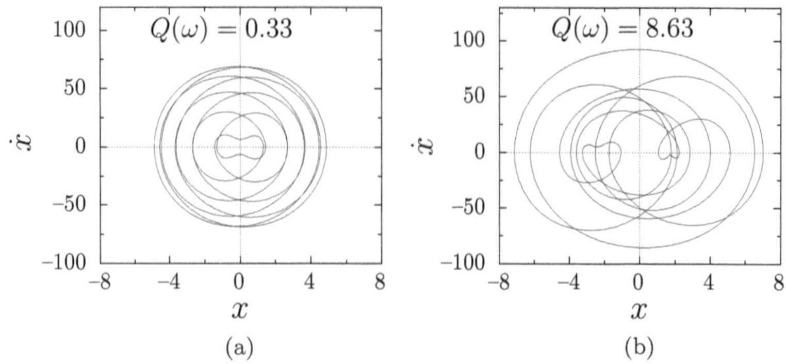

Fig. 2.6. Two coexisting periodic orbits for $g = 474$ with highly different values of $Q(\omega)$. The system is a single-well Duffing oscillator driven by the amplitude-modulated force. The parameter values of the system are as in Fig. 2.1.

where $\nu_{\rm m} \ll \nu_{\rm HF} \ll \nu_{\rm m}$ has been reported. This system is approximately modeled by the Duffing oscillator [52].

In Fig. 2.2(b), we need to account for the decrease in the value of $Q(\omega)$ when g is increased from 503.5 to 585. For $g > 400$, Q of X does not match with Q of x. Therefore, we consider the change in the phase portrait of the system (2.2). Figure 2.7 shows the phase portrait for four values of g in the interval [503.5, 585]. For $g = 503.5$, the orbit in the x–$\dot{x}$ plane consists of three parts. There is a part of the orbit enclosing both solid circles. In addition, we can clearly notice two other parts of the orbit — one enclosing left solid circle only and another right solid circle only. This is similar to an orbit moving around two equilibrium states. The distance between these two points decreases (and Q also decreases) with an increase in the value of g from 503.5. This is clearly evident from Figs. 2.7(a)–2.7(c). As shown in Fig. 2.7(d) at $g = 585$, the two points merged together at the origin, and the orbit moves around it like an orbit in a single-well potential. Q becomes a local minimum at this value of g. The values of $x_{\rm min}$, $x_{\rm max}$, $\dot{x}_{\rm min}$ and $\dot{x}_{\rm max}$ for the orbits for various values of g shown in Fig. 2.7 are almost the same. However, Q decreases from a higher value to a small value when g is varied from 503.5 to 585.

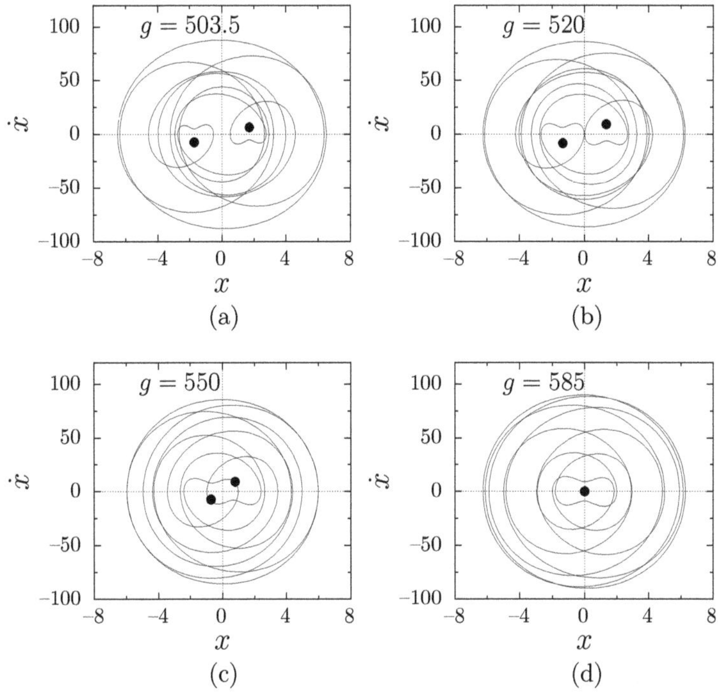

Fig. 2.7. Phase portrait of the system (2.2) for four values of g. The system is a single-well Duffing oscillator driven by the amplitude-modulated force. The parameter values of the system are as in Fig. 2.1.

2.5. Double-Well Duffing Oscillator

The potential V given by Eq. (2.3) is a symmetric double-well form for $\omega_0^2 < 0$ and $\beta > 0$. First, we consider the system Eq. (2.2) with a double-well potential and without the high-frequency force $g\cos(\Omega - \omega)t$.

2.5.1. *System with the force* $F_2(t) = f\cos\omega t + g\cos(\Omega + \omega)t$

For the double-well system, we can follow the theoretical treatment employed for the single-well system in Sec. 2.3.1. For a detailed analysis, see [4]. Now, in Eqs. (2.4)–(2.11), $\omega_0^2 < 0$. For the double-well case, the equilibrium points are given by

$$(X_0^*, \dot{X}_0^*) = (0, 0), \quad (X_\pm^*, \dot{X}_\pm^*) = (\pm\sqrt{-C_1/\beta}, 0) \tag{2.21}$$

with

$$C_1 = -|\omega_0^2| + \frac{3\beta g^2}{2(\Omega+\omega)^4}\,. \tag{2.22}$$

We find that for $g < g_{\mathrm{c}}$

$$g_{\mathrm{c}} = (\Omega+\omega)^2\sqrt{\frac{2|\omega_0^2|}{3\beta}}, \tag{2.23}$$

where there are three equilibrium points with $(X_0^*, \dot{X}_0^*)$ being unstable, while $(X_\pm^*, \dot{X}_\pm^*)$ being stable. For $g > g_{\mathrm{c}}$, only $(X_0^*, \dot{X}_0^*)$ exists and is stable. This means $V_{\mathrm{eff}}(X)$ given by Eq. (2.7) with $\omega_0^2 < 0$ is a double-well form for $g < g_{\mathrm{c}}$ and becomes a single-well form for $g > g_{\mathrm{c}}$.

The resonant frequency ω_{r}^2 given by Eq. (2.9) becomes

$$\omega_{\mathrm{r}}^2 = \begin{cases} 2|\omega_0^2| - \dfrac{3\beta g^2}{(\Omega+\omega)^4} & \text{for } g < g_{\mathrm{c}} \\ -|\omega_0^2| + \dfrac{3\beta g^2}{2(\Omega+\omega)^4} & \text{for } g > g_{\mathrm{c}}. \end{cases} \tag{2.24}$$

When the parameter g is varied, the resonance conditions are $S_g = 4\left(\omega_{\mathrm{r}}^2 - \omega^2\right)\omega_{\mathrm{r}}\omega_{\mathrm{rg}} = 0$ and $S_{gg} > 0$. From $\omega_{\mathrm{r}}^2 - \omega^2 = 0$, the values of g at which $Q(\omega)$ becomes maximum are obtained as

$$g_{\mathrm{VR}}^{(1)} = (\Omega+\omega)^2\sqrt{\frac{2|\omega_0^2| - \omega^2}{3\beta}}, \quad 2|\omega_0^2| > \omega^2, \tag{2.25}$$

$$g_{\mathrm{VR}}^{(2)} = (\Omega+\omega)^2\sqrt{\frac{2(|\omega_0^2| + \omega^2)}{3\beta}}\,. \tag{2.26}$$

Thus, for $2|\omega_0^2| > \omega^2$, resonance occurs at two values of g. Only one resonance occurs for $2|\omega_0^2| < \omega^2$.

Figure 2.8 presents the variation of the theoretical and the numerically computed $Q(\omega)$ and ω_{r}^2 against the control parameter

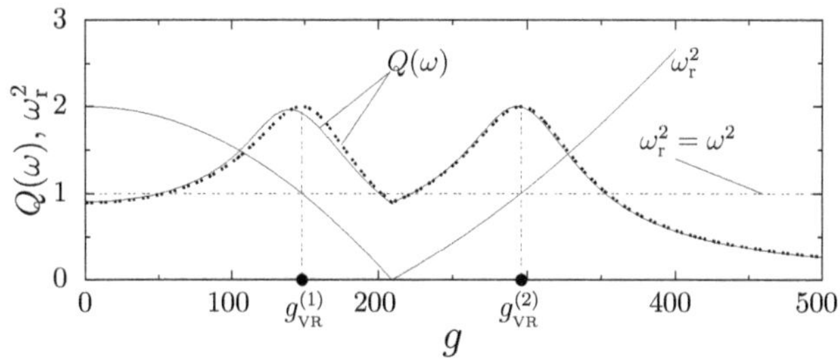

Fig. 2.8. $Q(\omega)$ and ω_r^2 vs. g for the Duffing oscillator (2.2) in the absence of the high-frequency force $g\cos(\Omega - \omega)t$ and for $\omega_0^2 = -1$, $\beta = 1$, $d = 0.5$, $f = 0.1$, $\omega = 1$ and $\Omega = 15$. Continuous and dashed lines represent numerically computed Q and theoretically computed Q. The horizontal dashed line corresponds to $\omega_r^2 = \omega^2 = 1$. Resonances occur at g for which $\omega_r^2 = \omega^2$. The two solid circles on the g-axis mark the values of g_{VR}.

g for $\omega_0^2 = -1$, $\beta = 1$, $d = 0.5$, $f = 0.1$, $\omega = 1$ and $\Omega = 15$. There are two resonances because $2|\omega_0^2| > \omega^2$. Q becomes maximum at two values of g at which ω_r^2 matches with $\omega^2 = 1$. At both resonances $Q = 1/(d\omega) = 2$.

2.5.2. *Vibrational resonance with two high-frequency forces*

Now, we also include the second high-frequency force $g\cos(\Omega - \omega)t$, so that the system is driven by the amplitude-modulated force. For the double-well case, the equations for X and ψ are given by Eq. (2.16) with $\omega_0^2 < 0$. Here also the theoretical $Q(\omega)$ obtained with the approximations $\ddot{\psi} = f\cos(\Omega + \omega)t + g\cos(\Omega - \omega)t$ and $\psi^2 = \langle \psi^2 \rangle$ does not closely match with the numerically computed $Q(\omega)$. Therefore, we consider Eq. (2.16) without the above approximations for ψ and ψ^2. We fix the parameter values as $\omega_0^2 = -1$, $\beta = 1$, $d = 0.5$, $f = 0.1$, $\omega = 1$ and $\Omega = 15$.

Figures 2.9(a) and 2.9(b) present a numerically computed Q of x and X as a function of g for $\omega = 1$ (corresponding to the case $2|\omega_0^2| > \omega^2$) and 1.5 (corresponding to $2|\omega_0^2| < \omega^2$), respectively. In both cases Q of X are in very good agreement with that of x for the considered range of values of g. The presence of an additional

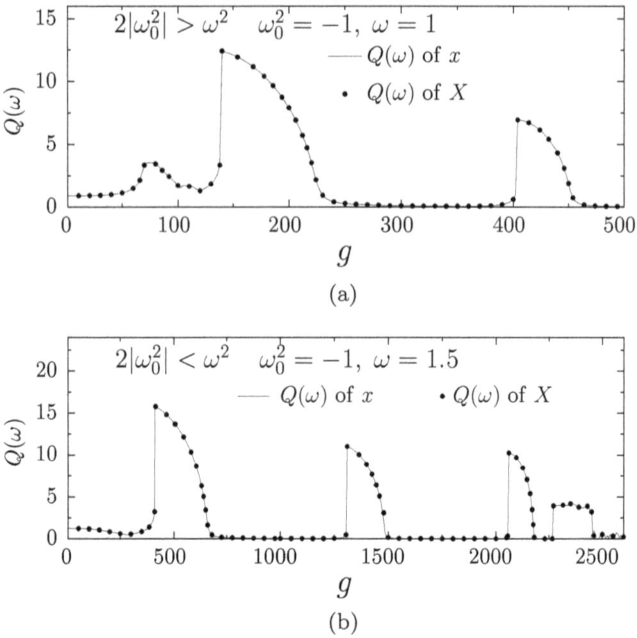

Fig. 2.9. Numerically computed $Q(\omega)$ of x (from Eq. (2.2)) and $Q(\omega)$ of X (from Eq. (2.16)) for the double-well Duffing oscillator system subjected to the amplitude-modulated force for (a) $2|\omega_0^2| > \omega^2$ with $\omega_0^2 = -1$ and $\omega = 1$ and (b) $2|\omega_0^2| < \omega^2$ with $\omega_0^2 = -1$ and $\omega = 1.5$. Other parameter values are $\beta = 1$, $d = 0.5$, $f = 0.1$ and $\Omega = 15$.

second high-frequency force $g\cos(\Omega - -\omega)t$ induces a higher number of resonances compared to the case of only one high-frequency force. For both $\omega = 1$ and $\omega = 1.5$, the resonance peaks are found for g values beyond the range covered in Fig. 2.9. Similar to the case of a single-well potential system, in the double-well case also for various range of values of g it is found that the response amplitude $Q > 1/(d\omega)$, the maximum possible value of Q in the system driven by one high-frequency force.

The double-well potential Duffing oscillator can admit a variety of bounded motions including motions confined to the left-well alone, the right-well alone and cross-well motion with a trajectory visiting both wells. For $\omega = 1$ and for $g < 120$, there are two coexisting orbits in the x–$\dot{x}$ plane as well as in the X–$\dot{X}$ plane. We consider

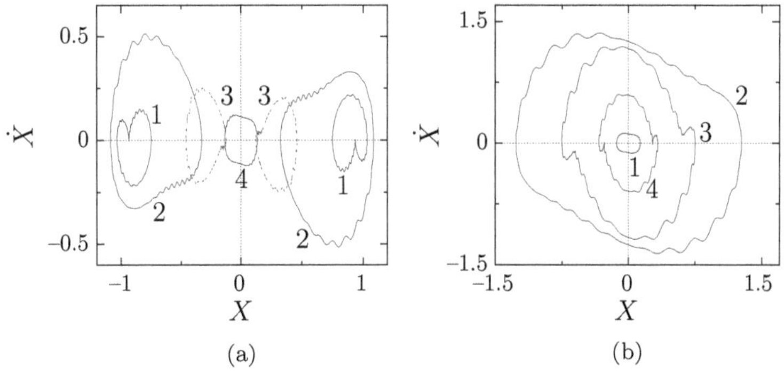

Fig. 2.10. Phase portrait of the slow motion of the double-well Duffing oscillator subjected to the amplitude-modulated force for the case of $\omega_0^2 = -1$, $\beta = 1$, $d = 0.5$, $f = 0.1$, $\omega = 1$ and $\Omega = 15$. (a) For the orbits $1-4$, the values of g being 50, 74.5, 115 and 120, respectively. (b) The values of g for the orbits 1–4 are 120, 139, 200 and 220, respectively.

here the slow variable. The centers of these two orbits are $X_+^{\rm c} > 0$ and $X_-^{\rm c} < 0$. As g increases from 0, both orbits expand and $X_\pm^{\rm c}$ move towards the origin. As the width of the orbit, defined as $X_{\rm w} = X_{\rm max} - X_{\rm min}$, where $X_{\rm max}$ and $X_{\rm min}$ are the maximum and minimum values of $X(t)$, respectively, increases, then the amplitude of X or the response amplitude $Q(\omega)$ increases with increasing values of g. This happens for $g \leq 74.5$. In Fig. 2.10, a phase portrait of a slow motion is plotted for several values of g. The orbits for $g = 50$ and 74.5 in this figure clearly show the increase in $X_{\rm w}$ and $X_\pm^{\rm c}$ approach the origin. At $g = 74.5$, Q reaches a maximum value 3.57. As g is further increased, $X_\pm^{\rm c}$ move towards the origin but $X_{\rm w}$ decreases leading to decrease Q. This happens for $74.5 < g < 120$. An example is shown in Fig. 2.10(a) for $g = 115$. At $g = 120$ both $X_+^{\rm c}$ and $X_-^{\rm c}$ merge together into the origin as shown in Fig. 2.10(a). At this value of g, both Q and $X_{\rm w}$ reach a minimum value.

For $g > 120$, there is no orbit confined to the region $X < 0$ alone and $X > 0$ alone. A second resonance occurs at $g = 139$. Figure 2.10(b) presents a phase portrait of a slow motion for a few values of g chosen on either side of $g = 139$. For $g = 120$, $X_{\rm w}$ is very small and is large and maximum for $g = 139$. A decrease in $X_{\rm w}$ is clearly seen in Fig. 2.10(b) for $g = 200$ and 220. In all cases, the center of the orbits is the origin. Similar behavior of X happens for g values around other resonances.

The orbits shown in Fig. 2.10 for various values of g clearly indicate that X cannot be approximated as $Q(\omega)f\cos(\omega t+\phi)$. It also has an slow–fast component as noted in the case of the single-well potential Duffing oscillator.

Hysteresis and a jump phenomenon are also found on the response amplitude curve when g is varied in the forward and reverse directions. Figure 2.11 presents Q, X_{w} and a bifurcation phenomenon

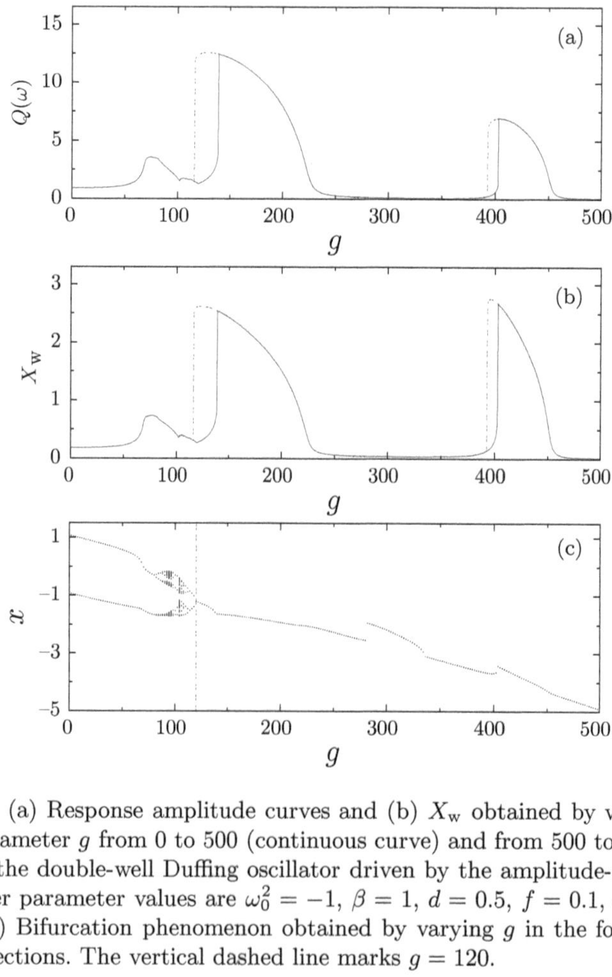

Fig. 2.11. (a) Response amplitude curves and (b) X_{w} obtained by varying the control parameter g from 0 to 500 (continuous curve) and from 500 to 0 (dashed curve) for the double-well Duffing oscillator driven by the amplitude-modulated force. Other parameter values are $\omega_0^2=-1$, $\beta=1$, $d=0.5$, $f=0.1$, $\omega=1$ and $\Omega=15$. (c) Bifurcation phenomenon obtained by varying g in the forward and reverse directions. The vertical dashed line marks $g=120$.

with g. We numerically calculated X_{w} with step size 0.25. For each value of g, 500 values of X_{w} corresponding to each period $T = 2\pi/\omega$ of the low-frequency driving force over the time interval $[0, 500T]$ are calculated. Then, the average value of X_{w} is obtained. We can clearly notice the similarities in the variation of Q and X_{w} in Figs. 2.11(a) and 2.11(b). In the bifurcation diagram for $g < 120$ (indicated by the vertical dashed line), the upper (lower) level points represent the values of x collected after every period T of the low-frequency force for the orbits existing in the region $x > 0$ ($x < 0$). For $g \in [82, 114]$, a period-doubling route to chaos and reverse bifurcation to period-T orbit occur.

2.6. Conclusion

Vibrational resonance in the undamped Duffing oscillator driven by the amplitude-modulated force is reported. The amplitude-modulated force can be recast into the form $F_3(t) = f\cos\omega t + g\cos(\Omega+\omega)t + g\cos(\Omega-\omega)t$. We compared the VR induced by the two high-frequency forces with that of the single high-frequency force. The VR phenomenon induced by the amplitude-modulated force has many interesting differences over the case of only one high-frequency force. Particularly, a higher number of resonance peaks, an enhanced response amplitude $Q(\omega)$ and a nondecaying behavior of $Q(\omega)$ even for large values of the control parameter g are realized. It is noteworthy to mention that there is no restriction on the number of resonance peaks with the system driven by the amplitude-modulated force. The slow variable X in the decomposition of the state variable x into $X+\psi$, where ψ is the fast variable, is not simply of the form $A_{\mathrm{L}}\cos(\omega t+\phi)$ but is modulated by a slow–fast motion as well.

Investigation of the effect of a second high-frequency force in different kinds of systems and the influence on other physically interesting forces including frequency-modulated force, pulse-modulated force and multiexternal forces may give rise fascinating results.

Acknowledgments

This work was supported by the Spanish State Research Agency (AEI) and the European Regional Development Fund (ERDF) under Project No. FIS2016-76883-P.

References

[1] S. Rajasekar and M. A. F. Sanjuan, *Nonlinear Resonances* (Springer, New York, 2016).

[2] P. S. Landa and P. V. E. McClintock, Vibrational resonance, *J. Phys. A: Math. Gen.* **33**(45) (2000) L433–L438.

[3] M. Gittermann, Bistable oscillator driven by two periodic fields, *J. Phys. A: Math. Gen.* **34**(24) (2001) L355–L357.

[4] I. I. Blekhman and P. S. Landa, Conjugate resonances and bifurcations in nonlinear systems under biharmonical excitation, *Int. J. Nonlin. Mech.* **39**(3) (2004) 421–426.

[5] A. A. Zaikin, L. Lopez, J. P. Baltanas, J. Kurths and M. A. F. Sanjuan, Vibrational resonance in a noise-induced structure, *Phys. Rev. E* **66**(1) (2002) 011106.

[6] E. Ullner, A. Zaikin, J. Garcia-Ojalvo, R. Bascones and J. Kurths, Vibrational resonance and vibrational propagation in excitable systems, *Phys. Lett. A* **312**(5–6) (2003) 348–354.

[7] B. Deng, J. Wang and X. Wei, Effect of chemical synapse on vibrational Resonance in coupled neurons, *Chaos* **19**(1) (2009) 013117.

[8] C. Stan, C. P. Cristescu, D. Alexandroaei and M. Agop, Stochastic resonance and vibrational resonance in an excitable system: The golden mean barrier, *Chaos, Solitons Fractals* **41**(2) (2009) 727–734.

[9] S. Jeyakumari, V. Chinnathambi, S. Rajasekar and M. A. F. Sanjuan, Single and multiple vibrational resonance in a quintic oscillator with monostable potentials, *Phys. Rev. E* **80**(4) (2009) 046608.

[10] S. Jeyakumari, V. Chinnathambi, S. Rajasekar and M. A. F. Sanjuan, Analysis of vibrational resonance in a quintic oscillator, *Chaos* **19**(4) (2009) 043128.

[11] U. E. Vincent, T. O. Roy-Layinde, O. O. Popoola, P. O. Adesina and P. V. E. McClintock, Vibrational resonance in an oscillator with an asymmetrical deformable potential, *Phys. Rev. E* **98**(6) (2018) 062203.

[12] S. Rajasekar, K. Abirami and M. A. F. Sanjuan, Novel vibrational resonance in multistable systems, *Chaos* **21**(3) (2011) 033106.

[13] T. O. Roy-Layinde, J. A. Laoye, O. O. Popoola and U. E. Vincent, Analysis of vibrational resonance in bi-harmonically driven plasma, *Chaos* **26**(9) (2016) 093117.

[14] J. P. Baltanas, L. Lopez, I. I. Blechman, P. S. Landa, A. Zaikin, J. Kurths and M. A. F. Sanjuan, Experimental evidence, numerics, and theory of vibrational resonance in bistable systems, *Phys. Rev. E* **67**(6) (2003) 066119.

[15] V. N. Chizhevsky and G. Giacomelli, Experimental and theoretical study of the noise-induced gain degradation in vibrational resonance, *Phys. Rev. E* **70**(6) (2004) 062101.

[16] V. N. Chizhevsky, Analytical study of vibrational resonance in an overdamped bistable oscillator, *Int. J. Bifur. Chaos* **18**(6) (2008) 1767–1773.

[17] V. M. Gandhimathi and S. Rajasekar, Vibrational and stochastic resonances in two coupled overdamped anharmonic oscillators driven by an amplitude modulated force, *Phys. Scr.* **76**(6) (2007) 693–698.

[18] F. Ci-Jun and L. Xian-Bin, Theoretical analysis on the vibrational resonance in two coupled overdamped anharmonic oscillators, *Chin. Phys. Letts.* **29**(5) (2012) 050504.

[19] P. Sarkar and D. Shankar Ray, Vibrational antiresonance in nonlinear coupled systems, *Phys. Rev. E* **99**(5) (2019) 052221.

[20] J. H. Yang and H. Zhu, Vibrational resonance in Duffing systems with fractional-order damping, *Chaos* **22**(1) (2012) 013112.

[21] J. H. Yang, M. A. F. Sanjuan, F. Tian and H. F. Yang, Saddle-node bifurcation and vibrational resonance in a fractional system with an asymmetric bistable potential, *Int. J. Bifur. Chaos* **25**(2) (2015) 1550023.

[22] T. Qin, T. Xie, M. Luo and K. Deng, Vibrational resonance in fractional-order overdamped multistable systems, *Chin. J. Phys.* **55**(2) (2017) 546–555.

[23] J. H. Yang, M. A. F. Sanjuan and H. G. Liu, Enhancing the weak signal with arbitrary high-frequency by vibrational resonance in fractional-order Duffing oscillators, *J. Comput. Nonlinear Dynam.* **12**(5) (2017) 051011.

[24] J. Yang, D. Huang, M. A. F. Sanjuan and H. Liu, Vibrational resonance in an overdamped system with a fractional order potential nonlinearity, *Int. J. Bifur. Chaos* **28**(7) (2018) 1850082.

[25] W. Guo and L. Ning, Vibrational resonance in a fractional order quintic oscillator system with time delay feedback, *Int. J Bifur. Chaos* **30**(2) (2020) 2050025.

[26] J. H. Yang and X. B. Liu, Controlling vibrational resonance in a multistable system by time delay, *Chaos* **20**(3) (2010) 033124.

[27] H. Yang and X. B. Liu, Controlling vibrational resonance in a delayed multistable system driven by an amplitude-modulated signal, *Phys. Scr.* **82**(2) (2010) 025006.

[28] C. Jeevarathinam, S. Rajasekar and M. A. F. Sanjuan, Theory and numerics of vibrational resonance in Duffing oscillators with time-delayed feedback, *Phys. Rev. E* **83**(6) (2011) 066205.
[29] A. Daza, A. Wagemakers, S. Rajasekar and M. A. F. Sanjuan, Vibrational resonance in a time-delayed genetic toggle switch, *Commun. Nonlinear Sci. Numer. Simul.* **18**(2) (2013) 411–416.
[30] C. Jeevarathinam, S. Rajasekar and M. A. F. Sanjuan, Effect of multiple time-delay on vibrational resonance, *Chaos* **23**(1) (2013) 013136.
[31] B. Deng, J. Wang, X. Wei, K. M. Tsang and W. L. Chan, Vibrational resonance in neuron populations, *Chaos* **20**(1) (2010) 013113.
[32] H. Yu, J. Wang, C. Liu, B. Deng and X. Wei, Vibrational resonance in excitable neuronal systems, *Chaos* **21**(4) (2011) 043101.
[33] Y. M. Qin, J. Wang, C. Men, B. Deng and X. Wei, Vibrational resonance in feedforward network, *Chaos* **21**(2) (2011) 023133.
[34] M. Uzuntarla, E. Yilmaza, A. Wagemakers and M. Ozer, Vibrational resonance in a heterogeneous scale free network of neurons, *Commun. Nonlinear Sci. Numer. Simul.* **22**(1–3) (2015) 367–374.
[35] C. Han, Y. Qin, Q. Qin, R. Wang, M. Lu, J. Zhao and Y. Che, Vibrational resonance without tuning in a neuronal parallel array, *Physica A* **523** (2019) 204–210.
[36] V. N. Chizhevsky, E. Smeu and G. Giacomelli, Experimental evidence of vibrational resonance in an optical system, *Phys. Rev. Lett.* **91**(22) (2003) 220602.
[37] V. N. Chizhevsky, Experimental evidence of vibrational resonance in a multistable system, *Phys. Rev. E* **89**(6) (2014) 062914.
[38] R. Jothimurugan, K. Thamilmaran, S. Rajasekar and M. A. F. Sanjuan, Experimental evidence for vibrational resonance and enhanced signal transmission in Chua's circuit, *Int. J. Bifur. Chaos* **23**(11) (2013) 1350189.
[39] A. Abusoua and M. F. Daqaq, Experimental evidence of vibrational resonance in a mechanical bistable twin-well oscillator, *J. Comput. Nonlinear Dynam.* **13**(6) (2018) 061002.
[40] Y. Ren and F. Duan, Theoretical and experimental implementation of vibrational resonance in an array of hard limiters, *Physica A* **456** (2016) 319–326.
[41] M. Coccolo, G. Litak, J. M. Seoane and M. A. F. Sanjuan, Energy harvesting enhancement by vibrational resonance, *Int. J. Bifur. Chaos* **24**(6) (2014) 1430029.
[42] J. Gao, J. Yang and S. Liu, Experimental application of vibrational resonance on bearing fault diagnosis, *J. Braz. Soc. Mech. Sci. Eng.* **41** (2019) 6.

[43] K. Abirami, S. Rajasekar and M. A. F. Sanjuan, Vibrational resonance in the Morse oscillator, *Pramana J. Phys.* **81**(1) (2013) 127–141.

[44] J. Juutilainen and R. de Seze, Biological effects of amplitude modulated radiofrequency radiation, *Scan. J. Work Env. Health* **24**(4) (1998) 245–254.

[45] M. Siewe Siewe, C. Tchawoua and S. Rajasekar, Homoclinic bifurcation and chaos in ϕ^6-Rayleigh oscillator with three wells driven by an amplitude modulated force, *Int. J. Bifur. Chaos* **21**(6) (2011) 1583–1593.

[46] L. E. Medina and W. M. Grill, Nerve excitation using an amplitude modulated signal with kilohertz-frequency carrier and non-zero offset, *J. Neuro Eng. Rehab.* **13** (2016) 63.

[47] Y. A. Amer, S. M. Ahmed, M. M. Dahshan and N. M. Ali, Dynamical chaos in ϕ^6-Rayleight oscillator with three wells driven an amplitude modulated force, *J. Adv. Maths.* **12**(8) (2016) 6545–6551.

[48] Y. Shi, Melnikov analysis of chaos and heteroclinic bifurcation in Josephson system driven by an amplitude-modulated force, *Int. J. Dynam. Control* **6**(2) (2018) 589–600.

[49] Y. Yu, Q. Wang, Q. Bi and C. W. Lim, Multiple-S-shaped critical manifold and jump phenomena in low frequency forced vibration with amplitude modulation, *Int. J. Bifur. Chaos* **29**(5) (2019) 1930012.

[50] G. Duffing, *Erzwungene Schwingungen bei Veränderlicher Eigenfrequenz und ihre Technische Bedeutung* (Sammlung Vieweg. Vieweg & Sohn, Braunschweig, 1918).

[51] J. Miles, Resonance and symmetry breaking for a Duffing oscillator, *SIAM J. Appl.* **49**(3) (1989) 968–981.

[52] A. Chowdhury, M. G. Clerc, S. Barbay, I. Robert-Philip and R. Brave, Weak signal enhancement by nonlinear resonance control in a forced nano-electromechanical resonator, *Nat. Commun.* (2020); doi: 10.1038/s41467-020-15827-3.

https://doi.org/10.1142/9789811221903_0003

Chapter 3

Generation of Self-excited and Hidden Multiscroll Attractors in Multistable Systems

R. J. Escalante-González*,‡ and E. Campos†,§

*Instituto Tecnológico de San Luis Potosí,
Av. Tecnologico S/N, UPA C.P. 78437,
Soledad de Graciano Sánchez, San Luis Potosí, Mexico
†Instituto Potosino de Investigación Científica y Tecnológica
A.C., División de Control and Sistemas Dinámicos,
Camino a la Presa San Jose 2055, Lomas 4 sección,
San Luis Potosi, Mexico 78216, Mexico
‡rodolfo.eg@slp.tecnm.mx
§eric.campos@ipicyt.edu.mx

In this work, we present approaches to generate self-excited and hidden multiscroll attractors by defining a vector field on $\mathbb{R}^3$ with an even number of equilibria. The vector field is defined by affine linear systems such that each equilibrium point is a saddle point. Hence, the space is partitioned in hyperbolic set. We start by generating a self-excited multiscroll attractor based on heteroclinic orbits. Interesting phenomena appear when the equilibria are separated by pairs, firstly, the system presents only one basin of attraction which is divided accordingly with the separation of the equilibria, and the coexistence of different self-excited double-scroll attractors arise. At a certain separation of equilibria, a hidden multiscroll attractors emerges. In this process of separation of equilibria, the resulting basin of attraction presents a widening.

3.1. Multiscroll Attractor

In general, a scroll attractor is such that the trajectories wrap or unwrap around a region of the phase space. According to [1], an attractor with three or more scrolls in the attractor is considered

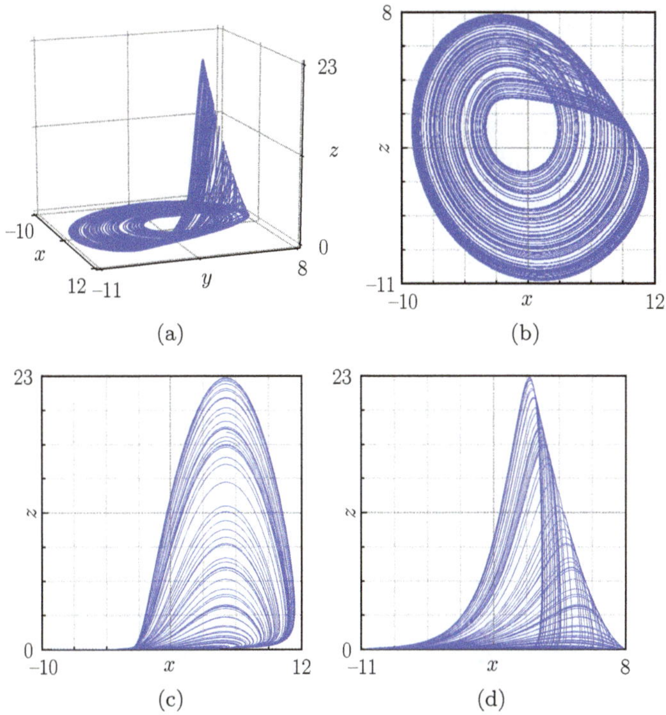

Fig. 3.1. Rössler attractor in (a) the phase space (x, y, z), and its projections onto the planes: (b) (x, y), (c) (x, z) and (d) (y, z).

a multiscroll attractor. The well-known Rössler attractor shown in Fig. 3.1 is an example of a single-scroll attractor whose system description is given by the following mathematical model:

$$\begin{aligned} \dot{x} &= -(y + z), \\ \dot{y} &= x + 0.2y, \\ \dot{z} &= 0.2 + z(x - 5.7), \end{aligned} \tag{3.1}$$

where x, y and z are the states of the system. The Rössler attractor belongs to the class of attractors known as self-excited because arbitrary neighborhoods around the equilibria intersect the basin of attraction of the system. The equilibria of the system are given by $(0.0070, -0.0351, 0.0351)$ and $(5.6929, -28.4648, 28.4648)$. The equilibria give a lot of information about the dynamics of the system.

The nonlinearity of this class of systems is given by the multiplication of two states. On the other hand, there are systems that generate the nonlinearity by piecewise linear systems.

The Chua's circuit [2] is a well-known piecewise linear system (PWL) which exhibits a double-scroll attractor as shown in Fig. 3.2. The equations of the Chua's circuit are as follows:

$$\begin{aligned}\dot{x} &= \alpha(-x + y - f(x)),\\ \dot{y} &= x - y + z,\\ \dot{z} &= -\beta y,\end{aligned} \tag{3.2}$$

where x, y and z are the states of the system. Here, α and β are system's parameters. The piecewise nonlinear function $f(x)$ is

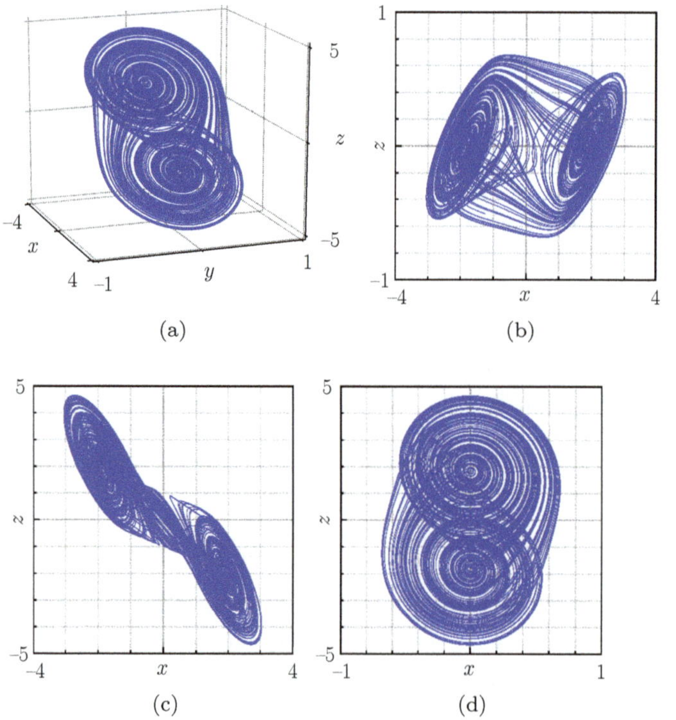

Fig. 3.2. Double-scroll attractor exhibited by the system (3.2) for $\alpha = 10$, $\beta = 14.87$, $m_0 = -0.68$, $m_1 = -1.27$ in (a) (x, y, z), (b) (x, y), (c) (x, z) and (d) (y, z).

given by

$$f(x) = m_0 x + \frac{1}{2}(m_1 - m_0)(|x+1| - |x-1|). \tag{3.3}$$

There are some works that are based on Chua's circuit but allows the generation of more scrolls, as the system reported by Suykens *et al.* [3]. They proposed a generalization of the Chua's circuit in order to construct n-scroll attractors. The generalization is as follows:

$$\begin{aligned} \dot{x} &= \alpha(y - h(x)), \\ \dot{y} &= x - y + z, \\ \dot{z} &= -\beta y, \end{aligned} \tag{3.4}$$

with

$$h(x) = m_{2q-1}x + \frac{1}{2}\sum_{i=1}^{2q-1}(m_{i-1} - m_i)(|x - c_i| - |x - c_i|), \tag{3.5}$$

where q is a positive integer, and α, β, q, m_k and c_l are system's parameters. An example with five scrolls is shown in Fig. 3.3.

There are also other approaches that are not based on Chua's circuit as the system reported by Yalçin *et al.* [4]. This approach uses systems of the form:

$$\dot{\mathbf{x}} = A\mathbf{x} + B\Phi(\mathbf{x}) \tag{3.6}$$

with

$$A = \begin{bmatrix} 0 & 1 & 0 \\ 0 & 0 & 1 \\ -a & -a & -a \end{bmatrix}, \quad B = \begin{bmatrix} 0 & 0 & 0 \\ 0 & 0 & 0 \\ 0 & 0 & a \end{bmatrix}, \tag{3.7}$$

$$\Phi = \begin{bmatrix} 0 \\ 0 \\ f_1(x) \end{bmatrix}, \tag{3.8}$$

$$f_1(x) = \sum_{i=1}^{M_x} g_{\frac{(-2i+1)}{2}}(x) + \sum_{i=1}^{N_x} g_{\frac{(2i-1)}{2}}(x) \tag{3.9}$$

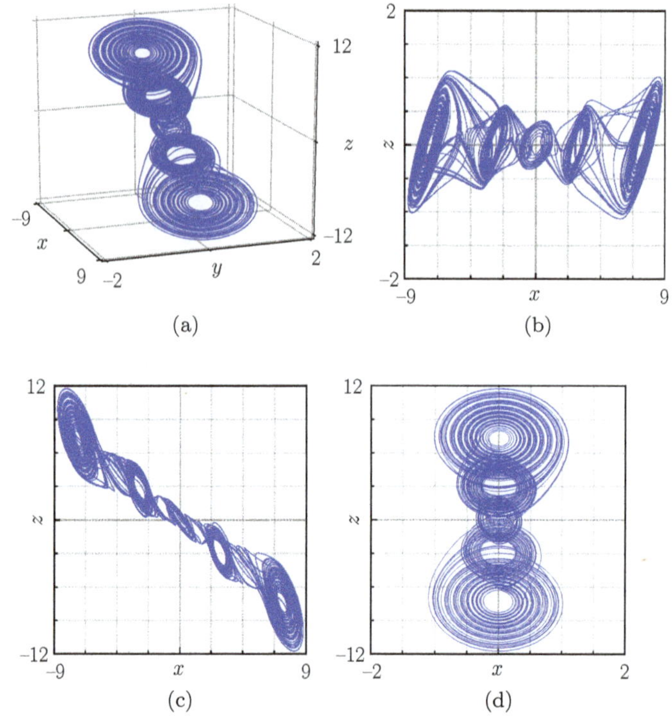

Fig. 3.3. Attractor with five scrolls exhibited by the system (3.4) with $\alpha = 9$, $\beta = 14.286$, $q = 3$, $m = [m_0;\ m_1; \ldots;\ m_{2q-1}] = [0.9/7;\ -3/7;\ 3.5/7;\ -2.7/7;\ 4/7;\ -2.4/7]$, $c = [c_1;\ c_2; \ldots; c_{2q-1}] = [1;\ 2.15;\ 3.6; 6.2; 9]$ in the phase space (a) (x, y, z), and its projections onto the plane: (b) (x, y), (c) (x, z) and (d) (y, z).

and

$$g_\theta(\zeta) = \begin{cases} 1, & \zeta \geq \theta\ \theta > 0, \\ 0, & \zeta < \theta\ \theta > 0, \\ 0, & \zeta \geq \theta\ \theta < 0, \\ -1, & \zeta < \theta\ \theta < 0, \end{cases} \tag{3.10}$$

where $\mathbf{x} = (x, y, z)^{\mathrm{T}} \in \mathbb{R}^3$ and $\zeta \in \mathbb{R}$.

The number of scrolls in the attractor generated by the system given by (3.6) and (3.7) is $M_y + N_y + 1$. To illustrate the approach, we consider the parameters $a = 0.8$, $M_y = 0$ and $N_y = 2$. Then, the system (3.6) exhibits a double-scroll attractor as shown in Fig. 3.4.

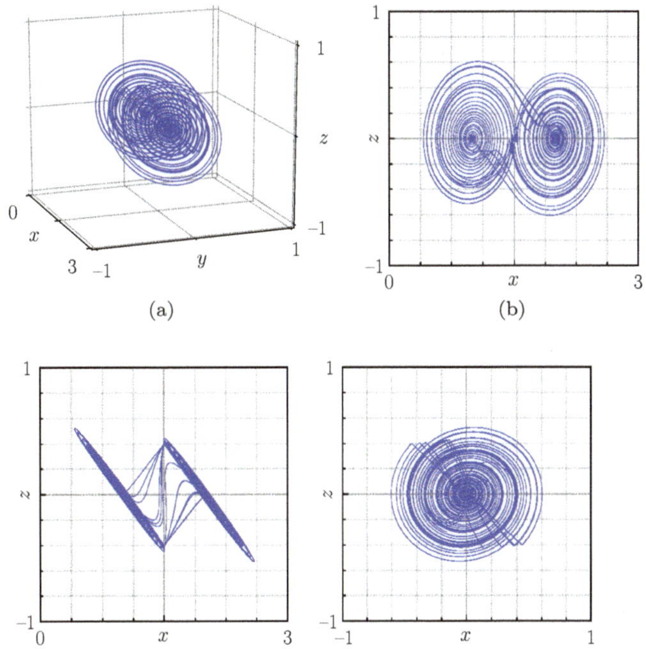

Fig. 3.4. Attractor exhibited by the system (3.6)–(3.8) with $a = 0.8$, $M_y = 0$ and $N_y = 2$ in the phase space (a) (x, y, z), and its projections onto the planes: (b) (x, y), (c) (x, z) and (d) (y, z).

Multiscroll attractors where the scrolls form a line are also known as one-directional (1D) grid-scroll attractors. When the scrolls are formed in two or three directions, then the attractors are known as two-dimensional (2D) or three-dimensional (3D) grid-scroll attractors, respectively. An approach for 2D grid-scroll attractors is also reported in [4]. For a 2D grid, the same system (3.6) is used with

$$A = \begin{bmatrix} 0 & 1 & 0 \\ 0 & 0 & 1 \\ -a & -a & -a \end{bmatrix}, \quad B = \begin{bmatrix} -1 & 0 & 0 \\ 0 & 0 & 0 \\ 0 & 0 & a \end{bmatrix} \tag{3.11}$$

and

$$\Phi = \begin{bmatrix} f_1(y) \\ 0 \\ f_2(x) \end{bmatrix} \tag{3.12}$$

with

$$f_1(y) = \sum_{i=1}^{M_y} g_{\frac{(-2i+1)}{2}}(y) + \sum_{i=1}^{N_y} g_{\frac{(2i-1)}{2}}(y), \tag{3.13}$$

$$f_2(x) = \sum_{i=1}^{m-1} \beta g_{p_i}(x), \tag{3.14}$$

where

$$p_i = M_y + 0.5 + (i-1)(M_y + N_y + 1) \tag{3.15}$$

$$\beta = M_y + N_y + 1. \tag{3.16}$$

The number of scroll in this approach is $m(M_y + N_y + 1)$ in an arrangement of $m \times (M_y + N_y + 1)$. To illustrate a 2D grid-scroll attractor, we consider the parameters $a = 0.8$, $M_y = 0$, $N_y = 2$ and $m = 3$. Then, the system (3.6) exhibits a 3×3 grid attractor as shown in Fig. 3.5.

For a 3D grid-scroll attractor, the same system (3.6) is used with

$$A = \begin{bmatrix} 0 & 1 & 0 \\ 0 & 0 & 1 \\ -a & -a & -a \end{bmatrix}, \quad B = \begin{bmatrix} -1 & 0 & 0 \\ 0 & -1 & 0 \\ 0 & 0 & a \end{bmatrix}, \quad \Phi = \begin{bmatrix} f_1(y) \\ f_1(z) \\ f_3(x) \end{bmatrix}, \tag{3.17}$$

where

$$f_1(z) = \sum_{i=1}^{M_z} g_{\frac{(-2i+1)}{2}}(z) + \sum_{i=1}^{N_z} g_{\frac{(2i-1)}{2}}(z) \tag{3.18}$$

and

$$f_3(x) = \sum_{l=1}^{k-1} \gamma g_{n_l}(x), \tag{3.19}$$

where

$$n_l = \rho + 0.5 + (l-1)(\rho + \varsigma + 1),$$

$$\gamma = \rho + \varsigma + 1$$

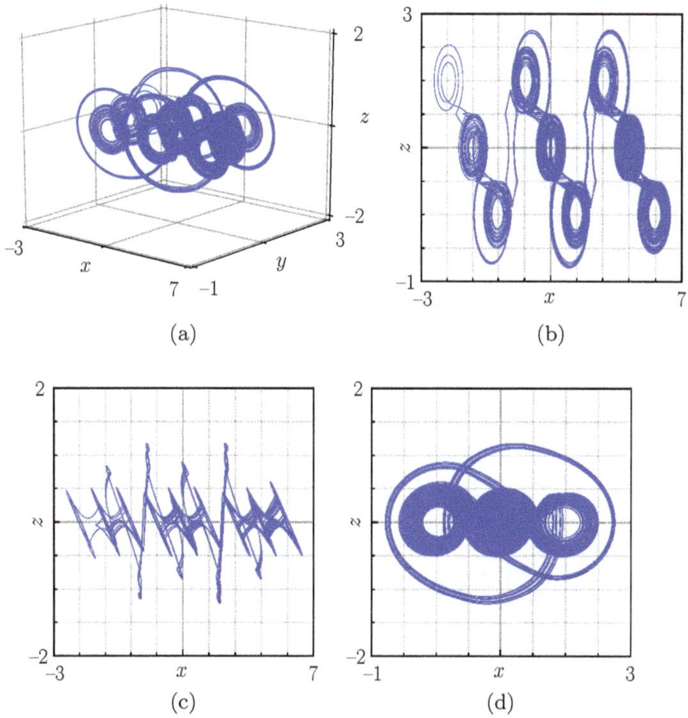

Fig. 3.5. Attractor exhibited by the system given by (3.6), (3.11) and (3.12) with $a = 0.8$, $M_y = 0$, $N_y = 2$ and $m = 3$ in the phase space (a) (x, y, z), and its projections onto the planes: (b) (x, y), (c) (x, z) and (d) (y, z).

with

$$\rho = \left| \min_{i,j} \{u_i^{eq,y} + u_j^{eq,z}\} \right|,$$

$$\varsigma = \left| \max_{i,j} \{u_i^{eq,y} + u_j^{eq,z}\} \right|,$$

and $u^{eq,y}$ and $u^{eq,z}$ are the vectors related to the equilibria for the variables y and z.

The equilibria are obtained from:

$$\begin{aligned} x + y + z &= f_3(x), \\ y &= f_1(y), \\ z &= f_1(z), \end{aligned} \tag{3.20}$$

where the y and z states are as follows:

$$u_i^{eq,y} = [-M_y; \ldots; -1; 0; 1; \ldots; N_y], \tag{3.21}$$

$$u_i^{eq,z} = [-M_z; \ldots; -1; 0; 1; \ldots; N_z]. \tag{3.22}$$

Thus, the approach produces $k \times (M_y + N_y + 1) \times (M_z + N_z + 1)$-scroll grid attractors. As an example of a 3D grid, we consider the parameters $a = 0.8$, $M_y = 0$, $N_y = 1$, $M_z = 0$, $N_z = 1$ and $k = 4$. Then, the system (3.6) exhibits a $4 \times 2 \times 2$ grid attractor as shown in Fig. 3.6.

Another generalization of the dimensionless Chua's circuit was proposed by Tang *et al.* [5]. The approach allows the generation of n-scroll attractors by using a sine function. The modified system is

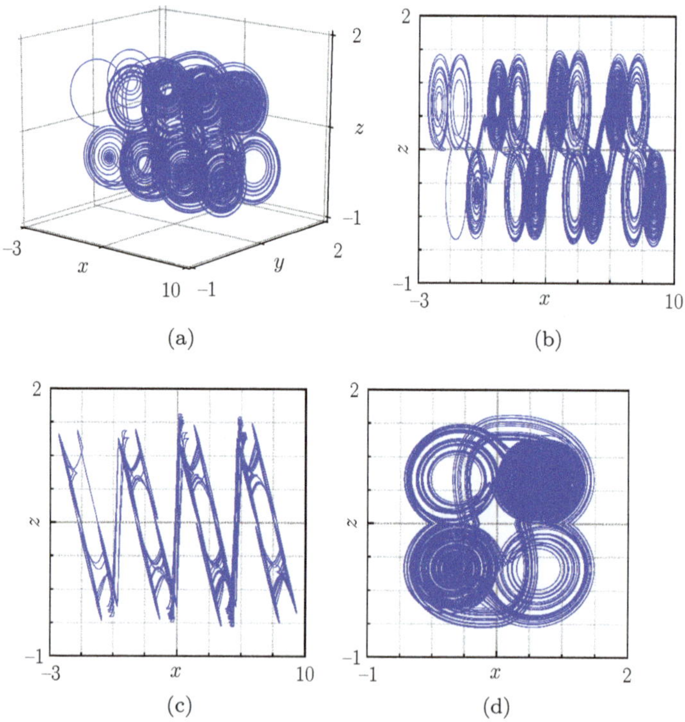

Fig. 3.6. Attractor exhibited by the system (3.6), (3.17) and (3.18) with $a = 0.8$, $M_y = 0$, $N_y = 1$, $M_z = 0$, $N_z = 1$ and $k = 4$ in the phase space (a) (x, y, z), and its projections onto the planes: (b) (x, y), (c) (x, z) and (d) (y, z).

as follows:

$$\begin{aligned}\dot{x} &= \alpha(y - f(x)),\\ \dot{y} &= x - y + z,\\ \dot{z} &= -\beta y\end{aligned} \tag{3.23}$$

with

$$f(x) = \begin{cases} \frac{b\pi}{2a}(x - 2ac) & \text{if}\, x \geq 2ac,\\ -b\sin\left(\frac{\pi x}{2a} + d\right) & \text{if} -2ac < x < 2ac,\\ \frac{b\pi}{2a}(x + 2ac) & \text{if}\, x \leq 2ac.\end{cases} \tag{3.24}$$

An n-scroll attractor can be generated by using the following relationship:

$$n = c + 1, \quad d = \begin{cases} \pi & \text{if } n \text{ is odd},\\ 0 & \text{if } n \text{ is even}.\end{cases} \tag{3.25}$$

For instance, using the parameters $\alpha = 10.814$, $\beta = 14.0$, $a = 1.3$, $b = 0.11$, $c = 2$ and $d = \pi$, a 5-scroll attractor shown in Fig. 3.7 is exhibited by the system (3.23).

Another approach to construct grid scroll attractors was proposed in [6]. This approach is based on the use of Unstable Dissipative System (UDS) defined as follows.

Definition 1 ([6]). Given by $\dot{\mathbf{x}} = A\mathbf{x} + B$ in $\mathbb{R}^3$ and eigenvalues λ_i with $i = 1, \ldots, 3$, we say the system is a UDS type I if $\sum_{i=1}^{3} \lambda_i < 0$, and two λ_i's are complex conjugate eigenvalues with positive real part $\Re\{\lambda_i\} > 0$ and the other is a negative real number.

The approach is a system of the form:

$$\dot{\mathbf{x}} = A\mathbf{x} + B(\mathbf{x}), \tag{3.26}$$

where $\mathbf{x} = (x_1, x_2, x_3)^{\mathrm{T}}$ is the state vector, $A \in \mathbb{R}^{3\times 3}$ is a linear operator and $B \in \mathbb{R}^3$ is a piecewise constant vector given by

$$B(\mathbf{x}) = \begin{cases} \mathrm{B}_1 & \text{if}\mathbf{x} \in D_1\\ \vdots & \quad\vdots\\ \mathrm{B}_k & \text{if}\mathbf{x} \in D_k,\end{cases} \tag{3.27}$$

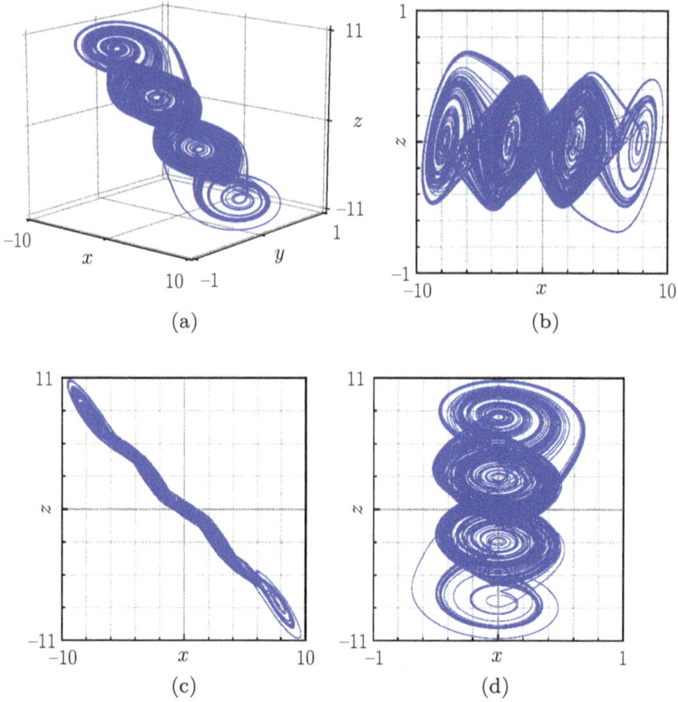

Fig. 3.7. Attractor exhibited by the system (3.23) with $\alpha = 10.814$, $\beta = 14$, $a = 1.3$, $b = 0.11$ and $c = 3$ in the phase space (a) (x, y, z), and its projections onto the planes: (b) (x, y), (c) (x, z) and (d) (y, z).

where $\mathbb{R}^3 = \bigcup_{i=1}^{k} D_i$. The linear operator $A \in \mathbb{R}^{3\times 3}$ is given by

$$A = \begin{pmatrix} 0 & 1 & 0 \\ 0 & 0 & 1 \\ -\alpha_{31} & -\alpha_{32} & -\alpha_{33} \end{pmatrix}, \tag{3.28}$$

where $\alpha_{31}, \alpha_{32}, \alpha_{33} \in \mathbb{R}$.

Thus, the equilibria of the system (3.26) are $\mathbf{x}_i^* = (x_{1i}^*, x_{2i}^*, x_{3i}^*)^{\mathrm{T}} = -A^{-1}B$ with $i = 1, \ldots, k$. The location of the equilibria depends on the vectors B_i:

$$B_i = \begin{pmatrix} b_{1i} \\ b_{2i} \\ b_{3i} \end{pmatrix} = \begin{pmatrix} -\beta_1 \\ -\beta_2 \\ \alpha_{31}\beta_3 + \alpha_{32}\beta_1 + \alpha_{33}\beta_2 \end{pmatrix}. \tag{3.29}$$

Thus, $\beta_1 = x_{2i}^*$, $\beta_2 = x_{3i}^*$ and $\beta_3 = x_{1i}^*$.

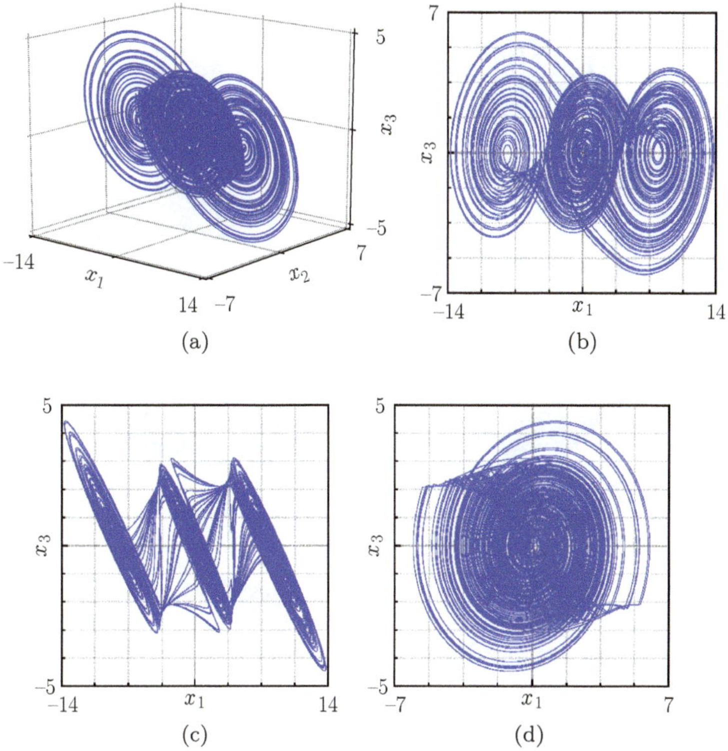

Fig. 3.8. Attractor of the system (3.26), (3.27), (3.28) and (3.30) in the phase space (a) (x, y, z), its projections onto the planes: (b) (x, y), (c) (x, z) and (d) (y, z).

In this way, the approach can be used to produce 1D, 2D and 3D grid-scroll attractors. As an example, we consider the 1D grid-scroll attractor as shown in Fig. 3.8 of the system (3.26) with the parameters $\alpha_{31} = \alpha_{32} = \alpha_{33} = -0.6$, $\beta_1 = \beta_2 = 0$ and with β_3 as follows:

$$\beta_3 = \begin{cases} 8 & \text{if } x_1 \geq 3.8, \\ 0 & \text{if } -3.8 < x_1 < 3.8, \\ -8 & \text{if } x_1 \leq -3.8. \end{cases} \tag{3.30}$$

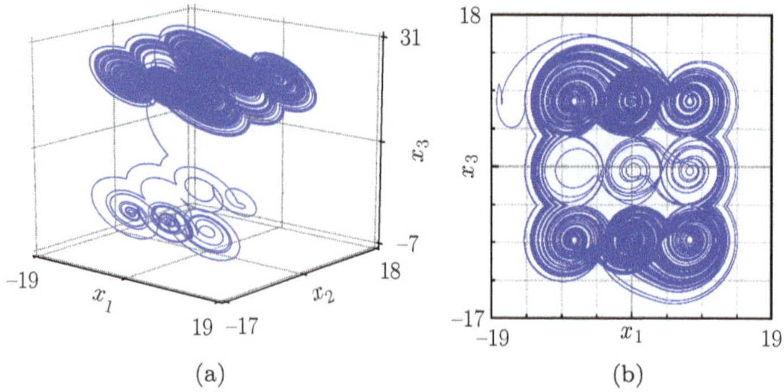

Fig. 3.9. Three-dimensional grid-scroll attractor in the phase space (a) (x, y, z), and its projections onto the planes: (b) (x, y), (c) (x, z) and (d) (y, z).

In Fig. 3.9, a 3D grid-scroll attractor is shown with a distribution of $3 \times 3 \times 3$ scrolls using the parameters $\alpha_{31} = \alpha_{32} = \alpha_{33} = -0.6$, β_3 as in (3.30) and β_2, β_1 as follows:

$$\beta_1 = \begin{cases} 8 & \text{if } x_2 \geq 4, \\ 0 & \text{if } -4 < x_2 < 4, \\ -8 & \text{if } x_2 \leq -4, \end{cases} \tag{3.31}$$

$$\beta_2 = \begin{cases} 24 & \text{if } x_3 \geq 18, \\ 12 & \text{if } 6 < x_3 < 18, \\ 0 & \text{if } x_3 \leq 6. \end{cases} \tag{3.32}$$

3.2. Hidden Attractors

We can find two classes of attractors according to [7]: self-excited attractors and hidden attractors. Self-excited attractors are those excited from unstable equilibria, where the basin of attraction intersect with an arbitrarily small open neighborhood of equilibria [8]. Attractors are hidden when their basin of attraction does not contain neighborhoods of the equilibrium points.

Self-excited attractors can be located numerically using initial conditions in the unstable manifold and in the vicinity of the unstable equilibrium points, however, this is not possible with hidden attractors, since the basin of attraction does not intersect with the vicinity of the equilibrium points.

A failure to consider the hidden attractors in a control system can lead to unexpected results. A known case where a hidden oscillations were not considered and lead to unexpected results is the crash of the aircraft YF-22 Boeing in 1992 [9].

In [10], Leonov *et al.* discussed the application of a special analytical–numerical algorithm for hidden attractor localization. They also demonstrated the existence of a hidden attractor in Chua's circuit.

According to the definition of hidden attractors, an attractor in a system without equilibria is considered a hidden attractor. To illustrate this class of systems, we consider the system proposed by Wei [11]:

$$\begin{aligned} \dot{x} &= -y, \\ \dot{y} &= cx + z, \\ \dot{z} &= ay^2 + xz - d. \end{aligned} \tag{3.33}$$

When $a = 3$, $c = 1$ and $d = 0$, system (3.33) is the Sprott D system [12], however when $a = 2$, $c = 1$ and $d = 0.35$, the system presents no equilibria and exhibits a hidden attractor shown in Fig. 3.10.

In [13], a numerical search was performed to produce a catalog of 17 systems without equilibria with chaotic flows called NE systems. One system from this catalog with a hidden chaotic attractor is the

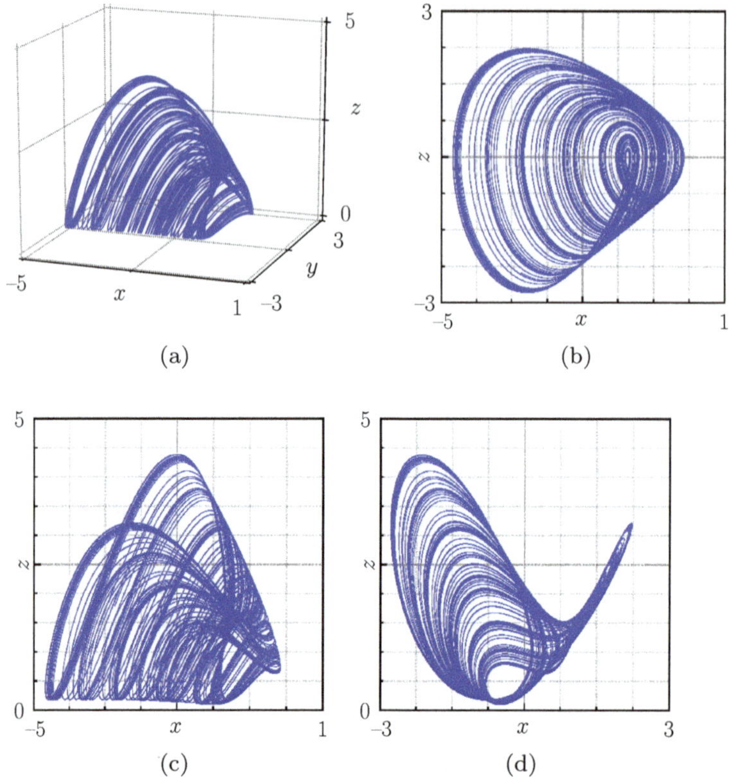

Fig. 3.10. Attractor of the system (3.33) in the phase space (a) (x, y, z), its projections onto the planes: (b) (x, y), (c) (x, z) and (d) (y, z). Initial condition $(-1.6, 0.82, 1.9)^{\mathrm{T}}$.

system NE8 whose description is as follows:

$$
\begin{aligned}
\dot{x} &= y, \\
\dot{y} &= -x - yz, \\
\dot{z} &= xy + 0.5x^2 - 1.3.
\end{aligned}
\tag{3.34}
$$

The hidden attractor exhibited by the system NE8 is shown in Fig. 3.11.

Hidden hyperchaotic attractor in systems without equilibria has also been reported. One example is the system reported by Wang

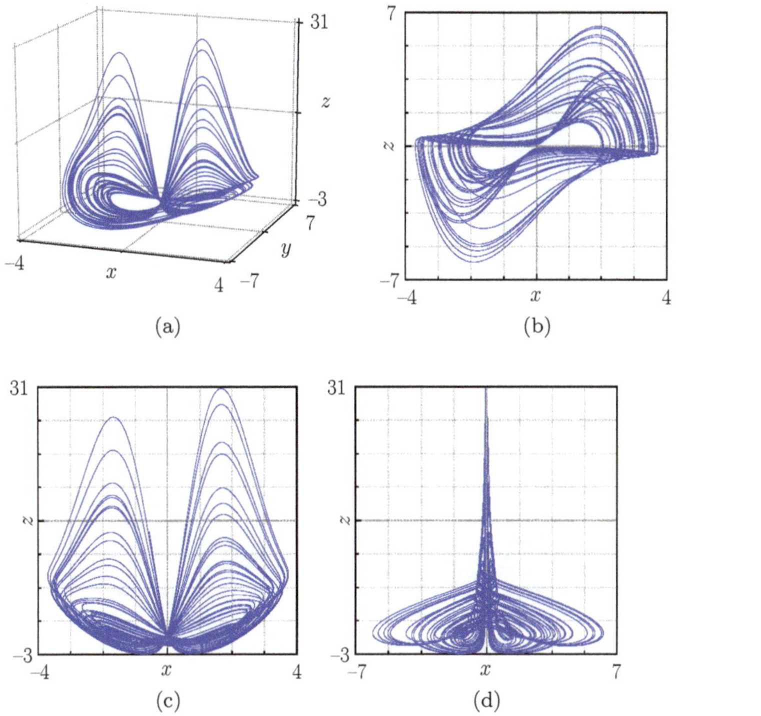

Fig. 3.11. Attractor of the system (3.34) in the phase space (a) (x, y, z), its projections onto the planes: (b) (x, y), (c) (x, z) and (d) (y, z). Initial condition $(0, 0.1, 0)^{\mathrm{T}}$.

et al., which is a four-dimensional system described as follows:

$$\begin{aligned} \dot{x} &= y, \\ \dot{y} &= -x + yz + axzw, \\ \dot{z} &= 1 - y^2, \\ \dot{w} &= z + bxz + cxyz. \end{aligned} \tag{3.35}$$

A simulation for the parameters $a = 8$, $b = -2.5$, and $c = -30$, with the initial condition $(0.1, 0.1, 0.1, 0.1)^{\mathrm{T}}$, is shown in Fig. 3.12.

The first piecewise linear system without equilibria with a hidden hyperchaotic attractor was reported by Li *et al.* This system is an approximated version of a modified diffusionless Lorenz system to which was added a linear feedback. The system is described as

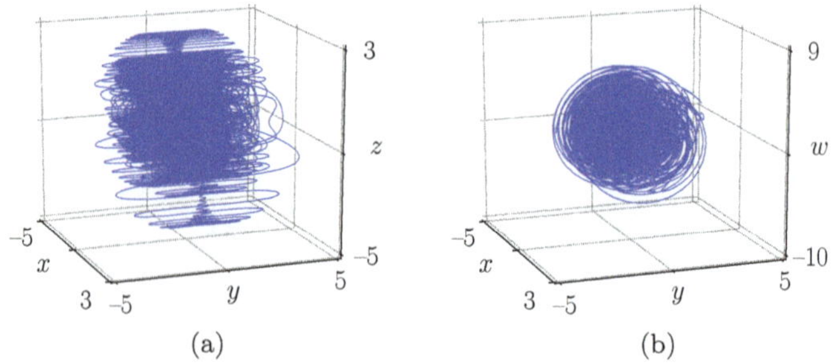

Fig. 3.12. Attractor of the system (3.35) projected onto (a) (x, y, z) and (b) (x, y, w). Initial condition $(0.1, 0.1, 0.1, 0.1)^{\mathrm{T}}$.

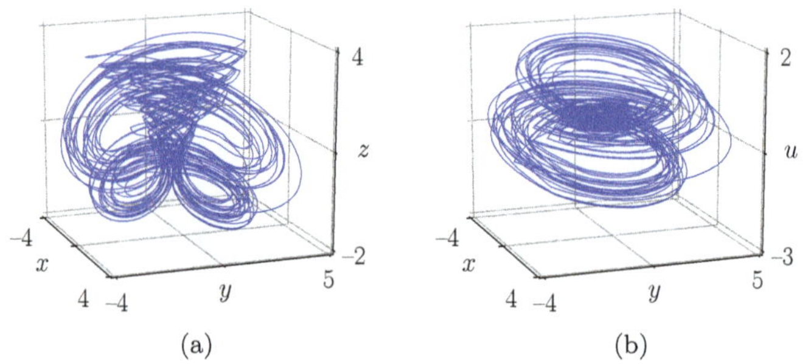

Fig. 3.13. Attractor of the system (3.36) projected onto (a) (x, y, z) and (b) (x, y, u). Initial condition $(0.1, 0.1, 0.1, 0.1)^{\mathrm{T}}$.

follows:

$$\begin{aligned}\dot{x} &= y - x,\\ \dot{y} &= -z\,\mathrm{sign}(x) + u,\\ \dot{z} &= |x| - a,\\ \dot{w} &= -by.\end{aligned} \tag{3.36}$$

The hidden hyperchaotic attractor is shown in Fig. 3.13. An interesting feature of this attractor is its global basin of attraction.

Hidden attractors can also be found in systems with equilibria. For instance, Wang and Chen reported a system with a chaotic attractor and a stable equilibrium point [14]. The system description is as follows:

$$\begin{aligned}\dot{x} &= yz + 0.006,\\ \dot{y} &= x^2 - y,\\ \dot{z} &= 1 - 4x.\end{aligned} \tag{3.37}$$

The hidden attractor is shown in Fig. 3.14.

Molaie *et al.* [15] have performed a systematic computer search using the Routh–Hurwitz stability criterion and found 23 systems with chaotic flows that were called SE systems. All the systems found present quadratic nonlinearities and a stable equilibrium point. One

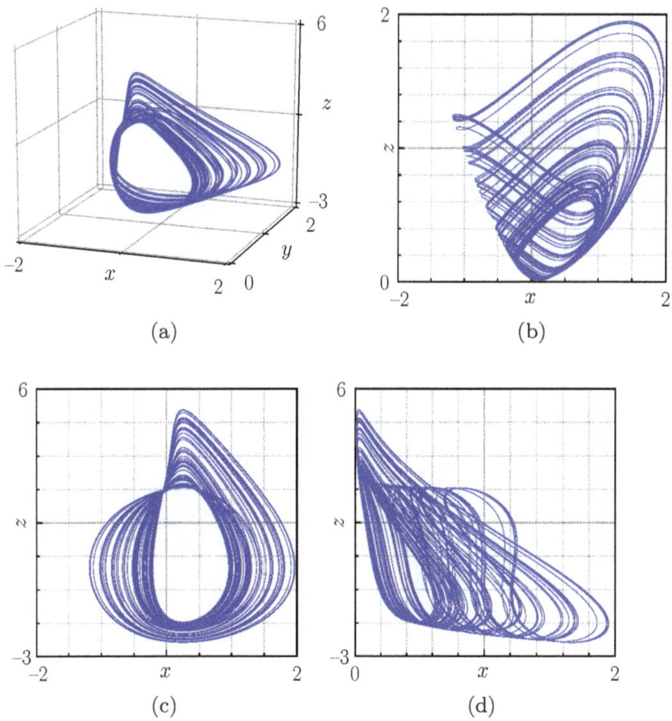

Fig. 3.14. Attractor of the system (3.37) in the phase space (a) (x, y, z), its projections onto the planes: (b) (x, y), (c) (x, z) and (d) (y, z). Initial condition $(1, 1, 1)^{\mathrm{T}}$.

example from this list is the SE8 system which is described as follows:

$$\begin{aligned}\dot{x} &= y,\\ \dot{y} &= -x + yz,\\ \dot{z} &= -z - 0.7x^2 + y^2 - 0.1.\end{aligned} \tag{3.38}$$

The hidden attractor exhibited by the SE8 system is shown in Fig. 3.15.

In [16], Jafari and Sprott introduced a new category of systems with a line equilibrium and a chaotic attractor. Although in such systems the basin of attraction may intersect the line equilibrium in some sections, the attractors are hidden from a computational point of view, i.e., knowledge about equilibria does not help in their

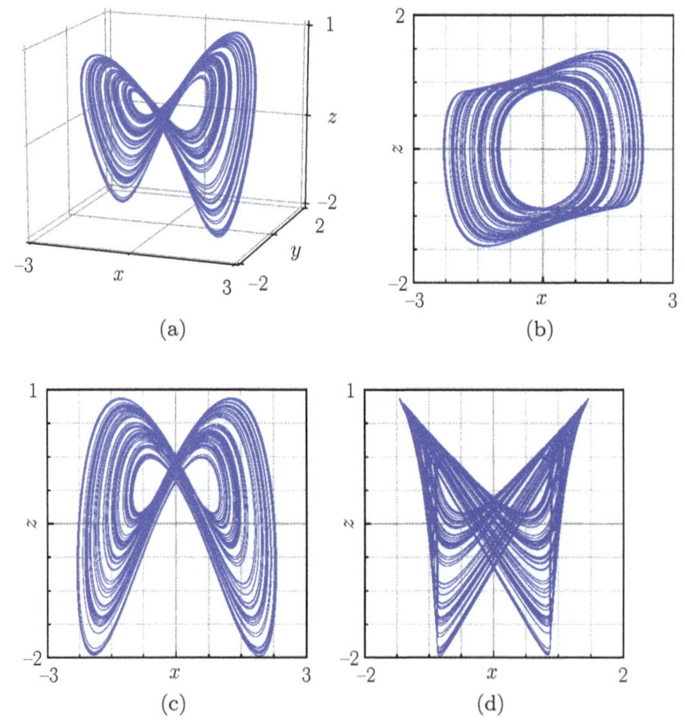

Fig. 3.15. Attractor of the system (3.38) in the phase space (a) (x, y, z), its projections onto the planes: (b) (x, y), (c) (x, z) and (d) (y, z). Initial condition $(0, 0.9, 0)^{\mathrm{T}}$.

localization. Using a systematic computer search, they found nine systems of this class. One of them, the LE3 system, is described as follows:

$$\begin{aligned} \dot{x} &= y, \\ \dot{y} &= -x + yz, \\ \dot{z} &= x^2 - 18xy - xz. \end{aligned} \tag{3.39}$$

The hidden attractor of the system LE3 is shown in Fig. 3.16.

In [17], a new class of piecewise linear system without equilibria with hidden scroll attractors was proposed. The construction of this type of systems is based on the use of linear operators without

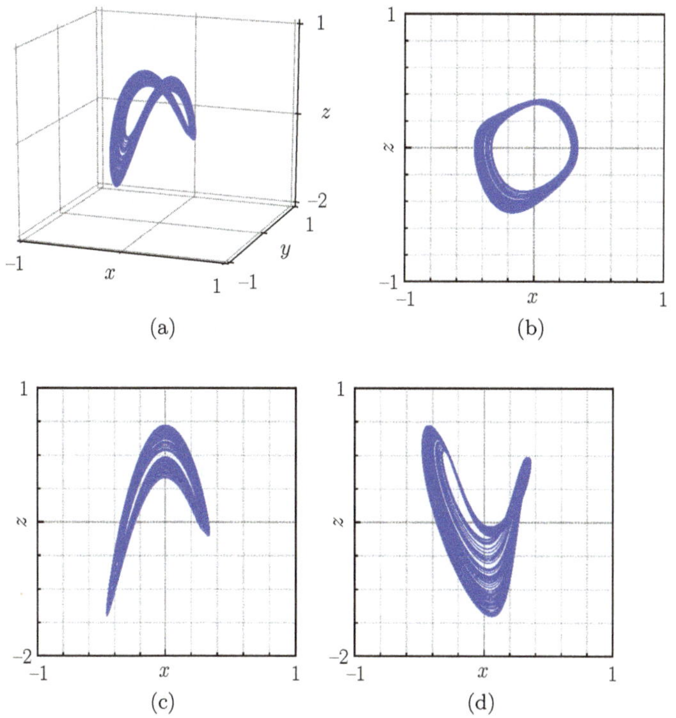

Fig. 3.16. Attractor of the system (3.39) in the phase space (a) (x, y, z), its projections onto the planes: (b) (x, y), (c) (x, z) and (d) (y, z). Initial condition $(0, -0.4, 0.5)^{\mathrm{T}}$.

inverse. An example of this class of systems is as follows:

$$\dot{\mathbf{x}} = \begin{cases} A\mathbf{x} + V + W_1 & \text{if } \mathbf{x} < S_1, x_1 < 0, \\ A\mathbf{x} + V + W_2 & \text{if } \mathbf{x} < S_1, \mathbf{x} < S_2, x_1 \geq 0, \\ A\mathbf{x} + W_1 & \text{if } \mathbf{x} \in S_1, x_1 < 0, \\ A\mathbf{x} + W_2 & \text{if } \mathbf{x} \in S_1, \mathbf{x} < S_2, x_1 \geq 0, \\ A\mathbf{x} - V + W_1 & \text{if } \mathbf{x} > S_1, \mathbf{x} < S_2, x_1 < 0, \\ A\mathbf{x} - V + W_2 & \text{if } \mathbf{x} > S_1, \mathbf{x} < S_2, x_1 \geq 0, \\ A\mathbf{x} + V + W_3 & \text{if } \mathbf{x} < S_3, \mathbf{x} \geq S_2, x_1 < 1, \\ A\mathbf{x} + V + W_4 & \text{if } \mathbf{x} < S_3, \mathbf{x} \geq S_2, x_1 \geq 1, \\ A\mathbf{x} + W_3 & \text{if } \mathbf{x} \in S_3, \mathbf{x} \geq S_2, x_1 < 1, \\ A\mathbf{x} + W_4 & \text{if } \mathbf{x} \in S_3, x_1 \geq 1, \\ A\mathbf{x} - V + W_3 & \text{if } \mathbf{x} > S_3, \mathbf{x} \geq S_2, x_1 < 1, \\ A\mathbf{x} - V + W_4 & \text{if } \mathbf{x} > S_3, x_1 > 1, \end{cases} \tag{3.40}$$

where

$$A = \begin{pmatrix} 0.5 & -10 & 0 \\ 10 & 0.5 & 0 \\ 0 & 0 & 0 \end{pmatrix} = (a_1, a_2, a_3), \quad V = \begin{pmatrix} 0 \\ 0 \\ 5 \end{pmatrix}, \tag{3.41}$$

and $W_i = k_i a_1$ with the parameters k_i given in Table 3.1.

The switching surfaces are given by the planes $S_1 = \{x \in \mathbb{R}^3 : x_3 = 0\}$, $S_2 = \{x \in \mathbb{R}^3 : x_1 + x_3/2 = 1\}$ and $S_3 = \{x \in \mathbb{R}^3 : x_3 = 2\}$. Each S_i divides $\mathbb{R}^n$ in two sets: the notation $x \geq S_i$ is used when x is in the set pointed by the vector $(1, 0, 1/2)^{\mathrm{T}}$ or in S_i and $x < S_i$ if it is in the opposite set.

The hidden double-scroll attractor is shown in Fig. 3.17.

Table 3.1. Vectors W_i for $i = 1, \ldots, 4$.

k_1	k_2	k_3	k_4
–0.1	0.1	–1.1	–0.9

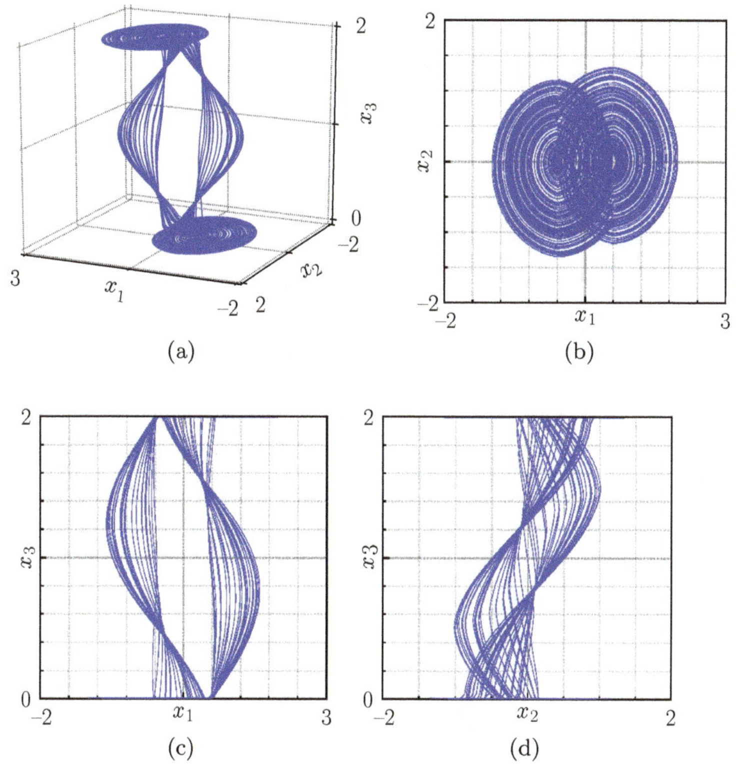

Fig. 3.17. Attractor of the system (3.40) in the phase space (a) (x, y, z), its projections onto the planes: (b) (x, y), (c) (x, z) and (d) (y, z). Initial condition $(0, 0, 0)^{\mathrm{T}}$.

This idea has been extended to a class of systems with grid scroll attractors in [18]. More work on the hidden attractors can be found in [18–28].

3.3. Emergence of a Hidden Double-Scroll Attractor

In [29], the generation of scroll attractors in a multistable systems was observed in a study on the widening of the basin of attraction. In the study, the considered system was given by

$$\dot{x} = Ax + B \tag{3.42}$$

with

$$A = \begin{pmatrix} 0 & 1 & 0 \\ 0 & 0 & 1 \\ -1.5 & 1 & 1 \end{pmatrix}, \quad B = \begin{pmatrix} 0 \\ 0 \\ B_c \end{pmatrix} \tag{3.43}$$

and

$$B_c(x_1) = \begin{cases} \delta + 1 & \text{if} \quad 2/3(\delta + 1/2 \leq x_1, \\ \delta & \text{if} \quad 0 \leq x_1 < 2/3(\delta + 1/2), \\ -\delta & \text{if} \quad -2/3(\delta + 1/2) \leq x_1 < 0, \\ -\delta - 1 & \text{if} \quad x_1 < -2/3(\delta + 1/2). \end{cases} \tag{3.44}$$

In this system, the parameter δ determines the separation of the equilibria. For example, with $\delta = 3$, there are two double-scroll attractors as in Fig. 3.18(a). If the parameter δ is increased, the two

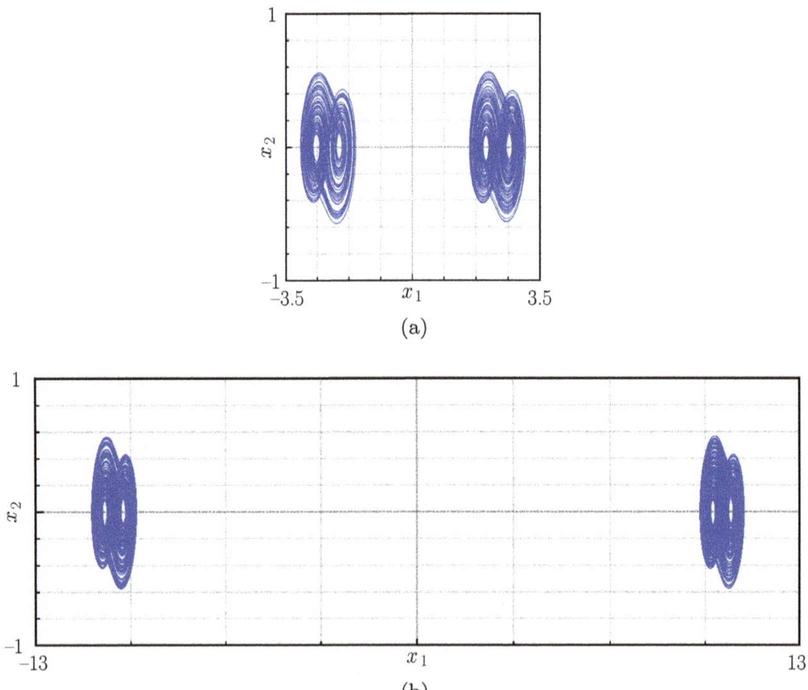

Fig. 3.18. Projections of two double-scroll attractors onto the plane (x, y) for (a) $\delta = 5$ and (b) $\delta = 15$.

double-scroll attractors are separated from each other. For instance, in Fig. 3.18(b) the parameter $\delta = 15$ allows the two double-scroll attractors to be separated more from each other than in the case $\delta = 3$.

If we continue to increase the parameter $\delta = 55$, then a hidden double-scroll attractor appears as in Fig. 3.19(a). The hidden attractor is colored red, while the two self-excited attractors are colored blue (see Fig. 3.19(b)). Each self-excited attractor is located in the center of the hidden attractor scrolls.

This phenomenon of the emergence of a hidden double-scroll attractor was studied in [30] in a class of piecewise linear systems. In order to introduce an approach to generate grid scroll hidden attractors, we will consider the class of systems proposed in [30].

Consider a finite partition of the space $X \subset \mathbb{R}^3$ $P = \{P_1, \ldots, P_\eta\}$ $(\eta > 1)$ such that $X = \bigcup_{1 \le i \le \eta} P_i$, and $P_i \cap P_j = \emptyset$ for $i \neq j$. Switching surfaces SW are located between adjacent atoms, i.e., for each pair of adjacent atoms P_i and P_j, $i \neq j$, $\mathrm{SW}_{i,j} = \mathrm{cl}(P_i) \cap \mathrm{cl}(P_j)$ is a switching surface, where $\mathrm{cl}(P_i)$ denotes the closure of a set P_i.

Now, it is possible to define a dynamical system $T : X \to X$ as follows:

$$\dot{\mathbf{x}} = A\mathbf{x} - f(\mathbf{x})B, \tag{3.45}$$

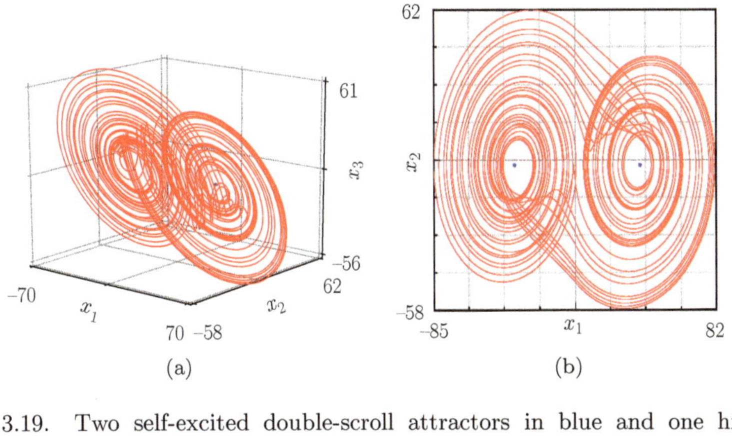

Fig. 3.19. Two self-excited double-scroll attractors in blue and one hidden double-scroll attractor in red in (a) the phase space (x, y, z) and (b) projections of the attractors onto the plane (x, y).

where $\mathbf{x} = (x_1, x_2, x_3)^{\mathrm{T}} \in \mathbb{R}^3$ is the state vector, and $A = \{\alpha_{ij}\} \in \mathbb{R}^{3\times 3}$ is a linear operator given by

$$A = \begin{pmatrix} \frac{a}{3} + \frac{2c}{3} & b & \frac{2c}{3} - \frac{2a}{3} \\ -\frac{b}{3} & a & \frac{2b}{3} \\ \frac{c}{3} - \frac{a}{3} & -b & \frac{2a}{3} + \frac{c}{3} \end{pmatrix}, \tag{3.46}$$

where $a, b \in \mathbb{R}^+$ and $c \in \mathbb{R}^-$. Here, A has a negative real eigenvalue $\lambda_1 = c$ with the eigenvector v_1 and a pair of complex conjugate eigenvalues with positive real part $\lambda_2 = a + ib$ and $\lambda_3 = a - ib$ with the eigenvectors v_2 and v_3, respectively.

The eigenvectors are as follows:

$$v_1 = \begin{pmatrix} 1 \\ 0 \\ \frac{1}{2} \end{pmatrix}, \quad v_2 = \begin{pmatrix} 0 \\ -1 \\ 0 \end{pmatrix}, \quad v_3 = \begin{pmatrix} -1 \\ 0 \\ 1 \end{pmatrix}. \tag{3.47}$$

Moreover, $B = (\beta_1, \beta_2, \beta_3)^{\mathrm{T}} \in \mathbb{R}^3$ is a constant vector. The functional $f : X \to \mathbb{R}$ is such that $f(\mathbf{x})B$ is a constant vector in each atom P_i. Thus, in each atom, there exists an equilibrium point $\mathbf{x}^*_{eq_i} = (x^*_{1_{eq_i}}, x^*_{2_{eq_i}}, x^*_{3_{eq_i}})^{\mathrm{T}} = -f(\mathbf{x})A^{-1}B$, with $i = 1, \ldots, \eta$.

In each atom P_i, there exist a stable manifold given by $W^s_{\mathbf{x}^*_{eq_i}} = \{\mathbf{x} + \mathbf{x}^*_{eq_i} : \mathbf{x} \in \mathrm{span}\{v_1\}\}$ and an unstable manifold given by $W^u_{\mathbf{x}^*_{eq_i}} = \{\mathbf{x} + \mathbf{x}^*_{eq_i} : \mathbf{x} \in \mathrm{span}\{v_2, v_3\}\}$ related to the saddle equilibrium point.

Let $P = \{P_1, P_2\}$ be a partition with two atoms and the constant vector $B \in \mathbb{R}^3$ given by

$$B = \begin{pmatrix} -\frac{a}{3} - \frac{2c}{3} \\ \frac{b}{3} \\ \frac{a}{3} - \frac{c}{3} \end{pmatrix}, \tag{3.48}$$

and the functional f given by

$$f(\mathbf{x}) = \begin{cases} -\alpha, & \mathbf{x} \in P_1, \\ \alpha, & \mathbf{x} \in P_2 \end{cases} \tag{3.49}$$

with $0 < \alpha \in \mathbb{R}$. The equilibria are located at $\mathbf{x}^*_{eq_1} = (-\alpha, 0, 0)^{\mathrm{T}}$ and $\mathbf{x}^*_{eq_2} = (\alpha, 0, 0)^{\mathrm{T}}$. The stable and unstable manifolds are as follows:

$$W^s_{\mathbf{x}^*_{eq_1}} = \{\mathbf{x} \in \mathbb{R}^3 | x_1 + \alpha = 2x_3, x_2 = 0\}, \tag{3.50}$$

$$W^u_{\mathbf{x}^*_{eq_1}} = \{\mathbf{x} \in \mathbb{R}^3 | x_1 + x_3 = -\alpha\}, \tag{3.51}$$

$$W^s_{\mathbf{x}^*_{eq_2}} = \{\mathbf{x} \in \mathbb{R}^3 | x_1 - \alpha = 2x_3, x_2 = 0\}, \tag{3.52}$$

$$W^u_{\mathbf{x}^*_{eq_2}} = \{\mathbf{x} \in \mathbb{R}^3 | x_1 + x_3 = \alpha\}. \tag{3.53}$$

The intersection points between the stable and unstable manifolds of adjacent atoms are as follows:

$$\mathbf{x}_{in_1} = \mathrm{cl}(W^s_{\mathbf{x}^*_{eq_1}}) \cap \mathrm{cl}(W^u_{\mathbf{x}^*_{eq_2}}) = \left(\frac{\alpha}{3}, 0, \frac{2\alpha}{3}\right)^{\mathrm{T}}, \tag{3.54}$$

$$\mathbf{x}_{in_2} = \mathrm{cl}(W^s_{\mathbf{x}^*_{eq_2}}) \cap \mathrm{cl}(W^u_{\mathbf{x}^*_{eq_1}}) = \left(-\frac{\alpha}{3}, 0, -\frac{2\alpha}{3}\right)^{\mathrm{T}}. \tag{3.55}$$

Assumption 3.1 ([30]). When the parameters a and b of the linear operator (3.46) control the oscillation around the equilibrium point $\mathbf{x}^*_{eq_i}$, we consider $b/a > 10$.

To illustrate the class of dynamical system given by (3.45), (3.46), (3.48) and (3.49), we define a specific system by considering parameters $a = 0.2$, $b = 5$, $c = -3$, $\alpha = 1$ with the switching surface SW$= \{\mathbf{x} \in \mathbb{R}^3 : 2x_1 - x_3 = 0\}$. The defined dynamical system has a heteroclinic loop as in Fig. 3.20(a) and exhibits a double-scroll attractor as shown in Fig. 3.20(b).

Now, let us consider a partition of the phase space with more atoms $P = \{P_1, P_2, P_3, P_4\}$ and the functional is defined as follows:

$$f(\mathbf{x}) = \begin{cases} -\alpha - \gamma, & \mathbf{x} \in P_1, \\ \alpha - \gamma, & \mathbf{x} \in P_2, \\ -\alpha + \gamma, & \mathbf{x} \in P_3, \\ \alpha + \gamma, & \mathbf{x} \in P_4, \end{cases} \tag{3.56}$$

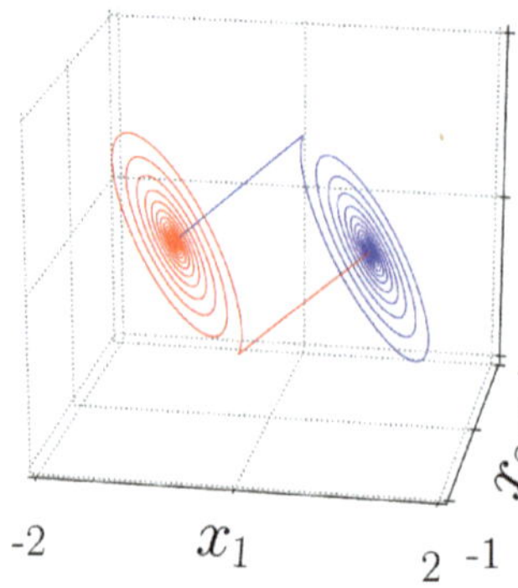

(a)

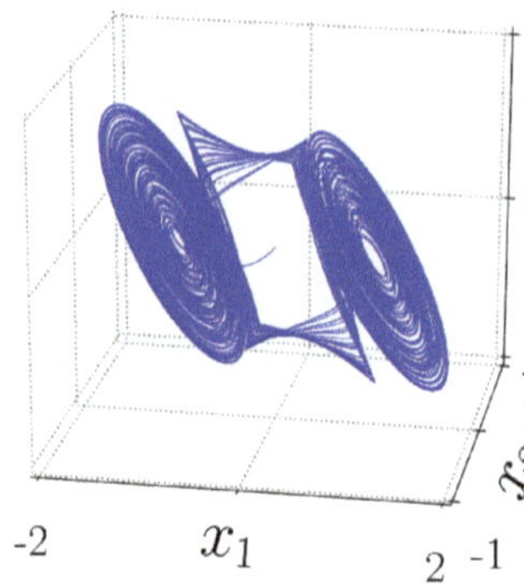

(b)

Fig. 3.20. (a) The heteroclinic loop of the system (3.45), (3.46), (3.48) and (3.49) with the switching surface $\{x \in \mathbb{R}^3 : 2x_1 - x_3 = 0\}$ and the parameters $a = 0.2, b = 5, c = -3, \alpha = 1$. (b) A double-scroll attractor that emerges from an heteroclinic loop using the following initial condition $\mathbf{x}_0 = (0, 0, 0)^{\mathrm{T}}$.

where $0 < \alpha, \gamma \in \mathbb{R}$ and $\gamma > \alpha$. The equilibria are located on the x_1-axis at:

$$\mathbf{x}^*_{eq_1} = \begin{bmatrix} -(\gamma + \alpha) \\ 0 \\ 0 \end{bmatrix}, \quad \mathbf{x}^*_{eq_2} = \begin{bmatrix} -(\gamma - \alpha) \\ 0 \\ 0 \end{bmatrix}, \quad \mathbf{x}^*_{eq_3} = \begin{bmatrix} (\gamma - \alpha) \\ 0 \\ 0 \end{bmatrix},$$

$$\mathbf{x}^*_{eq_4} = \begin{bmatrix} (\gamma + \alpha) \\ 0 \\ 0 \end{bmatrix}, \tag{3.57}$$

so $\mathbf{x}^*_{eq_1} \in P_1$, $\mathbf{x}^*_{eq_2} \in P_2$, $\mathbf{x}^*_{eq_3} \in P_3$ and $\mathbf{x}^*_{eq_4} \in P_4$.

The distance between the self-excited attractors should be big enough to allow the existence of a hidden double-scroll attractor so we make the following assumption.

Assumption 3.2 ([30]). We consider $\frac{\gamma}{\alpha} \geq 10$ to locate the self-excited attractors far from each other.

The three switching surfaces are defined as follows:

$$\begin{aligned} \mathrm{SW}_{1,2} &= \mathrm{cl}(P_1) \cap \mathrm{cl}(P_2) = \{\mathbf{x} \in \mathbb{R}^3 : 2x_1 - x_3 = -2\gamma\}, \\ \mathrm{SW}_{2,3} &= \mathrm{cl}(P_2) \cap \mathrm{cl}(P_3) = \{\mathbf{x} \in \mathbb{R}^3 : 2x_1 - x_3 = 0\}, \\ \mathrm{SW}_{3,4} &= \mathrm{cl}(P_3) \cap \mathrm{cl}(P_4) = \{\mathbf{x} \in \mathbb{R}^3 : 2x_1 - x_3 = 2\gamma\}, \end{aligned} \tag{3.58}$$

which fulfill that

$$\begin{aligned} &\mathrm{SW}_{i,(i+1)} \cap \{\mathbf{x} \in \mathbb{R}^3 : x_3 > 0\} \in P_i, \\ &\mathrm{SW}_{i,(i+1)} \cap \{\mathbf{x} \in \mathbb{R}^3 : x_3 \leq 0\} \in P_{i+1}. \end{aligned} \tag{3.59}$$

This selection of switching surfaces and the condition (3.59) ensure the existence of the four heteroclinic orbits.

To illustrate the system with a four atoms partition, we consider the parameters $a = 0.2, b = 5, c = -3, \alpha = 1$ and $\gamma = 10$. Then, the system presents two self-excited attractors which are shown in red in Fig. 3.21(a). The system does not exhibit a hidden double-scroll attractor, but transitory double scroll. Figure 3.21(a) shows (in blue) the trajectory for the initial condition $\mathbf{x_0} = (0, 0, 0)^T$ for a time $t \in [0, 150]$ a.u. (arbitrary units) and it can be seen how the big double scroll is just a transitory behavior.

The study in [30] revealed that the transitory behavior is observed for any selection of γ and α, which indicates that the separation between self-excited double-scroll attractors is not sufficient to

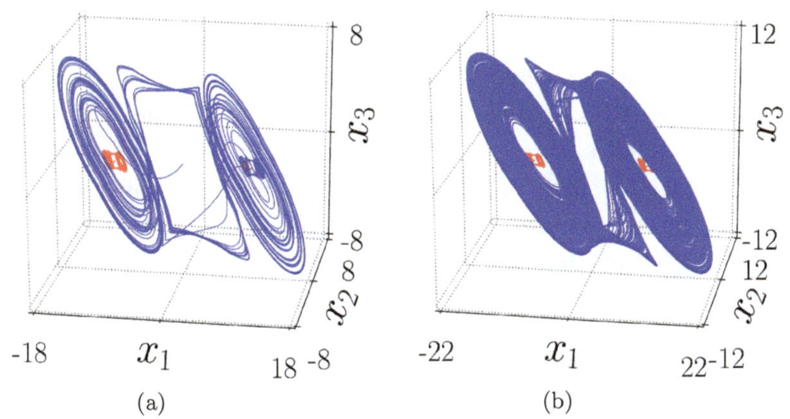

Fig. 3.21. Simulation of the system given by (3.45), (3.46), (3.48), (3.56) with $a = 0.2, b = 5, c = -3, \alpha = 1, \gamma = 10$ and $\alpha = 1$ for the initial condition $x = (0, 0, 0)^{\mathrm{T}}$ in blue. In (a) the surfaces are defined by (3.58) and (3.59), the simulation is shown for $t \in [0, 300]$ a.u. (arbitrary units), the trajectory converges to one of the self-excited attractors in red. In (b), the surfaces are defined by (3.60) and the simulation is shown for $t \in [50000, 51000]$ a.u., the hidden attractor emerges.

produce a hidden double-scroll attractor. It also revealed that the absence of a hidden attractor is related to the switching plane $\mathrm{SW}_{2,3}$, therefore the following switching surfaces is proposed:

$$\begin{aligned}\mathrm{SW}_{1,2} &= \mathrm{cl}(P_1) \cap \mathrm{cl}(P_2) = \{\mathbf{x} \in \mathbb{R}^3 : 2x_1 - x_3 = -2\gamma, x_3 < 0\},\\ \mathrm{SW}_{2,3} &= \mathrm{cl}(P_2) \cap \mathrm{cl}(P_3) = \{\mathbf{x} \in \mathbb{R}^3 : x_1 = 0\},\\ \mathrm{SW}_{3,4} &= \mathrm{cl}(P_3) \cap \mathrm{cl}(P_4) = \{\mathbf{x} \in \mathbb{R}^3 : 2x_1 - x_3 = 2\gamma, x_3 > 0\}.\end{aligned} \tag{3.60}$$

With the new switching surface $\mathrm{SW}_{2,3}$, the intersections $\mathrm{cl}(W^u_{\mathbf{x}^*_{eq2}}) \cap W^s_{\mathbf{x}^*_{eq3}} = \emptyset$ as well as $\mathrm{cl}(W^u_{\mathbf{x}^*_{eq3}}) \cap W^s_{\mathbf{x}^*_{eq2}} = \emptyset$. Moreover, the distance from $\mathrm{cl}(W^u_{\mathbf{x}^*_{eq2}}) \cap \mathrm{SW}_{2,3}$ to $W^s_{\mathbf{x}^*_{eq3}} \cap \mathrm{SW}_{2,3}$ and by symmetry from $\mathrm{cl}(W^u_{\mathbf{x}^*_{eq3}}) \cap \mathrm{SW}_{2,3}$ to $W^s_{\mathbf{x}^*_{eq2}} \cap \mathrm{SW}_{2,3}$ allow the trajectories to cross $\mathrm{SW}_{2,3}$ far from the stable manifolds and this lead to the existence of the hidden double-scroll attractor.

To illustrate the idea, we consider the system given by (3.45), (3.46), (3.48), (3.56) with the parameters $a = 0.2, b = 5, c = -3, \alpha = 1, \gamma = 10$, $\alpha = 1$ and the switching surfaces given by (3.60). In this case, the trajectory with initial condition $\mathbf{x_0} = (0,0,0)^{\mathrm{T}}$ does not converge to a self-excited attractor but the hidden attractor. Figure 3.21(b) shows the trajectory in blue for a time $t \in [50000, 51000]$ a.u.

Based on this idea of increasing the number of atoms in the partition, it is possible to extend the construction to multiscroll attractors, self-excited and hidden attractors. For instance, the partition for a triple-scroll attractor is $P = \{P_1, \ldots, P_6\}$ and a set of switching surfaces is given by

$$\begin{aligned}\mathrm{SW}_{1,2} &= \mathrm{cl}(P_1) \cap \mathrm{cl}(P_2) = \{\mathbf{x} \in \mathbb{R}^3 : 2x_1 - x_3 = -4\gamma, x_3 < 0\},\\ \mathrm{SW}_{2,3} &= \mathrm{cl}(P_2) \cap \mathrm{cl}(P_3) = \{\mathbf{x} \in \mathbb{R}^3 : x_1 = -\gamma\},\\ \mathrm{SW}_{3,4} &= \mathrm{cl}(P_3) \cap \mathrm{cl}(P_4) = \{\mathbf{x} \in \mathbb{R}^3 : 2x_1 - x_3 = 0, x_3 > 0\},\\ \mathrm{SW}_{4,5} &= \mathrm{cl}(P_4) \cap \mathrm{cl}(P_5) = \{\mathbf{x} \in \mathbb{R}^3 : x_1 = \gamma\},\\ \mathrm{SW}_{5,6} &= \mathrm{cl}(P_5) \cap \mathrm{cl}(P_6) = \{\mathbf{x} \in \mathbb{R}^3 : 2x_1 - x_3 = 4\gamma, x_3 > 0\},\end{aligned} \tag{3.61}$$

with

$$f(\mathbf{x}) = \begin{cases} -\alpha - 2\gamma, & \mathbf{x} \in P_1, \\ \alpha - 2\gamma, & \mathbf{x} \in P_2, \\ -\alpha, & \mathbf{x} \in P_3, \\ \alpha, & \mathbf{x} \in P_4, \\ -\alpha + 2\gamma, & \mathbf{x} \in P_3, \\ \alpha + 2\gamma, & \mathbf{x} \in P_4. \end{cases} \tag{3.62}$$

The equilibrium points are at the x_1-axis with $x_{1eq_1}^* = -2\gamma - \alpha$, $x_{1eq_2}^* = -2\gamma + \alpha$, $x_{1eq_3}^* = -\alpha$, $x_{1eq_4}^* = \alpha$, $x_{1eq_5}^* = 2\gamma - \alpha$, $x_{1eq_6}^* = 2\gamma + \alpha$.

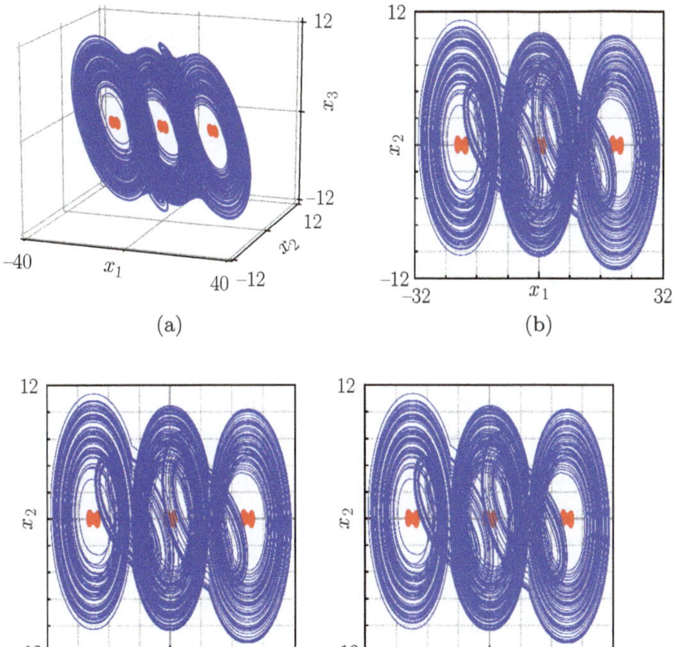

Fig. 3.22. Attractors of the system (3.45), (3.46), (3.48) and (3.62) with the switching surfaces given by (3.61) and the parameters $\gamma = 10$, $\alpha = 1$, $a = 0.2$, $b = 5$ and $c = -3$ in (a) (x, y, z), (b) (x, y), (c) (x, z) and (d) (y, z).

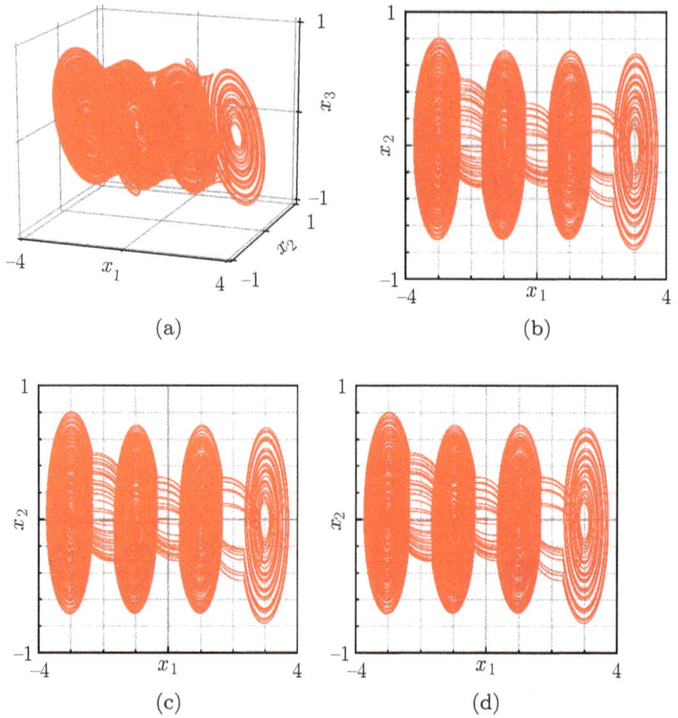

Fig. 3.23. Attractor of the system (3.45), (3.46), (3.48) and (3.62) with the switching surfaces given by (3.61) and the parameters $\gamma = 1$, $\alpha = 1$, $a = 0.2$, $b = 5$ and $c = -3$ in (a) (x, y, z), (b) (x, y), (c) (x, z) and (d) (y, z).

Therefore, by choosing the parameters $\gamma = 10$, $\alpha = 1$, $a = 0.2$, $b = 5$ and $c = -3$, the system (3.45), (3.46), (3.48) and (3.62) with the switching surfaces given by (3.61) generates three self excited double-scroll attractors and one triple-scroll hidden attractor as shown in Fig. 3.22. Also, a single-self-excited multiscroll attractor can be generated by choosing the parameters $\gamma = 1$, $\alpha = 1$, $a = 0.2$, $b = 5$ and $c = -3$ as shown in Fig. 3.23.

3.4. Generalization of Hidden Multiscroll Attractors

In this section, a brief description of a system that generates a higher number of scrolls reported in [31] is given by considering a dynamical system $T : X \to X$ whose dynamic is as follows:

$$\dot{\mathbf{x}} = A\mathbf{x} - AB(\mathbf{x}), \tag{3.63}$$

where $\mathbf{x} = (x_1, x_2, x_3)^{\mathrm{T}} \in \mathbb{R}^3$ is the state vector, $A = \{\alpha_{ij}\} \in \mathbb{R}^{3\times 3}$ is the linear operator given by (3.46) and $B : X \to X$ is a vector-valued function such that $AB(x)$ is a constant vector in each atom P_i. The saddle equilibrium point of each atom is given by $\mathbf{x}^*_{eqi} = (x^*_{1_{eq_i}}, x^*_{2_{eq_i}}, x^*_{3_{eq_i}})^{\mathrm{T}} = B(\mathbf{x})$, with $i = 1, \ldots, \eta$. Moreover, $B(\mathbf{x})$ is a vector-valued function defined as follows:

$$B(\mathbf{x}) = \begin{pmatrix} B_1 \\ B_2 \\ B_3 \end{pmatrix} = \begin{pmatrix} \alpha g(2(x_1 - f_1(x_1)) - x_3, x_3) + f_1(x_1) \\ 0 \\ 0 \end{pmatrix}, \tag{3.64}$$

where

$$f_1(x_1) = \sum_{j=1}^{N_{x_1}} \gamma u(x_1 + 2\gamma(j-1) - \gamma(N_{x_1} - 1)), \tag{3.65}$$

with $N_{x_1} \in \mathbb{Z}^+$. The parameters α and γ fulfill Assumption 3.2. The function $u(y)$ is the Heaviside step function:

$$u(y) = \begin{cases} 1 & \text{if } y \geq 0, \\ -1 & \text{if } y < 0, \end{cases} \tag{3.66}$$

and g is a step function defined as follows:

$$g(y, z) = \begin{cases} 1 & \text{if } y > 0 \text{ and } z \geq 0, \\ -1 & \text{if } y \leq 0 \text{ and } z \geq 0, \\ 1 & \text{if } y \geq 0 \text{ and } z < 0, \\ -1 & \text{if } y < 0 \text{ and } z < 0. \end{cases} \tag{3.67}$$

It can be seen that $g(y, z)$ is equal to $u(y)$ when $z < 0$ while for $z \geq 0$ it is similar to $u(y)$, however, 0 is mapped to 1 instead of -1.

In order to explain $B(\mathbf{x})$, let us separate B_1 and analyze each term and its effect in the sum, i.e., the term $\alpha g(2(x_1 - f_1(x_1)) - x_3, x_3)$ and the term $f_1(x_1)$.

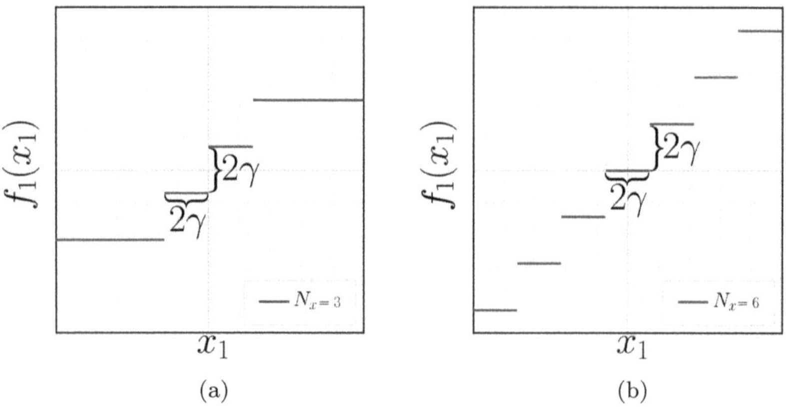

Fig. 3.24. The plot of the function $f_1(x_1)$ for (a) $N_{x_1} = 3$ and (b) $N_{x_1} = 6$.

Firstly, we consider the function $f_1(x_1)$. Then, its plot resembles a stair centered at the origin whose plateaus are of 2γ in height and width. In Fig. 3.24, we show two examples by considering $N_{x_1} = 3$ and $N_{x_1} = 6$.

Then, $f_1(x_1)$ generates N_{x_1} switching planes of the form $\{\mathbf{x} \in \mathbb{R}^3 : x_1 = \epsilon, \epsilon \in \mathbb{R}\}$ parallel to the plane x_2–x_3. Thus, $f_1(x_1)$ generates a partition $R = \{R_1, \ldots, R_{N_{x_1}+1}\}$ of X.

Now, we consider the term $\alpha g(2(x_1 - f_1(x_1)) - x_3, x_3)$. Then, this term generates a switching plane $\{\mathbf{x} \in R_i : 2(x_1 - f_1(x_1)) - x_3 = 0\}$ for $i = 1, \ldots, N_x + 1$. Since $f_1(x_1)$ takes $N_{x_1} + 1$ values, then $\alpha g(2(x_1 - f_1(x_1)) - x_3, x_3)$ generates $N_{x_1} + 1$ switching planes, one for each element of the partition R. Thus, the elements R_i are split and the partition $P = \{P_1, \ldots, P_{2N_{x_1}+2}\}$ is generated.

In this way, $B(\mathbf{x})$ locates two equilibrium points with a separation of 2α in the middle of each element R_i (along x_1) for $i = 1, \ldots, N_{x_1} + 1$. In the partition P, $B(\mathbf{x})$ locates an equilibrium point in each P_i for $i = 1, \ldots, 2N_{x_1} + 2$.

The equilibrium points are located along the x_1-axis and are given in N_{x_1} pairs, each pair of close equilibrium points has a separation of 2α. Let us denote the middle point of the line that joins a pair μ of nearby equilibria by cp_μ with $\mu = 1, \ldots, (N_{x_1} + 1)$. Then, the distance from cp_i to cp_{i+1} is 2γ.

The equilibrium points are located along the x_1-axis as follows:

$$\begin{aligned}
x^*_{1eq_1} &= (0)2\gamma - \gamma(N_{x_1}) - \alpha & x^*_{1eq_2} &= (0)2\gamma - \gamma(N_{x_1}) + \alpha \\
x^*_{1eq_3} &= (1)2\gamma - \gamma(N_{x_1}) - \alpha & x^*_{1eq_4} &= (1)2\gamma - \gamma(N_{x_1}) + \alpha \\
x^*_{1eq_5} &= (2)2\gamma - \gamma(N_{x_1}) - \alpha & x^*_{1eq_6} &= (2)2\gamma - \gamma(N_{x_1}) + \alpha \\
&\vdots & &\vdots \\
x^*_{1eq_{2N_{x_1}-1}} &= (N_{x_1} - 1)2\gamma - \gamma(N_{x_1}) - \alpha & x^*_{1eq_{2N_{x_1}}} &= (N_{x_1} - 1)2\gamma - \gamma(N_{x_1}) + \alpha \\
x^*_{1eq_{2N_{x_1}+1}} &= (N_{x_1})2\gamma - \gamma(N_{x_1}) - \alpha & x^*_{1eq_{2N_{x_1}+2}} &= (N_{x_1})2\gamma - \gamma(N_{x_1}) + \alpha.
\end{aligned} \tag{3.68}$$

The switching planes that are located in the middle of the self-excited attractors are given by

$$\begin{aligned}
\mathrm{SW}_{1,2} &= \{\mathbf{x} \in \mathbb{R}^3 : 2(x_1 - (x^*_{1eq_1} + \alpha)) - x_3 = 0, \\
&\qquad x_1 < x^*_{1eq_2} + (\gamma - \alpha)\} \\
\mathrm{SW}_{3,4} &= \{\mathbf{x} \in \mathbb{R}^3 : 2(x_1 - (x^*_{1eq_3} + \alpha)) - x_3 = 0, x^*_{1eq_3} \\
&\qquad - (\gamma - \alpha) \leq x_1 < x^*_{1eq_4} + (\gamma - \alpha)\} \\
&\ \ \vdots \\
\mathrm{SW}_{2N_{x_1}+1,2N_{x_1}+2} &= \{\mathbf{x} \in \mathbb{R}^3 : 2(x_1 - (x^*_{1eq_{N_{x_1}+1}} + \alpha)) - x_3 = 0, \\
&\qquad x^*_{eq_{2N_{x_1}+1}} - (\gamma - \alpha) \leq x_1\}.
\end{aligned} \tag{3.69}$$

The other switching surfaces are as follows:

$$\begin{aligned}
\mathrm{SW}_{2,3} &= \{\mathbf{x} \in \mathbb{R}^3 : x = x^*_{1eq_2} + (\gamma - \alpha)\} \\
\mathrm{SW}_{4,5} &= \{\mathbf{x} \in \mathbb{R}^3 : x = x^*_{1eq_4} + (\gamma - \alpha)\} \\
&\ \ \vdots \\
\mathrm{SW}_{2N_{x_1}-2,2N_{x_1}-1} &= \{\mathbf{x} \in \mathbb{R}^3 : x = x^*_{1eq_{2N_{x_1}-2}} + (\gamma - \alpha)\} \\
\mathrm{SW}_{2N_{x_1},2N_{x_1}+1} &= \{\mathbf{x} \in \mathbb{R}^3 : x = x^*_{1eq_{2N_{x_1}}} + (\gamma - \alpha)\}.
\end{aligned} \tag{3.70}$$

As an example of the construction, let us consider the parameters $a = 0.2$, $b = 5$, $c = -7$ and $N_{x_1} = 1$. Then, the system presents two self-excited double-scroll attractors and a hidden double-scroll attractor which is shown in Fig. 3.25(a). As a second example, let us consider $N_{x_1} = 4$. In this case, the system presents five self-excited double-scroll attractors and a hidden 5-scroll attractor as shown in Fig. 3.25(b). In this manner, the number of scrolls is equal to the number of self-excited double-scroll attractors which is $N_{x_1} + 1$.

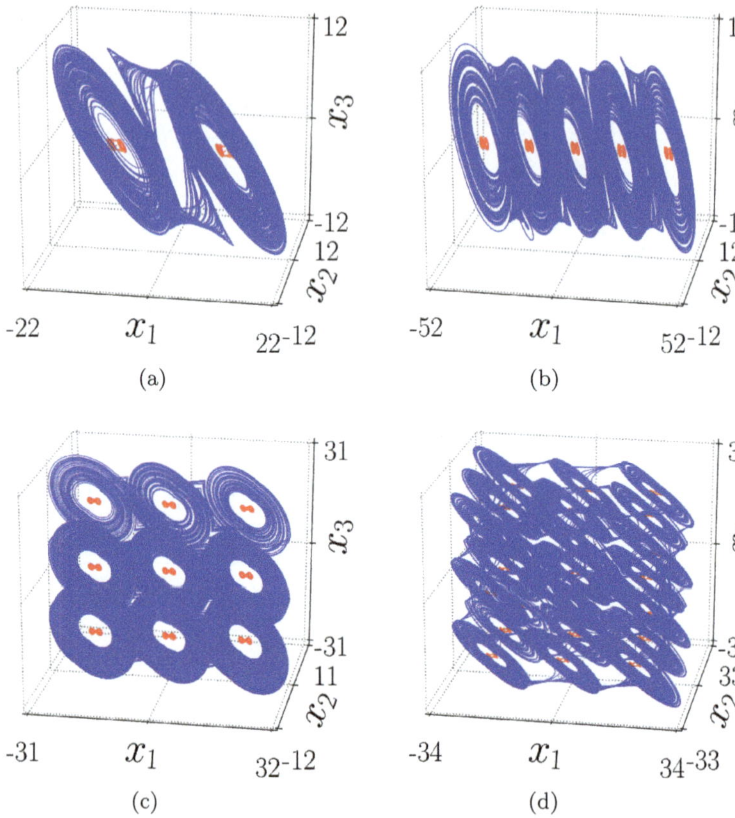

Fig. 3.25. Simulation of the hidden attractor (in blue) and the self-excited attractors (in red) exhibited by the system (3.63) with parameters $a = 0.2$, $b = 5$, $c = -7$. In (a) and (b), $B(\mathbf{x})$ is given by (3.64) with $N_{x_1} = 1$ and $N_{x_1} = 4$, respectively. In (c), $B(\mathbf{x})$ is given by (3.71) with $E_2 = 0$, $E_3 = 1$, $N_{x_1} = 2$ and $N_{x_3} = 2$. In (d), $B(\mathbf{x})$ is given by (3.71) with $E_2 = 1, E_3 = 1, N_{x_1} = 2$, $N_{x_2} = 2$, $N_{x_3} = 2$ and $w = 0.2$.

3.5. Extension for 2D and 3D Grid Scroll Hidden Attractors

The class of systems presented in Sec. 3.4 is 1D grid scroll hidden attractors. Now, the aforementioned class of systems is extended to generate 2D and 3D grid scroll hidden attractors. The idea consists in adding switching surfaces which are parallels to the planes x_1–x_2 and x_1–x_3. This leads to the modification of $B(\mathbf{x})$. For simplicity, the functions $f_i(\cdot)$ are written as f_i:

$$B(\mathbf{x}) = \begin{pmatrix} f_4 + f_1 - \dfrac{wf_2}{\gamma} \\ f_2 \\ f_3 + \dfrac{wf_2}{\gamma} \end{pmatrix}, \tag{3.71}$$

where

$$f_2 = E_2 \left(\sum_{k=1}^{N_{x2}} \gamma u(x_2 + 2\gamma(k-1) - \gamma(N_{x_2} - 1)) \right), \tag{3.72}$$

$$f_3 = E_3 \left(\sum_{l=1}^{N_{x3}} \gamma u((x_3 - \frac{wf_2}{\gamma}) + 2\gamma(l-1) - \gamma(N_{x_3} - 1)) \right), \tag{3.73}$$

$$f_1 = \sum_{j=1}^{N_{x1}} \gamma u((x_1 + \frac{wf_2}{\gamma}) + 2\gamma(j-1) - \gamma(N_{x_1} - 1)), \tag{3.74}$$

$$f_4 = \alpha g \left(2\left(x_1 + \frac{wf_2}{\gamma} - f_1 \right) - \left(x_3 - \frac{wf_2}{\gamma} - f_3 \right), \right.$$
$$\left. \left(x_3 - \frac{wf_2}{\gamma} - f_3 \right) \right), \tag{3.75}$$

where $w \geq 0$, $E_2, E_3 \in \{0, 1\}$ and $N_{x1}, N_{x2}, N_{x3} \in \mathbb{Z}^+$.

Let us define the equilibria found in Sec. 3.4 as $\mathbf{x}_{eq_j}^{*1D} = (x_{1eq_j}^{*1D}, x_{2eq_j}^{*1D}, x_{3eq_j}^{*1D})^T$ for $j = 1, \ldots, 2N_{x_1} + 2$. Then, the new equilibria for the new $B(\mathbf{x})$ in (3.71) are found from:

$$\mathbf{x}_{jkl}^{*3D} = \mathbf{x}_{eq_j}^{*1D} + \begin{pmatrix} 0 \\ 0 \\ (E_1 l)2\gamma - \gamma N_{x3} \end{pmatrix} + \begin{pmatrix} 0 \\ (E_2 k)2\gamma - \gamma N_{x_2} \\ 0 \end{pmatrix}$$
$$+ \begin{pmatrix} -\dfrac{wf_2(y)}{\gamma} \\ 0 \\ \dfrac{wf_2(y)}{\gamma} \end{pmatrix}, \tag{3.76}$$

where $k = 1, \ldots, 2N_{x_2} + 2$ and $l = 1, \ldots, 2N_{x_3} + 2$.

For instance, we consider $a = 0.2$, $b = 5$, $c = -7$, $w = 0.2$ with $E_2 = 0$, $E_3 = 1$, $N_{x_1} = 2$ and $N_{x_3} = 2$ for a 2D grid scroll hidden attractor as shown in Fig. 3.25(c) and $E_2 = 1$, $E_3 = 1$, $N_{x_1} = 2$, $N_{x_3} = 2$ and $N_{x_3} = 2$ for a 3D grid scroll hidden attractor as shown in Fig. 3.25(d).

In this chapter, we presented a review of multiscroll attractors and addressed different systems to generate a hidden multiscroll attractor based on PWL systems. A class of systems with 1D, 2D and 3D scroll attractors is generated around multiple self-excited attractors. This class of systems presents the coexistence of self-excited double-scroll chaotic attractors and a hidden multiscroll attractor.

Acknowledgment

E. Campos acknowledges CONACYT for the financial support through project no. A1-S30433.

References

[1] E. Campos-Cantón, J. G. Barajas-Ramírez, G. Solís-Perales and R. Femat, Multiscroll attractors by switching systems, *Chaos* **20**(1) (2010) 013116.
[2] T. Matsumoto, A chaotic attractor from Chua's circuit, *IEEE Trans. Circuits Syst.* **31**(12) (1984) 1055–1058. doi: 10.1109/TCS.1984.1085459.
[3] J. A. Suykens, A. Huang and L. O. Chua, A family of n-scroll attractors from a generalized Chua's circuit, *Arch. Elektron. Uber.* **51**(3) (1997) 131–138.
[4] M. E. Yalçin, J. A. K. Suykens and J. Vandewalle, Families of scroll grid attractors, *Int. J. Bifurc. Chaos* **12**(1) (2002) 23–41.
[5] K. W. Tang, G. Chen, G. Q. Zhong and K. Man, Generation of n-scroll attractors via sine function, *IEEE Trans. Circuits Syst. I: Fundamental Theory Appl.* **48**(11) (2001) 1369–1372.
[6] E. Campos-Cantón, Chaotic attractors based on unstable dissipative systems via third-order differential equation, *Int. J. Mod. Phys. C* **27**(1) (2016) 1650008.
[7] G. Leonov, N. Kuznetsov and V. Vagaitsev, Localization of hidden Chua's attractors, *Phys. Lett. A* **375**(23) (2011) 2230–2233.

[8] D. Dudkowski, S. Jafari, T. Kapitaniak, N. V. Kuznetsov, G. A. Leonov and A. Prasad, Hidden attractors in dynamical systems, *Phys. Rept.* **637** (2016) 1–50.
[9] B. Andrievsky, N. Kuznetsov, G. Leonov and A. Pogromsky, Hidden oscillations in aircraft flight control system with input saturation, *IFAC Proc.* **46**(12) (2013) 75–79.
[10] G. Leonov, N. Kuznetsov and V. Vagaitsev, Localization of hidden Chua's attractors, *Phys. Lett. A* **375**(23) (2011) 2230–2233.
[11] Z. Wei, Dynamical behaviors of a chaotic system with no equilibria, *Phys. Lett. A* **376** (2011) 102–108.
[12] J. C. Sprott, Some simple chaotic flows, *Phys. Rev. E* **50**(2) (1994) 647–650.
[13] S. Jafari, J. Sprott, and S. M. R. H. Golpayegani, Elementary quadratic chaotic flows with no equilibria, *Phys. Lett. A* **377** (2013) 699–702.
[14] X. Wang and G. Chen, A chaotic system with only one stable equilibrium, *Commun. Nonlinear Sci. Numer. Simul.* **17**(3) (2012) 1264–1272.
[15] M. Molaie, S. Jafari, J. C. Sprott and S. M. R. H. Golpayegani, Simple chaotic flows with one stable equilibrium, *Int. J. Bifurc. Chaos* **23**(11) (2013) 1350188, doi: 10.1142/S0218127413501885.
[16] S. Jafari and J. Sprott, Simple chaotic flows with a line equilibrium, *Chaos. Solitons Fractals* **57** (2013) 79–84.
[17] R. J. Escalante-González, E. Campos-Cantón and M. Nicol, Generation of multi-scroll attractors without equilibria via piecewise linear systems, *Chaos* **27**(5) (2017) 053109, doi: 10.1063/1.4983523.
[18] R. J. Escalante-González and E. Campos-Cantón, A class of piecewise linear systems without equilibria with 3-d grid multiscroll chaotic attractors, *IEEE Trans. Circuits Syst. II: Express Briefs* **66**(8) (2019) 1456–1460. doi: 10.1109/TCSII.2018.2886526.
[19] D. Cafagna and G. Grassi, Chaos in a new fractional-order system without equilibrium points, *Commun. Nonlinear Sci. Numer. Simul.* **19** (2014) 2919–2927 .
[20] Z. Wang, S. Cang, E. O. Ochola and Y. Sun, A hyperchaotic system without equilibrium, *Nonlinear Dynam.* **69** (2012) 531–537.
[21] C. Li, J. C. Sprott, W. Thio and H. Zhu, A new piecewise linear hyperchaotic circuit, *IEEE Trans. Circuits Syst. Part II: Express Briefs* **61**(12) (2014) 977–981.
[22] J. M. Munoz-Pacheco, E. Zambrano-Serrano, C. Volos, S. Jafari, J. Kengne and K. Rajagopal, A new fractional-order chaotic system with different families of hidden and self-excited attractors, *Entropy* **20**(8) (2018). doi: 10.3390/e20080564.

[23] X. Hu, C. Liu, L. Liu, J. Ni and S. Li, Multi-scroll hidden attractors in improved Sprott A system, *Nonlinear Dynam.* **86**(3) (2016) 1725–1734. doi: 10.1007/s11071-016-2989-5.
[24] X. Hu, C. Liu, L. Liu, Y. Yao and G. Zheng, Multi-scroll hidden attractors and multi-wing hidden attractors in a 5-dimensional memristive system, *Chinese Phys. B* **26**(11) (2017) 110502.
[25] F. R. Tahir, S. Jafari, V.-T. Pham, C. Volos and X. Wang, A novel no-equilibrium chaotic system with multiwing butterfly attractors, *Int. J. Bifurc. Chaos* **25**(4) (2015) 1550056.
[26] M. A. Kiseleva, N. V. Kuznetsov and G. A. Leonov, Hidden attractors in electromechanical systems with and without equilibria, *IFAC-PapersOnLine* **49**(14) (2016) 51–55.
[27] M.-F. Danca, N. Kuznetsov and G. Chen, Unusual dynamics and hidden attractors of the Rabinovich–Fabrikant system, *Nonlinear Dynam.* **88**(1) (2017) 791–805.
[28] R. J. Escalante-González and E. Campos-Cantón, Generation of chaotic attractors without equilibria via piecewise linear systems, *Int. J. Mod. Phys. C* **28**(01) (2017) 1750008. doi: 10.1142/S0129183117500085.
[29] L. Ontañón-García and E. Campos-Cantón, Widening of the basins of attraction of a multistable switching dynamical system with the location of symmetric equilibria, *Nonlinear Anal. Hybrid Syst.* **26** (2017) 38–47.
[30] R. J. Escalante-González and E. Campos-Cantón. Coexistence of hidden attractors and self-excited attractors through breaking heteroclinic-like orbits of switched systems (2019).
[31] R. J. Escalante-González and E. Campos-Cantón, Multistable systems with hidden and self-excited scroll attractors generated via piecewise linear systems, *Complexity* (2020).

https://doi.org/10.1142/9789811221903_0004

Chapter 4

Dynamics of the Dipole-Segment with Equal Masses and Arbitrary Rotation

Antonio Elipe*, Alberto Abad* and Alessandra F. S. Ferreira†

*Instituto Universitario de Matemáticas and Aplicaciones, Universidad de Zaragoza, 50009, Zaragoza, Spain

†Departamento de Matemática, UNESP, Brazil

Minor bodies in the Solar system usually have very irregular shape and its rotation differs much from one body to other. In recent years, there is a great interest on these bodies because there are planned missions to orbit around them and even to land on them. Hence, it is very important for mission planning to have a model, the simpler possible, that represents the gravity field of irregular bodies. Motivated by the shape of 216-Kleopatra, we propose the so-called dipole-segment system, which consists of two spherical masses placed at the endpoints of a finite massive straight segment. The gravity function of this body may be expressed in closed form, with no need of expansions in series. The problem depends on three parameters, namely, the mass ratio of the spherical masses (μ), the mass of the segment (μ_s) and the rotation (κ) of the system. In this work, we fix two of the parameters ($\mu = 0.5$, $\mu_s = 0.3$) and κ is the only free parameter considered. For this problem, we compute the equilibria, their stability and find families of periodic orbits depending on the evolution of the free parameter. The dynamics is very rich and it opens the way to new studies from the dynamical systems point of view.

4.1. Introduction

In recent years, there is an increasing interest in missions to minor bodies in the Solar system, in particular to small asteroids and comets nuclei. One of the main features of such bodies is their irregular shape and mass distribution as well as their period of rotation which usually is quite different from one body to another. Because of these facts, in order to study possible trajectories around then, it is necessary to determine their potential function and one of the first difficulties found is that the classical procedures like expanding the gravity field in power series, unless the body be near spheric, are no longer valid, for either these expansions are not convergent in some regions or it is necessary to deal with so many terms that make difficult to have a deep insight of the behavior of the orbits. This is the reason to searching for new models which may represent with certain accuracy the potential function and that be as much simple as possible in order to carry out obtain results with a reasonable amount of computational effort. Although these models may not represent exactly the potential function, no doubt they are very useful to start a preliminary orbit design study. Once selected the orbits more appropriated for the mission, it is a matter of refining then by introducing real data of the body to have a realistic orbit.

One way of having an accurate approximation to irregular bodies is given by Werner in his Ph.D. dissertation [1], where he covers the body with a polyhedron which faces are triangles. This work was continued in collaboration with Scheeres [2]. Although accurate, the model is rather involved and also it is not clear to obtain the gradient at the vertices or at the edges where the function is not derivable. Another approach is the one taken by Prieto-Llanos and Gómez-Tierno [3] who represent the body by two-fixed centers in arbitrary rotation (taken as a parameter). This model fits well with bodies made of the aggregation of two solids, like comets nuclei or asteroids like 25143 Itokawa. Recently, this model has been extended by Palacios *et al.* [4] by adding a third central body in such a way that the three bodies are aligned while uniformly rotating.

A different model was proposed by Halamek in his thesis [5] for highly elongated bodies, like for instance 1620 Geographos asteroid. The model consists of replacing the elongated body by a finite straight segment. The main advantage of the approach is that the potential function may be expressed in closed form as a logarithmic function of intrinsic quantities. Halameck's work was extended in the Riaguas' thesis [6] and in other works carried out by our research group [7–12], where we found families of periodic orbits, regions of chaoticity, nonlinear stability analysis of the equilibria, and even we determined the nonintegrability of the system.

However, most of minor bodies in the Solar system are not almost rectilinear but present some bulge, thus, Beiter and collaborators [13, 14] extended the straight segment model by adding another perpendicular segment in order to fit elongated two-axial and a three-axial ellipsoid.

But the more asteroids we know, the more surprises bring to us concerning its shape. There is a conspicuous body, 216-Kleopatra, a ham-bone-shaped asteroid (see Fig. 4.1) which suggests that a different approximation to model it should be used. With the previous models used, it seems quite natural to merge two of them, the fixed two-centers bodies and the rectilinear segment. This is precisely what Zeng *et al.* [16] proposed. For this problem, they compute the Hamiltonian, found numerically the equilibria and their stability (again numerically) and applied the model to some elongated asteroids.

Based on our experience with this type of problems, we attack the Dipole-segment (DS) problem from scratch in a forthcoming paper [17], that here after will be quoted as Paper A. We analytically determine the potential, its gradient, the equilibria and their stability, and find families of periodic orbits. The DS problem, as we shall see in the next section, depends on three free parameters, namely the mass ratio μ of the dipole components, the segment mass μ_s, and a third parameter κ related with the angular velocity of rotation of

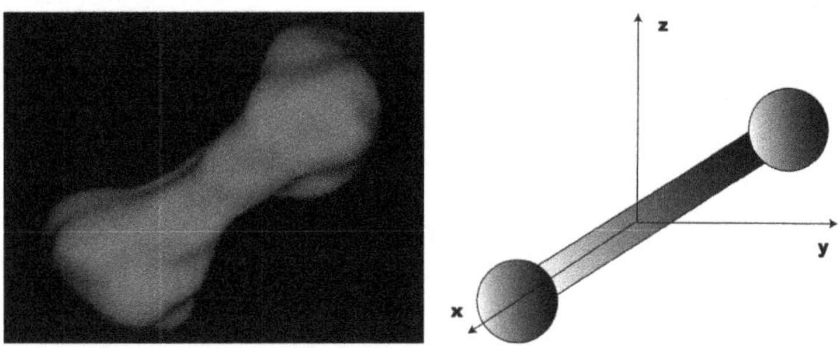

Fig. 4.1. Left: Image of the 216-Kleopatra asteroid, by Ostro *et al.* [15]. Right: Scheme of the Dipole-Segment (DS) system.

the system. Because of this triple dependence, it is rather complex to make a whole analysis of the previous items, in particular evolution of periodic orbits. This is the reason why in the above-mentioned Paper A [17], we fix two parameters, the angular rotation $\kappa = 1$ and $\mu = 0.5$, and compute with respect to the remaining parameter μ_s the evolution of families of periodic orbits.

In this paper, we proceed in similar way as in the above work, but now we still keep fix the mass ratio μ of the primaries which we shall assume to be identical each other, and also we fix $\mu_s = 0.3$, and we take the rotation parameter κ as the only free parameter.

4.2. Potential of the Dipole-Segment Problem

The Dipole-Segment (DS) system consists of two spherical bodies of masses m_1 and m_2 placed each one at the endpoints of a segment of mass m with constant linear mass distribution and length ℓ. Let us consider an inertial reference frame $OXYZ$ with origin at the center of mass of the DS. Besides, we consider a rotating frame $Oxyz$ with the axis Oz coinciding with OZ. We assume the DS is placed on the Ox-axis which is rotating with constant angular velocity ω (see Fig. 4.1, right). We call dipole-segment (DS) problem to the motion of an infinitesimal mass moving under the gravitational attraction of this body.

Let us define

$$M = m_1 + m_2 + m, \quad \mu = \frac{m_2}{m_1 + m_2}, \quad (1 - \mu) = \frac{m_1}{m_1 + m_2}, \quad \mu_s = \frac{m}{M},$$

and denote by

$$(-\ell_1, 0, 0), \quad (\ell_2, 0, 0), \quad \text{with } \ell_1 + \ell_2 = \ell$$

the coordinates of masses m_1 and m_2. With these, we have

$$\frac{m_1}{M} = (1 - \mu)(1 - \mu_s), \quad \frac{m_2}{M} = \mu(1 - \mu_s),$$

expressions that we will use later on.

Note also that when $\mu_s = 0$, there is no segment and we have the classical Two-Center or Dipole Problem; if $\mu_s = 1\,(\Leftrightarrow m_1 = m_2 = 0)$ we only have the segment; if $\mu = 0\,(\Leftrightarrow m_2 = 0)$ there is only the mass m_1 at the left extreme of the segment, while if $\mu = 1\,(\Leftrightarrow m_1 = 0)$ there is only the mass m_2 at the right end of the segment.

Because we assume that the DS body is placed on the Ox-axis, the coordinates of its center of mass in the synodic frame are $(x_0, 0, 0)$, where

$$\begin{aligned} x_0 &= \frac{1}{M}\left[-m_1\ell_1 + m_2\ell_2 + \int_{-\ell_1}^{\ell_2} \rho x\, dx\right] \\ &= \frac{1}{M}\left[-m_1\ell_1 + m_2\ell_2 + \frac{m}{2}(\ell_2 - \ell_1)\right]. \end{aligned}$$

But we also make the assumption that the origin of the reference frame is placed at the center of mass, hence $x_0 = 0$ and then,

$$\begin{aligned} \frac{\ell_2}{\ell} &= \frac{1}{M}\left(m_1 + \frac{m}{2}\right) = (1-\mu)(1-\mu_s) + \frac{1}{2}\mu_s, \quad \text{and} \\ \frac{\ell_1}{\ell} &= \mu(1-\mu_s) + \frac{1}{2}\mu_s. \end{aligned}$$

The DS is made of three masses, hence it potential is the addition of the three potentials created by each body. Taking for the segment the potential given by Riaguas [7], the potential may be expressed as

$$U = -\mathcal{G}M\left(\frac{(1-\mu)(1-\mu_s)}{r_1} + \frac{\mu(1-\mu_s)}{r_2} + \frac{\mu_s}{\ell}\log\left[\frac{r_1 + r_2 + \ell}{r_1 + r_2 - \ell}\right]\right),$$

where r_1 and r_2 are the distances from the particle to the spherical bodies at the endpoints of the segment

$$r_1^2 = (x + \ell_1)^2 + y^2 + z^2, \quad r_2^2 = (x - \ell_2)^2 + y^2 + z^2.$$

In the synodic reference system, the equations of the motion are

$$\mathbf{r}'' + 2\,\boldsymbol{\omega} \times \mathbf{r}' + \boldsymbol{\omega} \times (\boldsymbol{\omega} \times \mathbf{r}) = -\nabla U,$$

where the derivatives $(\Box')$ are computed with respect to the time s.

Now, we take as unit of time $1/\omega$ and as unit of length the length of the segment ℓ. With this, the equations of motion become

$$\begin{aligned} \ddot{x} - 2\dot{y} &= x - \frac{1}{\omega^2}\frac{\partial U}{\partial x} = \frac{\partial \Omega}{\partial x}, \\ \ddot{y} + 2\dot{x} &= y - \frac{1}{\omega^2}\frac{\partial U}{\partial y} = \frac{\partial \Omega}{\partial y}, \\ \ddot{z} &= -\frac{1}{\omega^2}\frac{\partial U}{\partial z} = \frac{\partial \Omega}{\partial z}, \end{aligned} \tag{4.1}$$

where $\dot{\square}$ denotes the derivative with respect to t and Ω is the effective potential

$$\Omega = \frac{1}{2}(x^2 + y^2) - \frac{1}{\omega^2}U.$$

It is easy to find that these equation have the Jacobian constant C

$$C = 2\Omega - (\dot{x}^2 + \dot{y}^2 + \dot{z}^2), \tag{4.2}$$

as usually happens in systems formulated by a synodic frame.

With the choice of length unit, while Eq. (4.1) remain unaltered, the effective potential becomes (see [7, 9] for details)

$$\Omega = \frac{1}{2}(x^2 + y^2) + \kappa \left(\frac{(1-\mu)(1-\mu_s)}{r_1} + \frac{\mu(1-\mu_s)}{r_2} + \mu_s \log\left[\frac{r_1 + r_2 + 1}{r_1 + r_2 - 1}\right]\right),$$

and the dimensionless parameter κ is

$$\kappa = \frac{\mathcal{G}M}{\omega^2 \ell^3} \in (0, \infty),$$

that is to say, the ratio of the gravitational acceleration to centrifugal acceleration.

Note also that $\kappa < 1$ means fast rotation of the segment, whereas $\kappa > 1$ means slow rotation (see [9]). As Prieto-Llanos and Gómez-Tierno [3] pointed out, the case $\kappa \geq 1$ will be more frequent, because in case of fast rotations it would be necessary cohesive strength to avoid body disruption under forces of inertia unbalanced by body self-gravitation. However, it is also interesting to analyze the dynamics in case of fast rotations.

4.3. Equilibria for Two Equal Masses $m_1 = m_2$ ($\mu = 1/2$)

To begin with, and since there are three parameters, in this chapter we shall consider $\mu = 1/2$, that is, the two masses placed at the endpoints of the segment are equal.

Jacobi constant (4.2) gives information on the dynamics of the problem, since form it, $C \leq 2\Omega$. In Fig. 4.2, we present the zero velocity curves for several values of $\kappa = \{0.5, 1.0, 5.0\}$ and $z = 0$. From the plots, it seems that there are four equilibria, two on the x-axis (apparently unstable) and another two on the y-axis. Besides, the equilibria move away form the origin as the parameter κ increases. We also plot the zero-velocity surface in Fig. 4.3 for two values of $\kappa = 1$, 19, which may help in having a guess on the evolution of the equilibria for different values of the parameter.

The equations of motion are

$$\ddot{x} - 2\dot{y} = \frac{\partial \Omega}{\partial x}, \quad \ddot{y} + 2\dot{x} = \frac{\partial \Omega}{\partial y}, \quad \ddot{z} = \frac{\partial \Omega}{\partial z}, \tag{4.3}$$

with the effective potential

$$\Omega = \frac{1}{2}(x^2 + y^2) + \kappa \left(\frac{(1-\mu_s)}{2r_1} + \frac{(1-\mu_s)}{2r_2} + \mu_s \log \left[\frac{r_1 + r_2 + 1}{r_1 + r_2 - 1} \right] \right).$$

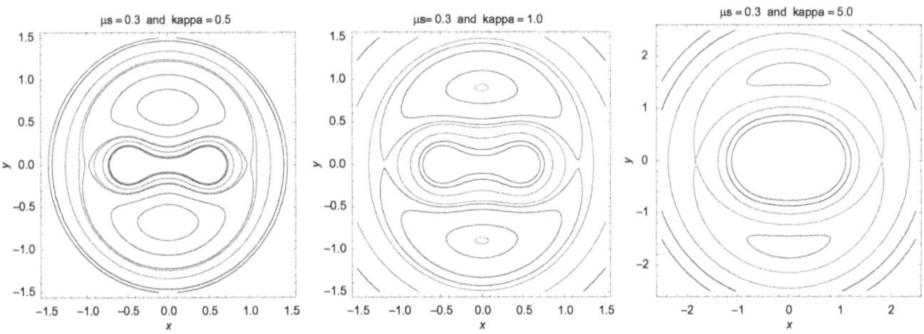

Fig. 4.2. Zero velocity curves for $\kappa = \{0.5, 1.0, 5.0\}$. Note that the equilibria move away from the origin as κ increases.

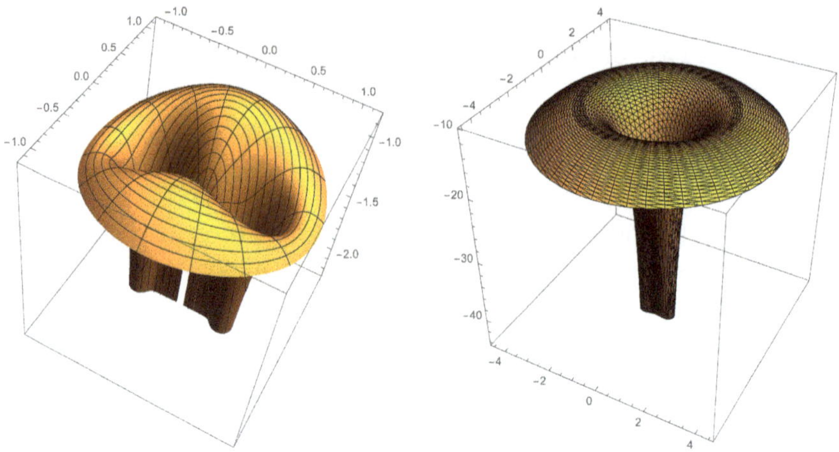

Fig. 4.3. Zero velocity surface for $\mu = 0.5$, $\mu_s = 0.3$ and $\kappa = 1$ (left) and $\kappa = 19$ (right).

Following Riaguas notation [6], let us introduce the auxiliary quantities

$$\begin{aligned} s = r_1 + r_2 &= \sqrt{(x+1/2)^2 + y^2 + z^2} + \sqrt{(x-1/2)^2 + y^2 + z^2}, \\ d = r_1 - r_2 &= \sqrt{(x+1/2)^2 = y^2 + z^2} - \sqrt{(x-1/2)^2 + y^2 + z^2}, \\ p = r_1 r_2 &= \sqrt{(x+1/2)^2 + y^2 + z^2}\sqrt{(x-1/2)^2 + y^2 + z^2}, \end{aligned}$$

with this, the equations of motion become

$$\begin{aligned} \ddot{x} - 2\dot{y} &= \frac{\partial \Omega}{\partial x} = \frac{\kappa(1-\mu_s)}{2}\left(\frac{(x+1/2)}{r_1^3} + \frac{(x-1/2)}{r_2^3}\right) - \left(1 - \frac{2\kappa\mu_s}{s\,p}\right) x, \\ \ddot{y} + 2\dot{x} &= \frac{\partial \Omega}{\partial y} = \left[\frac{\kappa(1-\mu_s)}{2}\left(\frac{1}{r_1^3} + \frac{1}{r_2^3}\right) - \left(1 - \frac{2\kappa\mu_s s}{(s^2-1)p}\right)\right] y, \quad (4.4) \\ \ddot{z} &= \frac{\partial \Omega}{\partial z} = \kappa\left[\frac{(1-\mu_s)}{2}\left(\frac{1}{r_1^3} + \frac{1}{r_2^3}\right) + \frac{2\kappa\mu_s s}{(s^2-1)p}\right] z. \end{aligned}$$

Equilibria are found by zeroing the right-hand side of Eq. (4.4). Since in the third equation, the coefficient of z is always positive, then $z = 0$, and hence, equilibria are always on the plane xy.

From second equation of (4.4), equilibria will be in the forms either $y = 0$ and x such that

$$\frac{\kappa(1-\mu_s)}{2}\left(\frac{(x+1/2)}{r_1^3}+\frac{(x-1/2)}{r_2^3}\right)-x\left(1-\frac{2\kappa\mu_s}{sp}\right)=0,$$

or

$$y \neq 0, \text{ and } \begin{cases} \dfrac{\kappa(1-\mu_s)}{2}\left(\dfrac{1}{r_1^3}+\dfrac{1}{r_2^3}\right)-\left(1-\dfrac{2\kappa\mu_s\, s}{(s^2-1)p}\right)=0, \\ \dfrac{\kappa(1-\mu_s)}{2}\left(\dfrac{(x+1/2)}{r_1^3}+\dfrac{(x-1/2)}{r_2^3}\right) \\ -x\left(1-\dfrac{2\kappa\mu_s}{sp}\right)=0. \end{cases}$$

Then, the possible equilibria are either on the x-axis and we shall denote them *collinear* equilibria, or on the xy-plane that we will denote as *triangular* equilibria. The problem is symmetric with respect to the two axes x and y because the two masses of the dipole are equal, hence we can reduce our analysis to only the first quadrant $\{(x,y) \in \mathbb{R}^2\,;\, x, y \geq 0\}$. Thus, the triangular points are on the y-axis and at times we will denote them as isosceles points.

In Paper A [17] we proved that there are always two collinear equilibria for all values of the parameters μ, μ_s, κ, and also there are two triangular equilibria for all μ, μ_s and $\kappa > 1/8$. For $\kappa \leq 1/8$, there are no triangular equilibria.

The above equations that provide the equilibria are rather involved, but may be solved numerically (as a matter of fact, even in the Restricted Three-Body Problem the position of the equilibria must be numerically computed). In Fig. 4.4 (left), we put the evolution of the coordinate x for collinear points and on the right part the coordinate y of the isosceles points when $\kappa \in (0, 10]$.

In which respect to the stability, we compute the variational matrix of Eq. (4.4)

$$\begin{pmatrix} 0 & 0 & 1 & 0 \\ 0 & 0 & 0 & 1 \\ \Omega_{xx} & \Omega_{xy} & 0 & 2 \\ \Omega_{xy} & \Omega_{yy} & -2 & 0 \end{pmatrix}, \tag{4.5}$$

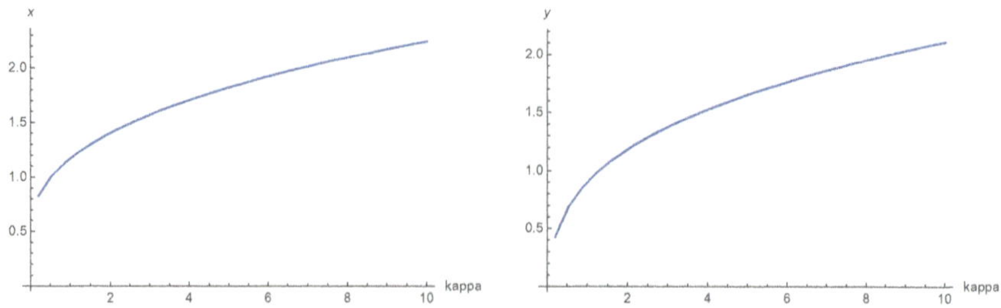

Fig. 4.4. Evolution of coordinates x of collinear (left) and y of triangular (right) equilibria with respect to $\kappa \in [0, 10]$ and $\mu = 0.5$, $\mu_s = 0.3$.

where the partial second derivatives are evaluated at the equilibrium point. With the parameters $\mu = 0.5$, $\mu = 0.3$ and κ free, when the matrix (4.5) is evaluated at the collinear points, its eigenvalues are $\pm bi$, $\pm a$, with $i = \sqrt{-1}$ and $b, A \in \mathbb{R}$, thus, since there is one eigenvalue with real part positive, these points are unstable for all κ.

In the case of the isosceles points, for $0.125 < \kappa < 18.1152$, the eigenvalues are of the form $\pm a \pm bi$, with $a, b \in \mathbb{R}\backslash 0$, that is, the eigenvalues are complex, but two of them have positive real part, hence, they are unstable. But for $\kappa > 18.1152$ all the eigenvalues are pure imaginary, and consequently they become linearly stable through a Hopf bifurcation. A detailed proof of this change of stability is given in Paper A [17].

4.4. Families of Symmetric Periodic Orbits for Values of κ

As it was pointed out by Poincaré [18], periodic orbits form the skeleton of any dynamical system, and in autonomous Hamiltonians, periodic orbits appear in families, depending on parameters. In this chapter, we will consider those periodic orbits that are symmetric with respect to the x-axis. The main feature of such an orbit is that it cuts perpendicularly this axis twice per period.

4.4.1. *The grid-search method*

Although there are many ways in computing periodic orbits, we shall use the so-called *grid-search method*, because it gives reliable results, is very intuitive, is systematic, in the sense that it scan all the domain, and we had good results in previous works, and it is specially useful in two degrees of freedom Hamiltonian problems. Next, we briefly describe the method. The interested reader may find more details on the literature, for instance in [4, 19, 20].

Let us assume that at an initial instant t_0 the initial conditions of a symmetric periodic orbit with period T are $(x(t_0), y(t_0), \dot{x}(t_0), \dot{y}(t_0)) = (x_0, 0, 0, \dot{y}_0)$. Then, because of the symmetry, at the next cut with the x-axis at the instant $t = t_0 + T/2$, it must be

$$(x(t_0+T/2), y(t_0+T/2), \dot{x}(t_0+T/2), \dot{y}(t_0+T/2)) = (x_{T/2}, 0, 0, \dot{y}_{T/2}).$$

Hence, we have to check whether the relations

$$y(x_0, 0, 0, \dot{y}_0; t_0 + T/2) = 0, \quad \text{and} \quad \dot{x}(x_0, 0, 0, \dot{y}_0; t_0 + T/2) = 0,$$

are fulfilled, and in that case, the orbit is periodic and symmetric with the initial conditions $(x_0, 0, 0, \dot{y}_0)$.

To continue the family of periodic orbits, by using the Jacobi's integral (4.2) we have that $\dot{y}_0^2 = 2\Omega(x_0, 0) - C$ and thus, we will consider the initial conditions on the plane (x, C). We build a regular grid on this plane (x_i, C_i), $1 \leq i \leq N \in \mathbb{N}$. From this mesh, we take two points (x_i, C_k) and (x_{i+1}, C_k) and for these two initial conditions we integrate the equations of motion (4.3) until the instant in which the trajectories intersect the x-axis

$$(x_i, 0, 0, \dot{y}_i \,|\, C_k;\, t = 0) \longmapsto (x_i, 0, \dot{x}_i, \dot{y}_i \,|\, C_k;\, t = T_i/2),$$
$$(x_{i+1}, 0, 0, \dot{y}_{i+1} \,|\, C_k;\, t = 0) \longmapsto (x_{i+1}, 0, \dot{x}_{i+1}, \dot{y}_{i+1} \,|\, C_k;\, t = T_{i+1}/2).$$

Next step is to check whether the product $\dot{x}_i(T_i/2) \cdot \dot{x}_{i+1}(T_{i+1}/2)$ is positive or not. If not, by continuity, there is an intermediate point $x_i^* \in (x_i, x_{i+1})$ in the grid such that $\dot{x}^*(T_i^*/2) = 0$, that is, the trajectory crosses orthogonally the x-axis, and hence, it is a symmetric periodic orbit. We store these initial conditions. For finding this point x_i^*, we use Brent's method due to its interesting properties, it is fast and always converges [21]. In the event the former product be positive, we move to the next point in the grid and repeat the procedure. Once checked every x_i on the row C_k, we proceed with next row C_{k+1} until every point of the mesh is tested. For the numerical integration, we use a Taylor series method [22, 23] which allows us to have extended precision, very convenient for bifurcation cases or for finding asymptotic points as we will see later on.

Note that since the body is the segment $[-1/2, 1/2]$, the initial coordinate is $|x_0| > 1/2$.

4.5. Families Evolution in Function of the Rotation Parameter κ

As it was already mentioned, we shall fix $\mu = 0.5$ and $\mu_s = 0.3$, and take the rotation parameter κ free. As the total mass is the unit, this means that the mass of the segment is 0.3, and each mass of the components of the dipole is equal to 0.35.

4.5.1. *Case k* = 1

Let us start with case $\kappa = 1$. In Fig. 4.5, we show the families of symmetric periodic orbits for this value of κ. This plot is very similar to the one appearing in Fig. 4 of Paper A (for $\mu_s = 0.5$). There are five families: m, l, ub, k, and va. Families m, l, k are the same as the ones in the classical Copenhagen problem which corresponds to our parameters $\mu = 0.5$, $\mu_s = 0$, and $\kappa = 1$. Families ub and va are originated as the union of families u and b and v and a of the same problem, respectively. The evolution of the families to reach this situation is described in Paper A.

In what follows of the section, we will show the evolution of these families and how other new families are generated as parameter κ varies. With this goal, we divide the above figure in three zones, namely the one containing family m, and zones Z_d and Z_i.

Family m is made of periodic orbits surrounding the dipole (see Fig. 4.6) and remains unaltered for whatever value of κ.

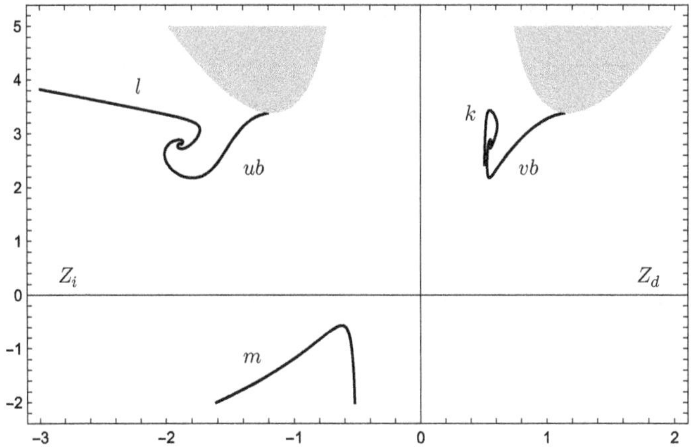

Fig. 4.5. Families of symmetric periodic orbits in the case $\kappa = 1, \mu = 0.5, \mu_s = 0.3$. Each point on the plane (x, C) corresponds to a periodic orbit.

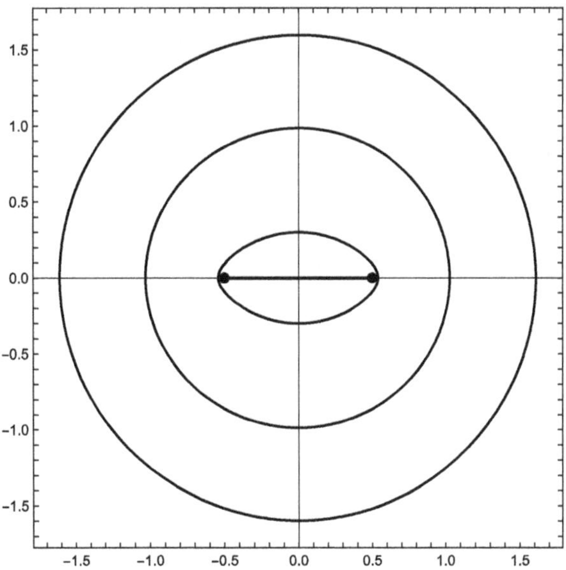

Fig. 4.6. Some orbits belonging to family m.

4.5.2. *Families evolution in Zone Z_i*

In Zone Z_i, placed on the upper-left part of Fig. 4.5, for whatever value of $\kappa < 1$ both families l and ub remain unaltered, with an asymptotic point (except for $\kappa < 1/8$ where triangular points do not exist; see Paper A for details) in which both families end up. Let us see what happens when $\kappa > 1$ increases.

As we can see on Fig. 4.7, when the parameter reaches the value $\kappa = 3$, family ub splits into two different families which are slight modifications of the classical families u and b. But besides, there is a new family.

In Fig. 4.8, we can see the families of the cases $\kappa = 4, 5, 8$, and 10. In the upper part of it ($\kappa = 4$ and 5), we do not show family b which is similar to the one in case $\kappa = 3$. We can see how in this zone Z_i the number of families increases, first as separated families, then, some families cross each other and lastly (for $\kappa = 8$) they originate two closed families which mutually intersect. For $\kappa = 10$, these two intersecting families do not exist. In Fig. 4.9, we plot some samples of periodic orbits of this zone.

We also may see the presence of tiny families (see the dotted rectangle in Fig. 4.8) which start at the end of family b and that are very close to it.

Finally, for greater values of κ (slower rotations) the number of families again decreases as we may see in Fig. 4.10, where we show the results for $\kappa = 60$ (left) and $\kappa = 100$ (right). For $\kappa > 100$, the number and shape of the families is always the same.

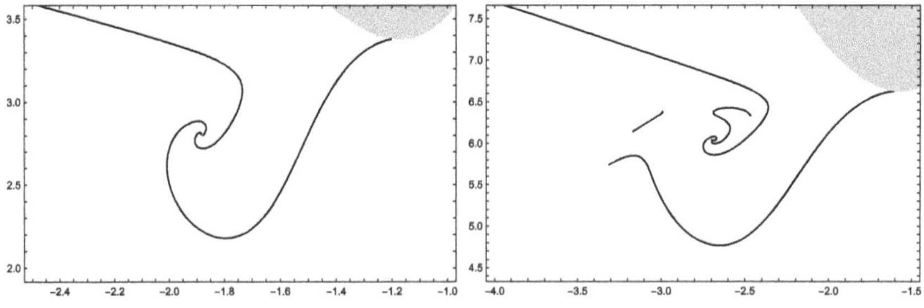

Fig. 4.7. Zone Z_i fo cases: $\kappa = 1$ (left) y $\kappa = 3$ (right).

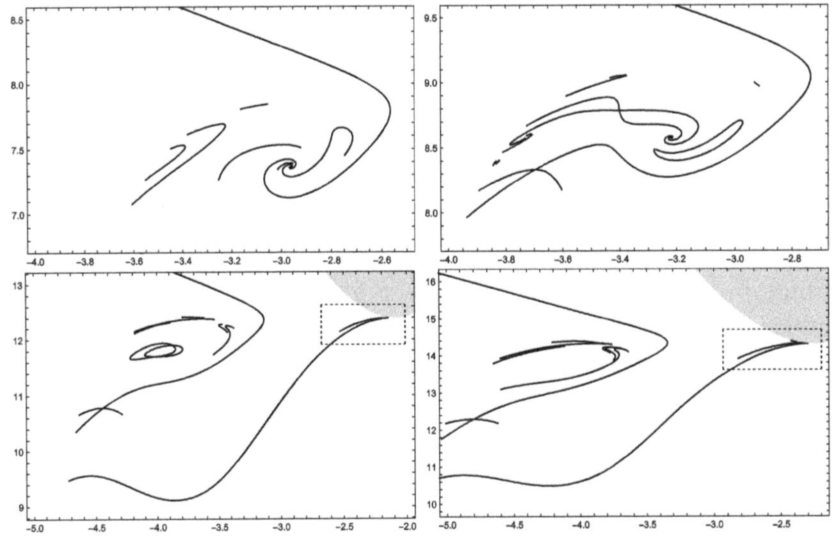

Fig. 4.8. Zone Z_i for cases: $\kappa = 4$ (top-left), $\kappa = 5$ (top-right), $\kappa = 8$ (bottom-left), and $\kappa = 10$ (bottom-right).

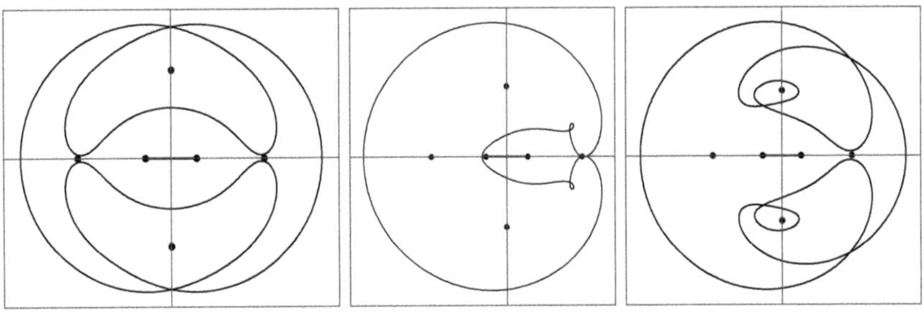

Fig. 4.9. Orbits of the central part of zone Z_i for $\kappa = 5$.

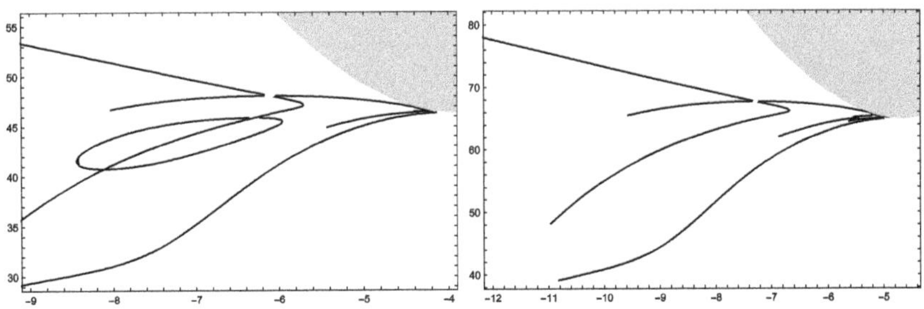

Fig. 4.10. Zone Z_i for cases: $\kappa = 60$ (left) y $\kappa = 100$ (right).

4.5.3. *Families evolution in Zone Z_d*

The behavior of families in this zone for $\kappa < 1$ is slightly different to the one of Zone Z_i. Indeed, family va persists for whatever value of κ, whereas family k is shortening by the side opposite to the asymptotic point where both families converge to, and eventually, for a value $\kappa \approx 0.03$, this family disappears.

In Fig. 4.11, we see the beginning of the evolution when κ increases from $\kappa = 1$. In the transition from $\kappa = 1$ to $\kappa = 3$, the family va splits into two parts. In a similar way, it happened with family ub in Zone Z_i. Besides two new families appear, one clearly visible between the new family v and family k, and the other just above the new curve a on the far right (inside the rectangle). In Fig. 4.12, we present several orbits of these families. These orbits are symmetric each other.

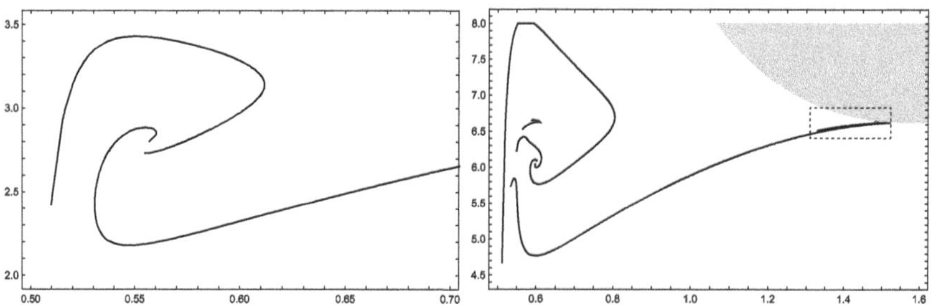

Fig. 4.11. Zone Z_d for cases: $\kappa = 1$ (left) y $\kappa = 3$ (right).

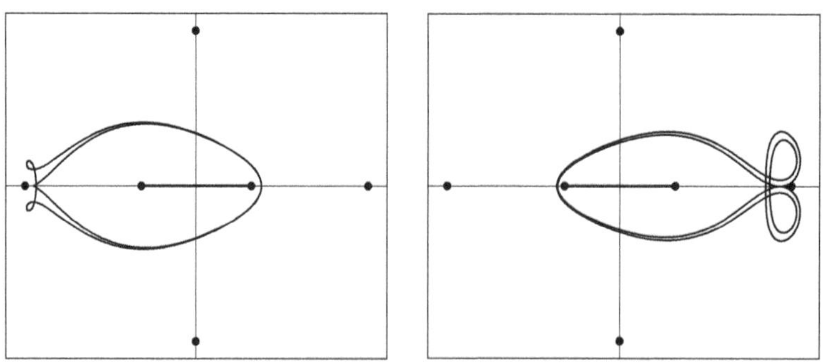

Fig. 4.12. Orbits of the new families when $\kappa = 3$.

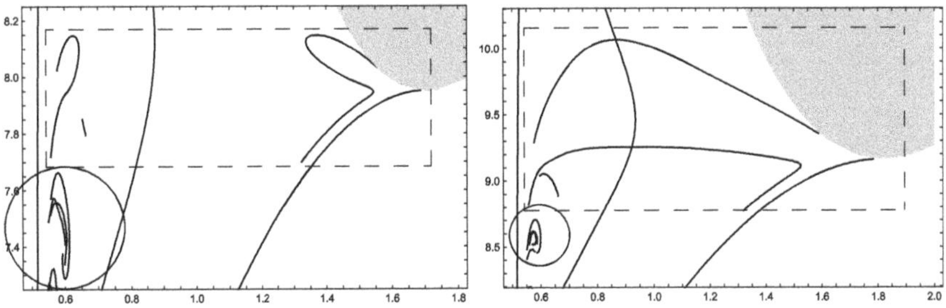

Fig. 4.13. Zone Z_d for cases: $\kappa = 4$ (left) and $\kappa = 5$ (right).

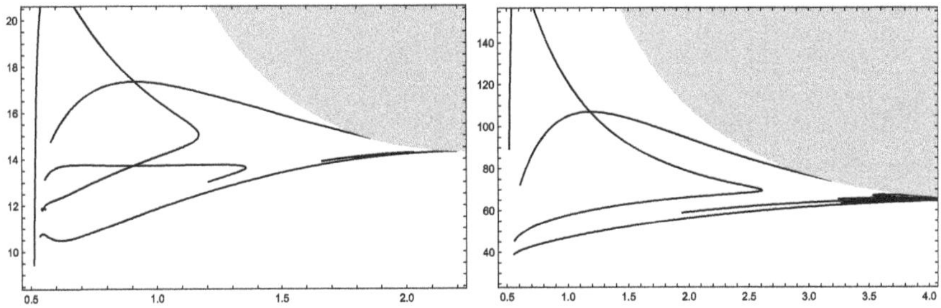

Fig. 4.14. Zone Z_d for $\kappa = 10$ (left) and $\kappa = 100$ (right).

In Fig. 4.13, we show the transition from $\kappa = 4$ to $\kappa = 5$. Curves inside the dashed rectangle corresponding to from $\kappa = 4$ (left part of the figure) are approaching each other until they merge and give rise to two new curves almost parallel which eventually separate one from other as we see for $\kappa = 5$ (dashed rectangle on the right part of Fig. 4.13). Besides in the area inside the circle, there appear a set of families similar to the ones shown in Zone Z_i (Fig. 4.9). These families do not exist for $\kappa > 10$.

Finally, for the case $\kappa = 10$, that is, for a value much lesser than the ones used in Zone Z_i, the number of families dramatically reduces and for very high values ($\kappa > 100$), families remain almost equal in number and shape. For the case $\kappa = 100$ and beyond, there are tiny families (inside the rectangle on the right part of Fig. 4.14) which end up at the extreme of family a or are vey close to it. On the left part of Fig. 4.15, there is a magnification of the mentioned rectangle.

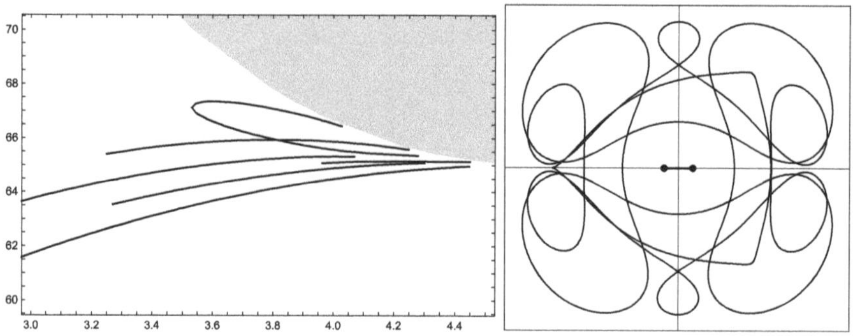

Fig. 4.15. (Left) Area inside the rectangle in Fig. 4.14 corresponding to $\kappa = 100$. (Right) Some orbits of families in this region.

On the right part of Fig. 4.15, we plot several orbits belonging to these families.

4.6. Conclusions

We study the motion of a particle around the dipole-segment problem, which consists of two masses attached to the endpoints of a massive rectilinear segment, in such a way that the system is uniformly rotating. This model is used to represent the gravity potential of some asteroids, like 256-Kleopatra, whose shape is quite similar to the DS body.

In our study, we consider that the two masses of the dipole segment are equal (0.35 mass unit) and that the mass of the segment is close to them (0.3 mass unit) whereas the rotation parameter κ is left as a free parameter. By using the grid-search method for computing symmetrical periodic orbits, we find families of periodic orbits, and we see that these families highly depend on this rotation parameter. We find that there are values at which there is a great variety of families, whereas at other, specially for fast rotations, the number of families decreases and remains very similar even for extreme fast rotations.

Acknowledgments

This work was supported by the Spanish Ministry of Economy (projects ESP2017-87113-R) and by the Aragon Government and European Social Fund (groups E24_20R). A. F. S. Ferreira was supported by São Paulo Research Foundation (FAPESP), grants 2019/15180-0 and 2016/23542-1.

References

[1] R. A. Werner, The gravitational potential of a homogeneous polyhedron or don't cut corners, *Celestial Mech. Dynam. Astron.* **59**(3) (1994) 253–278.
[2] R. A. Werner and D. J. Scheeres, Exterior gravitation of a polyhedron derived and compared with harmonic and mascon gravitation representations of asteroid 4769 castalia, *Celestial Mech. Dynam. Astron.* **65**(3) (1997) 313–344.
[3] T. Prieto-Llanos and M. A. Gómez-Tierno, Stationkeeping at libration points of natural elongated bodies, *J. Guidance, Control Dynam.* **17**(4) (1994) 787–794.
[4] M. Palacios, M. Arribas, A. Abad and A. Elipe, Symmetric periodic orbits in the Moulton–Copenhagen problem, *Celestial Mech. Dynam. Astron.* **131** (2019) 16.
[5] P. Halamek, Motion in the potential of a thin bar, Ph.D. dissertation, University of Texas at Austin (1988).
[6] A. Riaguas, Dinámica orbital alrededor de cuerpos celestes con forma irregular, Ph.D. dissertation, Universidad de Zaragoza, Spain (1999) (in Spanish).
[7] A. Elipe and A. Riaguas, Nonlinear stability under a logarithmic gravity field, *Int. Math. J.* **3**(4) (2003) 435–453.
[8] A. Elipe, V. Lanchares, T. López-Moratalla and A. Riaguas, Nonlinear stability in resonant cases: A geometrical approach, *J. Nonlinear Sci.* **11**(3) (2001) 211–222.
[9] A. Riaguas, A. Elipe and T. López-Moratalla, Nonlinear stability of the equilibria in the gravity field of a finite straight segment, *Celestial Mech. Dynam. Astron.* **81**(3) (2001) 235–248.
[10] A. Riaguas, A. Elipe and M. Lara, Periodic orbits around a massive straight segment, *Celestial Mech. Dynam. Astron.* **73**(1–4) (1999) 69–178.

[11] M. Arribas and A. Elipe, Non-integrability of the motion of a particle around a massive straight segment, *Phys. Lett. A* **281**(2–3) (2001) 142–148.
[12] A. Elipe and M. Lara, A simple model for the chaotic motion around (433), *J. Astronautical Sci.* **51**(4) (2003) 391–404.
[13] P. Bartczak and S. Breiter, Double material segment as the model of irregular bodies, *Celestial Mech. Dynam. Astron.* **86**(2) (2003) 131–141.
[14] P. Bartczak, S. Breiter and P. Jusiel, Ellipsoids, material points and material segments, *Celestial Mech. Dynam. Astron.* **96**(1) (2006) 31–48.
[15] https://apod.nasa.gov/apod/ap000510.html.
[16] X. Zeng, Y. Yu Zhang and X. Liu, The dipole segment model for axisymmetrical elongated asteroids, *Astrono. J.* **155** (2018) 85.
[17] A. Elipe, A. Abad, M. Arribas, A. F. S. Ferreira and R. V. de Moraes, Symmetric periodic orbits in the dipole-segment problem for two equal masses, *Astron. J.* **161** (2021) 274.
[18] H. Poincaré, *Les Méthodes Nouvelles de la Mécanique Céleste* (Gauthier-Villars et fils, Paris, 1892).
[19] A. Elipe, M. Arribas and T. J. Kalvouridis, Periodic solutions in the planar $(n+1)$ ring problem with oblateness, *J. Guidance, Control Dynam.* 30(6) (2007) 1640–1648.
[20] R. Barrio and F. Blesa, Systematic search of symmetric periodic orbits in 2DOF Hamiltonian systems, *Chaos, Solitons Fractals* **41**(2) (2009) 560–582.
[21] R. P. Brent, An algorithm with guaranteed convergence for finding a zero of a function, *Computer J.* **14**(4) (1971) 422–425.
[22] A. Abad, R. Barrio, F. Blesa and M. Rodríguez, TIDES: A free software based on the Taylor series method, *Monograf. Real Acad. Cien. Zaragoza* **35** (2011) 83–95.
[23] M. Rodríguez, A. Abad, R. Barrio and F. Blesa, TIDES: Algorithm 924: TIDES, a Taylor series integrator for differential equations, *ACM Trans. Math. Softw.* **39** (2012) 5.

https://doi.org/10.1142/9789811221903_0005

Chapter 5

The Interaction of Two Vortices Near a Boundary in Rotating Stratified Incompressible Flows

Xavier Carton*, Fabius Cedric Kouogang Tchuenkam and Armand Vic

LOPS/IUEM/UBO,
Rue Dumont D'Urville, 29280 Plouzané, France
**xcarton@univ-brest.fr*

A mathematical theory of two-dimensional vortices, represented by point charges, was developed at the end of the 19th century. Recently, additional physical effects, typical of ocean dynamics, have been added to the point-vortex model; they comprise density stratification, beta effect (the variation of the Coriolis effect with latitude), the presence of a boundary or of bottom topography. Here, we briefly review point-vortex dynamics theory. Then, we show how it can be applied to the study of oceanic vortices. We analyze the evolution of the North Brazil Current rings, idealized as point vortices, in a stratified rotating fluid, along a straight or angular wall, representing the northern coast of South America. We determine the vortex evolutions and the transitions between them. We conclude on some perspectives in this research field.

5.1. Introduction

Vortices are ubiquitous and energetic components of the oceanic and atmospheric flows. They exist at all scales, from the thousands of kilometers (e.g., atmospheric anticyclones and depressions), to the centimetric scale (e.g., in the turbulent dissipation of energy). Vortices play an essential role in the ocean dynamics at the mesoscale, defined by the range 20–200 km in horizontal sizes, and 2–20 days in internal timescales. These horizontal scales are similar to the internal radius of deformation (R_d), a scale at which an equilibrium exists between the Coriolis force due to the rotation of the Earth, and the buoyancy forces due to the stable density stratification. Smaller vortices ($R < R_d$) are called submesoscale vortices.

Vortices are long-lived, horizontally recirculating fluid motions, strongly constrained by the Earth rotation and the density stratification. Their horizontal velocity is often larger than that of other flow components, and than their own vertical velocity. Consequently, fluid particles follow closed and mostly horizontal trajectories in oceanic vortices.

Mesoscale vortices have long since been observed in the ocean. The Gulf Stream rings (large mesoscale vortices), which detach from the Gulf Stream offshore of Cape Hatteras, have been identified as early as the times of Benjamin Franklin [1, 2]. With the progress of *in-situ* sensors, these vortices were sampled more thoroughly [3–5]. With the birth of satellite radiometry (in the 1960s), synoptic views of these vortices were obtained [6] (see also Fig. 5.1).

Vortices deeper and smaller than the Gulf Stream rings were observed only later and sometimes, they were discovered by chance. One example of such small and deep vortices are the meddies (Mediterranean Water outflow eddies). The first paper reporting the measurement of a meddy in the Eastern North Atlantic Ocean was published only in 1984 [7]. It took some time and effort to identify the signature of these eddies in the sea surface temperature or elevation; this signature allows the tracking of meddies drifting away from their

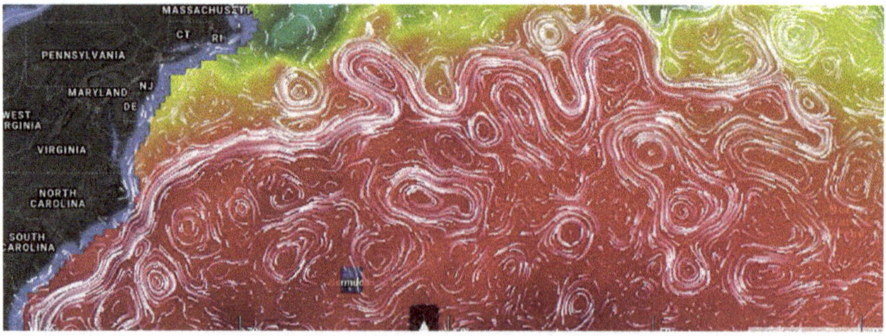

Fig. 5.1. The Gulf Stream (from the "Ocean Virtual Laboratory Web Portal (ESA, OceanDataLab)"; "https://ovl.oceandatalab.com".

formation site. Indeed, meddies form, close to the Iberian Peninsula, from the instability of the Mediterranean Water outflow in the North Atlantic Ocean. They are often generated with a companion cyclone with which they travel for a few weeks, and that they lose after collisions with other eddies [8] (see also Fig. 5.2). Globally, meddies travel southwestward from the Iberian Peninsula, forming thus a warm and salty plume at 1000 m depth in the Northeastern Atlantic Ocean.

Vortices at the surface of the ocean are seen as local bumps or troughs, in satellite measurements of the sea surface height. They can also be identified in maps of sea surface temperature, where they appear as warm or cold anomalies. Using a decade of sea surface height data (from ERS1-2 and T/P satellite measurements), statistics on the number and duration of surface vortices were derived [9]. Over this period, about 62,000 vortices (half of them anticyclones, half of them cyclones) were detected in all the Earth oceans, with lifetimes longer than a month.

Recently, a program of deep reaching vertical profilers of the ocean (the ARGO floats) provided a large database of temperature and salinity profiles over the whole globe. Detecting specific anomalies in these profiles indicates the presence of oceanic vortices. Following this principle, the contribution of oceanic vortices to the upper ocean circulation was calculated and was shown to be important [10].

Another energetic current than the Gulf Stream, which flows on the western side of the North Atlantic Ocean, and which produces vortices, is the North Brazil Current (NBC). This current heads northwestward along the northern coast of Brazil, retroflects and veers eastward, near 8°N. This abrupt change of direction requires that the current sheds vortices to conserve momentum [11, 12]. The shed vortices are anticyclones which can reach 200 km in radius. Once formed and detached from the NBC, these anticyclones drift northwestward along the northern coast of South America until they hit the Lesser Antilles Arc [13, 14] (see Fig. 5.3). This collision strongly modifies the structure of these vortices, the surface part of which is then often cut from their deeper part. A few mechanisms by which this may occur are studied in the present paper.

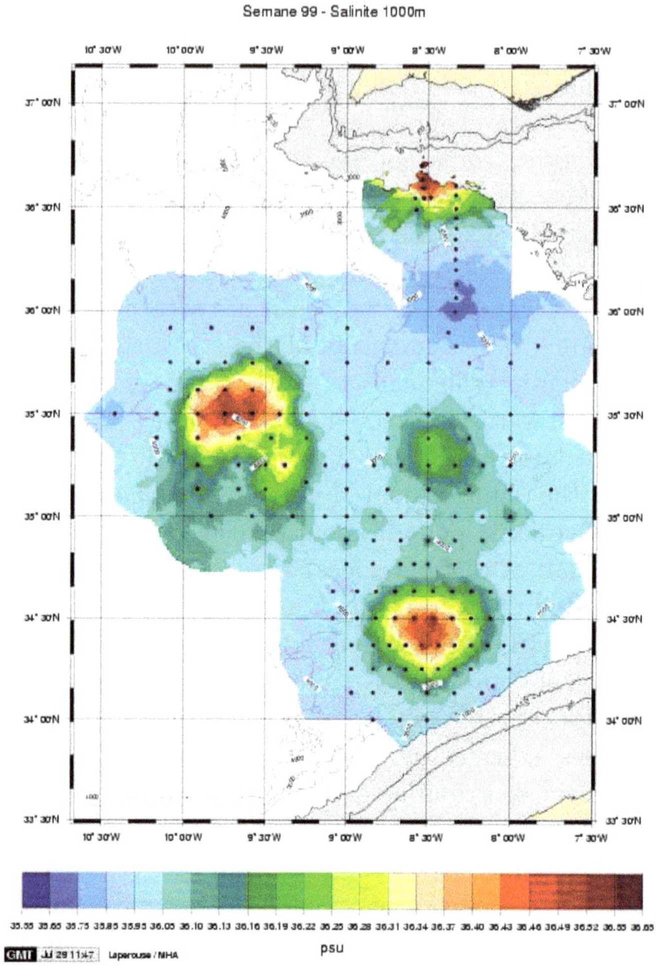

Fig. 5.2. Horizontal salinity map at 1000 m depth showing three eddies — two meddies (the saltiest two patches) and a cyclone (near them) — in the Gulf of Cadiz, between Portugal and Morocco. Data from the Semane 1999 experiment (carried out by SHOM, the French Navy Hydrographic and Oceanographic Service).

The important role of vortices in the circulation of the ocean has prompted many theoretical studies of their dynamics. Several review papers [15–18] and a book [19] provide a detailed overview of the research efforts of the last three decades on this subject, and of the

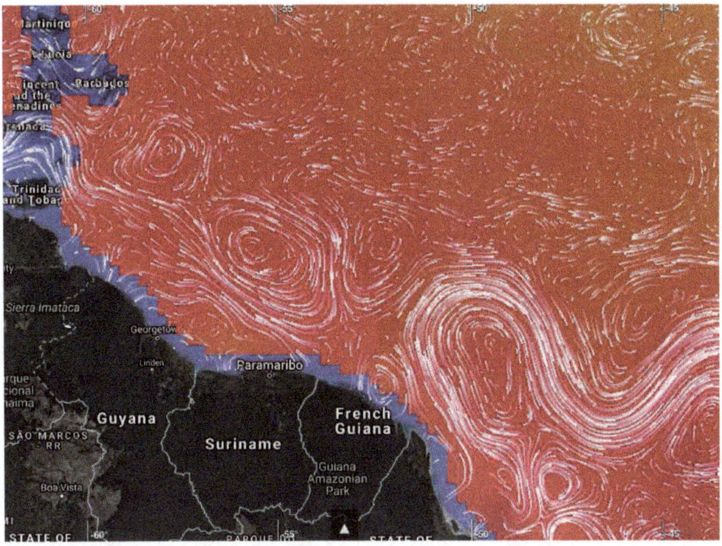

Fig. 5.3. Retroflection of the North Brazil Current, formation of a loop and of a detached anticyclonic eddy (from the "Ocean Virtual laboratory web portal (ESA, OceanDataLab)"; "https://ovl.oceandatalab.com".

various mathematical and numerical methods used. In particular, the two-dimensional and quasi-geostrophic fluid models are often used for the study of vortex dynamics because of the simple, linear relation between the conserved quantity in these models: (potential) vorticity, and the velocity field which advects it. The (potential) vorticity distribution can be chosen as pointwise, or extended, to idealize the observed vortices.

The goal of the present chapter is twofold. Firstly, starting from the earliest studies of point vortices, we will recall the mathematical equations governing their two-dimensional (2D) dynamics. Then, we will indicate how various physical effects (the rotation of the Earth, leading to the Coriolis force; the curvature of the Earth, leading to the so-called beta effect; the density stratification of the ocean; the bottom topography of the ocean; the presence of coasts) have progressively been added to these simple mathematical models.

Secondly, we will apply the model of quasi-geostrophic point vortices to the study of the North Brazil Current rings (vortices) along the northern coast of South America. We will idealize these vortices as point vortices in two layers of fluid, and we will simplify the coast firstly as a straight wall. Then, we will consider the Lesser Antilles Arc as a corner in the coastline, and we will evaluate the influence of this corner in the possible separation of vortices in the upper and lower layers (a process called vertical splitting, observed for the North Brazil Current rings). To study this problem, we will use both mathematical results for the point-vortex dynamics and a numerical model of quasi-geostrophic point-vortex dynamics. The various characteristics of the point-vortex motions (separation between the vortices, motion of their common center, distance to the coast) will be analyzed as time series.

The main results of this chapter are as follows: (1) for our case of application, the beta effect, the influence of bottom topography, a different coastline in the two layers, or an oscillatory sheared current, are all able to separate vertically the vortices of the two layer; (2) the point-vortex model encompassing all these effects is able to reproduce the global motion and evolution of real oceanic vortices (the North Brazil Current rings). Nevertheless, effects of vortex deformation, which are not taken into account with point vortices, are important in the context of vortex–vortex interaction. Thus, this study should be generalized with finite-area vortices in the future.

The present study will briefly recall the 2D model and the quasi-geostrophic models; then we will provide the equations governing point-vortex dynamics and we will explain how the various physical effects are taken into account in point-vortex models (Sec. 5.2). An application of a point-vortex model in a two-layer fluid, to the dynamics of the NBC vortices will be carried out in Sec. 5.3. Conclusions will then be drawn and openings will be provided (Sec. 5.4).

5.2. A Brief Review of Quasi-2D Flow and Vortex Models

Here, we remind the reader of the equations for 2D or quasi-2D incompressible fluid flows, and their solutions in terms of point vortices. It is of importance to note that in all these models, the flow evolution is governed by the conservation of a form of vorticity. Since the vorticity–streamfunction relation is a Laplace or a Helmholtz equation, vorticity is easily inverted into streamfunction; the velocity which advects the vorticity in time, derives from the streamfunction; thus, with the appropriate boundary conditions, the system of equations is closed mathematically. Characterizing the system via a unique function of space and time allows analytically tractable solutions for simple cases, and for more complex situations, the possibility to solve these equations numerically at a low-computational cost.

An example of such simple cases are vortex flows. For a single-isolated vortex, the flow has circular symmetry. On physical grounds, vortices are localized (the vorticity distribution has a compact support). Using the Gauss theorem, it is easily shown that the velocity field out of the vortex support can be computed from the total (area) integral of vorticity, called circulation. Therefore, a point-vortex and a finite-area vortex with identical circulations generate the same external flow. Thus point vortices are useful idealizations of real vortices for the computation of motion.

We start from the simple 2D fluid equations and point-vortex models. Then, we generalize these equations and models to incorporate more physical effects (pertinent to the ocean).

5.2.1. *Two-dimensional incompressible flows and vortex models*

5.2.1.1. *Two-dimensional Euler equations*

The two-dimensional (2D) Euler equations for incompressible, homogeneous, inviscid fluid motions[1] are composed of two momentum equations and of a continuity equation. The momentum equations express the balance between the horizontal acceleration and the horizontal pressure gradient normalized by the constant density of the fluid, and the continuity equation expresses the nondivergence of

[1]Here we consider that the fluid is contained in a horizontal plane with the vorticity vector along the vertical axis.

the horizontal velocity.

$$du/dt = -\partial_x p/\rho_0, \tag{5.1}$$

$$dv/dt = -\partial_y p/\rho_0, \tag{5.2}$$

$$\partial_x u + \partial_y v = 0, \tag{5.3}$$

where u, v denote the two components of the horizontal velocity (u pointing eastward along the x-axis, and v northward along the y-axis); p is pressure and ρ_0 is the fluid density. Taking the curl of the momentum equations, and using the continuity equation, leads to the conservation of the vorticity ζ:

$$d\zeta/dt = 0, \quad \zeta = \partial_x v - \partial_y u. \tag{5.4}$$

Since the vorticity vector is always vertical for 2D flows in the (x, y) plane, we use the vector norm as a conserved scalar quantity. Physically, vorticity is the self-rotation of each fluid particle. The continuity equation implies the existence of a streamfunction; the horizontal velocity is a two-dimensional curl of this streamfunction. The scalar vorticity is the two-dimensional Laplacian of this streamfunction:

$$u = -\partial_y \psi, v = \partial_x \psi, \zeta = \nabla^2 \psi. \tag{5.5}$$

The 2D Euler equations are thus composed of a nonlinear hyperbolic equation for the time evolution of vorticity, and of a linear elliptic equation relating vorticity to streamfunction. As a consequence, knowing the vorticity distribution, and the boundary conditions for streamfunction, the inversion of this elliptic equation provides the streamfunction necessary to evolve vorticity in time. As a first step, we will assume that the flow extends to infinity in both horizontal directions to perform this inversion. This closure of the problem shows that only the prescription of the initial vorticity, or of streamfunction, suffices to fully determine the flow evolution.

5.2.1.2. *Two-dimensional vortices*

The 2D Euler equations describe many fluid flows. Vortices often arise as a result of the instability of intense, sheared, flows. An isolated vortex is a simple, axisymmetric solution of the 2D Euler equation. As mentioned above, the external flows of a point-vortex and of a finite-area vortex are identical, if these two vortices have the same circulation. Therefore, point vortices are appropriate to study vortex motions. On the contrary, vortex deformation requires the use of finite-area vortices. Hereafter, we describe both types of vortices.

Consider first a vortex of finite extent; such a vortex is defined as a horizontally recirculating fluid motion; we assume that this vortex has a compact support, and that it is bounded by a closed material contour (that is, a contour advected with the fluid velocity). The Kelvin theorem [20] states that the vertical vorticity flux across the basic surface S of this finite-area vortex is conserved if the external forces, applied to the fluid, are either null or conservative (that is, with zero curl). This vorticity flux is also the circulation of the horizontal velocity around the material contour $C = dS$ bounding the vortex:

$$d\Gamma/dt = 0, \quad \Gamma = \int_S \zeta dx dy = \oint_C \vec{u} \cdot \vec{dl}. \tag{5.6}$$

To study the time evolution of the vortex, the prescription of the initial spatial distribution of vorticity is required. Vortex dynamics shows that the time evolution of the flow can be mathematically obtained in two cases:

(a) The vorticity distribution has a compact support and it is uniform over this domain. Particular cases of this are the Rankine vortex [21] (a disk of uniform vorticity) and the Kirchoff vortex [22] (an ellipse of uniform vorticity).

(b) The vorticity distribution is pointwise with finite circulation. Then, vorticity is a Dirac distribution. The associated stream-function is then simply the circulation multiplied by the Green's function

$$\psi(r) = \Gamma G(r), \quad G(r) = \ln(r)/(2\pi). \tag{5.7}$$

While finite-area vortices can deform under the action of a spatially nonuniform velocity field [23, 24], point vortices can only move. The velocity field around a point vortex is

$$u_r = 0, u_\theta = \Gamma/(2\pi r). \tag{5.8}$$

Considering a finite number of point vortices, the dynamical system governing their evolution is Hamiltonian [25]. For instance, with two vortices of circulation Γ lying at a distance $r_{12} = \sqrt{(x_1 - x_2)^2 + (y_1 - y_2)^2}$ from each other, the system has the Hamiltonian

$$H = \Gamma^2/(2\pi)\ln(r_{12}). \tag{5.9}$$

The integrability of point-vortex systems has been widely studied in 2D flows, leading to thorough review papers [26, 27].

5.2.2. *Quasi-two-dimensional fluid motion in stratified rotating fluids*

Two-dimensional point-vortex theory pertains only to the dynamics of vortices in a shallow, homogeneous ocean. In a stratified ocean, at the scale of oceanic vortices (about 100 km in horizontal size), the dynamics are quasi-2D due to the Earth rotation and to the density stratification. We show here how 2D point-vortex models can be generalized to a rotating stratified fluid, comparable to the Earth oceans.

Firstly, we will present the equations governing such fluid flows, called quasi-geostrophic flows for a continuously stratified, rapidly rotating fluid flow. Secondly, we will consider a two-layer model of such a fluid flow. Thirdly, we indicate how the various physical effects influencing oceanic vortex dynamics, can be included in a simple point-vortex model in the quasi-geostrophic model.

5.2.2.1. *The quasi-geostrophic model for a rapidly rotating, continuously stratified incompressible fluid*

In this subsection, we present the quasi-geostrophic model for rapidly rotating, strongly stratified fluid flows (e.g., those of vortices in the Earth oceans).

As a first step, we add global rotation to the 2D model. Consider a two-dimensional homogeneous and incompressible fluid in a *rotating* frame of reference, the Kelvin theorem generalizes as the Poincaré–Bjerkness theorem [25, 28].

$$d\Gamma_a/dt = 0, \quad \Gamma_a = \int (\zeta + 2\Omega)dxdy = \oint [\vec{u} + \vec{\Omega} \times \vec{r}] \cdot \vec{dl}. \quad (5.10)$$

The circulation is generalized as the area integral of the velocity curl plus twice the global rotation rate Ω. It is also the contour integral of the velocity plus the rotation rate times the radial vector. To distinguish between nonrotating and rotating frameworks, the adjective "relative" is associated to quantities (velocity, vorticity) in the former, and "absolute" in the latter.

Now let us add vertical density stratification to the fluid. The stratification and the global rotation render fluid motions mainly horizontal. This motion is governed by the so-called *quasi-geostrophic* equations [29]. These quasi-geostrophic equations hold when the Coriolis acceleration dominates the relative acceleration and when the flow is hydrostatic. Then a generalized form of vorticity, called potential vorticity, is conserved.

In a rotating framework with a vertical axis of global rotation, this conservation is written

$$dq/dt = 0, \quad q = \zeta + 2\Omega + \partial_z((4\Omega^2/N^2)\partial_z\psi), \quad (5.11)$$

where $N^2 = -(g/\rho_0)d\rho/dz$ is the stratification stability frequency or buoyancy frequency. If N^2 and Ω are constant, then one can scale the vertical coordinate via $Z = Nz/(2\Omega)$ and, we have the relation

$$q = \nabla^2_{3D}\psi + 2\Omega, \quad (5.12)$$

where the 3D Laplacian is taken in (x, y, Z) coordinates (in the particular case here, we assume that the fluid has a constant total thickness).

5.2.2.2. *The two-layer quasi-geostrophic model for the ocean*

On Earth, the effect of planetary rotation is not uniform with latitude. Planetary rotation participates in the total vorticity via the Coriolis parameter, which is the projection of this rotation on the local vertical axis: $f = 2\Omega\sin(\theta)$ where θ is the latitude. Locally, the Coriolis parameter f can be expanded linearly on the tangential plane to the Earth: $f = f_0 + \beta y$, where y is the North–South distance on this plane.

The density stratification in the ocean can be roughly approximated as two superimposed homogeneous layers (above and below the main pycnocline[2]). The two layers have thicknesses Δ_1 and Δ_2 at rest. The potential vorticity is conserved in each layer

$$dq_j/dt = 0, \quad j = 1, 2, \tag{5.13}$$

where j is the layer index, and

$$q_j = \nabla^2\psi_j + F_j(\psi_k - \psi_j) + f_0 + \beta y + f_0\delta_{j2}\mathrm{h_b}/\delta_2, \quad k = 3 - j, \tag{5.14}$$

where the Laplacian is two-dimensional. Here, we have assumed that the local ocean thickness Δ_l is not constant: $\Delta_l = \Delta - h_b$, where $\Delta = \Delta_1 + \Delta_2$ is the maximal depth of the ocean, and h_b is the elevation of the ocean bottom above its deepest point. The Kronecker symbol δ_{ij} indicates that the contribution of bottom elevation to potential vorticity concerns only the lower layer.

The two density layers are dynamically coupled via the raising and lowering of the density interface between them. This is represented mathematically by the layer coupling coefficient $F_j = f_0^2/(g'\Delta_j)$, where g' is the Earth gravity, reduced by the relative density difference between the layers. The potential vorticity–streamfunction relation is not diagonal in each layer. It is thus inverted via a diagonalization. The two vertical eigenvectors are called the barotropic and the baroclinic modes and they are defined by a linear combination of the layerwise quantities

$$\psi_{bt} = (\Delta_1/\Delta)\,\psi_1 + (\Delta_2/\Delta)\,\psi_2, \quad q_{bt} = (\Delta_1/\Delta)\,q_1 + (\Delta_2/\Delta)\,q_2, \tag{5.15}$$

$$\psi_{bc} = \psi_1 - \psi_2, \quad q_{bc} = q_1 - q_2. \tag{5.16}$$

[2]The pycnocline is the range of depths in the ocean at which the density of the seawater varies the most, vertically.

Then the potential vorticity–streamfunction relation is

$$q_{bt} = \nabla^2\psi_{bt} + f_0 + \beta y + f_0 h_b/H, \tag{5.17}$$

$$q_{bc} = \nabla^2\psi_{bc} - \psi_{bc}/R_d^2 - f_0 h_b/\Delta_2, \tag{5.18}$$

where $R_d = g'\Delta_1\Delta_2/(4\Omega^2\Delta)$ is the deformation radius.

We will use this two-layer quasi-geostrophic model in the following section to describe the motion of the oceanic vortices along a wall/coast.

5.2.2.3. *Point vortices in the two-layer quasi-geostrophic model*

Here, we show how physical effects which influence vortex dynamics in the ocean (such as beta-effect, irregular bottom topography, the presence of a coastline) can be included in a point-vortex model in the quasi-geostrophic dynamics.

Since the point-vortex charge Q_j (the layerwise equivalent of Γ) is the area integral of q_j we must let it vary with latitude to incorporate the effect of βy (the beta effect). This is not sufficient because fluid particles out of the vortex will also see their charge vary as they move. Therefore, we must add vortices with null initial charge ("ghost vortices") around our "active" vortices to take into account the neighboring fluid particles [30]. For vortices on the beta-plane, the value of β hereabove is multiplied by the elementary area separating the ghost vortices.

These "ghost" vortices can be generalized to take into account the effect of uneven bottom topography. This effect is called "the topographic beta-effect".

Finally, oceans being limited by horizontal boundaries, no cross-flow boundary conditions must be ensured. This is achieved via the addition of "image" vortices [31]. With a straight wall/coast, the image vortex is symmetrical to the real vortex, across the coast, and has an equal and opposite charge.

In two-layer or multi-layer quasi-geostrophic models, point-vortex interactions have extensively been studied [32–35]. Some of these studies analyzed the evolution of point vortices under the influence of an external flow, possibly leading to chaos [36–39].

5.2.2.4. *Numerical model for point vortices in the two-layer quasi-geostrophic dynamics*

When several vortices interact mutually, or with a coastline, only asymptotic solutions at large distances can be obtained (see, for instance, the solutions provided in the Appendix). When vortices are close to each other, or to the coast, their motion is then determined numerically.

The numerical code is based on a time iteration with the following steps:

- Knowing the position and potential vorticity of the two real vortices and of the ghost or image vortices, the streamfunction is calculated and the velocity of each point vortex is obtained.
- Each point vortex is advected horizontally using this velocity and a fourth-order Runge–Kutta scheme in time.
- On the beta plane, the potential vorticity of each vortex is updated from its position: $Q_{\text{new}} = Q_{\text{old}} + \beta(y_{\text{old}} - y_{\text{new}})$.
- Diagnostics (centroid position, distance to the coast) are calculated online.

In the numerical model, we normalize the length and timescales for simplicity; we set $R_d = R_1 = 1.0$ in the numerical model; this provides a length scale of about 120 km (the deformation radius in this region of the ocean varies between 100 and 150 km). In the reference case, the vortex circulation is set to unity (plus or minus unity). This provides a timescale of 0.36 day.

5.3. Interaction of Two Vortices Along a Wall/Coast in the Two-Layer Model

Here, we model the oceanic vortices along the northern coast of South America as point vortices, and we study their motion along a wall/coast. We vary several parameters of the model (layer thicknesses, beta effect, magnitude of bottom topography) or of the vortices (potential vorticity) to evaluate their impact on the vortex trajectories.

5.3.1. *Reference simulation*

The reference simulation is obtained by running the numerical model, initialized with two point vortices (one in each layer) along a straight wall (along the x-axis). These two point vortices lie at a normalized distance $d = 0.5$ from each other, and at a normalized distance $D = 1.0$ from the wall. These parameters correspond to observations of the North Brazil Current vortices which lie at 60 km from each other in the two layers and at 120 km from the coast.

The two layers have the same thickness $\Delta_1 = \Delta_2$. The two point vortices have the same charge $Q_1 = Q_2 = 1$.

A previous study [40] has provided a preliminary description of the motion of two such vortices along a straight wall (or coast), parallel to the x-axis, in the two layer quasi-geostrophic model. In particular, it has provided the mathematical formula for the along-wall translation velocity of their center of gravity (called "centroid"):

$$U = Q_{bt}/(4\pi D), \tag{5.19}$$

where D is the distance between this center of gravity and the wall. For identical vortices ($Q_1 = Q_2 = Q$), we have simply $U = Q/(4\pi D)$ which is the long-range velocity induced by the image vortex pair. This formula has also been used to check the accuracy of the numerical code.

The vortex pair trajectory is a cycloid (the addition of a translation due to the vortex images and of the co-rotation of the two vortices on the same side of the wall; see Fig. 5.4(a)). The extent of this cycloid depends on the initial orientation of the vortex pair (parallel or perpendicular to the wall), only by 25% in this reference case.

For vortices closer to the wall than in the reference case, the vortex motion in the translating frame of reference, increasingly deviates

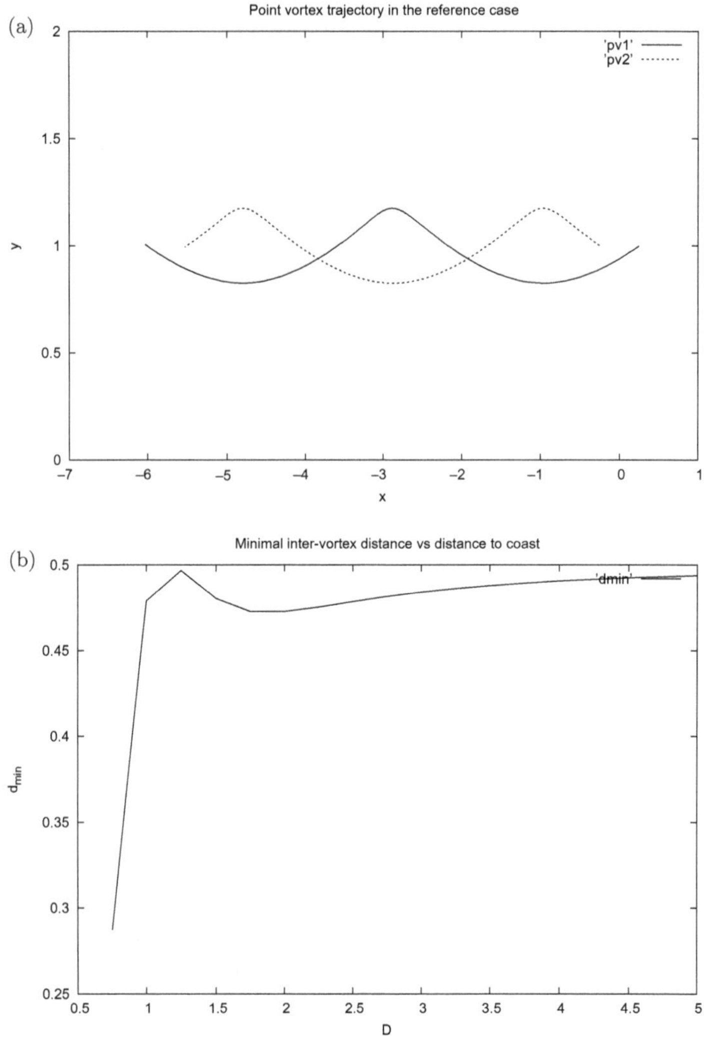

Fig. 5.4. (a) Trajectories in the plane of the two vortices along a zonal wall (parallel to Ox) in the reference configuration. (b) Minimal inter-vortex distance with respect to their distance to the wall.

from a circle (see Fig. 5.4(b)). For vortices more distant from the wall, the trajectory is very weakly dependent on the initial orientation.

Numerically, a critical distance of the two vortices to the wall was searched below which the two vortices would separate irreversibly, instead of co-rotating with a westward translation. Such a critical distance D_c was not found for $d = 1.0, 0.5$, at least for $D_c \geq 0.5$ (this minimal value was set by the oceanic observations). Analytically, the condition for vortex separation is $U \geq V_{12}$ where V_{12} is the velocity induced by one vortex of the pair on the other. This condition can be written, after simplifications, and considering only the long-range velocity generated by the image vortices

$$D < \frac{d}{1 - dK_1(d)} \tag{5.20}$$

using $R_d = 1$. For small d, the term $dK_1(d)$ goes as $\sqrt{d}\exp(-d)$; thus, for close vortices this implies that the wall must be at the same distance from each vortex, as its partner in the pair, which is not realistic.

Therefore, a spontaneous separation of the two vortices drifting along a straight wall was not found for physically relevant cases.

5.3.2. *Sensitivity study: Influence of the physical parameters*

Now, we evaluate how the vortex trajectories change when various physical parameters (layer thicknesses, vortex intensities, beta effect, bottom topography, coastline orientation) are modified. In particular, we look for cases where the two vortices would be irreversibly separated at the end of our numerical simulations.

5.3.2.1. *Varying the layer thicknesses and vortex intensities*

Here, we vary separately the layer thicknesses and the vortex intensities.

Firstly, a thicker lower layer ($\Delta_2/\Delta_1 = 3$) amplifies the loops of the upper layer trajectory and diminishes the lower layer loops as expected (figure not shown). The westward drift velocity is not changed. A lower layer vortex stronger than the upper one ($Q_2/Q_1 = 3$, with the same barotropic vorticity) has the same effect. This can be understood since the term $\Delta_j Q_j$ appears in the calculation of the velocities.

5.3.2.2. *Inclusion of the planetary and topographic beta effects*

Figure 5.5(a) shows the trajectory of the vortex pair when the planetary beta effect is added via the ghost vortices (this simulation is run for 1/3 of the total time of the reference case). The vortices also drift southward compared with this reference simulation; this southward drift of anticyclones due to beta effect only is well known. In the ocean, this drift would be arrested by the bottom topography near the coast. The zonal drift[3] is about the same in this case and for the same duration in the reference case. Note that the two vortices remain close to each other.

For a quantitative comparison with oceanic vortices, *in-situ* observations yield a drift velocity of about 0.2 m/s for the NBC rings, leading to a total distance of 216 km covered by these rings during the duration corresponding to this simulation; in the numerical model our point vortices drift covers 228 km. There error made here by this simple model is thus lower than 5%. Note also that, along the x-axis, the distance covered here is about 1/3 of the distance covered in the reference case, for 1/3 of the duration of this reference case. Therefore, the beta effect does not substantially modify the westward drift of the vortex pair induced by the image vortices. On the contrary, the influence of the beta effect is noticeable in the motion along the y-axis, inducing a 70 km drift towards the coast. This is clearly seen on Fig. 5.5(b).

Figure 5.5(c) presents the distance between the two vortices with respect to time. Without beta effect, this distance varies periodically with time. With beta effect this distance first increases and then decreases as the two vortices get closer to the wall/coast. This is due to the joint effect of their co-rotation and of the beta-drift.

We contrast now this simulation with a case where both planetary beta effect and topographic beta effect are present. In this region of the ocean, the topographic beta effect is opposite to the planetary beta effect and it can be up to ten times stronger locally (close to the

[3]Zonal means along the Ox-axis.

(a)

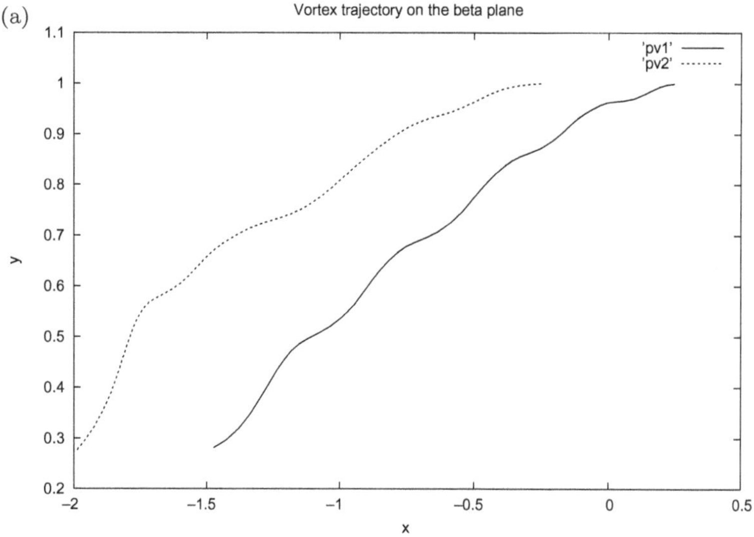

(b)

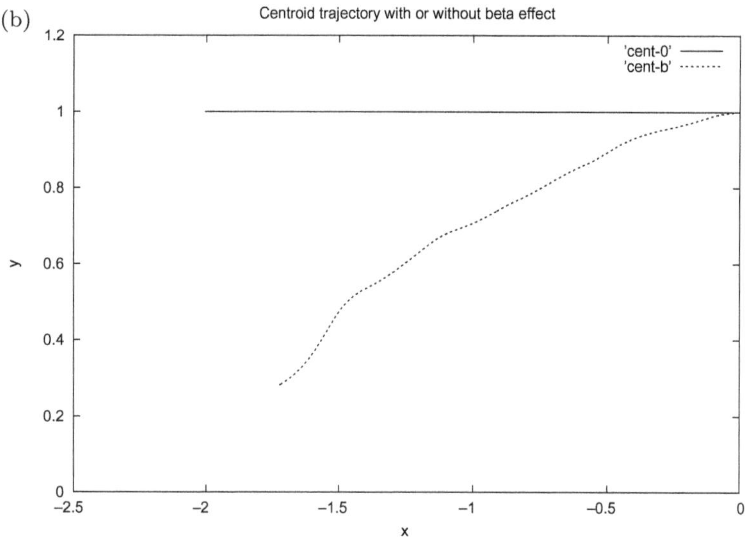

Fig. 5.5. (a) Trajectory of the two anticyclones on the (planetary) beta plane; (b) vortex centroid trajectory with or without beta effect; (c) inter-vortex distance with and without beta effect; (d) trajectories of the two point vortices when the total (planetary + topographic) beta effect in the lower layer is null, or opposite to that of the upper layer.

(c)

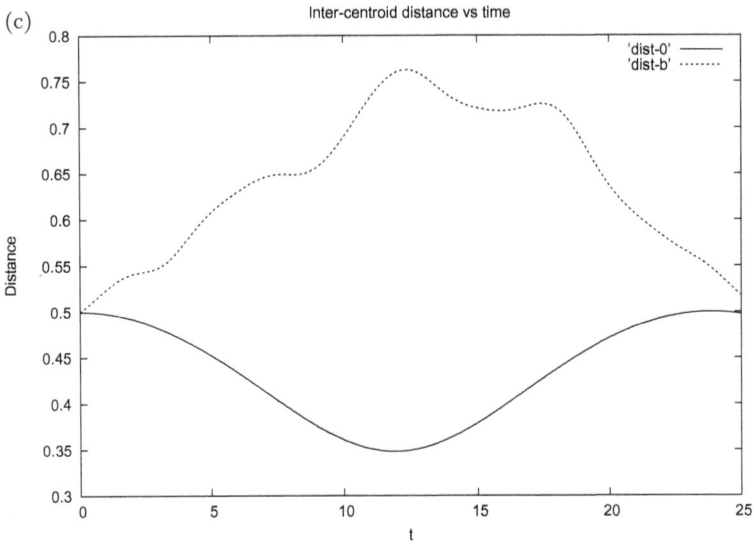

(d)

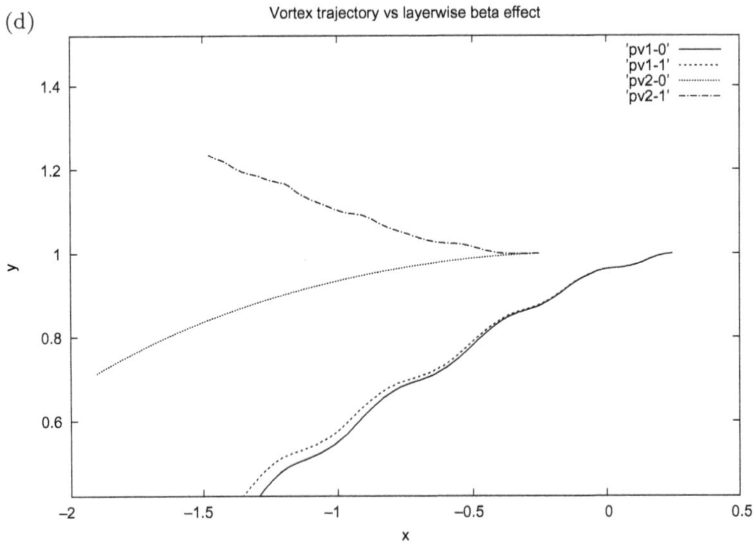

Fig. 5.5. (*Continued*)

coast). Therefore, we vary the relative intensity of the topographic and planetary beta effects with several numerical simulations. Figures 5.5(c) and 5.5(d) clearly show that:

- when planetary and topographic beta effects cancel out in the lower layer, the vortex in this layer drifts towards the wall under the influence of the upper layer vortex (via layer coupling);
- when the total lower layer beta effect is equal and opposite to that in the upper layer, the lower vortex trajectory is now away from the wall.

  As a conclusion to this analysis, the vertically different beta effects can be an efficient driver of vortex separation. Nevertheless, since the topography of the northern coast of South America is steep but localized, it will create a strong topographic beta effect only very close to the coast. Therefore, it is not likely to separate the two layerwise vortices far offshore.

For another mechanism able to achieve this separation, we now turn to the coastline orientation.

5.3.2.3. *Influence of the coastline orientation in both layers*

To maintain the number of parameters within a reasonable limit, and since the meridional extent[4] of the trajectories considered here is not large, we set the beta effect to zero.

Firstly, we consider that the wall (or coast in the ocean) forms a right angle at the origin of the plane, in both layers; secondly we will consider that this corner exists only in the lower layer and that the coast is straight (along Ox) in the upper layer. This latter case corresponds to the shallow passage between the Lesser Antilles Islands.

Considering first a right angle of the wall at the origin in both layers, we plot the two vortex trajectories on Fig. 5.6(a), for parameters corresponding to the reference case, and we add the path of their center of gravity (the "centroid"). This center follows

[4]Meridional means along the Oy direction.

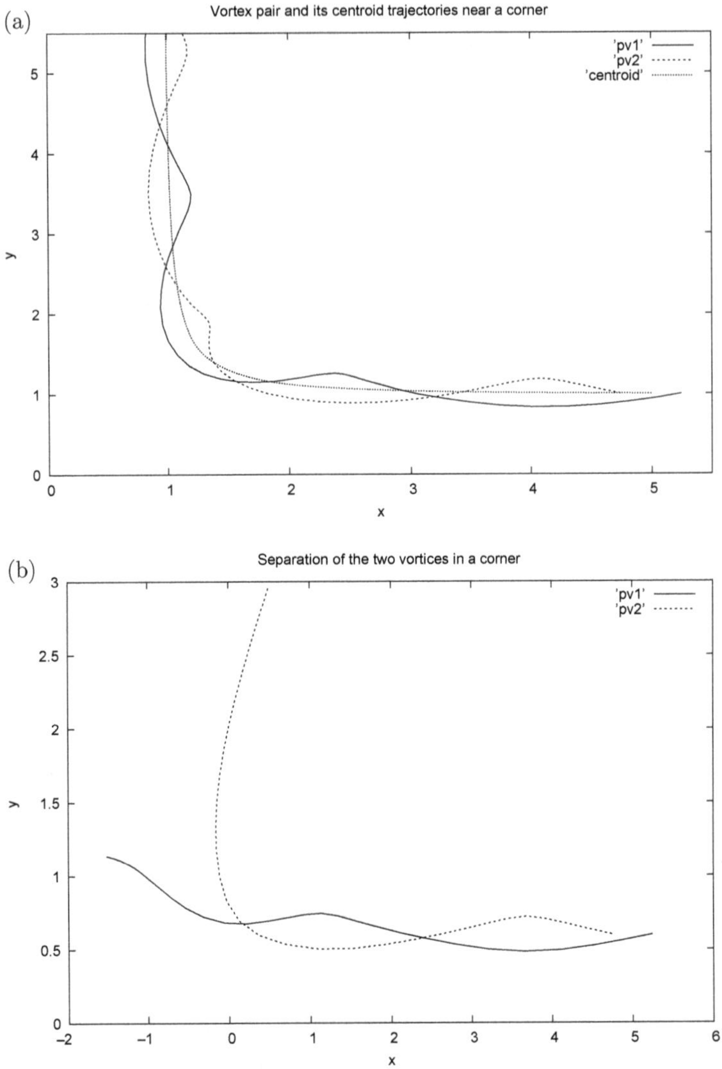

Fig. 5.6. (a) Trajectories of the two anticyclones and of their centroids when the coast/wall makes a right angle at the origin in the two layers; (b) trajectories of the two anticyclones when the coastline corner exists in the lower layer only; (c) trajectory of the two anticyclones (pv1 and pv2) in the presence of a cyclone (pv3); (d) regime diagram of the three vortices with respect to the position of the cyclone (abscissa) and of its strength (ordinate).

(c)

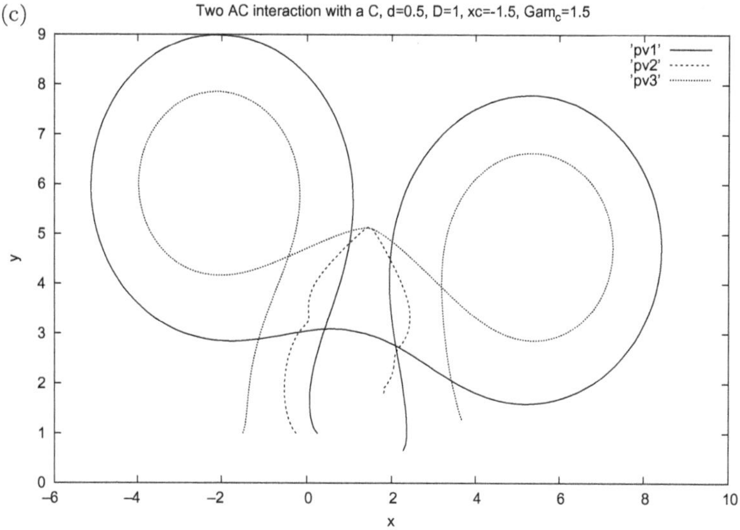

(d)

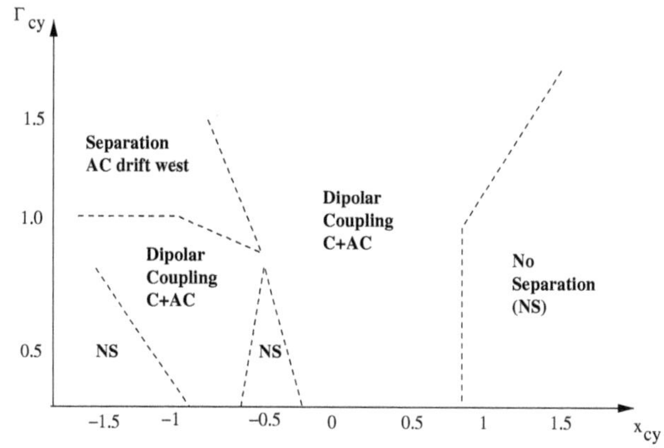

Fig. 5.6. (*Continued*)

a hyperbole as does the trajectory of a single point vortex near a right-angle corner.

Again the search for a separation of the vortex pair near the corner did not yield any realistic value for a minimal distance of these vortices to the wall.

Now we consider the case where the coastline orientation is only zonal (along Ox) in the upper layer, but has a right-angle corner at the origin of the plane in the lower layer. In this case, the upper layer vortex will drift westward along Ox due to the influence of its image vortex, while the lower layer vortex will veer to the North, also under the influence of its image vortex (see Fig. 5.6(b)).

Performing a series of simulations, setting $R_d = 1.0$ and $d = 0.5$, and varying D, we find that the critical distance to the wall for the separation of the upper and lower layer vortices, is $D_c = 0.9$ for $\Delta_1 = \Delta_2$. We find the same critical value D_c for $d = 1.0$ and $\Delta_1 = \Delta_2$, and also for $d = 0.5$ with $\Delta_2 = 3\Delta_1$.

Thus, we conclude that the separation of the two vortices can occur if they lie initially close to the wall, at a distance smaller than 110 km, in the real ocean. Indeed, the separation of upper and lower NBC rings, drifting close to the coast, has been observed near the Lesser Antilles Arc.

5.3.2.4. *Influence of a neighboring cyclone*

Cyclones are observed near the anticyclonic NBC rings. The ability of such cyclones to separate the upper and lower cores of a NBC ring is examined now. Many parameters are relevant to this problem: the relative strength of the cyclone, its position with respect to the anticyclones, its vertical structure and so on. To avoid a lengthy parameter search, we consider here a surface intensified cyclone ($Q_1(\mathrm{cy}) = Q_{\mathrm{cy}}, Q_2(\mathrm{cy}) = 0$) initially at the same latitude as the two anticyclones ($y_{\mathrm{cy}}(t = 0) = y_1(\mathrm{AC}_1, t = 0) = y_2(\mathrm{AC}_2, t = 0)$). We will only vary Q_{cy} and the longitude x_{cy} of this cyclone. The cyclone can thus lie to the east or to the west of the two anticyclones.

The cyclone has little influence on the anticyclones when either its strength (Q_{cy}) is weak or when its distance to the two anticyclones is large.

When the cyclone lies to the east and close to the anticyclones, it can pair with the closest one as a dipole, and then head northeastward (see Fig. 5.6(c)). This can also occur when the cyclone lies to the west of the anticyclones (that is, in the direction of their wall induced drift); then, a complex interaction, including leapfrogging, occurs. Finally, when the cyclone initially lies to the west of the anticyclones, and is strong, it can separate them; the anticyclones then drift westward, widely separated (see Fig. 5.6(d)).

5.3.2.5. *Influence of an oscillating, sheared, mean flow*

The influence of a mean flow on two vortices was studied in Carton *et al.* [40]. In that paper, this surrounding flow was steady. Here we evaluate the effect of an oscillatory mean flow on the two vortices; this mean flow has a period equal to the rotation period of the two vortices. The flow is along-wall, with a linear cross-wall variation. This variation is called the shear of the current.

For a shear of 0.03 (dimensionless value), the two anticyclones move westward, in co-rotation, while for a shear of 0.04 they finally separate (see Fig. 5.7). The smallness of this value (compared with Q_{AC}/d^2) indicates that the harmonic resonance is efficient to separate the two vortices without the need for a strong ambient current.

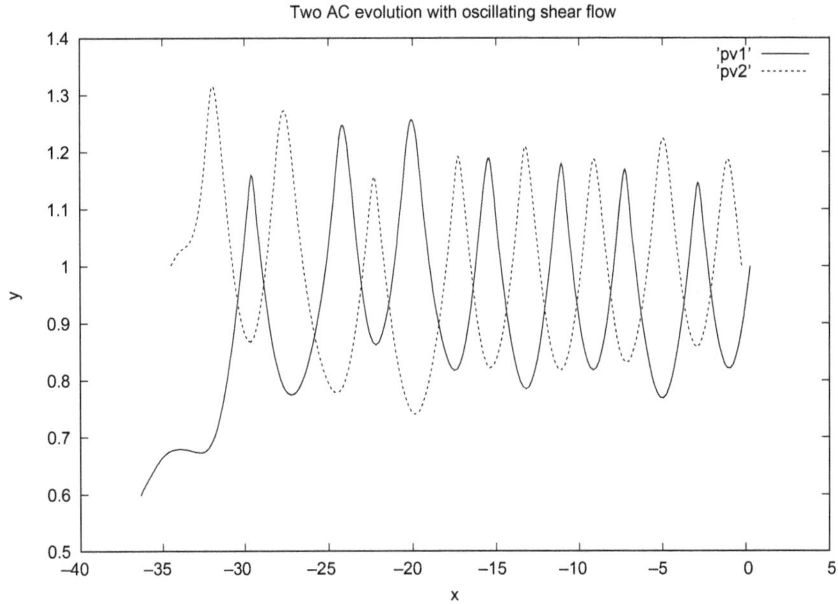

Fig. 5.7. Trajectories of the two anticyclones in the presence of a zonal (along Ox) shear current, oscillating with harmonic period.

5.4. Discussion, Conclusion

5.4.1. *Discussion*

In this chapter, we have shown how the point-vortex model evolved from the end of the 19th century to present. In particular, effects pertaining to geophysical flows were introduced in the point-vortex model: density stratification, Earth rotation (Coriolis effect) and curvature (planetary beta effect), the presence of boundaries, the effect of seafloor relief (topographic beta effect).

Then we applied this "geophysical" point-vortex model to the evolution of a pair of vortices along a straight wall/coast. This evolution is a cycloid in the absence of any other effect, when the two vortices are distant from the coast, at least as much as from each other. The planetary beta effect induces a southwestward drift of the two anticyclones. The northward sloping bottom of the sea induces a northwestward drift of the lower vortex.

In the presence of a corner of the coast, the two vortices veer northward. But they become finally separated if the coastal orientation changes direction only in the lower layer (e.g., the case of the Lesser Antilles Arc). This evolution corresponds to observations of the vertical splitting of the NBC rings.

Along the Northern coast of South America, cyclonic vortices are often observed in the vicinity of NBC rings. Therefore, the effect of a cyclone, either to the East or to the West of the pair of anticyclones, was investigated. It was shown that two main evolutions could occur: one where the cyclone pairs with one of the anticyclone and follows a complex trajectory, the other one where the cyclone initially separates the two anticyclones which then drift west, away from the cyclone. Finally, the effect of a time-periodic, sheared current, on the two anticyclones, was studied. This current induces a harmonic resonance which finally separates these anticyclones.

It must be noted that the vortex evolution was studied here with a two-layer point-vortex model; other models could have been used like a surface quasi-geostrophic model and a deep quasi-geostrophic, continuously stratified, model. It is expected that this would have provided qualitatively similar results, though with different thresholds in physical parameters between these evolutions. Since the real ocean currents in this region are frontal (i.e., density surfaces undergo strong variations of their vertical positions), it could also be useful to use semi-geostrophic point-vortex models which allow such frontal situations, contrary to quasi-geostrophic models which impose that isopycnal surfaces have weak vertical displacements.

5.4.2. *Conclusions*

Though our point-vortex model, and also our initial two-vortex configuration, are highly idealized compared with reality, they nevertheless show time evolutions qualitatively comparable with reality. Point-vortex models and mathematical models of vortex dynamics are thus an efficient tool for the exploration of the behavior of oceanic vortices. They also provide important information of the spreading of passive tracers around these vortices. These tracers can be pollutants, plastics, or oil slicks in the ocean. A natural extension of this study will use finite-area vortices in a two-layer quasi-geostrophic model; such vortices can deform, split or merge, and they are more resemblant to real vortices.

A following step in this study will use a three-dimensional ocean model, to determine all the effects which can lead to the separation of the two anticyclones. Other natural extensions of this short study are as follows:

- the study of point vortices which couple with the atmosphere and thus develop a source or a sink at their center; their resonance and evolution to chaos with an unsteady external flow should be studied;
- the study of chaotic dispersion of passive tracer in this case, and the coupling between the physical evolution of the vortices, and the evolution of ecosystems in their vicinity.

Finally, a validation of the three-dimensional model results on the *in-situ* and satellite observations will be necessary. In that respect, the sea surface temperature and elevation data will be used to track the various vortices in time. The vertical profiles of temperature and salinity obtained from the Argo profiling floats will help determine if the two cores of the NBC rings lie above each other. Finally, the hydrological data collected during the EUREC4A-OA experiment at sea, which form a dense set of observations, will provide a three-dimensional view of these vortices near the coast of South America. This analysis is forthcoming.

Appendix: Mathematical Solution of the Point-Vortex Motion in the Two-Layer Model

The velocities of the point vortices in the two-layer quasi-geostrophic model, with $Q_1 = Q_2 = \Gamma$, are given by the following relations in which vortex 1 is the real vortex in layer 1 with coordinates (x_1, y_1), vortex 2 is the real vortex in layer 2 with coordinates (x_2, y_2) and vortices 3 and 4 are respectively the images of vortices 2 and 1 across the coast (a straight wall along Ox). Thus, we make use of the symmetry properties to set $(x_3 = x_2, y_3 = -y_2, x_4 = x_1, y_4 = -y_1)$ in the calculations below (see also Fig. 5.8).

Vortices 5 and 6 are the images of 1 and 2 with respect to the wall along Oy, when present. Vortices 7 and 8 are the images of vortices 5 and 6 across the coast. Similarly, the symmetry properties are used to calculate the positions and distances for all vortices. We call r_{kl} the distance between vortices k and l. The Green's functions G depend on the layer i and j in which the vortices k and l reside. The terms $G'(r_{kl})$ below are in effect $G'_{ij}(r_{kl})$. The zonal and meridional velocities u_i and v_i are therefore

$$u_1 = -\frac{\Gamma}{2\pi}\left[\frac{(y_1 - y_2)}{r_{12}}G'(r_{12}) - \frac{(y_1 + y_2)}{r_{13}}G'(r_{13}) - \frac{2y_1}{r_{14}}G'(r_{14}) \right.$$
$$\left. - \frac{(y_1 - y_2)}{r_{16}}G'(r_{16}) - \frac{(y_1 + y_2)}{r_{17}}G'(r_{17}) - \frac{2y_1}{r_{18}}G'(r_{18})\right], \quad \text{(A.1)}$$

$$v_1 = \frac{\Gamma}{2\pi}\left[\frac{(x_1 - x_2)}{r_{12}}G'(r_{12}) - \frac{(x_1 - x_2)}{r_{13}}G'(r_{13}) - \frac{2x_2}{r_{15}}G'(r_{15}) \right.$$
$$\left. - \frac{(x_1 + x_2)}{r_{16}}G'(r_{16}) - \frac{(x_1 + x_2)}{r_{17}}G'(r_{17}) - \frac{2x_1}{r_{18}}G'(r_{18})\right], \quad \text{(A.2)}$$

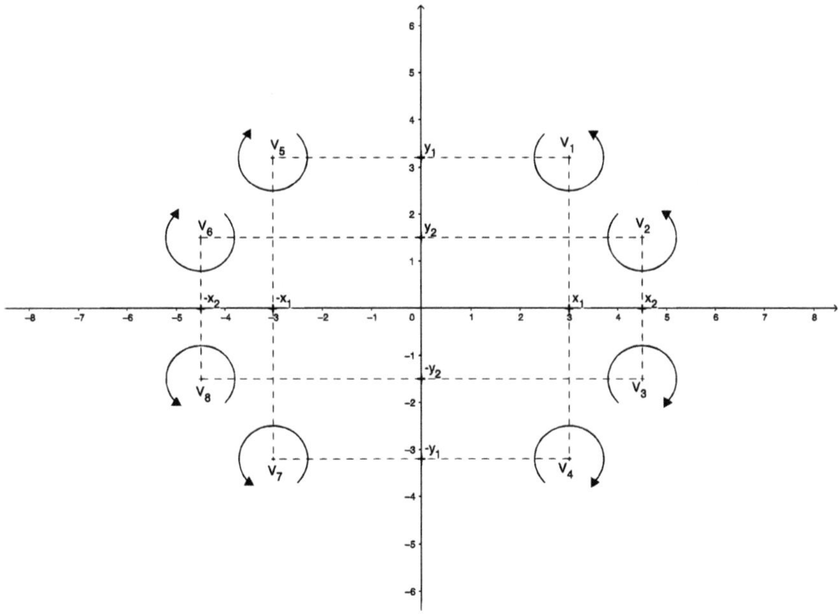

Fig. 5.8. Schematic configuration of the real vortices in the $x > 0, y > 0$ quadrant with their images across the Ox and Oy walls.

$$u_2 = -\frac{\Gamma}{2\pi}\left[\frac{(y_2 - y_1)}{r_{21}}G'(r_{21}) - \frac{(y_1 + y_2)}{r_{24}}G'(r_{24}) - \frac{2y_2}{r_{23}}G'(r_{23})\right.$$
$$\left. - \frac{(y_2 - y_1)}{r_{25}}G'(r_{25}) - \frac{2y_2}{r_{27}}G'(r_{27}) - \frac{(y_1 + y_2)}{r_{28}}G'(r_{28})\right], \quad \text{(A.3)}$$

$$v_2 = \frac{\Gamma}{2\pi}\left[\frac{(x_2 - x_1)}{r_{21}}G'(r_{21}) - \frac{(x_2 - x_1)}{r_{24}}G'(r_{24}) - \frac{(x_1 + x_2)}{r_{25}}G'(r_{25})\right.$$
$$\left. - \frac{2x_2}{r_{26}}G'(r_{26}) - \frac{2x_2}{r_{27}}G'(r_{27}) - \frac{(x_1 + x_2)}{r_{28}}G'(r_{28})\right]. \quad \text{(A.4)}$$

The layerwise Green's functions derivatives $G'_{ij}(r)$ are explicitly

$$G'_{11}(r) = \frac{1}{2\pi}\left[\frac{\Delta_1/\Delta}{r} + \frac{\Delta_2/\Delta}{R_d}K_1\left(\frac{r}{R_d}\right)\right], \quad \text{(A.5)}$$

$$G'_{12}(r) = \frac{1}{2\pi}\left[\frac{\Delta_2/\Delta}{r} - \frac{\Delta_2/\Delta}{R_d}K_1\left(\frac{r}{R_d}\right)\right], \quad \text{(A.6)}$$

$$G'_{22}(r) = \frac{1}{2\pi}\left[\frac{\Delta_2/\Delta}{r} + \frac{\Delta_1/\Delta}{R_d}K_1\left(\frac{r}{R_d}\right)\right], \tag{A.7}$$

$$G'_{21}(r) = \frac{1}{2\pi}\left[\frac{\Delta_1/\Delta}{r} - \frac{\Delta_1/\Delta}{R_d}K_1\left(\frac{r}{R_d}\right)\right] \tag{A.8}$$

with $R_d^2 = \frac{g'\Delta_1\Delta_2}{f_0^2\Delta}$ the deformation radius.

In the case of a coast formed of a single wall along Ox, these equations can be simplified with Taylor expansions in $\epsilon = d/D$ for vortices far away from the wall. In this case, the K_1 components of the velocity are negligible. We call $x_c = (x_1+x_2)/2, y_c = (y_1+y_2)/2, x_d = (x_1 - x_2)/2, y_d = (y_1 - y_2)/2$. We have $y_c = D/2$, $x_d = (d/2)\cos(\alpha)$, $y_d = (d/2)\sin(\alpha)$ and, with $Q_1 = Q_2 = \Gamma$,

$$u_c = \frac{\Gamma}{2\pi D}\,[2 - \epsilon^2\cos^2(2\alpha)], \quad v_c = 0, \tag{A.9}$$

$$u_d = -\frac{\Gamma\sin(\alpha)}{2\pi d}\,[1+\epsilon^2], \tag{A.10}$$

$$v_d = \frac{\Gamma\cos(\alpha)}{2\pi d}\,[1-\epsilon^2]. \tag{A.11}$$

The term in ϵ^2 in u_d, v_d leads to the deviation of the trajectory from a circle.

Acknowledgments

The authors acknowledge the support from the EUREC4A-OA international program in the initial design of these experiments. The authors dedicate this paper to Professor Miguel Sanjuan.

References

[1] B. Franklin, A letter from Dr. Benjamin Franklin to Mr. Alphonsus Le Roy, Member of Several Academies, at Paris. Containing sundry maritime observations, *Trans. Amer. Philos. Soc.* **2** (1786) 292–329.

[2] J. Williams, Memoir of Jonathan Williams on the use of the thermometer in discovering banks, soundings, etc., *Trans. Amer. Philos. Soc.* **3** (1793) 82–96.

[3] Anonymous, The Gulf Stream (report on P.E. Church's work), *Nature* **130** (1932) 162.

[4] P. E. Church, Surface temperatures of the Gulf Stream and the waters on its margin, *Bull. Amer. Met. Soc.* **17** (1935) 350–351.
[5] F. C. Fuglister and L. V. Worthington, Some results on a multiple ship survey of the Gulf Stream, *Tellus* **3** (1951) 1–14.
[6] F. C. Fuglister, Gulf Stream 60, *Progr. Oceanogr.* **3** (1963) 265–383.
[7] L. Armi and W. Zenk, Large lenses of highly saline Mediterranean Water, *J. Phys. Oceanogr.* **14** (1984) 1560–1576.
[8] P. L'Hegaret, X. Carton, I. Ambar, C. Menesguen, B. L. Hua, L. Cherubin, A. Aguiar, B. Le Cann, N. Daniault and N. Serra, Evidence of Mediterranean water dipole collision in the Gulf of Cadiz, *J. Geophys. Res. C* **119** (2014) 5337–5359.
[9] D. B. Chelton, M. G. Schlax, R. M. Samelson and R. A. de Szoecke, Global observations of large oceanic eddies, *Geophys. Res. Lett.* **34** (2007) L15606.
[10] K. Katsumata, Eddies observed by argo floats. Part I: Eddy transport in the upper 1000 dbar, *J. Phys. Oceanogr.* **36** (2016) 3471–3486.
[11] D. Nof and T. Pichevin, The retroflection paradox, *J. Phys. Oceanogr.* **26**(11) (1996) 2344–2358.
[12] T. Pichevin and D. Nof, The eddy cannon, *Deep-Sea Res.* **43**(9) (1996) 1475–1507.
[13] D. M. Fratantoni, W. E. Johns and T. L. Townsend, Rings of the North Brazil Current; their structure and behavior inferred from observations and a numerical simulation, *J. Geophys. Res. C* **100**(6) (1995) 10,633–10,654.
[14] D. M. Fratantoni and P. L. Richardson, The evolution and demise of North Brazil Current Rings, *J. Phys. Oceanogr.* **36** (2006) 1241–1264.
[15] J. C. McWilliams, Geostrophic vortices, in *Nonlinear Topics in Ocean Physics: Proceedings of the International School of Physics Enrico Fermi*, Vol. 109 (North-Holland, Amsterdam, 1991), pp. 5–50.
[16] X. Carton, Hydrodynamical modeling of oceanic vortices, *Surv. Geophys.* **22**(3) (2001) 179–263.
[17] X. Carton, Oceanic vortices, in *Fronts, Waves and Vortices*, Lecture Notes in Physics, Vol. 805 (Springer, 2010), pp. 61–108.
[18] G. J. F. van Heijst, Dynamics of vortices in rotating and stratified fluids in *Fronts, Waves and Vortices*, Lecture Notes in Physics, Vol. 805, (Springer, 2010), pp. 1–34.
[19] M. A. Sokolovskiy and J. Verron, *Dynamics of Vortex Structures in a Stratified Rotating Fluid*, Atmospheric and Oceanographic Sciences Library, Vol. 47 (Springer International Publishing, 2014), 382 pp.
[20] W. Thomson and Lord Kelvin, On vortex motion, *Trans. Roy. Soc. Edinb.* **25** (1869) 217–260.

[21] W. J. M. Rankine, *Manual of Applied Mechanics* (C. Griffen Co., London, 1858).
[22] G. Kirchhoff, *Vorlesungen ueber matematische Physic: Mechanik* (Teubner, Leipzig, 1876).
[23] M. E. Stern and G. R. Flierl, On the interaction of a vortex with a shear flow, *J. Geophys. Res. C* **92** (1987) 10,733–10,744.
[24] C. Eloy and S. Le Dizes, Stability of the Rankine vortex in a multipolar strain field, *Phys. Fluids* **13** (2001) 660.
[25] H. Poincaré, *Théorie des tourbillons* (Georges Carré éditeur, 1893).
[26] H. Aref, Integrable, chaotic, and turbulent vortex motion in two-dimensional flows, *Annu. Rev. Fluid Mech.* **15** (1983) 345–389.
[27] P. Newton, Point vortex dynamics in the post-Aref era, *Fluid Dyn. Res.* **46** (2014) 031401.
[28] V. Bjerknes, R. Rubenson and A. Lindstedt, *Ueber einen Hydrodynamischen Fundamentalsatz und seine Anwendung: besonders auf die Mechanik der Atmosphaere und des Weltmeeres* (Kungl. Boktryckeriet. PA Norstedt und Soener, 1898).
[29] J. G. Charney, On the scales of atmospheric motions, *Geofys. Publ.* **17** (1948) 3–17.
[30] N. J. Zabusky and J. C. McWilliams, A modulated point-vortex model for geostrophic, beta-plane dynamics, *Phys. Fluids* **25** (1982) 2175–2182.
[31] Anonymous, *Two-Dimensional Incompressible Irrotational Flow. Part A. Fluid Dynamics and Waves*, Lecture Notes, Vol. 2.1 (University of Oxford, OCIAM Mathematical Institute, 2014), available from https://courses.maths.ox.ac.uk/node/view_material/5464.
[32] X. Perrot and X. Carton, Point-vortex interaction in an oscillatory deformation field: Hamiltonian dynamics, harmonic resonance and transition to chaos, *Discrete Cont. Dyn. Syst. B* **11** (2011) 971–995.
[33] K. V. Koshel and E. A Ryzhov, Parametric resonance with a point-vortex pair in a nonstationary deformation flow, *Phys. Lett. A* **376** (2012) 744–747.
[34] E. A. Ryzhov and K. V. Koshel, Two-point-vortex evolution in an oscillatory shear flow with rotation, *Europhys. Lett.* **108** (2014) 24002.
[35] K. V. Koshel, E. A Ryzhov and X. J. Carton, Vortex interactions subjected to deformation flows: A review, *Fluids* **4** (2019) 14.
[36] E. A. Ryzhov, K. V. Koshel and X. J. Carton, Passive scalar advection in the vicinity of two point vortices in a deformation flow, *Eur. J. Mech. B Fluid* **34** (2012) 121–130.
[37] X. Perrot, X. Carton and A. Guillou, Geostrophic vortex alignment in external shear or strain, in *IUTAM Symposium on Turbulence in the Atmosphere and Oceans*, Vol. 28 (Springer–Verlag, Berlin, 2010), pp. 217–228.

[38] E. A. Ryzhov and K. V. Koshel, Resonance phenomena in a two-layer two-vortex shear flow, *Chaos* **26** (2016) 113–116.
[39] E. A. Ryzhov and K. V. Koshel, Parametric instability of a many point-vortex system in a multi-layer flow under linear deformation, *Reg. Chaotic Dyn.* **21** (2016) 254–266.
[40] X. Carton, B. Le Cann, A. Serpette and J. Dubert, Interactions of surface and deep anticyclonic eddies in the Bay of Biscay, *J. Mar. Sci.* **109–110** (2013) S45–S59.

https://doi.org/10.1142/9789811221903_0006

Chapter 6

Modal Asynchronicity in Pre-stressed Continuous Simply-Supported Beams with Transversal and Rotational Oscillators

Eduardo A. R. Ribeiro*,‡, Stefano Lenci† and Carlos E. N. Mazzilli*

*Escola Politécnica da Universidade de São Paulo.
Av. Prof. Luciano Gualberto,
Travessa 3, No. 380. 05508-010. São Paulo. Brazil
†Università Politecnica delle Marche
via Brecce Bianche 60131 Ancona, Italy
‡asceduardo@usp.br

The existence of structural systems where some vibration modes are perfectly localized/asynchronous (that is, where finite parts of the structure remain at rest along some modal responses) has been widely demonstrated in works published before. For discrete linear systems, we can resort to an inverse eigenvalue problem, where the eigenvectors are prescribed localized modes and the associated system parameters are the unknowns to be determined. In such cases, it is relatively simple to show that even very-simple models, with few degrees of freedom (DOFs), can exhibit modal localization. However, when it comes to continuous systems, the procedure is not so simple and our first efforts could solely entail nearly-localized modes. This barrier has been recently crossed, after we successfully derived localized modes in three continuous systems: two simply-supported beams with cantilever extensions and a tuning-fork-like structure. This was achieved after we fine-tuned the structures' discontinuities. Yet, in some cases calibrating the original parameters was not enough and the use of auxiliary systems (yielding new DOFs) was necessary. In this work, we propose another case study: A pre-stressed continuous simply-supported beam. We demonstrate that modal localization occurs after transversal and rotational oscillators are added and then parameter fine-tuning is performed.

6.1. Introduction

The classical modal response of a linear undamped structural model can be expressed as $\mathbf{w}_i(\mathbf{r},t) = f_i(\mathbf{r})\sin(\omega_i t + \varphi_i)$. The subscript i indicates that the model may display a variety of modal responses, according to the number of degrees of freedom (DOFs) it has. The ith *normal* vibration mode $f_i(\mathbf{r})$ depends solely on the system's material points position vector $\mathbf{r}$ in the reference configuration, while the sinusoidal function depends on the ith natural frequency ω_i (related to structural properties) and on a phase angle φ_i (dependent of the initial conditions to which the structure is subjected). As a consequence, all material points are governed by the same frequency and phase, and thus they are expected to vibrate in unison. Such modal responses are *synchronous* vibrations.

If the structure is subjected to proportional damping and linearity assumptions are still valid, the classical modes of vibration will remain the same, i.e., synchronous, although now the modal frequencies will be smaller and the modal amplitudes will exponentially decay with time. Yet, if the structure has nonproportional damping or if nonlinear effects must not be neglected, then the modal responses become potentially nonstationary, which means that the material points may no longer move in unison. Such nonstationary modes, referred to in the literature as nonsimilar [1–3], occur because modal displacements and velocities are out-of-phase. Hence, one might state that non-similar modes would be *asynchronous.* This interpretation is however questionable, since these modes are still governed by the same and single modal frequency.

In the literature, several definitions of asynchronous vibrations, not restricted to modal motions, can be found (for a detailed discussion on this, see [4]). In our understanding, vibrations in which material points do not move in unison, but are still governed by one single frequency, should *not* be told asynchronous. An altogether different case is that of vibrations characterized by different frequencies in different parts of the system's physical domain. In this sense, if one vibration mode is *localized,* that is, if a nonzero-measure region of the structure remains at rest during the modal response, we shall call it asynchronous, since the motionless part has a mathematically-null frequency, while the system's remaining part vibrates with a nonnull frequency. Furthermore, if we regard a generic free-vibration response composed of superimposed modes, from which at least one is localized, then the corresponding frequency spectra will be different throughout the system's physical domain. Hence, we choose to define asynchronous modes as localized modes, having as a consequence that if they are excited the overall system's response will be asynchronous.

The idea of localized free vibrations of structures, that is, which are localised despite not resorting to any kind of control, may sound unusual at first. However, it is not a novelty that vibrations can be passively controlled, by means, for example, of auxiliary systems added to the original structure. In fact, since Frahm proposed a "device for damping vibration of bodies" [5] and Den Hartog, Ormondroyd and Brock [6–8] established the theorical grounds of the now well-known theory of tuned-mass dampers (TMDs), extensive investigation has been made by different researchers and the literature survey now counts a large number of references on the subject, as addressed in [9].

It is known that, depending on the material properties of TMDs, the vibrations of the original structure may not only be reduced, but even annihilated at some points. Hence, if the parameters of these auxiliary devices are wisely chosen (that is, if they are *finely tuned*), then vibration can be reduced or suppressed at strategic points, which has obvious vibration-control applications. In this sense, Cha and others provide several studies on the use of finely-tuned oscillators for the sake of imposing nodes (collocated or not with the respective attachment points) at linear elastic beams [10–18]. For instance, in[10] it has been shown that an oscillator-chain can be used to simultaneously impose a node at different normal modes of a beam. In [11], it was shown that simple oscillators placed near to each other may not only entail nodes, but also nearly-localized vibrations. In [12, 13], the coupling of translational (TO) and rotational (RO) oscillators was demonstrated to be very efficient to produce fixed nodes with null deflection slope and also to *localize* vibrations with the aid of a few auxiliary systems.

In all cases, the addition of oscillators meant adding new DOFs to the original system, and localization for engineering purposes (that is, nearly-localized vibrations) appeared as a very-desirable, but not explicitly-pursued, collateral effect. However, it is reasonable to suppose that asynchronous vibrations can also be achieved without the necessity of adding new DOFs, provided that the original parameters are finely tuned. In this sense, Mazzilli and Lenci investigated the existence of *admissibility conditions for modal asynchronicity* (in short, admissibility conditions) in different structural systems. Resorting to the inverse eigenvalue problem, they showed that localized (and hence asynchronous) modes can be derived even in very simple linear systems. In other words, while in the regular modal analysis one calculates the modal frequencies and shapes for known structural parameters, this time the latter ones were calculated for imposed localized modes. Among the structural schemes regarded in their works, we mention Ziegler's columns subjected to circulatory forces [4], planar frames [19] and heavy-chains [20, 21]. The latter ones were later regarded under geometric nonlinearities [22, 23] and parametric excitation [24–26]. These systems were modeled with lumped masses and therefore they were all discrete, in the sense they were governed by ordinary differential equations (ODEs), although some of them were at least piecewise continuous [19]. Systems governed by partial differential equations (PDEs), and hence strictly continuous, were regarded in [27], where asynchronous modes were derived in pre-stressed beams on elastic foundation. However, such beams were modeled by means of an equivalent cable formulation (Bessel-like modes), and admissibility conditions could not guarantee rotation continuity at the interface between the vibrating region and the one at rest. Consequently, equivalent finite-element (FE) beam models could only reveal nearly-asynchronous modes.

Although Cha's and Mazzilli-Lenci's research lines might be distinct concerning their objectives and methods, it is clear that both lead to a fine-tuning problem. Therefore, we can combine them in order to impose, instead of nodes, motionless regions of *finite length* at linear elastic structures, with the aid of added oscillators. Bearing that in mind, we attached a TO at a simply-supported beam with a cantilever extension in [28] and we demonstrated that modal responses can be confined between the supports, provided that the TO be calibrated to yield null displacements along that extension. We also showed that, depending on the position of the TO, the resting part can also comprise part of the simply-supported span, although such highly-localized modes also have very high frequencies, which make them difficult to be realized in practice. The matter was again recast in [9], in which we readdressed the problem of [28] with three other study cases: a simply-supported beam with two cantilever extensions and two TOs, a tuning-fork-like structure and a cantilever with a TO and an RO attached at its free end. In that occasion, we demonstrated that: (i) both cantilevers will stay at rest if the TOs are finely-tuned and localization may also enter the simply-supported span, (ii) the tuning fork's stem will be at rest if their prongs (which can be understood, each one, as an equivalent TO–RO pair) are finely-tuned, and (iii) in some cases no admissibility conditions can be found, which is the case of the last model of that work. All the analytical results were confirmed by means of FE models.

The idea of combining the fine-tuning of the original parameters of a structure with the assemblage of additional DOFs is not new (see, for example, [29]). However, it is a novelty from the viewpoint of modal asynchronicity, since this method has proved to be very useful to obtain perfectly-asynchronous modes in continuous models. The reasoning is actually very simple: in order to guarantee a localized mode, the motionless part must be free of bending moments and shear forces; meanwhile, the vibrating part should obviously be subjected to such efforts. Hence, the only way to achieve that, disregarding the trivial solution, is by means of finely-tuned discontinuities along the system. If one structure does not contain any discontinuity, then simply tuning the original parameters will produce no effect and auxiliary devices should be added. Only then fine-tuning relationships can be sought.

In [9, 28] we confirmed the aforementioned argument by means of four case studies. Now we present a fifth one: a TO and an RO coupled to a pre-stressed simply-supported beam (see Fig. 6.1)). In the next sections, we formulate the mathematical model and present

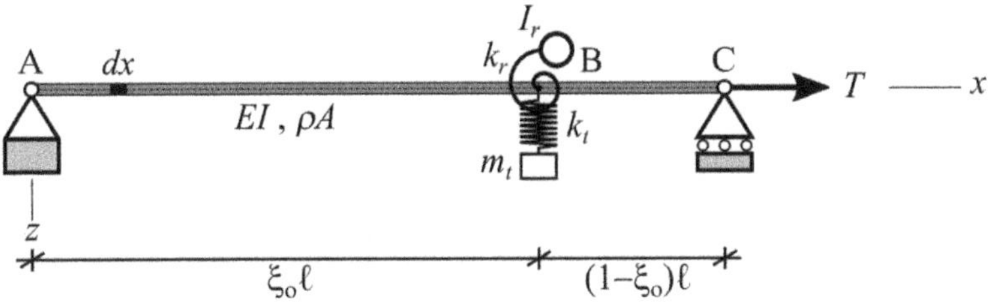

Fig. 6.1. Simply-supported beam with TO and RO attached.

a series of numerical simulations, which validate the analytical findings.

6.2. Mathematical Model

In this work, we study the structure shown in Fig. 6.1. It is a simply-supported beam with overall span length l, subjected to a constant axial load T, with uniform mechanical and geometrical properties, that is, with constant distributed mass ρA and flexural stiffness EI. To the bare beam, two auxiliary systems are attached, namely a TO with stiffness k_t and mass m_t, and an RO with stiffness k_r and rotary inertia I_r. Their position is defined by the dimensionless parameter ξ_0, such that $0 < \xi_0 < 1$, as seen in the figure.

6.2.1. *Beam*

The undamped linear free vibrations of the beam without the attached oscillators are governed by the following PDE [30]:

$$\rho A \frac{\partial^2 w}{\partial t^2} + EI \frac{\partial^4 w}{\partial x^4} - T \frac{\partial^2 w}{\partial x^2} = 0, \tag{6.1}$$

where $w(x, t)$ stands for the beam's transverse displacement at a given position x and time t. For a normalized formulation, we introduce the dimensionless time

$$\tau = \omega t, \tag{6.2}$$

where ω expresses a structure's natural (circular) frequency, and the dimensionless length

$$\xi = \frac{x}{\xi_0 l}. \tag{6.3}$$

We assume the validity of variable separation for the modal motion:

$$w(\xi, \tau) = D_1 f(\xi) \sin \tau, \tag{6.4}$$

where D_1 stands for the beam's translation amplitude. Hence, from (6.2) to (6.4), we rewrite (6.1) as

$$f^{iv} - \gamma f'' - \lambda f = 0, \tag{6.5}$$

where f expresses a vibration normal mode the structure may display. The following notation has been introduced:

$$\lambda = \frac{\rho A}{EI} \omega^2 (\xi_0 l)^4 \tag{6.6}$$

and

$$\gamma = \frac{T}{EI} (\xi_0 l)^2. \tag{6.7}$$

In (6.5) and in the sequel, $(\ )' = \frac{\partial}{\partial \xi}(\ )$ and $(\dot{\ }) = \frac{\partial}{\partial \tau}(\ )$ are derivatives with respect to ξ and τ. As it is well known, there is a countable infinity of frequencies and normal modes.

6.2.2. *Oscillators*

Regarding the attached oscillators, the motion of the TO is governed by the following ODE:

$$m_t \left(\frac{d^2 w_t}{dt^2} + \left. \frac{\partial^2 w}{\partial t^2} \right|_{x=\xi_0 l} \right) + k_t w_t = 0, \tag{6.8}$$

where the *relative* displacement of the TO reads

$$w_t(\tau) = D_1 f_t \sin \tau. \tag{6.9}$$

After the substitution of (6.2) into (6.4) and (6.9) into (6.8) and some manipulations, we find the dimensionless motion amplitude of the TO:

$$f_t = \frac{1}{\alpha_t - 1} f(1), \tag{6.10}$$

where

$$\alpha_t = \frac{k_t}{m_t \omega^2}. \tag{6.11}$$

On the other hand, the motion of the RO is governed by

$$I_r \left(\frac{d^2 \varphi_r}{dt^2} + \left. \frac{d^3 w}{\partial t^2 \partial x} \right|_{x=\xi_0 l} \right) + k_r \varphi_r = 0, \tag{6.12}$$

where

$$\varphi_r(\tau) = D_2 f_r \sin \tau \tag{6.13}$$

refers to the *relative* rotation of the RO and D_2 is its rotation amplitude. After the substitution of (6.2) into (6.4) and (6.13) into (6.12) and some manipulation, we find the dimensionless motion amplitude of the RO:

$$f_r = \frac{1}{\xi_0 l(\alpha_r - 1)} \frac{D_1}{D_2} f'(1), \tag{6.14}$$

where

$$\alpha_r = \frac{k_r}{I_r \omega^2}. \tag{6.15}$$

6.2.3. *Modal analysis*

The general solution of (6.5) is given by

$$f(\xi) = C_1 \sin(\beta_1 \xi) + C_2 \cos(\beta_1 \xi) + C_3 \sinh(\beta_2 \xi) + C_4 \cosh(\beta_2 \xi), \tag{6.16}$$

with

$$\beta_1 = \sqrt{\left(\frac{\gamma^2}{4} + \lambda \right)^{\frac{1}{2}} - \frac{1}{2}\gamma} \tag{6.17}$$

and

$$\beta_2 = \sqrt{\left(\frac{\gamma^2}{4} + \lambda \right)^{\frac{1}{2}} + \frac{1}{2}\gamma}. \tag{6.18}$$

The second derivative of $w(\xi, \tau)$ (related to the bending moment) and the third one (associated to the shear force), and hence f'' and f''', are discontinuous at ξ_0, due to the concentrated reactions from the oscillators. Therefore, f must be split in two parts:

$$f_1(\xi) = C_1 \sin(\beta_1 \xi) + C_2 \cos(\beta_1 \xi) + C_3 \sinh(\beta_2 \xi) + C_4 \cosh(\beta_2 \xi), \tag{6.19}$$

within $0 \le \xi \le 1$, and

$$f_2(\xi) = C_5 \sin(\beta_1 \xi) + C_6 \cos(\beta_1 \xi) + C_7 \sinh(\beta_2 \xi) + C_8 \cosh(\beta_2 \xi), \tag{6.20}$$

within $1 \le \xi \le \frac{1}{\xi_0}$.

Now, to calculate the constants C_1–C_8, we must take into account the structure's boundary and continuity/jump conditions. At the boundaries, we have:

$$\begin{aligned} &\text{(i)} \quad f_1(0) = 0, \text{(ii)} && f_1''(0) = 0, \\ &\text{(iii)} \quad f_2\left(\tfrac{1}{\xi_0}\right) = 0, \text{(iv)} && f_2''\left(\tfrac{1}{\xi_0}\right) = 0. \end{aligned} \tag{6.21}$$

At the oscillators' attachment point, two continuity conditions:

$$\text{(v)} \quad f_1(1) = f_2(1), \qquad \text{(vi)} \quad f_1'(1) = f_2'(1) \tag{6.22}$$

and two jump ones:

$$\begin{aligned} k_t w_t &= EI \left.\frac{\partial^3 w_R}{\partial x^3}\right|_{x=\xi_0 l} - EI \left.\frac{\partial^3 w_L}{\partial x^3}\right|_{x=\xi_0 l}, \\ -k_r \varphi_r &= EI \left.\frac{\partial^2 w_R}{\partial x^2}\right|_{x=\xi_0 l} - EI \left.\frac{\partial^2 w_L}{\partial x^2}\right|_{x=\xi_0 l}, \end{aligned} \tag{6.23}$$

where the subscripts L and R stand for the left and right limits at $x = \xi_0 l$. Such jump conditions may be rewritten with the aid of (6.4), (6.9), (6.10), (6.13) and (6.14), yielding

$$\text{(vii)} \; f_{1,2}(1) = \tfrac{\nu}{\alpha_t - 1} = f_2'''(1) - f_1'''(1) \tag{6.24}$$

and

$$\text{(viii)} \; -f_{1,2}'(1) \frac{\mu}{\alpha_r - 1} = f_2''(1) - f_1''(1), \tag{6.25}$$

in which $f_{1,2}$ and $f'_{1,2}$ stand for the fact that $f_1 = f_2$ and $f'_1 = f'_2$, and

$$\nu = \frac{k_t}{EI}(\xi_0 l)^3 \tag{6.26}$$

and

$$\mu = \frac{k_r}{EI}\xi_0 l \tag{6.27}$$

report the stiffness of each oscillator, for known beam parameters.

This system is therefore ruled by eight boundary and continuity/jump conditions, which in turn depend on eight constants, C_1–C_8. These equations can be cast into the matrix form:

$$[\boldsymbol{A}(\lambda, \gamma)]\{\boldsymbol{C}\} = \{\boldsymbol{0}\}, \tag{6.28}$$

where **C** collects the aforementioned constants. To have nontrivial solutions, the equation

$$\det \boldsymbol{A}(\lambda, \gamma) = 0 \tag{6.29}$$

must be satisfied. This provides infinitely many branches of solution $\lambda_n(\gamma)$. Each valid pair of parameters γ and λ is related to a set of constants C_1–C_8 (one remains undetermined), from which we obtain, by means of (6.19) and (6.20), the normal mode.

The procedure detailed above allows us to obtain any normal mode. However, in this work we are particularly interested in asynchronous (or localized) modal vibrations, so we need to consider extra/admissibility conditions, as stated in the introduction. If we want, for instance, to suppress any vibrations along the beam's right segment ($1 \leq \xi \leq \frac{1}{\xi_0}$), we must impose that

$$f_2(\xi) \equiv 0 \tag{6.30}$$

which means that all derivatives of (6.30) will also be identically null, and

$$\text{(iii)} \quad f_1(1) = 0, \quad \text{(iv)} \quad f'_1(1) = 0, \tag{6.31}$$

due to continuity conditions. Our problem is then reduced to four constants, C_1–C_4, and four "boundary" conditions. Gathering (i)–(iv) into (6.28) and solving (6.29) yields

$$\beta_1 \cos \beta_1 \sinh \beta_2 - \beta_2 \sin \beta_1 \cosh \beta_2 = 0, \tag{6.32}$$

which provides, with the aid of (6.17) and (6.18), λ_i for any arbitrary γ. In other words, (6.32) gives the natural frequencies ω_i of the asynchronous modes for a given axial tension T. The subscript i was purposely set to emphasize that, due to the periodic nature of function (6.32), for a given γ an infinite countable number of asynchronous modes exist, as it will be further detailed.

Equation (6.32) is sufficient to determine the frequencies of the asynchronous modes. Nevertheless, these results are valid only if the oscillators are properly tuned. In other words, the reactions coming from these auxiliary systems must be consistent with the beam's modal vibrations. In order to so impose it, we recall (6.24) and (6.25), which, due to the admissibility condition (6.30), are simplified to

$$f_1(1)\frac{\nu}{\alpha_t - 1} = -f_1'''(1) \tag{6.33}$$

and

$$f_1'(1)\frac{\mu}{\alpha_r - 1} = f_1''(1). \tag{6.34}$$

Since by the admissibility condition (6.30) and continuity ones (v) $f_1(1) = f_2(1) = 0$ and (vi) $f_1'(1) = f_2'(1) = 0$, in the expressions above, it is clear that if $k_t \neq m_t\omega^2$ and $k_r \neq I_r\omega^2$ (and hence α_t and $\alpha_r \neq 1$) then

$$f_1''(1) = 0 \tag{6.35}$$

and

$$f_1'''(1) = 0. \tag{6.36}$$

As a consequence, f_1 will be identically null and the entire beam will be at rest, which is not what we are looking for. Hence, the natural frequency of both oscillators for any asynchronous mode must be in resonance with the respective modal frequency. In mathematical terms, this means that

$$\alpha_t = \alpha_r = 1 \tag{6.37}$$

and hence

$$\omega^2 = \frac{k_t}{m_t} = \frac{k_r}{I_r}. \tag{6.38}$$

Such a condition is not unexpected. In fact, it is well known from the theory of TMDs that the maximum efficiency of an undamped oscillator is achieved when its natural frequency coincides with the main system's vibration frequency. When that occurs, the attachment point at the beam will have no translation (in the case of a TO) or no rotation (in the case of an RO). It comes from the admissibility conditions that the beam's translation and rotation must be null at $\xi = \xi_0$. Hence, in order to guarantee such conditions, both oscillators must be resonant with the asynchronous modal frequency.

With the aid of (6.6), (6.38) can be rewritten, yielding

$$k_t = \frac{EI}{l^4 \rho A} \frac{\lambda_i}{\xi_0^4} m_t \tag{6.39}$$

and

$$k_r = \frac{EI}{l^4 \rho A} \frac{\lambda_i}{\xi_0^4} I_r. \tag{6.40}$$

Equations (6.39) and (6.40) are the dimensional fine-tuning relationships for the TO and the RO, respectively. In each equation, for known modal frequency and structural parameters, one parameter between k_t and m_t is arbitrary and the other is given by (6.39), and one parameter between k_r and I_r is arbitrary and the other is given by (6.40).

Finally, in order to calculate the vibration amplitude of each added oscillator, we recall (6.23), which, regarding the aforementioned admissibility conditions, is reduced to

$$k_t w_t = -EI \left. \frac{\partial^3 w_L}{\partial x^3} \right|_{x=\xi_0 l} \text{ and } k_r \varphi_r = EI \left. \frac{\partial^2 w_L}{\partial x^2} \right|_{x=\xi_0 l}. \tag{6.41}$$

Substitution of (6.4), (6.9) and (6.13) into (6.41) leads to

$$f_t = -\frac{1}{\nu} f_1'''(1) \tag{6.42}$$

for the TO and

$$f_r = \frac{D_1}{D_2 \xi_0 l} \frac{1}{\mu} f_1''(1) \tag{6.43}$$

for the RO.

6.3. Results

In this section, we present the analytical and numerical results obtained from a series of simulations for the model of Fig. 6.1 and discuss relevant features of asynchronous modes. For the sake of an example, we have restricted our simulations to particular structural properties (see (6.44) and (6.45)). Nevertheless, one should not conclude that the features of asynchronous modes discussed in this section are restricted to the scenarios presented in the following. In fact, as observed from the bibliography addressed in the introduction and also from the dimensionless equations derived in the previous section, modal asynchronicity is observable in a variety of systems, provided that certain relationships among parameters are properly tuned.

In our simulations, we consider the following beam parameters:

$$EI = 1.3 \times 10^6\,\mathrm{N\,m^2},\ \rho A = 13\,\mathrm{kg\,m^{-1}} \text{ and } \ell = 10\,\mathrm{m} \tag{6.44}$$

and the following oscillators' parameters:

$$k_t = 5 \times 10^5\,\mathrm{Nm^{-1}} \text{ and } k_r = 5 \times 10^5\,\mathrm{N\,m\,rad^{-1}}. \tag{6.45}$$

It is noteworthy to mention that, since k_t and k_r are numerically the same, so are m_t and I_r, as seen from (6.38). The oscillator's relative position ξ_0 and the axial load T acting along the beam will be fixed or not, according to the objective of each simulation.

6.3.1. *Parametric analysis*

Since modes with large frequency are less important from a practical viewpoint, in our simulations we restrict our attention to those with lower frequencies. Recalling and rearranging (6.6), we obtain

$$\omega^2 = \frac{EI}{l^4 \rho A} \frac{\lambda}{\xi_0^4} \tag{6.46}$$

from which we conclude that an undesired modal frequency can be adjusted by means of: (i) varying λ (which means changing the modal shape) or (ii) varying ξ_0 (which means translating the oscillators along the beam). As we are going to see in the next simulations, the second alternative is preferable, especially if the high-frequency asynchronous mode is also the lowest solution (that is, if for the same set of admissibility conditions, that mode is the lowest asynchronous one). However, moving the oscillators must also be wisely done, otherwise the asynchronous mode will become too rigid (if $\xi_0 \ll 1$) or innocuous from the viewpoint of asynchronicity (if $\xi_0 \cong 1$).

In our first set of simulations, we obtain admissible values of λ for fixed $\xi_0 = 0.75$ and $T = 2500$ N (or $\gamma = 0.1082$). In practice, this means calculating the roots of (6.32) and defining the respective asynchronous modal shapes. These latter ones are reported in Fig. 6.2, where the first to fourth asynchronous modes are illustrated. In turn, Table 6.1 reports the derived analytical dimensionless parameters (λ, β_1 and β_2), the modal frequencies (obtained analytically and numerically with finite-elements models via ADINA [31]) and the mode numbers. Note that the correspondence between analytical

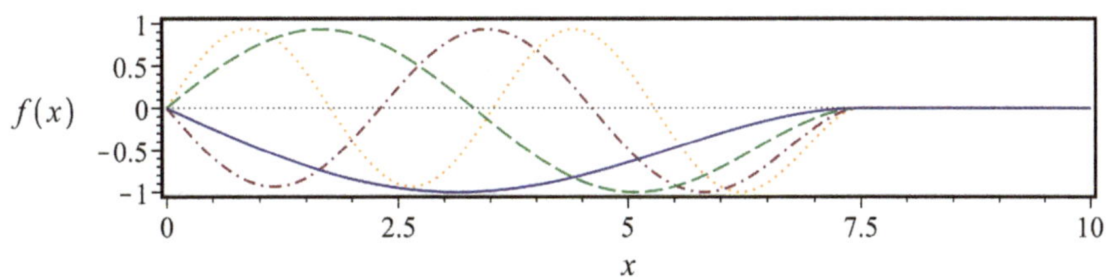

Fig. 6.2. Asynchronous modes for $\xi = 0.75$ and $T = 2500$ N. 1st mode: —, 2nd mode: - - - -., 3rd mode: -·-·-· 4th mode: · · · · · · · · · .

Table 6.1. Properties of asynchronous modes for $\xi_0 = 0.75$ and $T = 2500$ N.

i	λ_i	β_1	β_2	ω_{ana} (rad s^{-1})	ω_{num} (rad s^{-1})	Mode number
1	238.9663	3.9249	3.9386	86.91	86.89	3rd
2	2501.1276	7.0680	7.0757	281.15	281.04	4th
3	10877.7545	10.2099	10.2152	586.34	585.96	5th
4	31797.9361	13.3516	13.3557	1002.48	1001.64	7th

Table 6.2. Parameters of finely-tuned oscillators for $\xi_0 = 0.75$ and $T = 2500$ N.

i	m_t (kg) and I_r (kgm^2)	f_t	f_r
1	66.20	0.50	−0.9481
2	6.33	−2.92	3.0910
3	1.45	8.90	−6.5364
4	0.50	−20.76	11.6553

(ω_{ana}) and numerical (ω_{num}) frequencies validates the mathematical model and also localization findings.

Finely-tuned parameters for the TO and the RO are reported in Table 6.2. Both m_t and I_r are computed by the inverse of (6.39) and (6.40), and they are numerically equal since we chose $k_t = k_r$ in (6.45). Once the modal shape is defined, fine-tuning the oscillators means simply arbitrating adequate stiffness (k_t and k_r) and mass (m_t and I_r). For the fixed parameters of our simulations, we have

$$m_t = \frac{15820.3125}{\lambda_i} \text{ kg} \tag{6.47}$$

and

$$I_r = \frac{15820.3125}{\lambda_i} \text{ kg m}^2, \tag{6.48}$$

and thus m_t and I_r decrease by increasing i, so that "high" asynchronous modes require less mass from the added oscillators, which is welcomed (provided that such higher modes are not too rigid). Actually, from $i = 1$ we have $m_t = 66.20$ kg, which is about one half of the total weight of the beam ($\rho Al = 130$ kg), and thus is it not suitable from a practical point of view. From $i = 2$, onwards on the other hand, the added masses are realistic.

The motion amplitudes of the oscillators, computed by means of (6.42) and (6.43), are also important for practical applications. They are reported in Table 6.2 (normalized with respect to the motion amplitude of the beam). Here the opposite behavior is observed, namely f_t and f_r increase (in absolute value) with i, making "high" asynchronous modes unsuitable from a practical viewpoint. For example, f_t would be nearly 20 times the beam's maximum amplitude for $i = 4$.

Summing up, we see that there is a tradeoff: low added masses require "high" modes, while low vibration amplitudes of the attached oscillators require "low" modes, so that the optimal design, whichever definition of optimality is used, is expected to occur for "intermediate" modes. Another important observation coming from Table 6.2 is that always one of the two oscillators is in-phase ($f > 0$) while the other is out-of-phase ($f < 0$) with the motion of the beam.

Up to now, we have focused our attention on asynchronous (or localized) modes only. However, it is important to underline that they are just a subset of all structure's modes, and it is relevant to realize the mode number of the asynchronous mode within the list of all modes (asynchronous and synchronous). For $\xi_0 = 0.75$, $T = 2500\,\mathrm{N}$ and $i = 1$ (i.e., $m_t = 66.20\,\mathrm{kg}$ and $I_r = 66.20\,\mathrm{kgm}^2$, see Table 6.2), Fig. 6.3 displays the first five modes (computed numerically with ADINA), from which the third one is asynchronous. This analysis is repeated for all other cases of Tables 6.1 and 6.3, and the mode

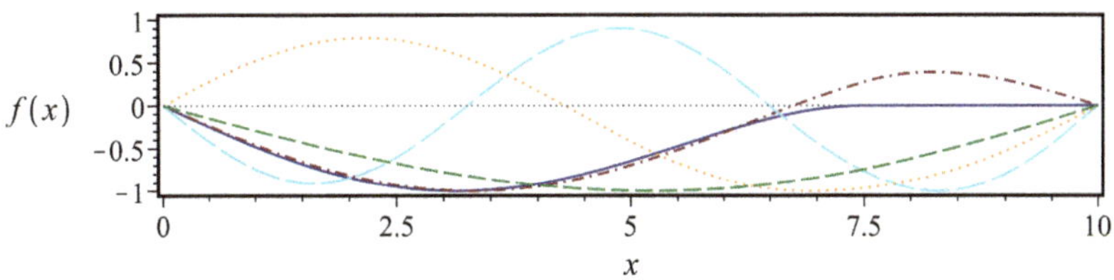

Fig. 6.3. Normal modes for $\xi_0 = 0.75$, $T = 2500$ N, $m_t = 66.20\,\mathrm{kg}$ and $I_r = 66.20\,\mathrm{kgm}^2$. (a) 1st mode ($\omega_1 = 24.63$ rad s^{-1}): - - - -, (b) 2nd mode ($\omega_2 = 74.72$ rad s^{-1}): -·-·-·, (c) 3rd mode (asynchronous, $\omega_3 = 86.89$ rad s^{-1}): ———, (d) 4th mode ($\omega_4 = 165.35$ rad s^{-1}): ·········, (e) 5th mode ($\omega_5 = 294.75$ rad s^{-1}).

number is reported in the last column of Table 6.1. We observe that the mode number increases with i, confirming, from another perspective, that "high" localized modes are meaningless from a practical point of view.

In the following, we evaluate the influence of the oscillators' position ($0.10 \leq \xi_0 \leq 0.90$) on the *first* asynchronous mode, for fixed $T = 2500\,\text{N}$. The modal shapes are depicted in Fig. 6.4, while the properties of each mode, and of the oscillators, are reported in Tables 6.3 and 6.4, respectively. Clearly, if we are interested in

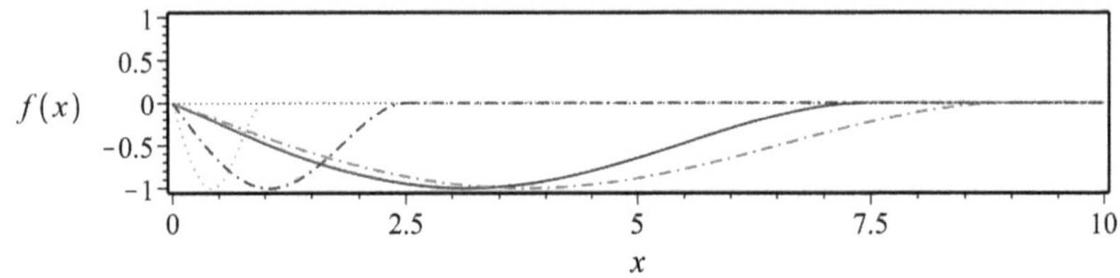

Fig. 6.4. First asynchronous mode for $T = 2500$ N and: (a) $\xi_0 = 0.10$: ·········, (b) $\xi_0 = 0.25$: -·-·-·, (c) $\xi_0 = 0.75$: ——— and (d) $\xi_0 = 0.90$: - - - -.

Table 6.3. Properties of the 1st asynchronous mode for $0.10 \leq \xi_0 \leq 0.90$ and $T = 2500$ N.

ξ_0	γ	β_1	β_2	λ_1	ω_{ana} (rad s^{-1})	ω_{num} (rad s^{-1})	Mode number
0.10	0.0019	3.9266	3.9268	237.7432	4875.89	4875.25	14th
0.25	0.0120	3.9264	3.9279	237.8594	780.33	780.16	6th
0.75	0.1082	3.9249	3.9386	238.9663	86.91	86.89	3rd
0.90	0.1558	3.9241	3.9439	239.5141	60.42	60.41	3rd

Table 6.4. Properties of finely-tuned oscillators for $0.10 \leq \xi_0 \leq 0.90$ and $T = 2500$ N.

ξ_0	m_t(kg) and I_r (kgm^2)	f_t	f_r
0.10	0.02	209.29	−53.2557
0.25	0.82	13.40	−8.5220
0.75	66.20	0.50	−0.9481
0.90	136.96	0.29	−0.6588

suppressing vibrations, small values of ξ_0 are preferable. However, like observed in other structures (see, for example, the first and second examples in [9]), as we confine the vibration within narrower regions, the vibrating frequency increases (together with the mode number), and these highly-confined modes are unlikely to occur in practice. On the other hand, ξ_0 cannot be too large either, otherwise the localized piece will be negligible (except if we are interested in suppressing vibration only in the neighborhood of one support). So, ξ_0 must be wisely chosen for the sake of realistic scenarios. Regarding the oscillators (see Table 6.4), as ξ_0 decreases, so does m_t and I_r (remember that k_t and k_r are constant), while f_t and f_r increase. However, as we compare these amplitudes, we observe that for lower ξ_0 the TO is more active than the RO, while for higher ξ_0 the latter one is more active. This is worth to be considered during the fine-tuning of the oscillators.

The influence of the axial load ($-2.5 \times 10^4 \leq T \leq 2.5 \times 10^5$ N — note that, for the sake of a reference number, Euler's static buckling load is $T_{\text{crit}} = -12.83 \times 10^4$ N) on the first asynchronous mode, for $\xi_0 = 0.75$, is reported in Tables 6.5 and 6.6. The modal shapes barely change within the regarded range and hence they are not depicted. As it is well known, because of the stiffening effect due to T, increasing the axial tension also increases the modal frequencies, although the changes are modest from a quantitative point of view (apart from the very-high value of $T = 2.5 \times 10^5$ N). Remarkably, in all cases,

Table 6.5. Properties of the 1st asynchronous mode for $\xi_0 = 0.75$ and $-2.5 \times 10^4\,\text{N} \leq T \leq 2.5 \times 10^5\,\text{N}$.

$T(N)$	γ	$\beta 1$	$\beta 2$	λ_1	ω_{ana} (rad s^{-1})	ω_{num} (rad s^{-1})	Mode number
−25000	−1.0817	3.9445	3.8049	225.2656	84.38	84.37	3rd
−2500	−0.1082	3.9284	3.9146	236.4756	86.45	86.44	3rd
0	0	3.9266	3.9266	237.7211	86.68	86.67	3rd
2500	0.1082	3.9249	3.9386	238.9663	86.91	86.89	3rd
25000	1.0817	3.9096	4.0456	250.1629	88.92	88.90	3rd
250000	10.8173	3.7883	5.0168	361.2052	106.85	106.81	3rd

Table 6.6. Properties of finely-tuned oscillators for $\xi_0 = 0.75$ and -2.5×10^4 N $\leq T \leq 2.5 \times 10^5$ N.

$T(N)$	m_t(kg) and I_r(kgm^2)	f_t	f_r
−25000	70.23	0.47	−0.9335
−2500	66.90	0.49	−0.9454
0	66.55	0.50	−0.9467
2500	66.20	0.50	−0.9481
25000	63.24	0.52	−0.9598
250000	43.80	0.72	−1.0711

the asynchronous modes are the third ones, i.e., there are only two synchronous modes with lower frequencies. Regarding the oscillators, as we increase T, it is noted a mild decrease in m_t and I_r (apart from $T = 2.5 \times 10^5$ N), showing that T has minor effects also in this respect.

6.3.2. *Sensitivity analysis*

It is well known that springs may have their stiffness affected by many practical factors, such as manufacturing processes, material imperfections, fatigue, etc. During the design stage, such effects are generally embodied by means of a tolerance range about the calculated stiffness. In order to evaluate how such a range would affect an asynchronous mode, in Figs. 6.5 to 6.7 we simulate scenarios in which: (i) the TO is detuned, (ii) the RO is detuned and (iii) both are detuned. The oscillators are modeled for a detuning range of ±10%, while all other parameters are kept fixed to the exact values needed for asynchronicity. Although perfect asynchronicity is lost, we still guarantee relatively small displacements in the region ideally at rest for perfect tuning.

6.3.3. *Considerations on a symmetric case*

Finally, we give special attention to the symmetric case $\xi_0 = 0.50$. For $T = 2500$ N (or $\gamma = 0.0481$), we have $\beta_1 = 3.9258$, $\beta_2 = 3.9319$, $\lambda = 238.2745$ and the natural frequency is given by $\omega_{\text{ana}} = 195.25\,\text{rad s}^{-1}$. The associated parameters of the oscillators, for the stiffnesses (6.45), are $m_t = 13.12$ kg and $I_r = 13.12$ kgm^2. This is

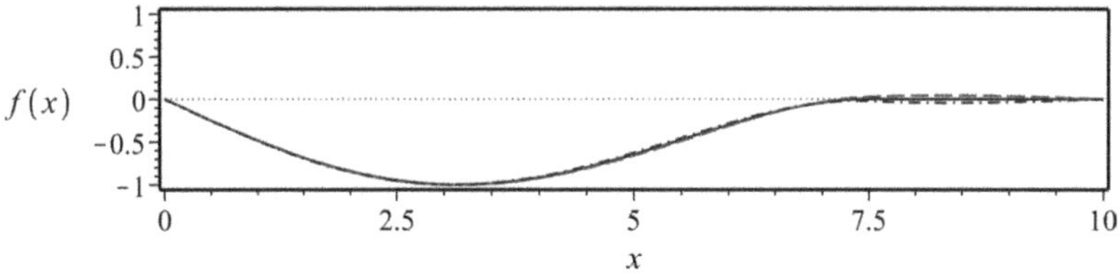

Fig. 6.5. Effect of detuning of the TO on the 1st asynchronous mode. (a) $k_t = 5 \times 10^5$ Nm^{-1}: —, (b) $k_t = 4.5 \times 10^5$ Nm^{-1}: - - - -. and (c) $k_t = 5.5 \times 10^5$ Nm^{-1}: -·-·-·.

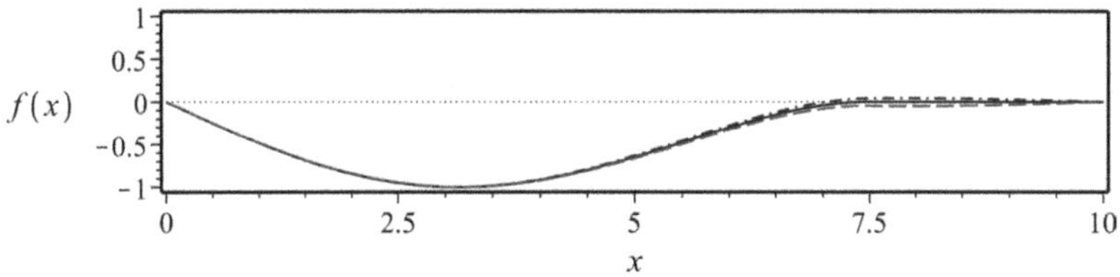

Fig. 6.6. Effect of detuning of the RO on the 1st asynchronous mode. (a) $k_r = 5 \times 10^5$ Nm rad^{-1}: —, (b) $k_r = 4.5 \times 10^5$ Nm rad^{-1}: - - - -. and (c) $k_r = 5.5 \times 10^5$ Nm rad^{-1}: -·-·-·.

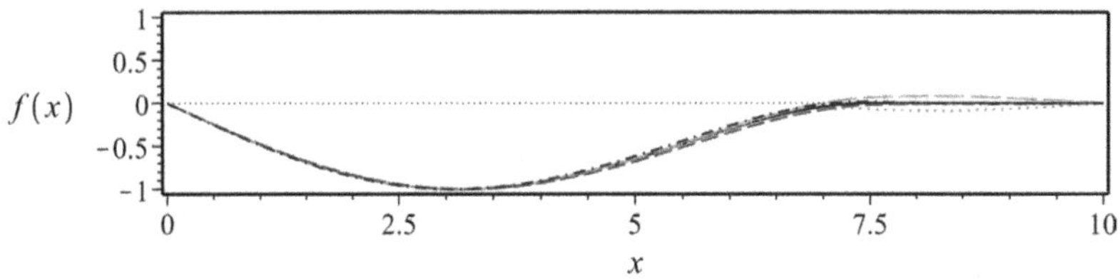

Fig. 6.7. Effect of detuning of both the TO and the RO on the 1st asynchronous mode. (a) $k_t = k_r = 5 \times 10^5$ Nm rad^{-1}: ———, (b) $k_t = k_r = 4.5 \times 10^5$ Nm rad^{-1}: - - - -. (c) $k_t = k_r = 5.5 \times 10^5$ Nm rad^{-1}: -·-·-·, (d) $k_t = 4.5 \times 10^5$ Nm rad^{-1} and $k_r = 5.5 \times 10^5$ Nm rad^{-1}: ·········, and (e) $k_t = 5.5 \times 10^5$ Nm rad^{-1} and $k_r = 4.5 \times 10^5$ Nm rad^{-1}: - - - -.

a special case since, because of the beam's symmetry with respect to the midpoint, the frequency has multiplicity 2 (i.e., they are two coinciding frequencies). This means that the eigenspace is 2D or, in other words, that there are at least two different normal modes associated to the same frequency, and any linear combination of these

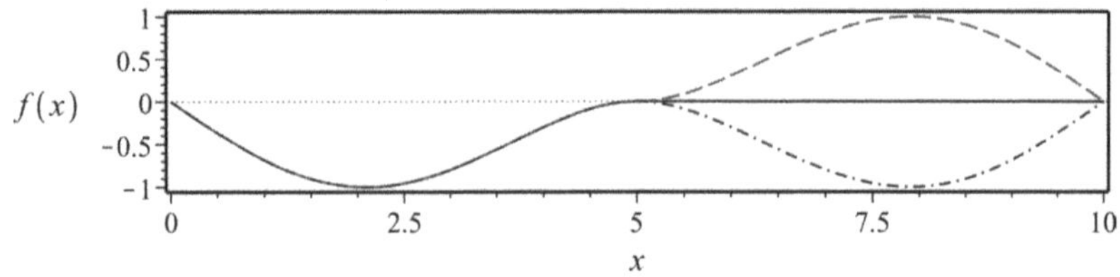

Fig. 6.8. Numerically-obtained modes with $\omega_{\mathrm{num}} = 195.21$ rad s^{-1}, for $\xi_0 = 0.50$ and $T = 2500$ N. (a) first of the two 3rd modes: - - - -., (b) second of the two 3rd modes: -·-·-· and (c) combination of both: —.

Table 6.7. Analytical and numerical results for $\xi_0 = 0.50$ and $T = 2500$ N for the oscillators.

Mode	f_t	f_r
First 3rd (numerical)	3.34	0
Second 3rd (numerical)	0	−4.2609
Average of the previous two	1.67	−2.1305
Asynchronous (analytical)	1.68	−2.1315

two modes is a normal mode, too. Actually, these two modes are not requested to be asynchronous (and actually they are synchronous): it is only needed that a linear combination of them be asynchronous.

This scenario is illustrated in Fig. 6.8, where the two numerically-obtained synchronous modes, corresponding to $\omega_{\mathrm{num}} = 195.21\,\mathrm{rad\,s^{-1}}$, are reported, together with their linear combination (it is just the average of them) providing the sought asynchronous mode. The same goes for the motion of the oscillators, whose amplitudes are compared in Table. 6.7.

To further confirm that we are dealing with a double frequency scenario, we report in Fig. 6.9 the function $f(\lambda) = \det \mathbf{A}(\lambda, 0.0481)$, whose zeroes give the natural frequencies (see (6.32)), and the first derivative of this function in Fig. 6.10. Clearly, for the third natural frequency we have coalesced zeroes (that is, $f = 0$ and $df/d\lambda = 0$), while for the first and second frequencies we have only isolated zeroes ($f = 0$, but $df/d\lambda \neq 0$; notice from the detail of Fig. 6.10 that, at the first frequency, $df/d\lambda \approx 0$, but still it is not null).

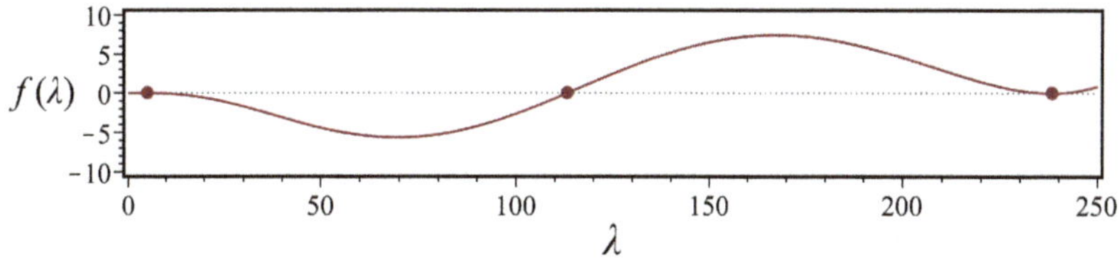

Fig. 6.9. Frequency function for $\xi_0 = 0.50$ and T = 2500 N.

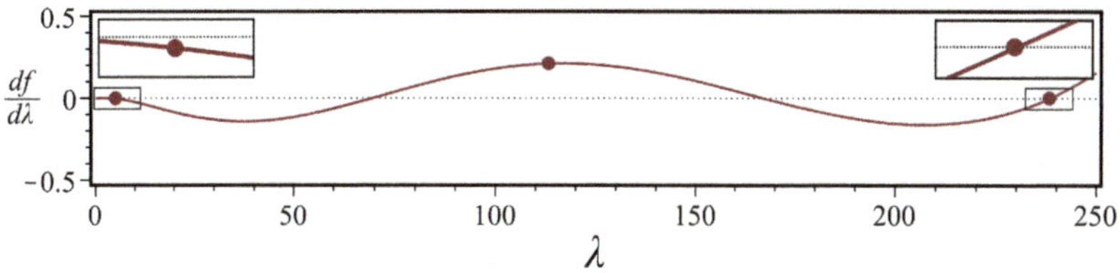

Fig. 6.10. First derivative of the frequency function for $\xi_0 = 0.50$ and $T =$ 2500 N.

6.4. Concluding Remarks

We have previously mentioned that localized (and hence asynchronous) modes can be observed in a variety of linear systems [4, 19–21], provided that some admissibility conditions are satisfied. We have also mentioned that some of these systems will still display asynchronicity for engineering purposes (that is, nearly-asynchronous modes) if geometric nonlinearities are taken into account [22, 23]. However, all these results referred to discrete systems (notice that in the continuous system of [27] we could only confirm near-asynchronous modes, as mentioned in the introduction). This barrier was crossed in [9, 28], where we demonstrated, by means of case studies, that continuous models of beams can display modal asynchronicity provided bending moments and shear forces are annihilated in the entire region at rest, during the dynamic response. This condition can only be achieved by means of finely-tuned discontinuities, which might come from new DOFs added to the original structure, from fine-tuning relationships among the system parameters or a combination of both. Now, we enhance the discussion of [9] by providing still another example of modal asynchronicity in a continuous system.

All the results presented so far are very interesting from a theoretical viewpoint, in the sense they explore a seldom-studied physical phenomenon. Yet, we stress that the motivation for studying asynchronous vibrations has at least three potential engineering applications: the enhancement of signal acquisition [32], vibration control and energy harvesting. For instance, in [25, 26] the electrical power of a parametrically-excited heavy-chain coupled to a piezoelectric device has been investigated under localized and nonlocalized scenarios, leading to more favorable results for localized vibrations, as far as energy harvesting is considered. It was also shown that, depending on the initial voltage applied to the structure, the energy harvester actually turns into an efficient vibration controller, suggesting that both applications can be exploited for the same system. Of course, there is still a long path to be followed towards the construction of efficient vibration controllers and energy harvesters using these ideas. However, we think that these incipient achievements let clear that the study of localized/asynchronous vibrations is worth making.

References

[1] A. F. Vakakis, Non-linear normal modes (NNMs) and their applications in vibration theory: An overview, *Mech. Syst. Sig. Process.* **11** (1997) 3–22.

[2] A. F. Vakakis, Non-similar normal oscillations in a strongly non-linear discrete system, *J. Sound Vib.* **158** (1992) 341–361.

[3] G. Kerschen, M. Peeters, J.C. Golinval and A.F. Vakakis, Nonlinear normal modes, part I: A useful framework for the structural dynamicist, *Mech. Syst. Sig. Process.* **23** (2009) 170–194.

[4] S. Lenci and C. E. N. Mazzilli, Asynchronous free oscillations of linear mechanical systems: A general appraisal and a digression on a column with a follower force, *Int. J. Non Lin. Mech.* **94** (2017) 223–234.

[5] H. Frahm, Device for damping vibrations of bodies (1911).

[6] J. P. Den Hartog, Mechanical Vibrations (McGraw-Hill, 1934).

[7] J. Ormondroyd and J. P. Den Hartog, The theory of the dynamic vibration absorber, *J. Appl. Mech.* **50** (1928) 9–22.

[8] J. E. Brock, A note on the damped vibration absorber, *J. Appl. Mech.* **284** (1946).

[9] E. A. R. Ribeiro, S. Lenci and C. E. N. Mazzilli, Modal localisation in continuous beam models: A discussion on fine-tuning and the use of auxiliary systems, *J. Sound Vib.* (2020).

[10] P. D. Cha and C. Pierre, Imposing nodes to the normal modes of a linear elastic structure, *J. Sound Vib.* **219** (1999) 669–687.

[11] P. D. Cha, Specifying nodes at multiple locations for any normal mode of a linear elastic structure, *J. Sound Vib.* **250** (2002) 923–934.

[12] P. D. Cha and X. Zhou, Imposing points of zero displacements and zero slopes along any linear structure during harmonic excitations, *J. Sound Vib.* **297**(1–2) (2006) 55–71.

[13] P. D. Cha and M. Chan, Mitigating vibration along an arbitrarily supported elastic structure using multiple two-degree-of-freedom oscillators, *J. Vib. Acoust.* **131** (2009) 031008.

[14] P. D. Cha, Enforcing nodes at required locations in a harmonically excited structure using simple oscillators, *J. Sound Vib.* **279** (3–5), (2005) 799–816.

[15] P. D. Cha, Imposing nodes at arbitrary locations for general elastic structures during harmonic excitations, *J. Sound Vib.* **272** (3–5), (2004) 853–868.

[16] P. D. Cha and K. Buyco, An efficient method for tuning oscillator parameters in order to impose nodes on a linear structure excited by multiple harmonics, *J. Vib. Acoust.* **137** (2015) 031018.

[17] P. D. Cha and C. Y. Chen, Quenching vibration along a harmonically excited linear structure using lumped masses, *J. Vib. Control* **17** (2010) 527–539.

[18] P. D. Cha and G. Ren, Inverse problem of imposing nodes to suppress vibration for a structure subjected to multiple harmonic excitations, *J. Sound Vib.* **290**(1–2) (2006) 425–447.

[19] S. Lenci and C. E. N. Mazzilli, Asynchronous modes of vibrations in linear conservative systems: An illustrative discussion of plane framed structures, *Meccanica* **52** (2017) 3131–3147.

[20] C. E. N Mazzilli and S. Lenci, Asynchronous modes of vibration in a heavy-chain model, in *17th Int. Symp. Dynamics Problems of Mechanics*, São Sebastião, Brazil (2017).

[21] C. E. N Mazzilli and S. Lenci, Asynchronous modes of vibration in a heavy-chain model, with linear and rotational springs, in *9th European Nonlinear Dynamics Conf.*, Budapest, Hungary (2017).

[22] E. A. R. Ribeiro, C. E. N. Mazzilli and S. Lenci, Influence of geometric non-linearities in the asynchronous modes of a heavy-chain model, in *2017 EMI Int. Conf.*, Rio de Janeiro, Brazil (2017).

[23] E. A. R. Ribeiro, C. E. N. Mazzilli and S. Lenci, Influence of geometric nonlinearities on the asynchronous modes of an articulated prestressed slender structure, *J. Vib. Acoust.* **141** (2018) 021007.

[24] B. A. P. Mendes, C. E. N. Mazzilli and E. A. R. Ribeiro, Parametric instability of a heavy-chain model with an asynchronous vibration mode, in *First Int. Nonlinear Dynamics Conf.*, Rome, Italy (2019).

[25] B. A. P. Mendes, C. E. N. Mazzilli and E. A. R. Ribeiro, Energy harvesting in a slender-rod model with modal asynchronicity, *Nonlinear Dyn.* **55** (2019) 1–14.

[26] B. A. P. Mendes, C. E. N. Mazzilli and E. A. R. Ribeiro, A piezoelectric vibration controller in a parametrically-excited system with modal localisation, in *Fourth Int. Conf. Recent Advances in Nonlinear Mechanics*, Łódź, Poland (2019).

[27] C. E. N. Mazzilli and E. A. R. Ribeiro, Asynchronous modes of beams on elastic media subjected to varying normal force: Continuous and discrete models, in *Symp. Exploiting Nonlinear Dynamics for Engineering Systems (IUTAM)*, Novi-Sad, Serbia (2018).

[28] E. A. R. Ribeiro, S. Lenci, B. A. P. Mendes and C. E. N. Mazzilli, Modal asynchronicity in a simply-supported beam with a cantilever extension, in *XXIV AIMETA Conf.*, Rome, Italy (2019).

[29] D. Richiedei, I. Tamellin and A. Trevisani, A general approach for antiresonance assignment in undamped vibrating systems exploiting auxiliary systems, in *Advances in Mechanism and Machine Science* (Springer 2019), pp. 4085–4094.

[30] R. W. Clough and J. Penzien, *Dynamics of Structures* (McGraw-Hill Education, 2003).

[31] J. K. Bathe, *Automatic Dynamic Incremental Nonlinear Analysis* (*ADINA*) (1986).

[32] N. Perchikov and O. V. Gendelman, Nonlinear dynamics of hidden modes in a system with internal symmetry, *J. Sound Vib.* **377** (2016) 185–215.

https://doi.org/10.1142/9789811221903_0007

Chapter 7

Weak Decay of Autocorrelations in Periodic Barrier Billiards

M. Courbage

Laboratoire Matière et Systèmes Complexes (MSC), UMR 7057 CNRS and Université de Paris, Case 7056, Bâtiment Condorcet, 10, rue Alice Domon et Léonie Duquet 75205 Paris Cedex 13, France

maurice.courbage@univ-paris-diderot.fr

Many studies of chaotic motions in recent decades have primarily concerned a form described by positive Lyapunov exponents. However, complex motions do not stop at this form, as many natural examples show us. Dynamic systems with zero Lyapunov exponents and weak decay of correlations, which are intermediate between systems with regular trajectories and chaotic systems with positive Lyapunov exponents, are present in billiards. We are up to investigate dynamic systems showing there is a gap between chaos and order.

7.1. Introduction

Billiards are simplified physical models of dynamics that allow us to understand the subtlety and complexity of dynamic systems. The complexity of the behavior of a natural system, generally described by dynamical variables, leads to memory effects with decreasing correlations. For a dynamical transformation T on the space X with an invariant probability measure μ, we say that two functions $f, g \in L^2(X, \mu)$ are mutually mixing (or have decay of correlations) if

$$\int_X f(T^n x) g(x) d\mu - \int_X f(x) d\mu \int_X g(x) d\mu \to 0 \quad \text{as } n \to \infty. \quad (7.1)$$

When $f = g$ we say that f is a self-mixing function (or has decay of auto-correlations).

Many studies of chaotic movements in recent decades have primarily concerned a form described by positive Lyapunov exponents. However, dynamical systems with zero Lyapunov exponents and some low rate of decay of correlations are present in billiards. Remind that the mutual *weak* mixing of two functions $f, g \in L^2(X, \mu)$, is defined by the property:

$$\lim_{n\to\infty} \frac{1}{n} \sum_{k=0}^{n-1} \left| \int_X f(T^n x) g(x) d\mu - \int_X f(x) d\mu \int_X g(x) d\mu \right| = 0. \quad (7.2)$$

When $f = g$ we say that f is self-weakly-mixing. The dynamical system (T, μ) is said to be weakly mixing if any two functions f and g of $L^2(X, \mu)$ are mutually weakly mixing. This is a weaker form of decay of correlations than the usual one above since (7.1) implies (7.2). These dynamical properties are related to Fourier spectrum of these functions as we shall see in the next section.

We must say that we distinguish the study of *mutually mixing functions* from the study of *mixing dynamical system.* This is pertinent when some functions represent physical variables with continuous spectrum as in our billiard. That case occurs especially in systems with mixed spectrum having both a discrete part and a continuous part. We show the presence of singularly continuous spectrum associated with a nondecaying correlations. Actually, we meet more and more dynamic physical systems with such spectra.

On the mathematical side, the interval exchange (IE) family of maps is a similar class of systems much studied in the mathematical literature. At the moment, the field of research on weak chaos widens and becomes very active in physics, mathematics and nonlinear science (see, e.g., [16, 18]).

The famous Sinaï billiard derived from the Lorentz gas model [4, 6, 10, 11] has been widely studied to explore the foundations of statistical mechanics. The diffusive nature of this gas was related to the Boltzmann equation [3, 13, 23].

Zaslavsky and others have carried out numerous studies on the anomalous transport of billiards as well as studies of deterministic diffusions from maps [14, 19, 22, 27].

7.2. Dynamical Properties of the Collision Map of Periodic Barrier Billiard Model

The barrier billiard has been proposed in [17] and studied in [24]. It consists in point-like particles moving freely on a plane and bouncing *elastically* off a periodic configuration of fixed plate barriers scatterers. A particle has a regular translation motion along an axis Ox, and because of the presence of a periodic array of slots through which the particle can escape along the orthogonal axis Oy from its original position and eventually comes back, it has a back-and-forth motion along Oy (Fig. 7.1). It follows that the motion along the x-direction is a one direction regular jump given by the velocity vector direction. The particle motion may be periodic or nonperiodic

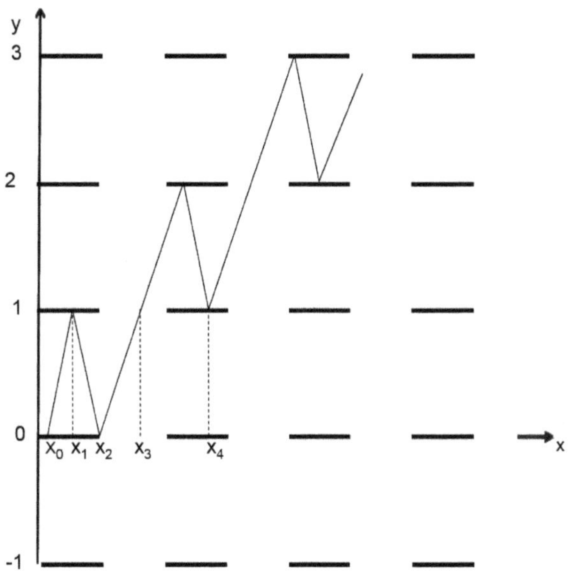

Fig. 7.1. The motion of a billiard particle.

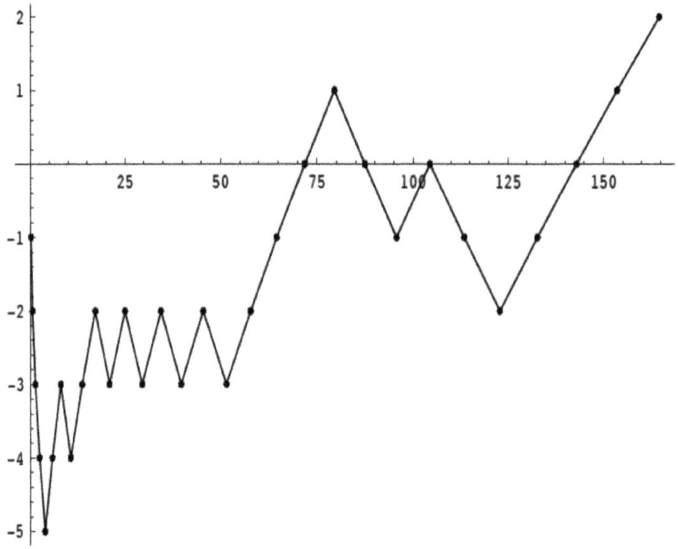

Fig. 7.2. The evolution of the y-coordinate as a function of n for irrational α. It shows a trajectory of a particle may be random in the plane when the angle α is an irrational number.

depending on whether the jump is a rational or an irrational number. In this last case, the particle motion is not periodic. One can see the obstacles lattice as cylindrical around the axis Oy, then such motion becomes quasi-periodic. The particle motion is apparently random and also unstable (see Fig. 7.2), but not exponentially since the divergence of neighboring trajectories is only linear. We shall study here the kind of chaotic motion it generates.

Let us be more specific for the sake of simplicity. The particle is elastically colliding with a set of plates bars of length $\beta < 1$ periodically distributed with a spatial unit period in both (x, y) directions. Let θ be the modulus of the angle between the obstacle and the outgoing velocity vector. The particle will move either always to the right or always to the left undergoing reflections, according to the value of the angle θ. Let us suppose that $0 < \theta < \frac{\pi}{2}$. At time $t = 0$, the coordinate of the particle is given by its position x_0 and the direction of its outgoing velocity. The position of the particle will increase between time $t = 0$ and time $t = 1$ by α along x-direction and by ± 1 along y-direction where $\alpha = (1/\tan\theta)$. The unit of time is then chosen as the time of the α jump along x. The trajectory of the particle will be specified at times $t = n$, by its x-coordinate x_n at the collision instant and the projection of the outgoing velocity vector along the y-direction, ϵ_n, which will change under a collision. Thus, $\epsilon_n = \pm 1$ depending on whether the velocity is upwards or downwards. We call ϵ_n the "*velocity direction*" of the particle at time n. From now on we specify $\beta = 1/2$. In order to represent the motion of the particle as a map, define the phase space of the collision map T by the coordinates (x, ϵ, θ). Let us introduce the function $\chi(x)$ as follows: $x \in \mathbb{T} = [0, 1[\to \{\pm 1\}$

$$\chi(x) = \begin{cases} -1 & \text{if } x \in [0, 1/2], \\ +1 & \text{if } x \in [1/2, 1]. \end{cases} \tag{7.3}$$

The transformation T maps the coordinates of the particle (x, ϵ, θ) at time $t = 0$ into the coordinates of the particle at time $t = 1$ according to the formula:

$$T(x, \epsilon, \theta) = (x + \alpha(\mathrm{mod} 1), \chi(x)\epsilon, \theta). \tag{7.4}$$

So the dynamics for the x-coordinate is a regular translation on the circle $\mathbb{T}$ called the rotation map of the circle and T maps the direction of the outgoing arrow to the next outgoing arrow. Obviously, there are a multitude of various trajectories, moreover all of them are unstable, since two neighboring trajectories will diverge, but the divergence rate increases at most linearly and not exponentially as in the Sinaï billiard. In order to study more deeply the dynamics, one needs first a statistical distribution, only stationary distributions can globally govern statistical information.

7.2.1. *The dynamics as a statistical process*

A stationary distribution over X should be invariant under the dynamic action. Such invariant probability distribution over X is immediately found as the product of the normalized Lebesgue measure over the variables (x, θ) and the probability measure $\left(\frac{1}{2}, \frac{1}{2}\right)$ for the coordinate ϵ.

The map T preserves this probability measure μ: $\mu(A) = \mu(T^{-1}A)$ for a subset A of X. The argument is as follows: take A as the product of an interval over x, say A_1, and some fixed ϵ, say $\epsilon = -1$. The measure $\mu(T^{-1}A)$ is then computed, using Fubini's theorem, successively first by fixing x, and integrating over ϵ, then integrating with respect to x. The measure of ϵ': $\chi(x)\epsilon' = -1$ is $1/2$ whatever is the value of x. Therefore, one obtains $\mu(T^{-1}A) = \frac{1}{2} \cdot l(A_1)$, since the translation by α preserves the length of A_1, that is $\mu(A)$. As θ is a constant of the motion, the collision map T is not ergodic. This also means that the time averages and ensemble averages are not equal for any dynamical variable (functions). For example, the time averages and ensemble averages are not equal for the characteristic function of a subset A

$$\chi_A(x, \epsilon, \theta) = \begin{cases} 1 & \text{if } (x, \epsilon, \theta) \in A, \\ 0 & \text{if } (x, \epsilon, \theta) \notin A, \end{cases}$$

where θ is limited to a strict subinterval of $[0, \pi]$. However, the system is ergodic if θ is fixed to an *irrational* single value, as follows from the well-known ergodicity of the irrational rotation of the circle. Let us now limit our dynamic system to the variables (x, ϵ).

The interesting process of successive reversals of the velocity directions of a particle is the apparently random aperiodic process. It varies with the initial conditions x_0 since, we have

$$(x_n, \epsilon_n) = (x_0 + n\alpha, \chi(x_0 + (n-1)\alpha) \cdots \chi(x_0)\epsilon_0). \tag{7.5}$$

They occur irregularly an infinite number of times and the series has an interesting Fourier spectrum.

7.2.2. *Continuous spectral measure of function f*

Remind that, for a discrete temporal evolution, the Fourier spectral measure of a dynamic variable represented by a function f is defined as a Fourier transform of a measure according to the formula:

$$\langle U^n f, f \rangle = \int_0^1 \exp(2i\pi n\lambda) d\sigma_f(\lambda) = \hat{\sigma}_f(n), \tag{7.6}$$

where U^n is the unitary evolution operator associated to T acting on the Hilbert space $\mathfrak{H} = L^2(X, \mu)$ such as

$$U^n f(x) = f(T^n x). \tag{7.7}$$

The autocorrelations coefficients of f,

$$\langle U^n f, f\rangle = \int_X f(T^n x) f(x) d\mu, \tag{7.8}$$

are called the Fourier coefficients of the spectral measure σ_f.

The spectral decomposition theorem for unitary operators stipulates that the Hilbert space decomposes into three orthogonal subspaces that are invariant under the action of U such that

$$\mathfrak{H} = \mathfrak{H}_d \oplus \mathfrak{H}_{\text{ac}} \oplus \mathfrak{H}_{\text{sc}}. \tag{7.9}$$

Here $\mathfrak{H}_d$ is the subspace of all eigenfunctions of U, $\mathfrak{H}_{\text{ac}}$ is the subspace of functions having spectral measures that are absolutely continuous with respect to μ and $\mathfrak{H}_{\text{sc}}$ is the subspace of functions having spectral measures that are singularly continuous with respect to μ. In other words,

- $f \in \mathfrak{H}_d$ if and only if its Fourier spectral measure is composed of discrete delta peaks;
- $f \in \mathfrak{H}_{\text{ac}}$ if and only if its Fourier spectral measure has density with respect to the Lebesgue measure on the circle: $d\sigma_f(\lambda) = \rho(\lambda)d\lambda$. Then, it follows from (7.6) that σ_f is the Lebesgue measure (i.e., $\rho = 1$) if $\hat{\sigma}_f(n) = 0, \forall n \neq 0$. If an absolutely continuous measure has density $\rho(\lambda)$ which can vanish only on a subset of zero Lebesgue measure, then it is said to be *equivalent to the Lebesgue measure*. This is the case, for instance, if the Fourier coefficients $\hat{\sigma}_f(n)$ decay exponentially rapidly as $|n| \to \infty$. In fact, the density $\rho(\lambda)$ given above as an entire series of $z = \exp(2i\pi\lambda)$ will be uniformly convergent, so that it has analytic continuation in some annulus around the unit circle. Therefore $\rho(\lambda)$ has only finite number of zeros;
- $f \in \mathfrak{H}_{\text{sc}}$ if and only if its Fourier spectral measure is continuous (i.e., single points have zero measure) and supported by a set of zero Lebesgue measure on the circle like the measure generated by the devil staircase.

We shall denote

$$\mathfrak{H}_{\mathfrak{c}} = \mathfrak{H}_{\mathrm{ac}} \oplus \mathfrak{H}_{\mathrm{sc}}. \tag{7.10}$$

A function $f \in \mathfrak{H}_c$ if and only if $\sigma_f(\lambda) = 0$, for all λ. Then, the measure σ_f is said *continuous.*

The relation of the Fourier spectrum to the decay of correlations reads thanks to the remarkable Wiener lemma which stipulates that:

$$\lim_{n\to\infty} \frac{1}{n} \sum_{k=0}^{n-1} |\hat{\sigma_f}(k)|^2 = \sum_i (\sigma_f(\lambda_i))^2, \tag{7.11}$$

where λ_i are the points of the circle such that $\sigma_f(\lambda_i) > 0$ (see, e.g., [5]). The left-hand side of the above limit is called *the Cesàro-mean of the sequence* $|\hat{\sigma_f}(k)|^2$. Let us remark here that *the convergence to zero of the above Cesàro-mean is equivalent to the same convergence for the sequence of the absolute values* $|\hat{\sigma_f}(k)|$ since the terms of that sequence are uniformly bounded by $\|f\|^2$.

Thus, the convergence to zero of the above Cesàro-mean gives a characterization of the continuity of the spectral measure of f:

> *A necessary and sufficient condition for the continuity of this spectral measure is the decay of the Cesàro-mean of the auto-correlation coefficients of f. That is also to say that the function f is self-weakly-mixing.*

It is well known that the convergence to zero of a numerical sequence implies the convergence to zero of its Cesàro-mean. The converse is not necessarily true.

In the case the measure has a density, $\rho \in L^1(S^1, d\lambda)$, (i.e., σ_f is absolutely continuous), it comes from (7.6) that

$$\langle U^n f, f\rangle = \int_0^1 \exp(2i\pi n\lambda)\rho(\lambda)d(\lambda). \tag{7.12}$$

Then, by the Riemann–Lebesgue lemma, $\langle U^n f, f\rangle$ decays to zero as $n \to \infty$. If, moreover, $\rho \in L^2(S^1, d\lambda)$, we may write

$$\rho(\lambda) = \sum_{n=-\infty}^{\infty} \hat{\sigma}_f(n) \exp(2i\pi n\lambda). \tag{7.13}$$

If the continuous spectral measure is singular, e.g., if the Cesàro-mean of the series converges to zero, but not to the coefficients themselves, we not have a decay of the f autocorrelations. A case study that concerns the velocity direction function $f = \epsilon$ in our billiard.

Let us apply this theorem for the collision map:

$$T(x, \epsilon,) = (x + \alpha (\mathrm{mod}\, 1), \chi(x)\epsilon). \tag{7.14}$$

First, the subspace of functions $f(x, \varepsilon)$ that depend only on x is invariant generated by all eigenfunctions of U. Denote it $\mathfrak{H}_1$. It follows from the definition of the scalar product of two functions, (f_1, f_2):

$$\langle f_1, f_2 \rangle = \int_X \overline{f_1(x, \varepsilon)} f_2(x, \varepsilon) d\mu(x, \varepsilon) \tag{7.15}$$

$$= \frac{1}{2} \int_0^1 \left[\overline{f_1(x, -1)} f_2(x, -1) + f_1(x, 1) f_2(x, 1) \right] dx. \tag{7.16}$$

Consider now the two-dimensional subspace of complex functions that depend only on ε, with the scalar product as above. It has the orthogonal basis $\{1, \varepsilon\}$. It comes that, for any given f, there are $g, h \in L^2(S^1, dx)$ such that

$$f(x, \varepsilon) = g(x) + \varepsilon h(x). \tag{7.17}$$

To verify directly this assertion, it is possible to determine $g(x)$ and $h(x)$ by solving two equations taking successively $\varepsilon = 1, \varepsilon = -1$ and obtaining:

$$g(x) = \frac{1}{2}(f(x, 1) + f(x, -1)), \tag{7.18}$$

$$h(x) = \frac{1}{2}(f(x, 1) - f(x, -1). \tag{7.19}$$

It follows from (7.17) that the subspace $\mathfrak{H}_2$ of the functions of the form $\varepsilon h(x)$ is the orthogonal complement of $\mathfrak{H}_1$. Moreover, the action of U restricted to $\mathfrak{H}_1$ (respectively, $\mathfrak{H}_2$) is given by $Ug(x) = g(x + \alpha)$ (respectively, $U(\varepsilon h(x)) = \varepsilon \chi(x) h(x + \alpha)$). It shows that $\mathfrak{H}_1$ and $\mathfrak{H}_2$ are invariant. Therefore, the spectrum of U restricted to $\mathfrak{H}_1$ is reduced to the set of eigenvalues of the unitary operator associated to the rotation map acting on $\mathfrak{H}_1$, $\{e^{2i\pi n\alpha}, n \in \mathbb{Z}\}$. The spectral measure associated to any function from $\mathfrak{H}_2$ is given through the Fourier coefficients: $\langle U^n(\varepsilon h), \varepsilon h\rangle$. An argument given by Oseledets [20] shows that this spectral measure is continuous, that is, $\varepsilon h(x)$ could not be an eigenfunction of U. Suppose, *ad absurdum*, that there exists $h(x)$ such that $\varepsilon h(x)$ is an eigenfunction for an eignvalue λ of modulus 1, that is,

$$U(\varepsilon h(x)) = h(x+\alpha)\chi(x)\varepsilon, \tag{7.20}$$

$$h(x+\alpha)\chi(x) = \lambda h(x) \tag{7.21}$$

almost everywhere. Then, translating by 1/2 the same equation:

$$\overline{h(x+1/2+\alpha)\chi(x+1/2)} = \bar{\lambda}\overline{h(x+1/2)} \tag{7.22}$$

and multiplying the last two equations and observing that $\chi(x)\chi(x+1/2) = -1$, we have

$$\overline{h(x+1/2+\alpha)}h(x+\alpha) = -\overline{h(x+1/2)}h(x), \tag{7.23}$$

which means that $\overline{h(x+1/2)}h(x)$ is an eigenfunction of the unitary operator of rotation map by α for the eigenvalue -1. A contradiction with the fact that α is irrational and U cannot have -1 as eigenvalue. This completes the spectral decomposition of T (7.14):

$$\mathfrak{H}_{\mathfrak{d}} = \mathfrak{H}_1, \tag{7.24}$$

$$\mathfrak{H}_{\mathfrak{c}} = \mathfrak{H}_2. \tag{7.25}$$

As seen above, the velocity direction in terms of n is equal to

$$\varepsilon_n(x,\varepsilon) = \chi(x+(n-1)\alpha) \times \cdots \times \chi(x)\varepsilon \tag{7.26}$$

and its spectral measure is continuous, that is, according to Wiener lemma, the Cesàro-mean of the velocity direction autocorrelations decays to zero. The question is therefore about the decay or not of the autocorrelation coefficients, themselves given by

$$\langle \varepsilon_n, \varepsilon\rangle = \int_0^1 \chi(x+(n-1)\alpha) \times \cdots \times \chi(x)dx. \tag{7.27}$$

A first answer has been given in [22] where it is shown that, when α is a Liouville irrational number, *the velocity direction autocorrelations will not decay to zero.* This is a case of singularly continuous spectrum and nonmixing function in the barrier billiard.

7.3. Concluding Remarks

We have shown that the process of reversing the direction of velocity in barrier billiards with irrational α can have a singularly continuous spectrum, that is, a decay of autocorrelations only on average without effective decay. These long-memory effects, observed similarly in other maps with zero Lyapunov exponents in the gap between order and chaos, can coexist with other strange aspects, as an exponential decay of correlations in zero entropy maps (other strange decay rates in a dynamical system have been found in finite ideal gas [12]). This gap is richer than it looks as illustrated by the emergence in this billiard of a slow, anomalous transport depending on the values of α. Note also that if the Fourier coefficients $\hat{\sigma}_f(n)$ decay exponentially rapidly as $|n| \to \infty$, the process is linearly unpredictable but not in the sense of KS-entropy.

The mathematical system of the interval exchange (IE), which is a big generalization of the rotation map, also produces singularly continuous spectrum. In this regard, let us quote that Avila and Forni [2] have shown that in general such a dynamic system, when it is not an irrational rotation, it is a weak mixing system. There is a huge literature on (IE) (see, for instance, the references in [1]).

In physics, singularly continuous spectrum has been found in relation with the substitutions dynamical systems used in condensed matter, especially, quasi-crystals (see, e.g., [21]), also in nonlinear physics, especially, dissipative hydrodynamics (see [25]). In the physicists language, it is called a fractal spectrum, interesting because of its "vague" memory effects.

George Zaslavsky highlighted in the barrier billiard a plausible diffusive behavior due to the irregular collisions of the particle and the presence of a corridor where the particle can escape very far without collisions, but also "stickiness" traps. Sinai billiards with infinite horizon have a similar structure except that the obstacle border is convex, which produces genuine chaos. In [22], the results of the simulations show a strong dependence of α on the transport behavior. Zaslavsky developed his scaling method [26] for estimating the mean displacement along y. This is what has been taken in [27]. In massive simulations, Zaslavsky and Edelman took an average over α in an interval. They suggested a fractional kinetics as a nonstationary statistical dynamics. From the point of view of ergodic theory, the addition of the variable θ makes the system nonergodic. This is of great importance and we will report on it in a separate publication.

Acknowledgments

The author thanks Drs S. M. Saberi, W. Ettoumi, M. Zaks and G. Zaslavsky for useful discussions. Thanks also to my colleagues from the Physics Department of Université de Paris for their support.

References

[1] E. H. El Abdalaoui, On the spectrum of α-rigid maps, *J. Dyn. Control Syst.* **15** (2009) 453.
[2] A. Avila and G. Forni, Weak mixing for interval exchange transformations and translation flows, *Ann. Math. Second Ser.* **165**(2) (2007) 637–664.
[3] C. Boldrighini, L. A. Bunimovich and Y. Sinai, On the Boltzmann equation for the Lorentz gas, *J. Statist. Phys.* **32**(3) (1983) 477–501.
[4] L. A. Bunimovich and Y. G. Sinaï, Statistical properties of Lorentz gas with periodic configuration of scatterers, *Commun. Math. Phys.* **78** (1981) 479–497.
[5] I. P. Cornfeld, S. V. Fomin and Ya. G. Sinai, *Ergodic Theory* (Springer, New York, 1981).
[6] M. Courbage, M. Edelman, S. M. Saberi Fathi and G. M. Zaslavsky, Problem of transport in billiards with infinite horizon, *Phys. Rev. E* **77** (2008) 1.
[7] M. Courbage, Notes on spectral theory, mixing and transport, in *Proc. Chaotic Dynamics and Transport in Classical and Quantum Systems*, P. Collet *et al.* (eds.) (Kluwer Academic Publishing, The Netherlands, 2005).
[8] M. Courbage and D. Hamdan, Unpredictability in some nonchaotic dynamical systems, *Phys. Rev. Lett.* **74** (1995) 5166–5169.

[9] M. Courbage and D. Hamdan, Decay of correlations and mixing properties in a dynamical system with zero entropy, *Ergod. Theory Dyn. Syst.* **17**(1) (1997) 87–103.

[10] G. Cristadoro, T. Gilbert, M. Lenci and D. P. Sanders, Measuring logarithmic corrections to normal diffusion in infinite-horizon billiards, *Phys. Rev. E* **90**(2) (2014) 022106.

[11] C. P. Dettmann, New horizons in multidimensional diffusion: The Lorentz gas and the Riemann hypothesis. *J. Statist. Phys.* **146** (2012) 181–204.

[12] A. K. Evans, The long-time behavior of correlation functions in dynamical systems, in *Stochastic and Chaotic Dynamics in the Lakes: STOCHAOS*, D. S. Broomhead, E. A. Luchinskaya, P. V. E. McClintock and T. Mullin (eds.), AIP Conference Proceedings, Vol. 502 (American Institute of Physics, 2000), 394 pp.

[13] G. Gallavotti, Divergences and the approach to equilibrium in the Lorentz and the wind-tree models, *Phys. Rev.* **185** (1969) 308.

[14] T. Geisel and S. Thomae, Anomalous diffusion in intermittent chaotic systems, *Phys. Rev. Lett.* **52** (1984) 1936.

[15] S. Gouëzel and I. Melbourne, Moment bounds and concentration inequalities for slowly mixing dynamical systems, *Electron. J. Probab.* **19**(93) (2014) 1–30.

[16] E. Gutkin and A. Katok, Weakly mixing billiards, in *Holomorphic Dynamics*, X. Gomez-Mont, J. A. Seade and A. Verjovski (eds.), Lecture Notes in Mathematics, Vol. 1345 (Springer, Berlin, 1988).

[17] J. H. Hannay and R. J. McCraw, Barrier billiards: A simple pseudointegrable system, *J. Phys. A* **23** (1990) 887–889.

[18] M. Horvat, M. Degli Esposti, S. Isola, T. Prosen and L. Bunimovich, On ergodic and mixing properties of triangle map, *Physica D* **238** (2009) 395–415.

[19] R. Metzler, J.-H. Jeon, A. G. Cherstvy and E. Barkai, Anomalous diffusion models and their properties: Non-stationarity, non-ergodicity, and ageing at the centenary of single particle tracking, *Chem. Phys.* **16** (2014) 24128–24164.

[20] V. I. Oseledets, The spectrum of ergodic automorphisms, *Dokl. Akad. Nauk SSSR* **168**(5) (1966) 1009–1011.

[21] A. S. Pikovsky, M. A. Zaks, U. Feudel and J. Kurths, Singular continuous spectra in dissipative dynamical systems, *Phys. Rev. E* **52** (1995) 285–296.

[22] S. M. Saberi Fathi, W. Ettoumi and M. Courbage, Transport in the barrier billiard, *Phys. Rev. E* **93** (2016) 062216.

[23] H. Spohn, Kinetic equations from Hamiltonian dynamics: Markovian limits, *Rev. Mod. Phys.* **52** (1980) 569–615.

[24] J. Wiersig, Singular continuous spectra in a pseudointegrable billiard, *Phys. Rev. E* **62** (2000) R21.
[25] M. A. Zaks and A. Nepomnyashchy, Subdiffusive and superdiffusive transport in plane steady viscous flows, *Proc. Natl. Acad. Sci. USA* **116**(37) (2019) 18245–18250.
[26] G. M. Zaslavsky, Renormalization group theory of anomalous transport in systems with Hamiltonian chaos, *Chaos* **4** (1994) 25.
[27] G. M. Zaslavsky and E. Edelman, Weak mixing and anomalous kinetics along filamented surfaces, *Chaos* **11** (2001) 295–305.

Part II

Stochastic Dynamics, Fractal Structure Analysis and Numerical Errors

https://doi.org/10.1142/9789811221903_0008

Chapter 8

Stochastic Sensitivity Analysis of Noise-Induced Phenomena in Discrete Systems

Irina Bashkirtseva and Lev Ryashko*

Ural Mathematical Center,
Institute of Natural Sciences and Mathematics,
Ural Federal University,
Lenina, 51, Ekaterinburg, 620000, Russia
*lev.ryashko@urfu.ru

A problem of the analysis of noise-induced phenomena in map-based stochastic systems is considered. We give an overview of the new general approach for the study of such phenomena on the base of stochastic sensitivity technique and confidence domain method. Mathematical background of the stochastic sensitivity analysis of equilibria, cycles, closed invariant curves, and chaotic attractors is presented. Constructive abilities of suggested analytical approach are illustrated for diverse stochastic phenomena in discrete systems with quadratic map and coupling.

8.1. Introduction

At present, mathematical models in the form of nonlinear differential or difference equations are widely used to study dynamical processes observed in different areas of science [1–7]. The presence of random disturbances is an inevitable attribute of the functioning of any real system. The interrelation of nonlinearity and stochasticity often leads to new phenomena that have no analogues in the original deterministic models. Even weak random disturbances in systems with strong nonlinearity can cause noise-induced transitions, stochastic resonance, stochastic bifurcations, noise-induced order-chaos transitions, stochastic excitability, etc. [8–19].

Direct numerical modeling remains the main tool for studying such nonlinear stochastic phenomena. Within the framework of this extremely costly method, it is difficult to obtain clear parametric descriptions of various stochastic regimes of the models under study. In order to carry out a detailed parametric analysis of probabilistic mechanisms of these new stochastic phenomena, the development of analytical approaches is required.

In the present paper, we focus on the study of randomly forced discrete-time dynamical systems. A rigorous mathematical description of the dynamics of probability distributions in map-based dynamical systems is given by functional equations with the Perron–Frobenius operators [10]. However, the analytical solution of such equations, even in the one-dimensional case, is possible only for specially selected examples. In these circumstances, approaches using analytical approximations and asymptotics are in demand.

In [20, 21], the constructive approach based on the stochastic sensitivity analysis was proposed for equilibria and cycles of map-based systems forced by random noise. Further, this approach was extended to systems with more complex quasiperiodic [22, 23] and chaotic [24–26] attractors. Based on the analytical stochastic sensitivity technique and confidence domain method, a wide range of research problems was solved: noise-induced transitions between attractors and their parts, stochastic excitement, backward stochastic bifurcations, noise-induced order-chaos transitions, etc. [27, 28]. Probabilistic mechanisms of spiking-bursting transformation in neuron models [29–32], phenomena of noise-induced extinction in population models [33–36], stochastic dynamics of business cycles [37, 38] and coupled systems [39, 40] were studied using this apparatus.

It should also be noted that, on the basis of the stochastic sensitivity technique, constructive control methods have been developed that provide required regimes with assigned probabilistic features in stochastic discrete systems [41–43].

This chapter aims to give a short overview of the stochastic sensitivity technique and confidence domains method for regular and chaotic attractors of general discrete-time stochastic systems. Along with the mathematical background, we give key examples and show how our approach works in the analysis of various stochastic phenomena.

8.2. Stochastic Sensitivity

Consider a nonlinear randomly forced discrete-time system

$$x_{t+1} = f(x_t, \eta_t), \quad \eta_t = \varepsilon \xi_t, \tag{8.1}$$

where x is an n-dimensional vector, $f(x, \eta)$ is a smooth n-dimensional vector-function, ξ_t is an m-dimensional uncorrelated random process with parameters $\mathrm{E}\xi_t = 0$, $\mathrm{E}\xi_t\xi_t^\top = V$, and ε is the scalar parameter of noise intensity.

Let $\bar{x}_t$ be a solution of the deterministic system

$$x_{t+1} = f(x_t, 0) \tag{8.2}$$

with the initial state $\bar{x}_0$. Denote by x_t^ε the solution of the system (8.1) with the initial data $x_0^\varepsilon = \bar{x}_0 + \varepsilon\nu$ where ν is n-dimensional random vector with parameters $\mathrm{E}\nu = 0$, $\mathrm{E}\nu\nu^\top = U$. The vector ν is uncorrelated with ξ_t.

In the study of the dispersion of random solutions x_t^ε of system (8.1) around the deterministic solution $\bar{x}_t$ we will use the asymptotics:

$$z_t = \lim_{\varepsilon \to 0} \frac{x_t^\varepsilon - \bar{x}_t}{\varepsilon}.$$

Consider the following linear extension system:

$$\begin{aligned} x_{t+1} &= f(x_t, 0), \\ z_{t+1} &= F(x_t)\, z_t + S(x_t)\, \xi_t. \end{aligned} \tag{8.3}$$

Here,

$$F(x) = \frac{\partial f}{\partial x}(x, 0), \quad S(x) = \frac{\partial f}{\partial \eta}(x, 0).$$

The $n \times n$-matrices $M_t = \mathrm{E} z_t z_t^\top$ of second moments satisfy the deterministic system

$$\begin{aligned} x_{t+1} &= f(x_t, 0), \\ M_{t+1} &= F(x_t) M_t F^\top(x_t) + S(x_t) V S^\top(x_t). \end{aligned} \tag{8.4}$$

For the fixed deterministic solution $\bar{x}_t$, second moments M_t are uniquely defined by equations

$$M_{t+1} = F(\bar{x}_t) M_t F^\top(\bar{x}_t) + S(\bar{x}_t) V S^\top(\bar{x}_t), \quad M_0 = U. \tag{8.5}$$

The sequence of matrices M_t defines the stochastic sensitivity of the deterministic sequence $\bar{x}_t$ and gives for small noise intensity the following approximation:

$$\mathrm{E}(x_t^\varepsilon - \bar{x}_t)(x_t^\varepsilon - \bar{x}_t)^\top \approx \varepsilon^2 M_t.$$

8.2.1. *Stochastic sensitivity of equilibria*

Let $\bar{x}$ be an exponentially stable equilibrium of the deterministic system (8.2). Due to stability of $\bar{x}$, it holds that $\rho(F(\bar{x})) < 1$, where $\rho(F)$ is the spectral radius of the matrix F.

For the solution $\bar{x}_t \equiv \bar{x}$, the system (8.5) can be rewritten as

$$M_{t+1} = F(\bar{x}) M_t F^\top(\bar{x}) + S(\bar{x}) V S^\top(\bar{x}), \quad M_0 = U. \tag{8.6}$$

Because of $\rho(F(\bar{x})) < 1$, the system (8.6) has a unique stable stationary solution $M_t \equiv M$, where the matrix M is governed by the following matrix equation

$$M = F(\bar{x}) M F^\top(\bar{x}) + Q, \quad Q = S(\bar{x}) V S^\top(\bar{x}). \tag{8.7}$$

The matrix M characterizes the stochastic sensitivity of the stable equilibrium $\bar{x}$.

8.2.2. *Stochastic sensitivity of k-cycles*

Let the deterministic system (8.2) have the exponentially stable k-cycle Γ with elements $\bar{x}_1, \ldots, \bar{x}_k$. Elements of this cycle are connected by the following relations:

$$f(\bar{x}_i, 0) = \bar{x}_{i+1}\,(i = 1, \ldots, k-1), \quad f(\bar{x}_k, 0) = \bar{x}_1.$$

The necessary and sufficient condition of the exponential stability of the cycle Γ is

$$\rho(F_k \cdot \ldots \cdot F_2 F_1) < 1, \quad F_t = F(\bar{x}_t). \tag{8.8}$$

For the cycle Γ, Eq. (8.5) is written as

$$M_{t+1} = F_t M_t F_t^\top + Q_t, \quad Q_t = S(\bar{x}_t)\, V\, S^\top(\bar{x}_t), \tag{8.9}$$

where matrices F_t, Q_t are k-periodic.

Due to the condition (8.8), this equation has a unique k-periodic solution W_t satisfying

$$W_{t+1} = F_t W_t F_t^\top + Q_t, \quad W_{k+1} = W_1. \tag{8.10}$$

The matrices W_t define the stochastic sensitivity of the elements $\bar{x}_t$ of the cycle Γ. For calculation of the stochastic sensitivity matrices $W_1, W_2, \ldots, W_k$, the following algorithm can be used.

The matrix W_1 is the solution of the matrix equation

$$W_1 = F W_1 F^\top + Q, \tag{8.11}$$

where $F = F_k \cdot \ldots \cdot F_2 F_1$, $Q = Q_k + F_k Q_{k-1} F_k^\top + \cdots + F_k \cdot \ldots \cdot F_2 Q_1 F_2^\top \cdot \ldots \cdot F_k^\top$. Other stochastic sensitivity matrices $W_2, \ldots, W_k$ can be found directly by the recurrent formula (8.10).

8.2.3. *Stochastic sensitivity of closed invariant curves*

In nonlinear discrete systems of dimension two and higher, regular oscillatory attractors can be more complicated. Here, a classic example is a closed invariant curve that appears via Neimark–Sacker bifurcation.

Let the deterministic system (8.2) have the exponentially stable closed invariant curve Γ. Such curve can be formed by solutions of the system (8.2) in different ways. We will consider cases when the curve Γ consists of equilibria, discrete limit cycles, or quasiperiodic solutions.

Let us study the asymptotics

$$z_t = \lim_{\varepsilon \to 0} \frac{x_t^\varepsilon - \bar{x}_t^\varepsilon}{\varepsilon},$$

where x_t^ε is the solution of stochastic system (8.1), and $\bar{x}_t^\varepsilon$ is the point of the curve Γ that is nearest to x_t^ε. Consider the corresponding system of stochastic linear extension:

$$\begin{aligned} x_{t+1} &= f(x_t, 0), \quad x \in \Gamma, \\ z_{t+1} &= P(x_{t+1}) \left[F(x_t)\, z_t + S(x_t)\, \xi_t \right], \quad z \in \mathbb{R}^n. \end{aligned} \tag{8.12}$$

Here,

$$F(x) = \frac{\partial f}{\partial x}(x, 0), \quad S(x) = \frac{\partial f}{\partial \eta}(x, 0),$$

$P(x)$ is the matrix of the projection onto a hyperplane $\Pi(x)$ that is orthogonal to the curve Γ at the point x.

For the second moments $M_t = \mathrm{E} z_t z_t^\top$, the following system can be written:

$$\begin{aligned} x_{t+1} &= f(x_t, 0), \\ M_{t+1} &= P(x_{t+1}) \left[F(x_t) M_t F^\top(x_t) + S(x_t) V S^\top(x_t) \right] P(x_{t+1}). \end{aligned} \tag{8.13}$$

Consider the case when the curve Γ is the family of k-cycles. Let $\bar{x}$ be an arbitrary point of the curve Γ and $\{\bar{x}_1, \bar{x}_2, \ldots, \bar{x}_k\}$ be the k-cycle γ starting from the point $\bar{x}$: $\bar{x}_{t+1} = f(\bar{x}_t, 0), \bar{x}_1 = \bar{x}$. From (8.13) for the cycle γ, it follows that

$$M_{t+1} = P_{t+1} \left[F_t M_t F_t^\top + Q_t \right] P_{t+1}, \tag{8.14}$$

where matrices

$$P_t = P(\bar{x}_t), \quad F_t = F(\bar{x}_t), \quad Q_t = S(\bar{x}_t)\, V\, S^\top(\bar{x}_t)$$

are k-periodic. Because of the exponential stability of Γ, the system (8.14) has a unique k-periodic solution W_t:

$$W_{t+1} = P_{t+1} \left[F_t W_t F_t^\top + Q_t \right] P_{t+1}, \quad W_{k+1} = W_1. \tag{8.15}$$

The set of matrices $\{W_1, \ldots, W_k\}$ defines stochastic sensitivity of the closed invariant curve Γ at the points $\{\bar{x}_1, \ldots, \bar{x}_k\} \subset \mathcal{X}$. Note that the first matrix W_1 can be found from the following equation

$$W_1 = FW_1F^\top + H,$$

where

$$F = P_1F_kP_kF_{k-1} \cdot \ldots \cdot P_2F_1, \quad H = H^{(k)},$$

$$H^{(0)} = 0, \quad H^{(j)} = P_{j+1}\left[F_jH^{(j-1)}F_j^\top + Q_j\right]P_{j+1}, \quad j = 1, \ldots, k.$$

Other matrices $W_2, \ldots, W_k$ are found recurrently from (8.15).

Due to presence of projection matrices in (8.15), the matrices W_t are singular ($\det W_t = 0$). So, in the two-dimensional case we can use the following representation: $W_t = m_tp_tp_t^\top$. Here, $m_t = p_t^\top W_tp_t$ and p_t is an orthonormal vector to Γ at the point $\bar{x}_t$. Thus, the stochastic sensitivity function $m(x)$ of the curve Γ at the point $\bar{x}$ has a value $m(\bar{x}) = m_1$.

In the case when the closed curve Γ is formed by the family of quasiperiodic solutions of the deterministic system (8.2), we can approximate Γ by an appropriate periodic sequence (k-cycle) and use the theory presented above. Note that the higher the order of this approximation, longer the period k of the cycles used.

Nonlinear systems can possess attractors formed by m separate closed invariant curves (m-torus). For this more general case, m-fold superposition of one-step mappings can be used. For this superposition, m-piece attractor splits into m coexisting 1-piece attractors. This allows us to use the stochastic sensitivity theory presented above for 1-piece closed invariant curves. Consider such a reduction for the case of the 2-torus.

The dynamical system obtained by two-fold stochastic map (8.1) is written as

$$x_{t+2} = f(f(x_t, \eta_t), \eta_{t+1}). \tag{8.16}$$

For new variables $y_t = x_{2t-1}$ and $\nu_{1,t} = \eta_{2t-1}$, $\nu_{2,t} = \eta_{2t}$, this system can be rewritten as

$$y_{t+1} = f(f(y_t, \nu_{1,t}), \nu_{2,t}). \tag{8.17}$$

Consider the function $\psi(y,\nu) = \varphi(\varphi(y,\nu_1),\nu_2)$ and $2m$-dimensional random process $\nu_t = (\nu_{1,t},\nu_{2,t})$ and rewrite the system (8.17) as

$$y_{t+1} = \psi(y_t,\nu_t). \tag{8.18}$$

As a result, the stochastic sensitivity analysis of the 2-piece closed invariant curve $\Gamma = \Gamma_1 \cup \Gamma_2$ in the initial stochastic system (8.1) is reduced to the study of the stochastic sensitivity of separate coexisting 1-piece invariant curves Γ_1 and Γ_2 in the stochastic system (8.18).

8.2.4. *Stochastic sensitivity of chaotic attractors*

Consider now how the general stochastic sensitivity technique is used in the case of chaotic attractors. Here, we restrict our representation by one-dimensional systems with the one-band chaotic attractor.

Let the deterministic system (8.2) have a chaotic attractor $\Gamma = (a,b)$. It is supposed that the function $f(x,0)$ has a single maximum at the point $c \in (a,b)$:

$$\max f(x,0) = f(c,0), \quad \frac{\partial f}{\partial x}(c,0) = 0.$$

Borders a and b of the chaotic attractor Γ are connected with the point c as follows: $b = f(c,0)$, $a = f(b,0) = f(f(c,0),0)$.

The stochastic sensitivity of the chaotic attractor Γ at the boundary points a and b is defined by the stochastic sensitivity of the corresponding points of the solution of system (8.2) passing through these borders [24]. So, let us consider the solution $\bar{x}_t$ of system (8.2) with the initial state $\bar{x}_1 = c$. The first iterations give us $\bar{x}_2 = f(\bar{x}_1,0) = b$, $\bar{x}_3 = f(\bar{x}_2,0) = a$. The stochastic sensitivity of the right border $x = b$ of the attractor Γ is equal to the stochastic sensitivity M_2 of the state $\bar{x}_2$ of the solution $\bar{x}_t$: $M(b) = M_2$. In a like manner, the stochastic sensitivity $M(a)$ of the left border a is equal to the stochastic sensitivity of $\bar{x}_3$: $M(a) = M_3$. It follows from the general system (8.4) that

$$M_2 = q_1 M_1 + s_1, \quad M_3 = q_2 M_2 + s_2, \tag{8.19}$$

where

$$q_1 = \left[\frac{\partial f}{\partial x}(c,0)\right]^2, \quad s_1 = s(c), \quad q_2 = \left[\frac{\partial f}{\partial x}(b,0)\right]^2, \quad s_2 = s(b),$$

where

$$s(x) = \frac{\partial f}{\partial \eta}(x,0) V \left[\frac{\partial f}{\partial \eta}(x,0)\right]^\top.$$

It follows from $\frac{\partial f}{\partial x}(c,0) = 0$ that

$$M(a) = \left[\frac{\partial f}{\partial x}(f(c,0),0)\right]^2 s(c) + s(f(c,0)), \quad M(b) = s(c). \quad (8.20)$$

Stochastic sensitivity technique presented here for one-band chaotic attractors can be extended to multi-band chaotic attractors.

8.3. Confidence Domains

Values of stochastic sensitivity can be effectively used for the approximation of dispersion of random states around deterministic attractors in the form of confidence domains.

8.3.1. *Confidence ellipsoid around the equilibrium*

Let M be the stochastic sensitivity matrix of the stable equilibrium $\bar{x}$ in system (8.1). Using this matrix, one can write the Gaussian approximation of the probability density

$$\rho(x) = \frac{1}{\sqrt{(2\pi\varepsilon^2)^n \det(M)}} \exp\left\{-\frac{1}{2\varepsilon^2}(x-\bar{x})^\top M^{-1}(x-\bar{x})\right\}.$$

For geometrical description of the spatial features of the dispersion of random states around the equilibrium $\bar{x}$, one can use confidence domains in the form of ellipsoids.

For general case, the confidence n-dimensional ellipsoid is written as

$$\left(x-\bar{x}, M^{-1}(x-\bar{x})\right) = \varepsilon^2 K(P), \quad (8.21)$$

where P is the fiducial probability. The function $K(P)$ is inverse to the function

$$P(K) = \frac{\Phi_n(K)}{\Phi_n(\infty)}, \quad \Phi_n(K) = \int_0^{\sqrt{K}} e^{-\frac{t^2}{2}} t^{n-1} dt.$$

For $n = 1$,

$$P(K) = \sqrt{\frac{2}{\pi}} \int_0^{\sqrt{K}} e^{-\frac{t^2}{2}} dt = \operatorname{erf}\left(\sqrt{\frac{K}{2}}\right), \quad \operatorname{erf}(x) = \frac{2}{\sqrt{\pi}} \int_0^x e^{-t^2} dt,$$

and the confidence interval is

$$(\bar{x} - r, \bar{x} + r), \quad r = \varepsilon\sqrt{2M}\, \operatorname{erf}^{-1}(P). \tag{8.22}$$

For two-dimensional case, we have

$$P(K) = 1 - \mathrm{e}^{-\frac{K}{2}}, \quad K(P) = -2\ln(1 - P),$$

and the confidence ellipse is defined by the equation

$$\frac{z_1^2}{\mu_1} + \frac{z_2^2}{\mu_2} = \varepsilon^2 K(P), \tag{8.23}$$

where $z_1 = (x - \bar{x}, u_1), z_2 = (x - \bar{x}, u_2)$ are coordinates of the ellipse in the basis of normalized eigenvectors u_1, u_2 of the matrix M, parameters μ_1, μ_2 are corresponding eigenvalues, and ε is the noise intensity. So, eigenvalues and eigenvectors of the stochastic sensitivity matrix M significantly affect the spatial configuration and size of the confidence ellipse.

For three-dimensional case, we have

$$P(K) = \sqrt{\frac{2}{\pi}} \left[\int_0^{\sqrt{K}} e^{-\frac{t^2}{2}} dt - \sqrt{K} e^{-\frac{K}{2}} \right] = \operatorname{erf}\left(\sqrt{\frac{K}{2}}\right) - \sqrt{\frac{2K}{\pi}} \mathrm{e}^{-\frac{K}{2}}.$$

The three-dimensional confidence ellipsoid is defined by the equation

$$\frac{z_1^2}{\mu_1} + \frac{z_2^2}{\mu_2} + \frac{z_3^2}{\mu_3} = \varepsilon^2 K(P), \tag{8.24}$$

where μ_1, μ_2, μ_3 are eigenvalues, u_1, u_2, u_3 are normalized eigenvectors of the stochastic sensitivity matrix M, and $z_1 = (x - \bar{x}, u_1)$, $z_2 = (x - \bar{x}, u_2)$, $z_3 = (x - \bar{x}, u_3)$.

8.3.2. *Confidence domains for k-cycle*

Let $M_1, \ldots, M_k$ be stochastic sensitivity matrices of states $\bar{x}_1, \ldots, \bar{x}_k$ of the k-cycle Γ of system (8.1). Using these matrices, one can write the approximation of the probability density

$$\rho(x) = \frac{1}{k} \sum_{t=1}^{k} \frac{1}{\sqrt{(2\pi\varepsilon^2)^n \det(M_t)}} \exp\left\{ -\frac{1}{2\varepsilon^2}(x - \bar{x}_t)^\top M_t^{-1}(x - \bar{x}_t) \right\}$$

of random distribution near Γ.

Depending on the dimension of the system, confidence domains for k-cycles have the following form.

For $n = 1$, the confidence domain is a set of k confidence intervals around states $\bar{x}_1, \ldots, \bar{x}_k$ of the k-cycle Γ.

For $n = 2$, these domains are confidence ellipses defined by equations:

$$\frac{z_1^2}{\mu_{1,t}} + \frac{z_2^2}{\mu_{2,t}} = -2\varepsilon^2 \ln(1 - P), \quad (t = 1, 2, \ldots, k). \tag{8.25}$$

Here, $z_1 = (x - \bar{x}_t, u_{1,t}), z_2 = (x - \bar{x}_t, u_{2,t})$ are coordinates of the ellipse in the basis of normalized eigenvectors $u_{1,t}, u_{2,t}$ of the matrix M_t, parameters $\mu_{1,t}, \mu_{2,t}$ are corresponding eigenvalues.

For $n = 3$, these domains are formed by confidence ellipsoids.

8.3.3. *Confidence domains for closed invariant curves*

Here, we restrict our discussion by two-dimensional case. Let Γ be a closed invariant curve of the system (8.2) for $n = 2$, and $m(x)$ is the stochastic sensitivity function defined at the points of Γ. Using $m(x)$, one can find a confidence interval on the line that is orthogonal to the closed invariant curve Γ at the point x. Boundaries of this interval can be written as

$$x_{1,2} = x \pm \varepsilon q \sqrt{2m(x)} p(x). \tag{8.26}$$

Here, the parameter $q = \mathrm{erf}^{-1}(P)$, $\mathrm{erf}(x) = \frac{2}{\sqrt{\pi}} \int_0^x e^{-t^2} dt$, and P is the fiducial probability. Under the variation of x along Γ, the family of these intervals forms the corresponding confidence band around Γ.

8.4. Examples

Consider one-dimensional model with the quadratic map

$$x_{t+1} = 1 - \mu x_t^2. \tag{8.27}$$

The system (8.27) has two equilibria

$$\tilde{x} = \frac{-1 - \sqrt{1+4\mu}}{2\mu}, \quad \bar{x} = \frac{-1 + \sqrt{1+4\mu}}{2\mu}.$$

The equilibrium $\tilde{x}$ is unstable for any $\mu > 0$. This equilibrium is shown by the lower red curve in Fig. 8.1. The equilibrium $\bar{x}$ is stable for any $0 < \mu < 0.75$. Passing $\mu = 0.75$ from left to right, this equilibrium loses its stability and a stable 2-cycle appears. Under increasing μ, the system possesses various attractors, both regular and chaotic. The corresponding Feigenbaum tree is shown by blue in Fig. 8.1. At the crisis bifurcation point $\mu = 2$, this tree touches the unstable equilibrium $\tilde{x}$ and disappears. Here, the basin of attractors of the Feigenbaum tree is shown by white. In the interval $0 < \mu < 2$, the lower border of this basin is $\tilde{x}$. The upper border of this basin (upper red curve) is $\hat{x} = \sqrt{(1 - \tilde{x})/\mu}$.

For all initial states which belong to light blue domain, we have $\lim_{t\to\infty} x_t = -\infty$.

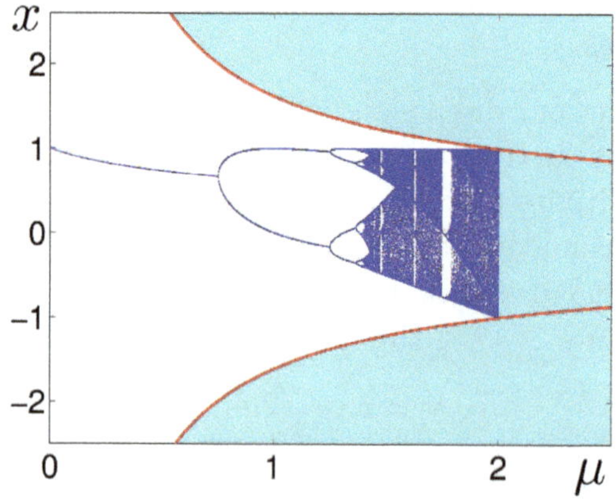

Fig. 8.1. Attractors of system (8.27).

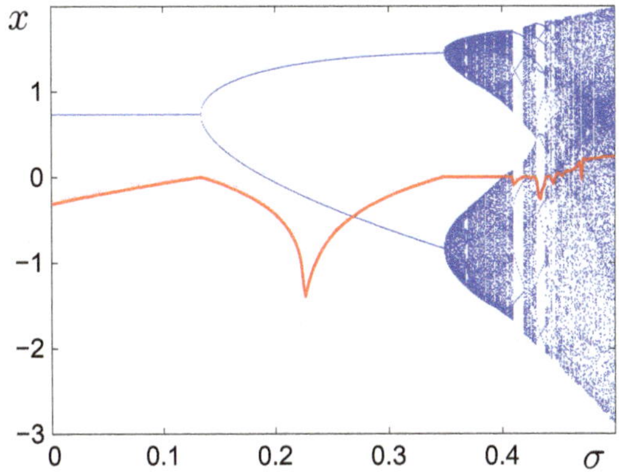

Fig. 8.2. Attractors of the system (8.28) with $\mu = 0.5$. Largest Lyapunov exponents are shown in red.

Consider now a system of two coupled quadratic maps:

$$\begin{aligned} x_{t+1} &= 1 - \mu x_t^2 + \sigma(y_t - x_t), \\ y_{t+1} &= 1 - \mu y_t^2 + \sigma(x_t - y_t), \end{aligned} \tag{8.28}$$

where σ is a coupling parameter. In what follows, we fix $\mu = 0.5$. Attractors of the system (8.28) with $\mu = 0.5$ are plotted in Fig. 8.2 vs. parameter σ. Here, a wide diversity of attractors is observed: equilibria, discrete cycles of various multiplicity, closed invariant curves, and chaotic attractors. Largest Lyapunov exponents for these attractors are shown in red.

To demonstrate stochastic sensitivity technique and confidence domains method, we will use the following stochastic system:

$$\begin{aligned} x_{t+1} &= 1 - \mu x_t^2 + \sigma(y_t - x_t) + \varepsilon\xi_{1,t}, \\ y_{t+1} &= 1 - \mu y_t^2 + \sigma(x_t - y_t) + \varepsilon\xi_{2,t}. \end{aligned} \tag{8.29}$$

Here, $\xi_{1,t}, \xi_{2,t}$ are uncorrelated scalar random Gaussian processes with parameters $\mathrm{E}\xi_{i,t} = 0$, $\mathrm{E}\xi_{i,t}^2 = 1$ $(i = 1, 2)$, and ε is the scalar parameter of noise intensity.

Let us consider a behavior of the stochastic system (8.29) for $\sigma = 0.2$. Corresponding deterministic system has the stable 2-cycle

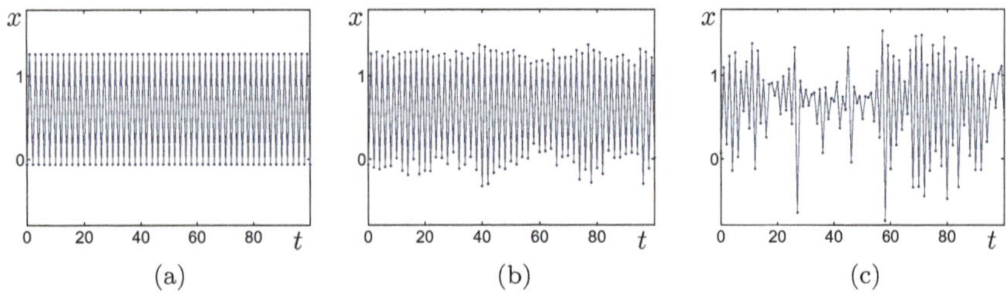

Fig. 8.3. Time series of (8.29) with $\sigma = 0.2$ and (a) $\varepsilon = 0$, (b) $\varepsilon = 0.05$, (c) $\varepsilon = 0.2$.

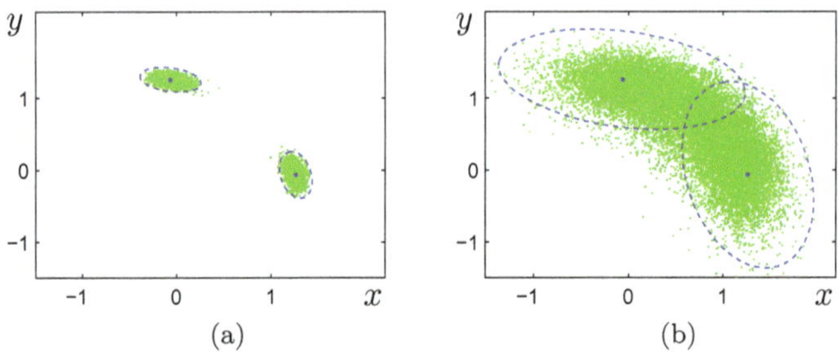

Fig. 8.4. Stochastic system (8.29) with $\sigma = 0.2$. Random states (green) and confidence ellipses for (a) $\varepsilon = 0.05$, (b) $\varepsilon = 0.2$.

(see periodic time series in Fig. 8.3(a) for $\varepsilon = 0$). Under increasing noise, these regular 2-periodic oscillations are destroyed (see Fig. 8.3(b) for $\varepsilon = 0.05$ and Fig. 8.3(c) for $\varepsilon = 0.2$).

Consider how this stochastic transformation can be studied by confidence domains method. In Fig. 8.4, states of the deterministic 2-cycle are shown by black dots, random states of system (8.29) are plotted in green, and confidence ellipses are shown by dashed. For $\varepsilon = 0.05$, random states are localized near states of the deterministic 2-cycle, and confidence ellipses are small and well separated from each other. Under increasing noise, the size of ellipses increases too. For $\varepsilon = 0.2$, random states are widely dispersed and stochastic mixing is observed. The intersection of confidence ellipses signals the onset of this new stochastic regime of the destruction of rhythmic oscillations.

Let us consider a behavior of the stochastic system (8.29) for $\sigma = 0.37$. For this parameter value, the deterministic system exhibits

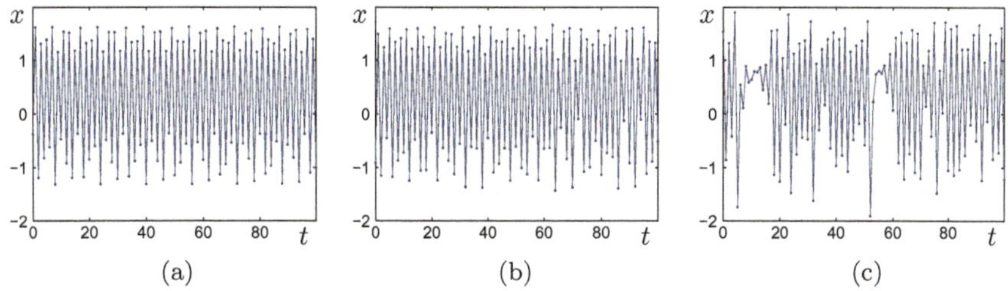

Fig. 8.5. Time series of system (8.29) with $\sigma = 0.37$ and (a) $\varepsilon = 0$, (b) $\varepsilon = 0.03$, (c) $\varepsilon = 0.15$.

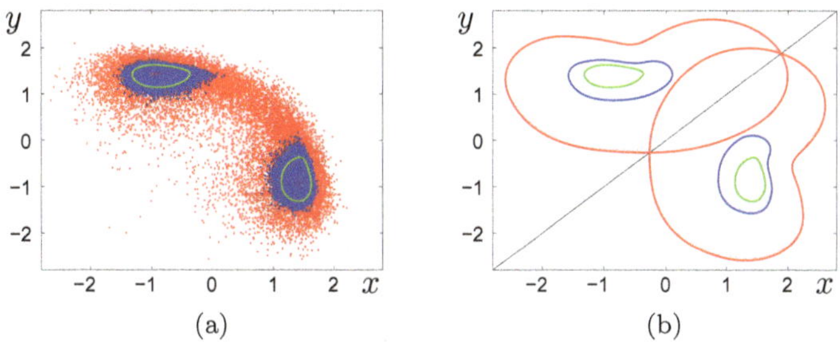

Fig. 8.6. Stochastic system (8.29) with $\sigma = 0.37$: (a) random states for $\varepsilon = 0$ (green), $\varepsilon = 0.03$ (blue), $\varepsilon = 0.15$ (red); (b) closed invariant curves of the deterministic 2-torus (green) and external borders of confidence bands for $\varepsilon = 0.03$ (blue), $\varepsilon = 0.15$ (red).

a stable 2-torus with two coexisting closed invariant curves (see for $\varepsilon = 0$ quasiperiodic time series in Fig. 8.5(a) and green closed curves in Fig. 8.6). As noise increases, this regular quasiperiodic regime is destroyed (see Fig. 8.5(b) for $\varepsilon = 0.03$ and Fig. 8.5(c) for $\varepsilon = 0.15$). Corresponding transformation of the random states distribution is shown in Fig. 8.6(a).

Following theory presented in Secs. 8.2.3 and 8.3.3, one can find stochastic sensitivity function for these closed invariant curves and construct corresponding confidence bands. External borders of these bands are shown in Fig. 8.6(b) for $\varepsilon = 0.03$ in blue and for $\varepsilon = 0.15$ in red. Here, the intersection of these confidence bands also signals the destruction of regular oscillations. Moreover, in this case, we observe a phenomenon of the noise-induced transition from order to chaos

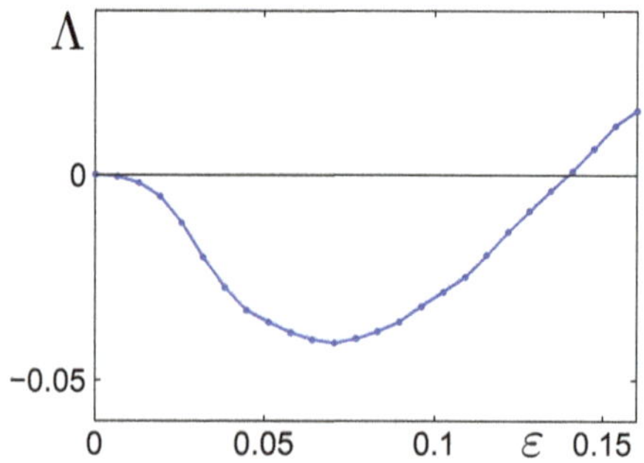

Fig. 8.7. Noise-induced chaos in stochastic system (8.29) with $\sigma = 0.37$. Largest Lyapunov exponent $\Lambda(\varepsilon)$.

(see Fig. 8.7). This is confirmed by the largest Lyapunov exponent which changes its sigh from minus to plus with increasing noise.

Finally, we will show how stochastic sensitivity analysis can be applied to the study of noise-induced transformation of chaotic attractors.

Consider a stochastic variant of one-dimensional system

$$x_{t+1} = 1 - \mu x_t^2 + \varepsilon \xi_t, \tag{8.30}$$

where ξ_t is uncorrelated scalar random Gaussian process with parameters $E\xi_t = 0$, $E\xi_t^2 = 1$. Random states for various values of noise intensity are shown in Fig. 8.8(a).

As can be seen, an increase of noise results in the shift to the left of μ-value of the crisis bifurcation. Consider details of this noise-induced effect and methods of its analysis with the help of the confidence intervals.

In Fig. 8.8(b) for $\mu = 1.9$, the deterministic chaotic attractor is shown in gray, random states of system (8.30) lying outside this chaotic attractor are shown in blue, the unstable equilibrium $\tilde{x}$ is shown by solid red line, and the lower border of the confidence intervals found by theory from the Sec. 8.2.4 is shown by red dashed line.

For weak noise, the dashed line well describes the deviation of random states (blue dots) from the deterministic chaotic attractor

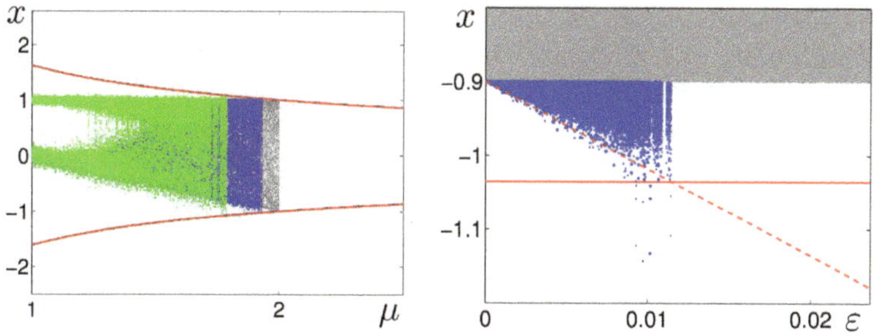

Fig. 8.8. Stochastic system (8.30): (a) states of attractors for $\varepsilon = 0$ (gray), $\varepsilon = 0.01$ (blue), $\varepsilon = 0.03$ (green); (b) noise-induced destruction for $\mu = 1.9$. Here, deterministic chaotic attractor is shown in gray, points of system (8.30) lying outside chaotic attractor are shown in blue, unstable equilibrium is shown by solid red, and lower border of the confidence intervals is shown by red dashed.

(gray). Under increasing ε, this deviation increases, and the intersection of the lower border of the confidence interval (dashed red) with the unstable equilibrium (solid red) well estimates the noise intensity corresponding to the noise-induced crisis.

Acknowledgment

The work was supported by RFBR (20-01-00165).

References

[1] J. Awrejcewicz, *Nonlinear Dynamics: New Theoretical and Applied Results* (Akademie Verlag, 1995).
[2] A. D. Bazykin, *Nonlinear Dynamics of Interacting Populations* (World Scientific, 1998).
[3] I. R. Epstein and J. A. Pojman, *An Introduction to Nonlinear Chemical Dynamics: Oscillations, Waves, Patterns, and Chaos* (Oxford University Press, 1998).
[4] E. M. Izhikevich, *Dynamical Systems in Neuroscience: The Geometry of Excitability and Bursting* (MIT Press, Cambridge, 2007).
[5] B. Ibarz, J. M. Casado and M. A. F. Sanjuán, Map-based models in neuronal dynamics, *Phys. Rep.* **501** (2011) 1–74.
[6] S. H. Strogatz, *Nonlinear Dynamics and Chaos With Applications to Physics, Biology, Chemistry, and Engineering* (Taylor & Francis Group, 2015).

[7] M. Edelman, E. E. N. Macau and M. A. F. E. Sanjuán, *Chaotic, Fractional, and Complex Dynamics: New Insights and Perspectives* (Springer, Berlin, 2018).
[8] W. Horsthemke and R. Lefever, *Noise-Induced Transitions* (Springer, Berlin, 1984).
[9] N. G. van Kampen, *Stochastic Processes in Physics and Chemistry* (North-Holland, Amsterdam, 1987).
[10] A. Lasota and M. C. Mackey, *Chaos, Fractals, and Noise: Stochastic Aspects of Dynamics* (Springer, Berlin, 1994).
[11] L. Arnold, *Random Dynamical Systems* (Springer-Verlag, Berlin, 1998).
[12] J. Gao, S. Hwang and J. Liu, When can noise induce chaos? *Phys. Rev. Lett.* **82**(6) (1999) 1132–1135.
[13] A. A. Zaikin, L. López, J. P. Baltanás, J. Kurths and M. A. F. Sanjuán, Vibrational resonance in a noise-induced structure, *Phys. Rev. E* **66** (2002) 011106.
[14] B. Lindner, J. Garcia-Ojalvo, A. Neiman and L. Schimansky-Geier, Effects of noise in excitable systems, *Phys. Rept.* **392** (2004) 321–424.
[15] V. S. Anishchenko, V. V. Astakhov, A. B. Neiman, T. E. Vadivasova and L. Schimansky-Geier, *Nonlinear Dynamics of Chaotic and Stochastic Systems. Tutorial and Modern Development* (Springer–Verlag, Berlin, 2007).
[16] M. McDonnell, N. Stocks, C. Pearce, and D. Abbott, *Stochastic Resonance: From Suprathreshold Stochastic Resonance to Stochastic Signal Quantization* (Cambridge University Press, Cambridge, 2008).
[17] J. M. Seoane, L. Huang, M. A. F. Sanjuán and Y.-C. Lai, Effect of noise on chaotic scattering, *Phys. Rev. E* **79** (2009) 047202.
[18] E. del Rio, M. A. Sanjuán and S. Elaskar, Effect of noise on the reinjection probability density in intermittency, *Commun. Nonlinear Sci. Numer. Simulat.* **17**(9) (2012) 3587–3596.
[19] P. Schuster, *Stochasticity in Processes: Fundamentals and Applications to Chemistry and Biology* (Springer, Berlin, 2016).
[20] I. Bashkirtseva, L. Ryashko and I. Tsvetkov, Sensitivity analysis of stochastic equilibria and cycles for the discrete dynamic systems, *Dynam. Contin. Discrete Impulsive Syst. Ser. A: Math. Anal.* **17** (2010) 501–515.
[21] I. Bashkirtseva and L. Ryashko, Stochastic sensitivity analysis of noise-induced intermittency and transition to chaos in one-dimensional discrete-time systems, *Physica A* **392**(2) (2013) 295–306.
[22] I. Bashkirtseva and L. Ryashko, Stochastic sensitivity analysis of the attractors for the randomly forced Ricker model with delay, *Phys. Lett. A* **378**(48) (2014) 3600–3606.

[23] I. Bashkirtseva and L. Ryashko, Stochastic sensitivity of the closed invariant curves for discrete-time systems, *Physica A* **410** (2014) 236–243.
[24] I. Bashkirtseva and L. Ryashko, Approximating chaotic attractors by period-three cycles in discrete stochastic systems, *Int. J. Bifurc. Chaos* **25**(10) (2015) 1550138.
[25] I. Bashkirtseva and L. Ryashko, Stochastic sensitivity of regular and multi-band chaotic attractors in discrete systems with parametric noise, *Phys. Lett. A* **381**(37) (2017) 3203–3210.
[26] I. Bashkirtseva and L. Ryashko, Stochastic sensitivity analysis of chaotic attractors in 2D non-invertible maps, *Chaos, Solitons Fractals* **126** (2019) 78–84.
[27] I. Bashkirtseva, E. Ekaterinchuk and L. Ryashko, Stochastic sensitivity analysis and noise-induced chaos in 2D logistic-type model, *Int. J. Bifurc. Chaos* **26**(04) (2016) 1650053.
[28] I. Bashkirtseva and Ryashko, Stochastic sensitivity analysis of noise-induced order-chaos transitions in discrete-time systems with tangent and crisis bifurcations, *Physica A* **467** (2017) 573–584.
[29] I. Bashkirtseva and L. Ryashko, Analysis of noise-induced chaos-order transitions in Rulkov model near crisis bifurcations, *Int. J. Bifurc. Chaos* **27**(3) (2017) 1730014.
[30] I. Bashkirtseva, V. Nasyrova and L. Ryashko, Analysis of noise effects in a map-based neuron model with Canard-type quasiperiodic oscillations, *Commun. Nonlinear Sci. Numer. Simulat.* **63** (2018) 261–270.
[31] I. Bashkirtseva, V. Nasyrova and L. Ryashko, Noise-induced bursting and chaos in the two-dimensional Rulkov model, *Chaos, Solitons Fractals* **110** (2018) 76–81.
[32] I. Bashkirtseva, V. Nasyrova and L. Ryashko, Stochastic spiking-bursting excitability and transition to chaos in a discrete-time neuron model, *Int. J. Bifurc. Chaos* **30**(10) (2020) 2050153.
[33] I. Bashkirtseva, Preventing noise-induced extinction in discrete population models, *Discrete Dynam. Nature Soc.* **2017** (2017) 9610609.
[34] I. Bashkirtseva and L. Ryashko, How environmental noise can contract and destroy a persistence zone in population models with Allee effect, *Theoret. Population Biol.* **115** (2017) 61–68.
[35] I. Bashkirtseva, Crises, noise, and tipping in the Hassell population model, *Chaos* **28** (2018) 033603.
[36] I. Bashkirtseva, L. Ryashko and B. Spagnolo, Combined impacts of the Allee effect, delay and stochasticity: Persistence analysis, *Commun. Nonlinear Sci. Numer. Simulat.* **84** (2020) 105148.

[37] I. Bashkirtseva, L. Ryashko and A. Sysolyatina, Analysis of stochastic effects in Kaldor-type business cycle discrete model, *Commun. Nonlinear Sci. Numer. Simulat.* **36** (2016) 446–456.

[38] J. Jungeilges, T. Ryazanova, A. Mitrofanova and I. Popova, Sensitivity analysis of consumption cycles, *Chaos* **28**(5) (2018) 055905.

[39] E. Ekaterinchuk, J. Jungeilges, T. Ryazanova and I. Sushko, Dynamics of a minimal consumer network with bi-directional influence, *Commun. Nonlinear Sci. Numer. Simulat.* **58** (2018) 107–118.

[40] I. Bashkirtseva and L. Ryashko, Stochastic deformations of coupling-induced oscillatory regimes in a system of two logistic maps, *Physica D* **411** (2020) 132589.

[41] I. Bashkirtseva and L. Ryashko, Stabilization of stochastic cycles and chaos suppression for nonlinear discrete-time systems, *Nonlinear Dynam.* **67** (2012) 2505–2517.

[42] I. Bashkirtseva and L. Ryashko, Attainability analysis in the problem of stochastic equilibria synthesis for nonlinear discrete systems, *Int. J. Appl. Math. Comput. Sci.* **23**(1) (2013) 5–16.

[43] I. Bashkirtseva, Method of stochastic sensitivity synthesis in a stabilisation problem for nonlinear discrete systems with incomplete information, *Int. J. Control* **90**(8) (2017) 1652–1663.

https://doi.org/10.1142/9789811221903_0009

Chapter 9

The Role of Noise in Chaotic Intermittency

Ezequiel del Rio*,‡ and Sergio Elaskar†,§

*Departamento de Física Aplicada. ETSI Aeronáutica y del Espacio. Universidad Politécnica de Madrid, Plaza Cardenal Cisneros 3. 28040 Madrid, Spain

†Departamento de Aeronáutica, FCEFyN and Instituto de Estudios Avanzados en Ingeniería y Tecnología CONICET and Universidad Nacional de Córdoba, Argentina

‡ezequiel.delrio@upm.es

§selaskar@unc.edu.ar

Intermittency is a particular route to the deterministic chaos characterized by spontaneous transitions between laminar and chaotic dynamics. The classical theory of intermittency was firstly introduced by Pomeau and Manneville in the context of the Lorenz system. Later, the classical theory was extended into a new intermittency theory that includes those cases that produce statistic behavior far from the classical predictions. This chapter is devoted to the additive noise effect in the maps that have intermittency. From classical point of view, two approaches addressing this problem are the most important ones: Fokker–Planck and renormalization group analysis. Both cases only consider the noise effect in the local map, that is, near to the unstable point where the laminar behavior takes place. Later, in a paper with Prof. Miguel A. F. Sanjuán the noise effect on the new intermittency theory was considered and it was shown that the noise on the chaotic region also affects the statistical dynamics in the laminar region. It has also been found good agreement between the new proposed noise theory and the numerical simulations. Recently, experimental evidences on the new intermittency theory have been found in continuous systems. The noise effect has also been investigated in these systems and some differences with theory predictions have been found hence open questions remain for future investigations.

9.1. Introduction

Intermittency is a particular route to the deterministic chaos characterized by spontaneous transitions between laminar and chaotic dynamics. This concept has been introduced by Pomeau and Manneville in the context of the Lorenz system [1]. Later, intermittency has been found in a variety of different systems including, for example, periodically forced nonlinear oscillators, Rayleigh–Bénard convection, derivative nonlinear Schrödinger equation, and in development of turbulence in hydrodynamics [2–4]. Proper qualitative and quantitative characterizations of intermittency based on experimental data are especially useful for studying problems with partial or complete lack of knowledge on exact governing equations, as it frequently happens e.g., in Economics, Biology, and Medicine [5, 6]. In this case, special attention has to be payed to the length of data sets required for robust estimation of the model parameters.

According to Pomeau and Manneville, intermittency can be classified into three types depending on the local geometry of their Poincaré map: type I for quadratic maps and type II and type III for cubic ones [7]. However, other nonlinear types had been reported [8–10], hence the local laminar dynamics of type I intermittency is determined by the one-dimensional (1D) map in the form:

$$x_{n+1} = \varepsilon + x_n + a\, x_n^p \quad \text{(type I)}, \tag{9.1}$$

where $a > 0$ accounts for the weight of the nonlinear component determined by the exponent p. The laminar behavior of type II and type III are described by the maps:

$$x_{n+1} = (1+\varepsilon)x_n + a x_n^p \quad \text{(type II)}, \tag{9.2}$$

$$x_{n+1} = -(1+\varepsilon)x_n - a x_n^p \quad \text{(type III)}. \tag{9.3}$$

In type II and type III, the fixed point of the system becomes unstable for positive values of the parameter ε. In type I, the instability is through a narrow corridor formed with the bisector line. Another condition for a 1D map to possess intermittency is to have a reinjection mechanism mapping back the trajectories from the chaotic zone into the laminar one. This mechanism is described by the so-called reinjection probability density (RPD), which is determined by the chaotic dynamics of the system itself. The function RPD together with the local map determines all the dynamical properties of the system. In general, it is difficult to get an analytical expression for the RPD, hence different approximation have been used. The most common approximation used to obtain many classical results on intermittency theory have been to consider the RPD uniform, and thus independent of the reinjection point [2, 7, 11–16]; however, it works fine in a few model cases only [11, 13, 15]. Another approach deals with the δ-function. It considers reinjection into a given point in the presence of noise [14, 17]. Recently, to describe the reinjection mechanism of a wide class 1D maps exhibiting intermittency a generalized RPD was introduced, a parametric power law function depending on a free parameter $m \in (0, 1)$. The generalized RPD includes the uniform reinjection as a particular case $m = 1/2$ [9, 18]. We showed that RPD function is determined by the Poincaré map in the neighbor of points with infinite or zero tangent [19, 20]. Based on this fact, an analytical procedure has been proposed to obtain the value of α from the analytical expression of the map [19]. For some celebrated classical maps having intermittency, this method predicts the complete statistical behavior already found by numerical investigation, including the fundamental characteristic of the intermittency, that is, the probability density of the length of laminar phase $\psi(l)$ where l approximates the number of iterations in the laminar region, i.e., the length of the laminar phase. Note that, as the function $\psi(l)$ can be estimated for time series, it is usually used to identify the intermittency type by comparing with the analytic estimation.

The new RPD has been found in many 1D maps, and the analytical predictions for RPD and for $\psi(l)$ have been numerically confirmed even in the pathological cases of intermittency described in the literature [21], hence the generalized reinjection probability density provides faithful description of anomalous and standard intermittencies in the unified framework [19, 20].

As the noise is always present in experiments, it is of a fundamental importance to know the effect of noise on the intermittency phenomenon, in particular on the RPD. There are many papers devoted to study noise effect on chaotic intermittency, by means of the normalization group analysis [8], or by using the Fokker–Planck equation [16, 17, 22, 23] among the others. In Sec. 9.2, we will resume the main results of these theories. Note that, in spite of the fact that the noise affects the whole region where the system dynamics takes place, classically, the researches are devoted to the noise effect on the laminar region of the 1D map, and there were no studies in classical theory of intermittency focused on the effect of noise on the RPD. In Sec. 9.4, we introduced the main results that we need coming from the new RPD described by a power law. This RPD introduces a novel scenario because, whereas the classical uniform reinjection should remain constant under a wide class of noise distributions, the new RPD can be affected by the noise. For 1D maps, an analytical approach to the noise reinjection probability density (NRPD) has been presented [24–27]. Note that the probability density of the laminar lengths $\psi(l)$ depends on the reinjection probability density, hence the NRPD also determines a noisy $\psi(l)$. In Sec. 9.5, we review the generalized noise theory based on the work with Prof. Sanjuan [24].

The theory of intermittency sketched before has been developed for 1D maps. However, some continuous systems that contract volume in phase spaces can be described by the 1D maps [28], hence, it was possible, to extend the new intermittency framework to continuous systems and experimental confirmation has been reported [29] as we will explain in Sec. 9.7.

9.2. Noise Effect: Classical Approach

In this section, we present the Fokker–Planck and renormalization group analysis results related to the noise effect on chaotic intermittency.

9.2.1. *Fokker–Planck approach*

In this subsection, the Fokker–Planck equation is used to analyze the noise effects in classical types of intermittency. Noise effects on type

I intermittency are studied in the next subsection. Type II and type III intermittencies are analyzed in Sec. 9.2.1.2.

9.2.1.1. *Type I intermittency*

In the noiseless case, Eq. (9.1) describes the local map for type I intermittency. To consider the noise, an additive noise term is included in the local map:

$$x_{n+1} = F(x_n) = \varepsilon + x_n + a\,x_n^2 + \sqrt{2D}\xi_n. \tag{9.4}$$

Here D is the spatial diffusion and ξ_n is a white Gaussian of zero mean and delta-correlated noise. The difference equation (9.4) can be approximated by the stochastic differential equation as follows [7]:

$$\frac{dx}{dt} = -\frac{dV(x)}{dx} + \sqrt{2D}\xi(t), \tag{9.5}$$

where $\xi(t)$ is considered white Gaussian of zero mean and delta-correlated in time, $\langle \xi(t) \rangle = 0$ and $\langle \xi(t')\xi(t) \rangle = \delta(t' - t)$, and the potential $V(x)$ is

$$V(x) = -\frac{1}{3}ax^3 - \varepsilon x + c. \tag{9.6}$$

Figure 9.1 shows the local map and the corresponding potential for values of $\varepsilon < 0$ where the local map is stable and presents two fixed points that meet in a single point for $\varepsilon = 0$. For $\varepsilon > 0$, the map does not have fixed points, and there is a narrow passage which is characteristic type I intermittency.

The extreme points of the potential in Eq. (9.6) correspond to the stable and unstable fixed points of the map, which are given by

$$x_s = -\sqrt{-\varepsilon/a} \quad \text{and} \quad x_u = \sqrt{-\varepsilon/a}. \tag{9.7}$$

Note that in the pure deterministic map, this scenario does not present intermittency because there is a stable fixed point. However, the noise perturbation can move the system outside on the potential well and then the trajectory goes through the laminar region as in the usual intermittency case.

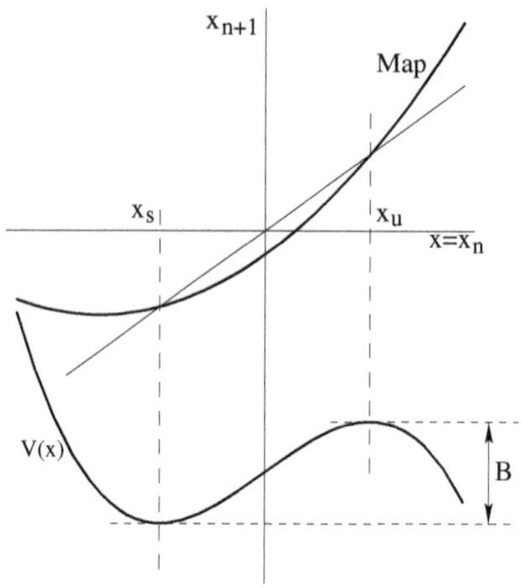

Fig. 9.1. Local map of Eq. (9.4) for $\varepsilon < 0$ and its associated potential equation (9.6). The potential barrier is also indicated.

We can apply the standard methodology in the stochastic theory Eq. (9.5) to obtain the corresponding backward Fokker–Planck equation (FPE)

$$\frac{\partial G(x,t)}{\partial t} = -\frac{dV(x)}{dx}\frac{\partial G(x,t)}{\partial x} + D\frac{\partial^2 G(x,t)}{\partial x^2}, \tag{9.8}$$

where $G(x,t)$ is the probability density of finding a particle at the position x in the time t. The average laminar length in the intermittency context corresponds here to the mean first-passage time (MFPT) over the potential barrier $(V(x_s) - V(x_u))$, that is the mean escaping time $T(x)$ defined by

$$T(x) = -\int_0^\infty \frac{\partial G(x,t)}{\partial t} t dt, \tag{9.9}$$

where $G(x,0) = 1$ and $\lim_{t\to\infty} G(x,t) = 0$ are boundary conditions, that is, all particles are placed in x at $t = 0$ and no particle can be found at x for long time enough. By integrating Eq. (9.8), we get a differential equation for $T(x)$ as follows:

$$0 = 1 - \frac{dV(x)}{dx} + D\frac{d^2 T(x)}{dx^2}. \tag{9.10}$$

Equation (9.10) is not integrable analytically and it requires Taylor expansion to get an estimation for MFPT in the following form (see, for more details, [20]):

$$T = \frac{2\pi}{\sqrt{V''(x_s)V''(x_u)}}\left(\frac{-B}{D}\right). \tag{9.11}$$

Here B is the potential barrier to spring over by the particle in the potential well. For the potential equation (9.6), we have (see Fig. 9.1)

$$B = V(x_u) - V(x_s) = \frac{2a}{3}\left(\frac{|\varepsilon|}{a}\right)^{\frac{3}{2}}, \tag{9.12}$$

and taking into account Eq. (9.7), we get the approximate estimation for the MFPT, that is, for $\bar{l}$ in the case of $\varepsilon < 0$,

$$\bar{l} \propto T = \frac{\pi}{\sqrt{a|\varepsilon|}}\exp\left(\frac{4}{3D\sqrt{a}}|\varepsilon|^{\frac{3}{2}}\right). \tag{9.13}$$

Note that to get Eq. (9.13) we have used the extreme points of the potential, hence Eq. (9.13) cannot be applied to the case $\varepsilon > 0$, where there are not extreme points. Therefore, the particle will cross the laminar region in a similar way than in the pure determinism case, because the map iterations will be dominant over the small noise perturbation. Note also that the limit $D \to 0$ corresponds to absence of random perturbation, then we have $T \to \infty$, and the particle is trapped in the potential well. From Eq. (9.13) we get for large $|\varepsilon| \gg 0$ that the second term is dominant, hence we have $\ln T \approx 1/D|\varepsilon|^{3/2}$. This approximation has been observed numerically in the two coupled Rössler oscillators [30]. The solution given by Eq. (9.13) is consistent with the result from the renormalization group analysis as we will see in Sec. 9.3.

9.2.1.2. *Type II and type III intermittencies*

For type II and type III intermittencies, the perturbed map by a random noise can be obtained from the noiseless map of Eqs. (9.2)

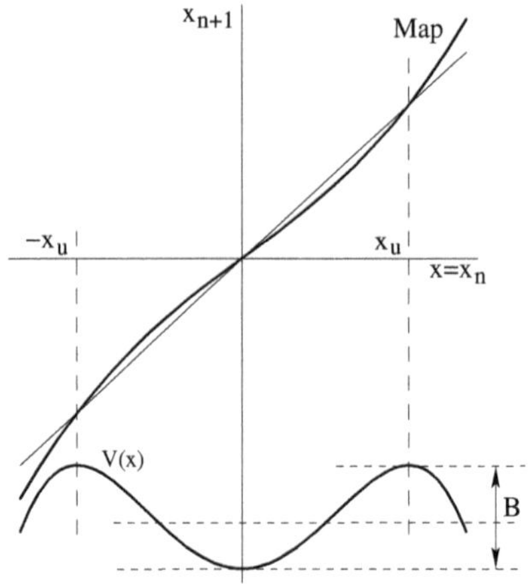

Fig. 9.2. Local map of Eq. (9.14) for $\varepsilon < 0$ and its associated potential Eq. (9.15). The potential barrier is also indicated.

and (9.3) as follows:

$$x_{n+1} = \mp(1+\varepsilon)x_n \mp a\, x_n^3 + \sqrt{2D}\xi_n. \tag{9.14}$$

The difference equation (9.14) can be approximated by Eq. (9.5) where the potential $V(x)$ is a symmetric function given by

$$V(x) = -\frac{1}{4}ax^4 - \frac{\varepsilon}{2}x^2 + c. \tag{9.15}$$

This function has two maxima and a single potential well at $(x = 0)$ (see Fig. 9.2).

Hence, after applying the previous procedure and taking into account that

$$B = V\left(-\sqrt{-\frac{\varepsilon}{a}}\right) - V(0) = \frac{\varepsilon^2}{4\,a}, \tag{9.16}$$

we can obtain

$$\bar{l} \propto T = \frac{\sqrt{2}\pi}{|\varepsilon|} \exp\left(\frac{1}{3D\,a}|\varepsilon|^2\right). \tag{9.17}$$

The previous approximation for $\bar{l}$ has been observed in an electronic circuit [32].

Notice that Eq. (9.11) is very similar to the Arrhenius formula for chemical reaction theory. As the approximation given by Eq. (9.11) depends only on the potential expression, it can be extended to other nonlinear terms in the local maps.

9.3. Renormalization Group and Scaling Theory

The Renormalization Group Theory (RGT) for deterministic chaos was firstly applied to the Feigenbaum cascade because the period-doubling cascade provides a mathematical structure to be renormalized.

A similar method has been applied to the local 1D map that has intermittency. It is important to note that the new general expression for the RPD is not considered in the RGT applied to intermittency, in particular, the role of the LBR [31].

Let us consider the next local map

$$x_{n+1} = F(x_n) = \varepsilon + x_n + a\,|x_n|^p + \sigma\xi_n, \tag{9.18}$$

where we rewrite the diffusion term as $\sigma = \sqrt{2D}$.

Suppose that we map the point x' l times getting x_l, that is, $x_l = F^l(x')$. Let $L(\varepsilon, \sigma)$ be the number of iteration of the map F needed by x' to exit from the laminar region. It is clear that if we used the map F^l instead F, this number should be reduced by a factor l. Let us consider a new set of parameters ε' and σ' mapping in a single iteration x' into x_l, hence we have

$$L(\varepsilon', \sigma') = \frac{1}{l}L(\varepsilon, \sigma). \tag{9.19}$$

Note that, for now, the relation between (ε', σ') and (ε, σ) is unknown, but let us introduce two new scaling exponents ρ and ν

$$\varepsilon' = l^\rho \varepsilon \quad \text{and} \quad \sigma' = l^\nu \sigma \tag{9.20}$$

so Eq. (9.19) can be rewritten as follows:

$$L(\varepsilon, \sigma) = lL(l^\rho \varepsilon, l^\nu \sigma). \tag{9.21}$$

If we multiply Eq. (9.21) by a common factor, we just rescale the number of iterations in the laminar region, hence we can choose a new scale: $l^{\rho}\varepsilon = 1$, so Eq. (9.21) becomes

$$L(\varepsilon,\sigma) = \varepsilon^{\frac{-1}{\rho}} Z\left(\frac{\sigma}{\varepsilon^{\frac{\nu}{\rho}}}\right), \tag{9.22}$$

where the new function Z is defined as $Z(\cdot) = L(1,\cdot)$.

Now, the main task is to evaluate the exponents ρ and ν. This activity can be done by solving and linearizing the renormalization group equation getting the following result (see [20]):

$$\rho = \frac{p}{p-1} \quad \text{and} \quad \nu = \frac{p+1}{p-1}. \tag{9.23}$$

Following a similar argument used to develop the RG in the period doubling context, we postulated the existence of a function $g(x)$ satisfying the following equation:

$$\frac{1}{\alpha} g(\alpha x) = g^{l}(x), \tag{9.24}$$

hence Eq. (9.22) becomes

$$L(\varepsilon,\sigma) = \varepsilon^{\frac{1-p}{p}} Z\left(\frac{\sigma}{\varepsilon^{\frac{p+1}{p}}}\right). \tag{9.25}$$

Notice that whereas the renormalization group has a clear meaning in the context of the Feigenbaum cascade, here the real meaning is a little different. Remember that the characteristic exponent β is defined for $\varepsilon \to 0$ as

$$\bar{l} \propto \varepsilon^{-\beta}, \tag{9.26}$$

and from Eq. (9.22) in the nonperturbed case ($\sigma = 0$), the function $L(\varepsilon, 0)$ determines how fast L increases as ε decreases and then, we must have

$$\beta = -\frac{p-1}{p}. \tag{9.27}$$

But, for $\sigma \neq 0$ it is difficult to determine the characteristic exponent because the function Z in Eq. (9.25) remains unknown.

Note also that for the particular value $p = 2$, Eq. (9.25) takes the specific form

$$L(\varepsilon, \sigma) = \frac{1}{\sqrt{\varepsilon}} Z \left(\frac{\sigma}{\varepsilon^{\frac{3}{2}}} \right) \tag{9.28}$$

so it is consistent with Eq. (9.13). It is interesting to emphasize that Eq. (9.13) was derived for $\varepsilon < 0$ whereas this is not the case for Eq. (9.28). In fact for $\varepsilon \gg 0$ and $\sigma > 0$ the scaling properties of the intermittency do not follow Eq. (9.28) and the deterministic component of the map is dominant over the small noise [30].

The cases of type II and type III present an additional difficulty because Eq. (9.22) for $\sigma = 0$ has the following form [33]:

$$L(\varepsilon, x_o, c) = 2^{-n} L(\lambda^n \varepsilon, \alpha^n x_o, \alpha^n c), \tag{9.29}$$

where x_0 is the starting point in the laminar region and c determines the end of the laminar region. The important feature of Eq. (9.29) is that the laminar region presents three characteristic exponents, that is $L(\lambda^n \varepsilon, \alpha^n x_o, \alpha^n c)$ remains constant under change of both $\alpha^n x_o$ and $\alpha^n c$. The Renormalization Group Theory cannot determine whether this is true or not. Note that in the previous sections, we devoted the classical noise theory applied to intermittencies, it only considers the noise on the local map.

9.4. Extended Intermittency Theory to a More General Reinjection

From early times it was observed that the classical intermittency theory was not enough to properly explain some experimental results [34] or numerical simulations [31]. The main problem was that originally the RPD was considered constant, and in general this is a very strong hypothesis. In absence of noise, it was observed that a generalized RPD given by the next power law

$$\phi(x) = b(x - \hat{x})^{\alpha} \tag{9.30}$$

is common to abroad class of 1D maps having type I, II, III and V intermittencies [9, 10, 18, 20]. Moreover, the RPD (9.30) has been observed in experiment with electronic circuits [29, 34]. Let us summarize the main results cumming from this new RPD that we need in the following sections. The RPD (9.30) has two free parameters $\hat{x}$ and α determined by the chaotic region dynamics. The parameter $\hat{x}$ is the lower bound of reinjection (LBR), that is $\hat{x} \leq x$, and it can strongly determines the dynamical behavior of the map [18, 21, 25, 26, 35]. The symmetry case, $x \leq \hat{x}$, can be taken into account by substituting in Eq. (9.30) $(x - \hat{x})$ by $(\hat{x} - x)$. As the following results can be extended in a natural way to the symmetry case, for sake of clarity, in this work we consider only the RPD given by Eq. (9.30). The exponent α is determined by the geometry of the map in a vicinity of a point with infinite or zero tangent, and b is determined by the normalization condition: $\int_{\hat{x}}^{c} b(x - \hat{x})^{\alpha} dx = 1$, where c determines the upper limit of the laminar zone. Note that the RPD (9.30) includes the constant reinjection approach as the particular case $\alpha = 0$. A simple methodology to evaluate the free parameters $\hat{x}$ and α from numerical or experimental data has been proposed by using the function $M(x)$ defined as

$$M(x) = \frac{\int_{x_s}^{x} \tau\, \phi(\tau)\, d\tau}{\int_{x_s}^{x} \phi(\tau)\, d\tau} \quad \text{if} \int_{x_s}^{x} \phi(\tau) d\tau \neq 0 \quad \text{or}$$

$$M(x) = 0 \quad \text{if} \int_{x_s}^{x} \phi(\tau) d\tau = 0, \tag{9.31}$$

where x_s is a suitable starting point having $x_s < x$ for all reinjected point x. The domain of definition of $M(x)$ is the laminar region. As $M(x)$ is defined by means of integrals, the effects coming from the statistical fluctuations are reduced, even for a relatively low number of data [21]. Moreover, note that for a given value of x, $M(x)$ is the average of reinjection points in the interval (x_s, x), hence, if we sort the reinjections according to the relation $x_j < x_{j+1}$, a simple estimation of the function $M(x)$ is obtained by

$$M(x_l) \approx \frac{\sum_{j=1}^{l} x_j}{l}, \tag{9.32}$$

which has been used in this work to evaluate the function $M(x)$ instead of using Eq. (9.31).

An important property of the function $M(x)$ is that for the RPD given by Eq. (9.30), it becomes linear as follows:

$$M(x) = \begin{cases} m(x - \hat{x}) + \hat{x} & \text{if } x > \hat{x}, \\ 0 & \text{otherwise,} \end{cases} \tag{9.33}$$

where the exponent α is determined by the slope m, hence the RPD is given by

$$\phi(x) = b(x - \hat{x})^{\alpha} \quad \text{where } \alpha = -\frac{1 - 2m}{1 - m}. \tag{9.34}$$

Note that by a simple least mean square fit, Eq. (9.33) provides the parameters $\hat{x}$ and α determining the RPD. Assuming $\alpha > -1$, that is $0 < m < 1$, the normalization condition converges, and for $m = 1/2$ the classical approach $\phi(x) = cte$ is recovered. On the other hand, the RPD (9.34) has two limit cases:

$$\phi_0(x) = \lim_{m \to 0} \phi(x) = \delta(x - \hat{x}), \quad \phi_1(x) = \lim_{m \to 1} \phi(x) = \delta(x - c). \tag{9.35}$$

Let us describe how the RPD affects the duration of the laminar phase described by the probability density of the laminar lengths $\psi(l)$. To determine $\psi(l)$ we follow the classical approach [7], and we approximate the dynamics of the local map by a differential equation. The appropriated equation to model the local behavior of the experiment described below is Eq. (9.3) with $p = 2$, so the approximate differential equation results

$$\frac{d|x|}{dl} = \varepsilon|x| + a|x|^2. \tag{9.36}$$

By solving the above equation, we obtain

$$l(|x|, c) = \int_{|x|}^{c} \frac{dz}{\varepsilon\, z + a\, z^2} = \frac{1}{\varepsilon}\left[\ln\left(\frac{c}{|x|}\right) - \ln\left(\frac{\varepsilon + a\, c}{\varepsilon + a\, |x|}\right)\right], \tag{9.37}$$

which is referred to a local behavior of the map in the neighborhood around of the unstable fixed point. Now, the probability density of the laminar lengths $\psi(l)$ is given by [7, 20]

$$\psi(l) = \phi(X(l,c)) \left| \frac{dX(l,c)}{dl} \right|, \tag{9.38}$$

where $X(l,c)$ is $l^{-1}(x,c)$, the inverse function of $l(x,c)$ in Eq. (9.37) with respect to its first argument, that is,

$$X(l,c) = \frac{\varepsilon}{(a+\varepsilon/c)\, e^{\varepsilon\, l} - a}. \tag{9.39}$$

Taking into account Eq. (9.36) and the above relations, we finally have

$$\psi(l) = \phi(X(l,c)) \left[aX(l,c)^2 + \varepsilon X(l,c) \right]. \tag{9.40}$$

An important difference with the classical approach is the factor $\phi(X(l,c))$ in Eq. (9.40), hence the $\psi(l)$ depends not only on the local laminar map described by $X(l,c)$ but also on the chaotic one, described by the factor $\phi(X(l,c))$. Classically, $\phi(X(l,c)) \approx$ constant was considered, so the first factor in Eq. (9.40) does not appear.

9.5. Effect of Noise on the RPD

In this section, we mainly follow the argument given in [24]. The figures used in this section come from the same publication.[1] As in the noiseless intermittency, in the noisy case, we also use the function $M(x)$ to investigate the noise RPD (NRPD) in systems with type II and type III intermittency.

To study type II intermittency, let us consider the following map (see Fig. 9.3):

$$x'_{n+1} = \begin{cases} F(x_n) + \sigma\xi_n, & x_n \le x_r, \\ (F(x_n) - 1)^{\gamma} + \sigma\xi_n, & x_n > x_r, \end{cases} \tag{9.41}$$

where ξ_n is a uniform distributed noise verifying $\langle \xi_m, \xi_n \rangle = \delta(m-n)$ and $\langle \xi_n \rangle = 0$. The noise strength is given by σ, and $F(x) = (1+\epsilon)x_n + (1-\epsilon)x_n^3$. We denote by x_r the root of the equation $F(x_r) = 1$.

[1]This article was published in *Commun. Nonlinear Sci. Numer. Simulat.*, **17**, del Rio, E., Sanjuán, M.A.F. & Sergio Elaskar, "Effect of noise on the reinjection probability density in intermittency", pp.3587–3596. Copyright Elsevier (2012).

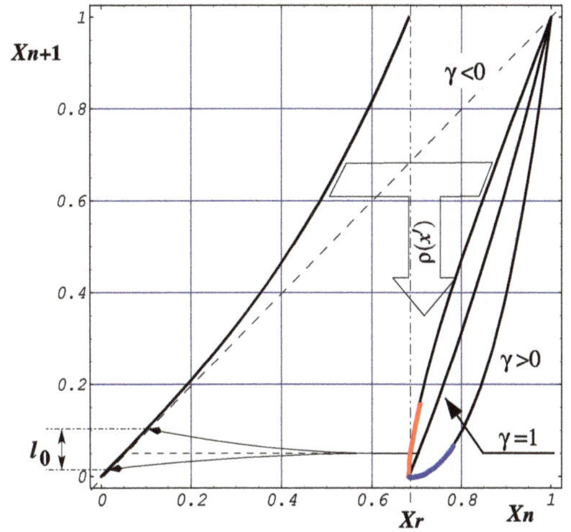

Fig. 9.3. Map (9.41) with type II intermittency. The arrows indicate the noiseless trajectories spreading into the laminar region into the interval l_0.

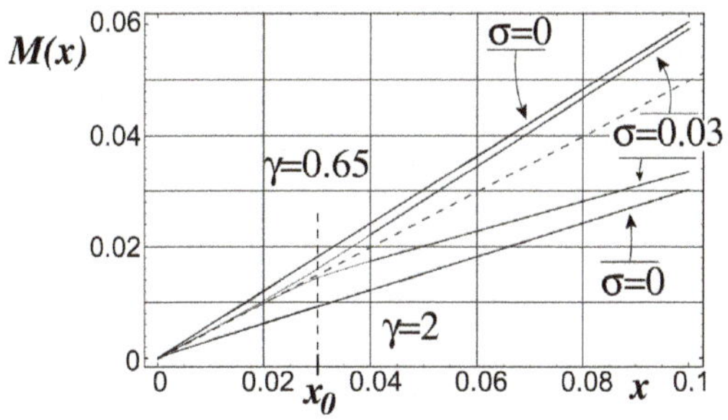

Fig. 9.4. Function $M(x)$ for the map (9.41) evaluating using Eq. (9.32). The dashed line with slope 1/2 shows the uniform reinjection case. The lines above the dashed one correspond to $\gamma = 0.65$ for two values of the noise strength as indicated. The same values of noise strength is used for the two lines below the dashed one, that correspond to $\gamma = 2$. For all the cases, $\varepsilon = 0.001$ and $c = 0.1$ are fixed. The solid lines show the corresponding least mean square fit, very close to the numerical simulations.

We evaluate the function $M(x)$, in the noisy and noiseless cases, and the results are shown in Fig. (9.4). Note that the function $M(x)$ looks like a smooth function even in the noisy case ($\sigma = 0.03$), due to the fact that the integrals in the definition of $M(x)$ filter the noise.

Whereas in the noiseless case, the function $M(x)$ can be approximated by a linear function [9], in the noisy case, the function $M(x)$ can be approximated by a piecewise linear function with different slopes, on each side of x_0 (see Fig. 9.4). The value x_0 depends on the noise strength σ. In the region $x < x_0$, the slope of $M(x)$ approaches 1/2, as we expect for the uniform reinjection. Note, however, that in the noiseless case there is no uniform reinjection, but there is a power law given by Eq. (9.34). On the side $x_0 < x$, the slope of $M(x)$ is different from the 1/2 taking a value close to the corresponding noiseless value. Note that for $\gamma = 0.65$ (the cases $\gamma > 1$ can be analyzed in a similar way) in the noisy case the slope is $m \approx 0.61$ which is very close to the noiseless case $m \approx 0.60$. Hence in that region the NRPD must be similar to the RPD function. This means that, by the analysis of the noisy data, we can predict the RPD function for the noiseless case. To do this, we proceed like in the noiseless case [9, 18], but considering only the data on the right side of x_0 in Fig. 9.4. That is, by least mean square analysis we can calculate the slope m in Eq. (9.33), which determines the reinjection function given by Eq. (9.34). This is shown in Fig. 9.5 for the same values of γ than in Fig. 9.4. That is, in Fig. 9.5(a), we have $\gamma < 1$ so that $\lim_{x\to 0} \phi(x) = 0$ and in Fig. 9.5(b), we have $\gamma > 1$ so that $\lim_{x\to 0} \phi(x) = \infty$ [9]. The result displayed in this figure shows that,

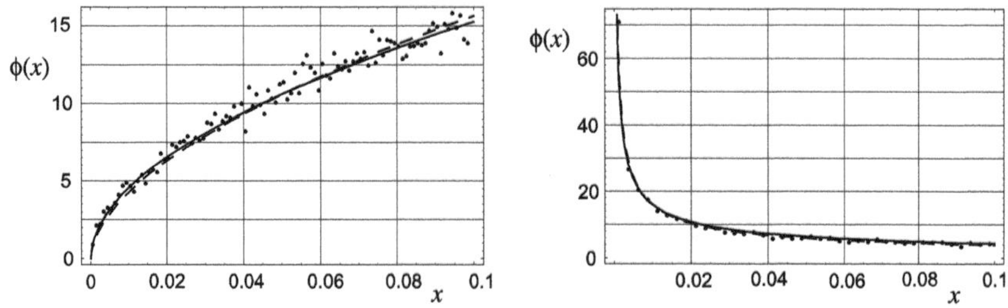

(a) In this case $\gamma < 1$. We fix $\gamma = 0.65$ (b) In this case $\gamma > 1$. We fix $\gamma = 2$

Fig. 9.5. RPD for the map (9.41). Dots correspond to numerical computation for $\sigma = 0$ whereas solid lines correspond to Eq. (9.34) using two values of m obtained by fitting data with $\sigma = 0$ (dashed line). For $\sigma \neq 0$ (solid line) we have used only the points $x > x_0$ in Fig. 9.4. The values of the parameter γ are the same that in Fig. 9.4.

in both cases, from the function $M(x)$ evaluated in the noisy cases it is possible to approximate the RPD in the noiseless case.

It is important to note that whereas the noise is applied to the whole map, the function $M(x)$ evidences that, on the right side of x_0 of Fig. 9.4, the reinjection function is robust against the noise but on the left side of x_0, the noise changes the RPD approaching it to the uniform reinjection, at least locally around $x = 0$.

In type III intermittency, there is a similar scenario. We use the map

$$x_{n+1} = F(x_n) = -(1+\varepsilon)\, x_n - a\, x_n^3 + d\, x_n^6\, \sin(x_n) + \sigma\xi_n, \quad (9.42)$$

where $\sigma\xi_n$ is the noise term.

As in the previous case, we start with a numerical evaluation of the function $M(x)$ by using the approximation given by Eq. (9.32). The results are shown in Fig. 9.6 for two values of the noise strength σ together with the noiseless case $\sigma = 0$. As in Fig. 9.4, for values close to the origin (on the left of the arrows), the function $M(x)$ approaches $M(x) \approx 0.5x$, but for points on the right-hand side of the arrows we have $M(x) \approx mx$ where the slope m is similar as to in the noiseless map. Note, however, that here the transition from the 0.5 slope to a slope close to the noiseless case takes place for bigger

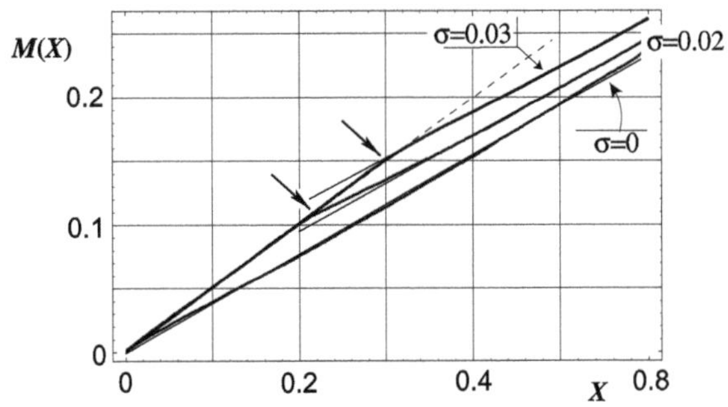

Fig. 9.6. Function $M(x)$ for the map (9.42) for three different values of the noise strength as indicated. Dots correspond to the numerical evaluation of $M(x)$ and solid lines show the corresponding least mean square fit that, according with their slopes ($m_{(\sigma=0)} = 0.364$, $m_{(\sigma=0.02)} = 0.375$ and $m_{(\sigma=0.03)} = 0.367$), can be considered as parallel lines. The dashed line slope is 1/2 and it corresponds to a uniform reinjection. The parameters are: $c = 0.8$, $a = 1.1$, $d = 1.35$, $\varepsilon = 10^{-4}$. Arrows show when $M(x)$ change the 1/2 slope.

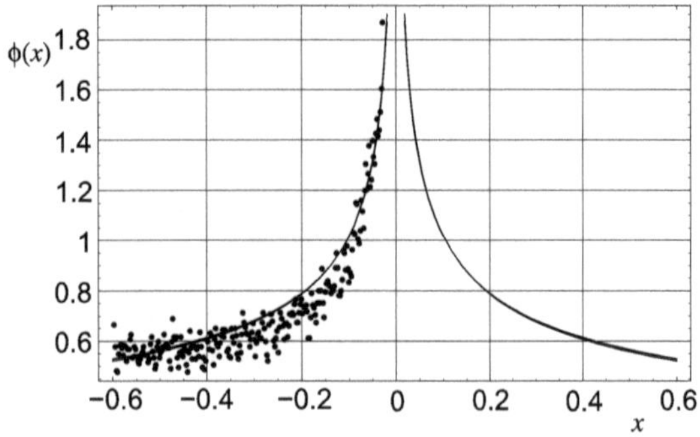

Fig. 9.7. RPD for the map (9.42). Dots correspond to numerical data. For clarity, we have omitted the points on the right-hand side. Solid lines correspond to Eq. (9.34) evaluated using the three values of m of Fig. 9.6, where the three of them are almost coincident. The parameters are: $c = 0.6$, $a = 1.1$, $d = 1.35$ $\varepsilon = 10^{-4}$.

values of x_0 than in the type II case. That is, the effect of noise on the function $M(x)$ is stronger than in the previous case.

As the three values of m estimated from data in Fig. 9.6 are similar, we can use each of them to describe the RPD of the noiseless case. Hence, as in the type II intermittency we can estimate the behavior of the noiseless map from noisy data, as shown in Fig. 9.7.

9.5.1. *NRPD in type II intermittency*

To get an analytical expression for the NRPD, denoted here by $\Phi(x)$, we analyze the effect of noise on the reinjection trajectories, as it is sketched in Fig. 9.3. A noiseless trajectory represented by a dashed line is perturbed by noise. As a consequence of this, the reinjection point must be ended inside of an interval represented by l_0. That is, the noiseless density $\phi'(x)$ should be transformed into a new density $\Phi(x)$ according to the convolution

$$\Phi(x) = \int \phi'(y) G(x - y, \sigma) dy, \tag{9.43}$$

where $G(x, \sigma)$ is the probability density of the noise term $\sigma\xi_n$ in Eq. (9.41).

Whereas the function $\phi'(x)$ is in general unknown, we have information on $\Phi(x)$ by means of the function $M(x)$ as was explained above. Furthermore, by the convolution properties, we expect $\Phi(x) \approx \phi'(x)$ in the region where the slope of $\phi'(x)$ is small. On the other hand, the slope of the noisy function $M(x)$ approaches the corresponding slope without noise, determining the density $\phi(x)$ in the noiseless case. These facts suggest that $\phi'(x) \approx \phi(x)$, where $\phi(x) = b|x|^\alpha$ is the RPD without noise, hence the parameters b and α can be taken as the same values that in the noiseless case. Observe that, although the noise is applied to the complete map, the associated power law to the effect of the nonlinearity observed in the noiseless map appears to be robust against noise. To verify this hypothesis, we introduce $\phi'(x) = b|x|^\alpha$ in Eq. (9.43). As noise source we used a random variable ξ in the interval $[-1, 1]$, hence its probability density G in Eq. (9.43) is given by

$$G(x, \sigma) = \frac{\Theta(x + \sigma) - \Theta(x - \sigma)}{2\sigma}, \tag{9.44}$$

where $\Theta(x)$ is the Heaviside step function. Finally, after integrating Eq. (9.43), we get the NRPD as

$$\Phi(x) = \frac{1}{c^{1+\alpha}} \frac{(|x| + \sigma)^{1+\alpha} - \mathrm{Sg}(|x| - \sigma)||x| - \sigma|^{1+\alpha}}{2\sigma}, \tag{9.45}$$

where $\mathrm{Sg}(x)$ is the sign function that extracts the sign from its argument. To compare Eq. (9.45) with the numerical simulations, we use the values of α determined by the values of m obtained from Fig. 9.4. The results are plotted in Fig. 9.8 for different values of σ, showing a good agreement between the numerical simulations and the analytical predictions. To plot the solid lines of Fig. 9.8, we can use either values of m obtained from noisy data of Fig. 9.4, or the corresponding value from the noiseless data because both values are approximately equal. However, it is not the case when $\gamma = 2$ and $\sigma = 0.03$. Here the noisy slope of $M(x)$ is $m_n \approx 0.268$, so the error relative to the noiseless value (0.30) is around 10%. This error makes a similar deviation between Eq. (9.45) and numerical data as shown in Fig. 9.9 by a dashed line. Nevertheless, we can compare Eq. (9.45)

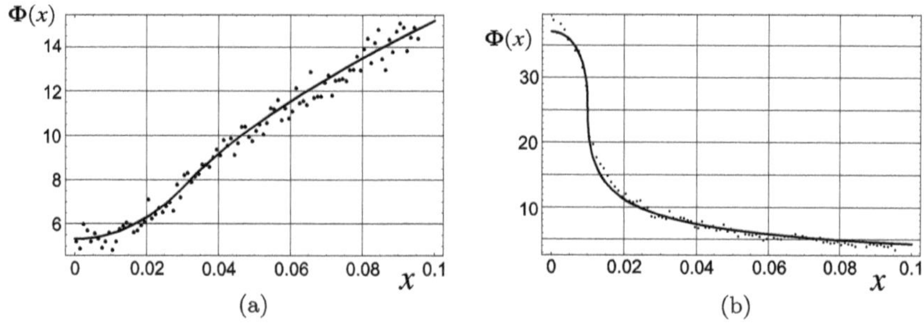

Fig. 9.8. NRPD for the map (9.41) using two set of parameter. Dots correspond to numerical data whereas the analytical approximation given by (9.45) is plotted as a solid line. (a) $c = 0.1$, $\gamma = 0.65$, $\sigma = 0.03$. (b) $c = 0.1$, $\gamma = 2$ and $\sigma = 0.01$.

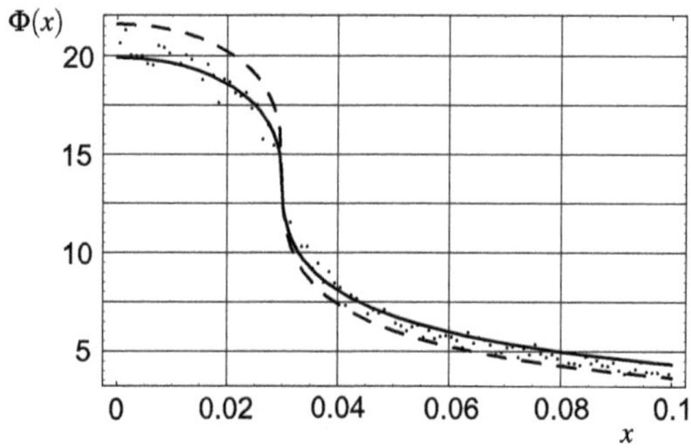

Fig. 9.9. NRPD for the map (9.41) for $c = 0.1$, $\gamma = 2$ and $\sigma = 0.03$. Dots correspond to the numerical values of NRPD. The dashed line correspond to the analytical approximation Eq. (9.45) using the value $m = 0.268$, obtained from the noisy case of Fig. 9.4. We also plot Eq. (9.45), solid line, but we used $m = 0.30$, as obtained from Eq. (9.47).

using a noiseless value of m with the numerical noisy data. This is shown in Fig. 9.9 by a solid line in good agreement with the numerical simulations. This proves that the hypothesis $\phi'(x) = b|x|^\alpha$ works well even when the high level of noise makes it significant differences between the slopes of $M(x)$ with and without noise.

Moreover, even in this case it is possible to estimate the function $M(x)$ for points $x > x_p$ far enough from the critical point $x = 0$, that is $x_p > x_0$ (see Fig. 9.4). To do this, we note that in the region $x > x_p$ the slope of $\phi'(x)$ is small, so according with Eq. (9.43) we expect

$\Phi(x) \approx \phi'(x)$, hence by using the definition of $M(x)$, (Eq. (9.31)) we have

$$M_n(x) \approx M(x) + \frac{\int_0^x \tau(\Phi(\tau) - \phi(\tau))\, d\tau}{\int_0^x \phi(\tau)\, d\tau} \quad \text{if } x > x_p, \tag{9.46}$$

where $M_n(x)$ and $M(x)$ are the noisy and noiseless functions, respectively, and we assume that the integral in the denominator does not depend on noise. With the approximation given by Eq. (9.32), we get for $x_l > x_p$

$$M(x_l) \approx M_n(x_l) - \frac{C}{l} \approx \frac{\sum_{j=1}^{l} x_j}{l} - \frac{C}{l}. \tag{9.47}$$

As $M(0) = 0$, we can choose the value C to approximate $M(0) \approx 0$. In our case, the value $C = 350$ gives the line $M(x_l) = 0.0004 + 0.298x_l$, with a small independent term. As we expected, this line has a slope very close to the noiseless case, $m \approx 0.301$. This means that even in this case with a very high value of the noise strength, the density $\phi(x) = b|x|^\alpha$ of the noiseless map is recovered.

9.5.2. *NRPD in type III intermittency*

The RPD, $\phi(x)$, for type III intermittency follows a power law depending on the neighborhood of the maximum and minimum of the map Eq. (9.42) [18]. In Fig. 9.10, the noiseless trajectory of a point starting near the maximum of the map it sketched by a dashed line. Because of the noise, this trajectory may spread over a region of some width, say l_0. Note that l_0 will be rescaled by a suitable factor K increasing the length up to $l_1 = Kl_0$ on the graph of the map. To get an analytical approximation for the NRPD, $\Phi(x)$, we consider the map (9.42) as a composition of the noiseless map

$$x'_n = -(1+\varepsilon)\, x_n - a\, x_n^3 + d\, x_n^6 \sin(x_n) \tag{9.48}$$

and the new map defined as

$$x_{n+1} = x'_n + \sigma\xi_n, \tag{9.49}$$

by just adding a noise to the variable. Now we focus on a point x_n lying close to the maximum of the map (9.48). As in the type II

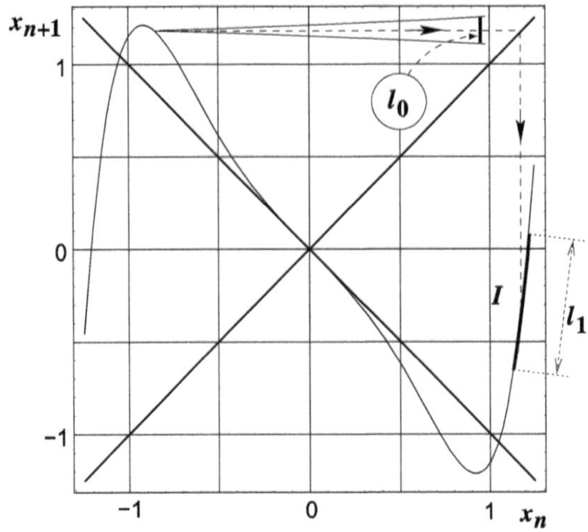

Fig. 9.10. Map of Eq. (9.42). Dashed line between the two solid lines indicate the effect of the map on a point near the maximum. These solid lines indicate the effect of the noisy map on the same point that will be mapped on the interval I on the graph of the map.

intermittency, if $\rho'(x)$ is the invariant density in that region due to the map (9.48), the effect on this density of the map (9.49) can be obtained according to

$$\rho(x) = \int \rho'(\tau)G(\tau - x, \sigma)d\tau, \tag{9.50}$$

where $\rho(x)$ is the invariant density observed in the interval l_0. Contrary to the type II case, here points on l_0 are not directly mapped on the laminar region, so to get the NRPD we must propagate the density $\rho(x)$ by the map (9.42). Then, we propagate $\rho(x)$ by the map (9.48) to reach for points in the interval I

$$\rho_I(x) = \frac{dF^{-1}(x)}{dx}\,\rho(F^{-1}(x)) \tag{9.51}$$

and later, we must propagate the density $\rho_I(x)$ by the map (9.49) to obtain the NRPD

$$\Phi(x) = \int \rho_I(y)G(y - x, \sigma)dy. \tag{9.52}$$

By using Eqs. (9.50) and (9.51), we get

$$\Phi(x) = \int\int \frac{dF^{-1}(y)}{dy}\,\rho'(\tau')G(\tau' - F^{-1}(y),\sigma)G(y-x,\sigma)\,dy\,d\tau' \tag{9.53}$$

and by rescaling the dummy variable $\tau' = F^{-1}(\tau)$ we can rewrite the last equation in a better form to be compared with the noiseless case as

$$\Phi(x) = \int\int \phi(\tau)G_F(\tau,y)G(y-x,\sigma)\,dyd\tau, \tag{9.54}$$

where we use the following definitions:

$$\phi(x) = \frac{dF^{-1}(x)}{dx}\,\rho'(F^{-1}(x)) \tag{9.55}$$

and

$$G_F(\tau,y) = \frac{dF^{-1}(y)}{dy}G(F^{-1}(\tau) - F^{-1}(y),\sigma). \tag{9.56}$$

Equation (9.55) shows the effect of the noiseless map on the density ρ'. By using the aforementioned hypotheses for type II intermittency, we can approximate ρ' by the same power law that in the noiseless map. Using the linear approximation $F(x) \approx F(x_{-1}) + K(x_{-1} - x)$, where $K = dF/dx$ corresponds to the mean slope of the curve I, the density ϕ in Eq. (9.55) can be approximated by a similar power law than ρ' [18], that is $\phi(x) \approx b|x|^{\alpha}$ even in the noisy case. By considering the previous linear approximation, we get for the function $G_F(\tau,y)$

$$G_F(\tau,y) \approx G(\tau - y, K\,\sigma). \tag{9.57}$$

In our case $K \approx 10$, so that the density $G(y-x,\sigma)$ in the integral (9.54) is very narrow as compared with the approximation for G_F given by Eq. (9.57), so that in the limit $G(y-x,\sigma) \approx \delta(y-x)$, we get for the NRPD

$$\Phi(x) = \int \phi(\tau)G(\tau - x, K\,\sigma)d\tau, \tag{9.58}$$

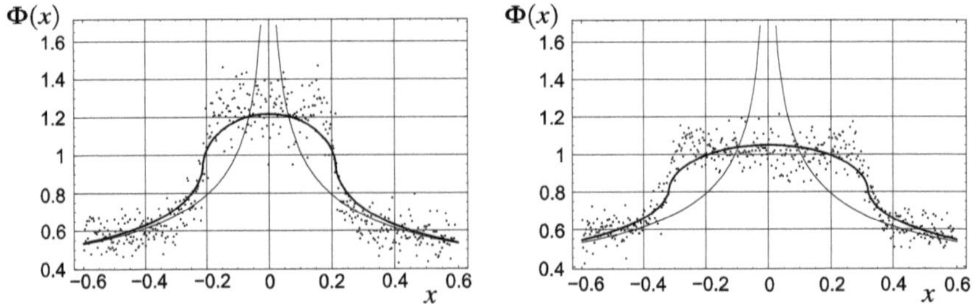

(a) $\sigma = 0.02$ (b) $\sigma = 0.03$

Fig. 9.11. NRPD for two values of the noise strength σ of Eq. (9.42). Dots refer to numerical simulation whereas solid lines are the plots of the analytical approximation to NRPD given by Eq. (9.59). Note that to illustrate the effect of the noise, the noiseless RPD is also plotted. As in Fig. 9.7, the parameters are: $c = 0.6$, $a = 1.1$, $d = 1.35$, $\varepsilon = 10^{-4}$.

a expression similar to Eq. (9.43) obtained to describe the noise in the type II intermittency. Hence, we can proceed in the same way that in the previous case,

$$\Phi(x) = \frac{1}{c^{1+\alpha}} \frac{(|x| + K\,\sigma)^{1+\alpha} - \mathrm{Sg}(|x| - K\,\sigma)||x| - K\,\sigma|^{1+\alpha}}{2K\,\sigma}. \quad (9.59)$$

Note that Eq. (9.59) is similar to Eq. (9.45) but here the factor K produces an amplification of the effect of the noise. This expression is plotted in Fig. 9.11 showing a good agreement with the numerical data.

9.6. Characteristic Relations

In absence of noise, the maps (9.41) and (9.42) exhibit the RPD given by Eq. (9.34). Recently, it has been shown that for this RPD, the critical exponent β of the characteristic relation is given by $\bar{l} \propto \epsilon^{\beta}$, where $\bar{l}$ is the average laminar length and it is determined by the slope of the function $M(x)$ as follows [9]:

$$\beta = \frac{\alpha + 2 - p}{p - 1} = \frac{1 + p(m-1)}{(p-1)(1-m)}. \quad (9.60)$$

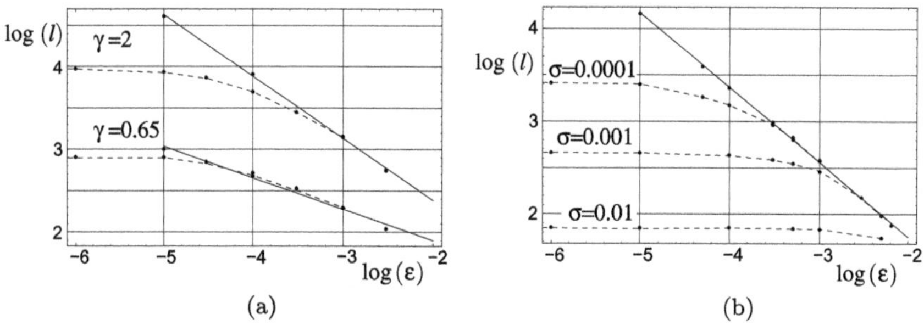

Fig. 9.12. Dots show numerical data and solid line represents the least squares straight fitting data corresponding to $\sigma = 0$. For two values of γ, the dashed lines join the noisy data ($\sigma = 10^{-4}$) reaching the corresponding saturation label for small values of ε. (a) Characteristic relations for the map (9.41) for $c = 0.1$ and two values of γ. (b) Characteristic relations for the map (9.42) for $c = 0.6$, $a = 1.1$ and $d = 1.35$.

A log–log plot of the characteristic relation $\bar{l} \propto \epsilon^{\beta}$ is shown by the solid lines in Figs. 9.12(a) and 9.12(b) when $\sigma = 0$. Note that for $\sigma \neq 0$, this characteristic relation and the numerical simulations are close for $\varepsilon > \sigma$. On the contrary, for low values of ε, ($\varepsilon < \sigma$), the time escape from the laminar region due to the dynamics of the map is different than the random escape. Hence, the escape speed is not practically influenced by a further decrease of ε because it has a small effect on the random escape. As a consequence, the average laminar length, $\bar{l}$, reaches a saturation value. Note that this saturation value depends on the value of the noise strength σ, that governs the random speed in the laminar region, but also it depends on the NRPD that is generated during the chaotic phase and governs the starting point in the local map. This is clear in Fig. 9.12(a), since the local map where the laminar phase takes place is the same for the two cases shown in the figure. The value of noise strength σ is also the same, and the only difference is the parameter γ that determines the RPD in the noiseless case. Hence, contrary to what has been suggested [17, 22], the average laminar length depends on the reinjection probability density.

9.7. Experimental Confirmation

In this section, we mainly follow the argument given in [29]. The figures used in this section come from the same publication.[2] An experimental confirmation of the noise theory explained in previous section has been presented recently [29, 36]. In any experiment, the data are affected by a noise intrinsic to measure procedure. However, our interest here in to study the change of the system behavior due to its internal noise. For this purpose, it was made an analog circuit described by a three-order differential equation. In Fig. 9.13, it is shown an experimental trajectory in the 3D phase space and the

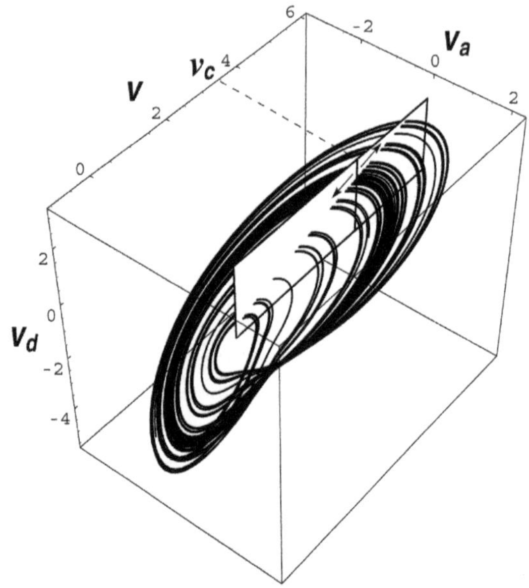

Fig. 9.13. Experimental time series for three experimental voltages, $v_a(t)$, $v(t)$ and $v_d(t)$. The rectangle on the plane $v_a = 0$ used to define the Poincaré is also shown. On this rectangle, it is displayed the vertical line that corresponds to $x_n = 0$ in dimensionless variables.

[2]This article was published in *Commun. Nonlinear Sci. Numer. Simulat.*, **64**, del Rio, E. & Sergio Elaskar, "Experimental evidence of power law reinjection in chaotic intermittency", pp.122–134. Copyright Elsevier (2018).

Poincaré plan used to define the Poincaré map that has type III intermittency.

Let us describe the procedure used to increase the noise in our dynamical system. In a first approximation, the internal noise strength in the circuit does not depend on the level of the signal. However, the circuit was designed looking for an interesting property for our purpose, that is, the signal level is proportional to an external constant voltage called v_0. This voltage does not change the dynamics of the system. Consequently, to decrease the signal-to-noise (S/N) ratio, we must reduce the voltage v_0 whereas the rest of the parameters remain without changing. With this procedure, we have used the natural noise in the system without an additional noise generator. Figure 9.14(a) shows two superimposed dimensionless data from experimental Poincaré maps corresponding to large S/N ($v_0 = 0.275V$) and low S/N ($v_0 = 0.05V$). Note that both maps differ approximately in a 10% due to internal component of the circuit [29].

Following the procedure of Sec. 9.4, we evaluate the function $M(x)$ for dimensionless experimental data corresponding to low and

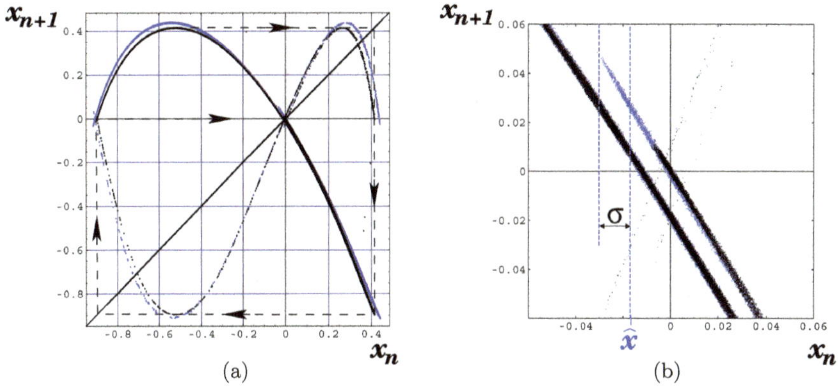

Fig. 9.14. Higher density of dots indicates the experimental Poincaré map. Dimensionless data that correspond to two experimental Poincaré maps with different S/N ratio; black for large and blue for low S/N ratio. (a) Continuous black lines show the bisector one and the straight line with slope means one passing thought the fixed point. Black arrow shows a trajectory going from the maximum to the laminar region. (b) Enlargement of subfigure (a) around the fixed points. In blue color, the noise effect with noise strength σ can be observed.

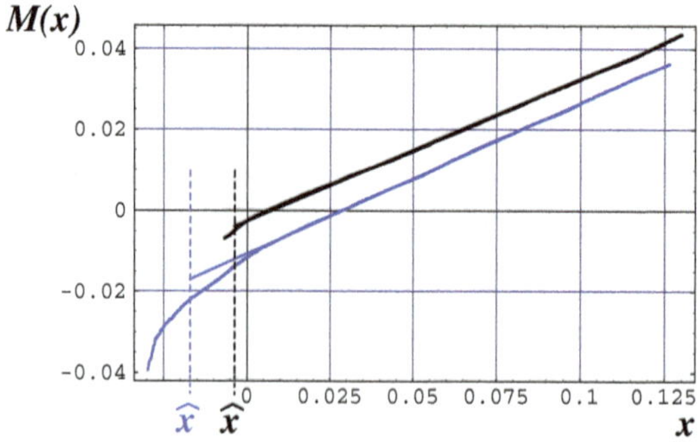

Fig. 9.15. Experimental functions $M(x)$ for the maps represented in Fig. 9.4. Black color indicates the case $v_0 = 0.275V$ whereas blue color refers to $v_0 = 0.05V$. Dots correspond to experimental data and solid lines show the corresponding least mean square fit. Vertical dashed line indicate the corresponding values for the LBRs.

large S/N ratio. The vertical dashed lines in Fig. 9.15 indicate the corresponding values $\hat{x}$ obtained by fitting the subset of experimental data with $x > 0$.

In view of Fig. 9.15, some experimental results can be commented. Firstly, we observe that for $x > 0$, that is for point far from $\hat{x}$, $M(x)$ is linear even in the case of large S/N relation (blue points in Fig. 9.15). Moreover, according with analytical predictions, the values of m that found are similar in both cases, as it is predicted in Sec. 9.5 (see Fig. 9.4 for $x > x_0$ and Fig. 9.6 for x on the right side of the arrows).

Secondly, in the scaled plot of Fig. 9.15, the knee sharp region around the corresponding values of $\hat{x}$ increases as the S/N ratio decreases, hence for the $v_0 = 0.05V$ (blue points in Fig. 9.15) we observe a largest number of reinjected points on the left of $\hat{x}$ than for $v_0 = 0.275V$ case. This phenomenon produces the blue tail observed in Fig. 9.14(b) on the left of $\hat{x}$. Both phenomena are in agreement theory explained in Sec. 9.5. Concerning with the large S/N ratio case ($v_0 = 0.05V$), the NRPD is given by Eq. (9.43), where $G(x - y, \sigma)$ is the probability density of the internal noise perturbing the experimental Poincaré map. This function is in general unknown, however we can approach by Eq. (9.44). Now, after evaluate the integral (9.43), the NRPD reaches

$$\Phi(x) = \begin{cases} 0, & x < \hat{x} - \sigma, \\ \dfrac{b}{2\sigma(\alpha+1)}(x - \hat{x} + \sigma)^{\alpha+1}, & \hat{x} - \sigma \le x \le \hat{x} + \sigma, \\ \dfrac{b\left[(x - \hat{x} + \sigma)^{\alpha+1} - (x - \hat{x} - \sigma)^{\alpha+1}\right]}{2\sigma(\alpha+1)}, & \hat{x} + \sigma < x. \end{cases} \tag{9.61}$$

Some comments concerning with Eq. (9.61) can be done. Whereas the expression (9.44) has been previously used to evaluate Eq. (9.43) in the context of 1D maps having type I, type II and type III intermittency [24–26], the result obtained here, Eq. (9.61), is different from the previous ones. This is because our Poincaré map is not a symmetric map as it happens in [24] for type III intermittency and, on the other hand, there are not applied the usual border conditions to keep data in the domain of definition of the map, as it happen in numerical investigations with 1D maps (see, for instance, [26]). In our experimental dynamical system, the domain of definition of the Poincaré map is not imposed and is determined by the system itself.

Equation (9.61) has three free parameters, α, $\hat{x}$ and σ, that we must estimate. It is important to emphasize that the parameters $\hat{x}$ and σ in Eq. (9.61) determine the noiseless RPD (see Eq. (9.30)). As we mention before, the slope of $M(x)$ is robust against the noise, hence we can still use Eq. (9.34) to obtain α. The case of $\hat{x}$ is different and there is not a general method to evaluate from the noisy $M(x)$. For instance, for a noisy symmetric reinjection there is not a simple way to estimate the noiseless $\hat{x}$ [24], whereas for the map without symmetry considered in [25], it is proposed an estimation for $\hat{x}$.

Concerning with our experiment, it is clear that $\hat{x}$ is located in the middle of the interval $[\hat{x}-\sigma, \hat{x}+\sigma]$, and according with Eq. (9.61) the NRPD takes particular values at the endpoints of the interval. That is, $\Phi(\hat{x} - \sigma) = 0$ and Φ presents a vertex point at $x = \hat{x} + \sigma$, hence we can estimate the value of σ and $\hat{x}$ from the experimental NRPD of Fig. 9.16 (dots). In this experiment, a good choice is $\hat{x} = \hat{x}_{0.05}$ and $\sigma \approx 0.013$ as it is indicated by vertical lines in Fig. 9.16.

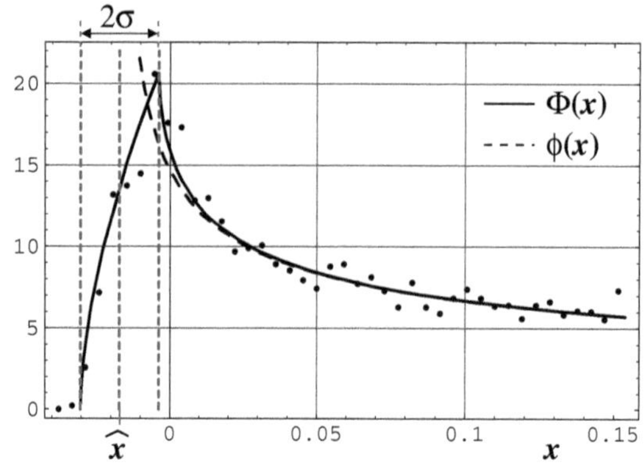

Fig. 9.16. NRPD for the map of Fig. 9.11. Dots correspond to experimental data. Solid lines correspond to the analytical expression (9.61) evaluated using the values of m and $\hat{x}$ of Fig. 9.15. Dashed curve indicates the ideal noiseless RPD.

Equation (9.61) is represented for this set of parameter values in good agreement with its experimental estimation. It is interesting to note that in this experiment the whole function $M(x)$ is robust against the noise, not just its slope. To show clearly the noise effect on the RPD, in Fig. 9.16 the RPD of Eq. (9.30) is superimposed in dashes line, in other words, the RPD that should be observed in an ideal noiseless experiment. In this case, the straight line $x = \hat{x}$ should be a vertical asymptote for $\phi(x)$.

9.8. Conclusions and Future Researches

In this chapter, we have reviewed the classical theory of noise applied to chaotic intermittency. Classically, additive noise applied to 1D maps was considered. Despite that the noise perturbs the whole map, only the noise effect on the local map was considered. We show the main results given by the Fokker–Planck and renormalization group analysis.

A decade ago, a new intermittency theory was proposed. The so-called pathological case, that is, maps that significantly deviated from the classical analytical previsions, were included in the new framework in a natural way. Also the classical results have been included as a particular case as well. Concerning to the noise effect, the new intermittency theory provides a new scenario; now the noise in the chaotic region may affect the statistic in the laminar region. This is because the RPD is not uniform and can be affected by the noise. In Sec. 9.5, based on a paper with Prof. Miguel A.F. Sanjuán [24], we review the new methodology to evaluate the noise effect on chaotic intermittency. It has been found good agreement between the new proposed noise theory and the numerical simulations in many 1D maps with type I, type II and type III intermittencies. We found that in the indirect reinjection, that is, in the cases that are necessary several map iterations to go from the extreme point of the map to the laminar region, the convolution (9.43) must be evaluated in each iteration needed to reach the laminar region, as it is shown in Sec. 9.5.2. Recently, experimental evidences on the new intermittency theory have been found in continuous systems. The theory was applied to a suitable Poincaré section in the phase space of a continuous experimental system. The noise effect also has been investigated in experimental systems and it was found good agreement with theory predictions explained in Sec. 9.5.2, but until now, some open questions remain for future investigations. For instance, in the numerical map it is necessary to propagate the effect of the noise by Eq. (9.43) for each iteration, whereas in the experimental Poincaré map, the experimental data is well fitted by a single propagation of the noise from the maximum of the map. This difference can be due to the fact that the natural noise disturbing the circuit is continuously applied in time, however this hypothesis has not worked in general [36, 37]. Hence an open question for future investigations is to determine in which case the theory of Sec. 9.5.2 works with just one convolution, or when it is necessary one convolution per map iteration from the maximum to the fixed points as it happened in numerical simulations for 1D maps [37].

Acknowledgments

This research is supported the Spanish Ministry of Science and Innovation (MICINN) under Project No. ESP2013-41078-R (EdR), and by National University of Córdoba under a SECyT Project (SE) and by Universidad Politécnica de Madrid.

References

[1] P. Manneville and Y. Pomeau, Different ways to turbulence in dissipative dynamical systems, *Physica D* **1**(1) (1980) 219–226.
[2] M. Dubois, M. Rubio and P. Berge, Experimental evidence of intermittencies associated with a subharmonic bifurcation, *Phys. Rev. Lett.* **51** (1983) 1446–1449.
[3] S. G. Stavrinides, A. N. Miliou, Th. Laopoulos and A. N. Anagnostopoulos, The intermittercy route to chaos of an electronic digital oscillator, *Int. J. Bifurc. Chaos* **18** (2008) 1561–1566.
[4] G. Krause, S. Elaskar and E. del Rio, Type I intermittency with discontinuous RPD in the truncation model of the derivative nonlinear Schrödinger equation, *Nonlinear Dynam.* **77** (2014) 455–466.
[5] J. Zebrowski and R. Baranowski, Type I intermittency in nonstationary systems models and human heart rate variability, *Physica A* **336** (2004) 74–83.
[6] A. Chian, *Complex Systems Approach to Economic Dynamics*, Lecture Notes in Economics and Mathematical Systems, Vol. 592 (Springer, Berlin, 2007), pp. 39–50.
[7] H. Schuster and W. Just, *Deterministic Chaos: An Introduction* (Wiley-VCH Verlag GmbH & Co. KGaA, Weinheim, 2005).
[8] J. E. Hirsch, M. Nauenberg and D. J. Scalapino, Intermittency in the presence of noise: A renormalization group formulation, *Phys. Lett. A* **87** (1982) 391–393.
[9] E. del Rio and S. Elaskar, New characteristic relations in type II intermittency, *Int. J. Bifurc. Chaos* **20** (2010) 1185–1191.
[10] E. del Rio, S. Elaskar and J. M. Donoso, Laminar length and characteristic relation in type I intermittency, *Commun. Nonlinear Sci. Numer. Simulat.* **19** (2014) 967–976.
[11] P. Manneville, Intermittency, self-similarity and $1/f$ spectrum in dissipative dynamical systems, *J. Phys.* **41** (1980) 1235–1243.
[12] A. Pikovsky, G. Osipov, M. Rosenblum, M. Zaks and J. Kurths, Attractor-repeller collision and eyelet intermittency at the transition to phase synchronization, *Phys. Rev. Lett.* **79** (1997) 47–50.
[13] C. M. Kim, O. J. Kwon, E.-K. Lee and H. Lee, New characteristic relations in type I intermittency, *Phys. Rev. Lett.* **73** (1994) 525–528.
[14] C. M. Kim, G. S. Yim, J. W. Ryu and Y. J. Park, Characteristic relations of type III intermittency in an electronic circuit, *Phys. Rev. Lett.* **80** (1998) 5317–5320.

[15] J. H. Cho, M. S. Ko, Y. J. Park and C. M. Kim, Experimental observation of the characteristic relations of type I intermittency in the presence of noise, *Phys. Rev. E* **65** (2002) 036222.
[16] W. H. Kye, S. Rim, C. M. Kim, J. H. Lee, J. W. Ryu, B. S. Yeom and Y. J. Park, *Phys. Rev. E* **68** (2003) 036203.
[17] A. A. Koronovskii and A. E. Hramov, Type II intermittency characteristics in the presence of noise, *Eur. Phys. J. B* **62** (2008) 447–452.
[18] S. Elaskar, E. del Rio and J. M. Donoso, Reinjection probability density in type III intermittency, *Physica A* **390** (2011) 2759–2768.
[19] E. del Rio, and S. Elaskar, On the theory of intermittency in 1D map, *Int. J. Bifurc. Chaos* **26** (2016) 1650228.
[20] S. Elaskar and E. del Rio, *New Advances on Chaotic Intermittency and its Applications* (Springer, 2017).
[21] E. del Rio, S. Elaskar and V. Makarov, Theory of intermittency applied to classical pathological cases, *Chaos* **23** (2013) 033112.
[22] J. E. Hirsch, B. A. Huberman and D. J. Scalapino, Theory of intermittency, *Phys. Rev. A* **25** (1982) 519–532.
[23] A. S. Pikovsky, Attractor-repeller collision and eyelet intermittency at the transition to phase synchronization, *J. Phys A* **16** (1983) L109.
[24] E. del Rio, M. A. F. Sanjuán and S. Elaskar, Effect of noise on the reinjection probability density in intermittency, *Commun. Nonlinear Sci. Numer. Simulat.* **17** (2012) 3587–3596.
[25] G. Krause, S. Elaskar and E. del Rio, Noise effect on statistical properties of type I intermittency, *Physica A* **402** (2013) 318–329.
[26] S. Elaskar, E. del Rio, G. Krause and A. Costa, Effect of the lower boundary of reinjection and noise in Type II intermittency, *Nonlinear Dynam.* **79**, (2015) 1411–1424.
[27] S. Elaskar, E. del Rio and A. Costa, Reinjection probability density for type III intermittency with noise and lower boundary of reinjection, *J. Comput. Nonlinear Dynam.* **12** (2017) 031020–031020.
[28] E. Ott, *Caos in Dynamical Systems* (Cambridge University Press, Cambridge, 2002).
[29] E. del Rio and S. Elaskar, Experimental evidence of power law reinjection in chaotic intermittency, *Commun. Nonlinear Sci. Numer. Simulat.* **64** (2018) 122–134.
[30] W. Kye and C. H. Kim, Characteristic relations of type I intermittency in the presence of noise, *Phys. Rev. E* **62** (2000) 6304–6307.
[31] O. Kwon, C. Kim, E. Lee and H. Lee, Effects of reinjection on the scaling property of intermittency, *Phys. Rev. E* **53** (1996) 1253–1256.
[32] W. Kye, S. Rim and C. H. Kim, Experimental observation of characteristic relations of type III intermittency in the presence of noise in a simple electronic circuit, *Phys. Rev. E* **68** (2003) 036203.

[33] H. Kodama, S. Sato and K. Honda, Reconsideration of the renormalization-group theory on intermittent chaos, *Phys. Lett. A* **157** (1991) 354–356.
[34] E. del Rio, M. G. Velarde and A. Rodríguez-Lozano, Long time data series and difficulties with the characterization of chaotic attractors: A case with intermitency III, *Chaos, Solitons Fractals* **4** (1994) 2169.
[35] C. M. Kim, G. S. Yim, Y. S. Kim, J. M. Kim and H. W. Lee, Experimental evidence of characteristic relations of type I intermittency in an electronic circuit, *Phys. Rev. E* **56** (1997) 2573–2577.
[36] E. del Rio and S. Elaskar, An experimental investigation on noisy intermittency, in *Proc. of the Internat. Conf. Theoretical Approaches in Non-Linear Dynamical Systems*, Lodz, Poland, December 2–5 (2019), pp. 97–109.
[37] E. del Rio and S. Elaskar, Experimental results versus computer simulations of noisy Poincaré maps in an intermittency scenario, *Regular Chaotic Dynam.* **25** (2020) 281–294.

https://doi.org/10.1142/9789811221903_0010

Chapter 10

Fractal Structures in a Binary Schwarzschild Black Hole System

E. E. de Souza Filho, A. C. Mathias and R. L. Viana
Department of Physics, Federal University of Paraná, 81530-990, Curitiba, Paraná, Brazil

In open Hamiltonian systems, the dynamics underlying chaotic scattering presents a number of fractal structures. One example is the deflection of light rays produced by a pair of supermassive (Schwarzschild) black holes. A light ray approaching this system can diverge to infinity, fall down into one of the black holes or can orbit around the black hole pair as periodic orbits. Using a two-dimensional area-preserving map proposed by Moura and Letelier, we investigated the escape basins and their fractal boundaries, using the corresponding basin and basin boundary entropies. We also varied the distance between the black holes, showing how it impacts in the complexity of the basin structure.

10.1. Introduction

Fractal structures appear often in chaotic systems, both dissipative and conservative ones [1]. In General Relativity, the geodesic motion in a curved space is a dynamical system of physical interest [2]. The presence of fractal structures in open chaotic systems is caused by the existence of an invariant nonattracting chaotic manifold, the so-called strange saddle [3], which influences the chaotic scattering of particles for example [4].

The chaoticity of geodesic motion is responsible for fractal distribution of the light ray scattering by a pair of black holes. This dynamics is described by the null geodesics equations [5]. One example is the Majumdar–Papapetrou binary black hole system [6, 7]. The investigation by Sanjuán and coworkers of fractal exit basins in a Majumdar–Papapetrou binary black hole shows that the escape basin boundaries are not just fractal but also display the stronger Wada property: any boundary point belongs to the boundary of at least two other basins [5]. Moura and Letelier have proposed a two-dimensional map to describe the scattering of light rays by a system of two static Schwarzschild black holes [8].

We investigate the fractality of the escape basin by using a scattering map [8]. The escape basins are known to play an important role in astrophysical investigations, since they are actually the so-called shadows of a black hole [9]. Thus, we quantitatively investigate the fractal nature of the structure of escape basin by computing the basin entropy and the basin boundary entropy [10, 11].

This chapter is organized as follows. In Sec. 10.2, we introduce the basic formulas for the scattering of a light ray by a spherically symmetric black hole. In Sec. 10.3, we present the two-dimensional map that describes the light ray deflection due to the binary black hole system discussing some of its dynamic properties. In Sec. 10.4, we present some numerical results of the escape basin as we vary the distance between black holes. In Sec. 10.5, we characterize the fractality of the escape basins using the basin entropy and the basin boundary entropy. Finally, in last section we report our conclusions.

10.2. Basic Equations

Assuming the spacetime metric $g_{\mu\nu}$ with signature $(-,+,+,+)$ and using Einstein's summation convention for repeated indexes, we define $c = G = 1$. In a curved spacetime, the light rays follow geodesics given by

$$g_{\mu\nu} dx^{\mu} dx^{\nu} = 0, \tag{10.1}$$

where we consider a metric for a symmetrically spherical and static spacetime with length element

$$ds^2 = A(r)dt^2 - B(r)dr^2 - C(r)\left(d\theta^2 + \sin^2\theta d\phi^2\right), \tag{10.2}$$

where $A(r)$, $B(r)$ and $C(r)$ define the metric produced by the black hole.

For a photon approaching the black hole from infinity with impact parameter b, the substitution of (10.2) into (10.1) gives the following general equation for the geodesics

$$\frac{A(r)B(r)}{[C(r)]^2}\left(\frac{dr}{d\phi}\right)^2 + \frac{1}{[U(r)]^2} = \frac{1}{b^2}, \tag{10.3}$$

where $[U(r)]^{-2} = A(r)/C(r)$. After approaching the black hole with a minimum distance r_0 the photon is deflected and emerges out in the other direction. From (10.3) this distance is called the critical impact parameter

$$b_c = \sqrt{\frac{C(r_0)}{A(r_0)}} \equiv \sqrt{\frac{C_0}{A_0}}. \tag{10.4}$$

On substituting (10.4) back into (10.3) it turns out that the angle of deflection is $\alpha = I(r_0) - \pi$, where

$$I(r_0) = \int_{r_0}^{\infty} dr \sqrt{\frac{B(r)}{C(r)}} \left(\frac{A_0}{A(r)}\frac{C(r)}{C_0}\right)^{-1/2}. \tag{10.5}$$

In the limit of strong gravitational fields, we can expand (10.5) so as to obtain

$$\alpha(r_0) = -a\ln\left(\frac{r_0}{r_m}\right) + c + o(r_0 - r_m) = -\bar{a}\ln\left(\frac{b}{b_m} - 1\right)$$
$$+\,\bar{c} + o(b - b_m), \tag{10.6}$$

where a, $\bar{a}$, c and $\bar{c}$ depend on the functions A, B and C, evaluated at the photosphere radius r_m. Similarly to (10.4) we have $b_m = \sqrt{C(r_m)/A(r_m)}$.

We shall consider Schwarzschild black holes, such that these functions are given, in a suitable system of units [12]

$$A(r) = 1 - \frac{2M}{r}, \tag{10.7}$$

$$B(r) = \left(1 - \frac{2M}{r}\right)^{-1} = \frac{1}{A(r)}, \tag{10.8}$$

$$C(r) = r^2. \tag{10.9}$$

The Schwarzschild metric has an event horizon, given by the radii where it diverges, corresponding to $r = 2M$, without loss of generality we will rewrite

$$R = \frac{r}{2M}. \tag{10.10}$$

Using (10.7)–(10.9) and computing the extreme of the effective potential results that the radius of the photosphere is $R_m = 3/2$.

From these results, Bozza [13] shows that the coefficients of the expansion for the deflection angle (10.6) are given by

$$\bar{a} = 1, \tag{10.11}$$

$$\bar{c} = -\pi + 0.9496 + \ln 6 \approx -0.4002, \tag{10.12}$$

$$b_m = \frac{3\sqrt{3}}{2} \approx 2.5981. \tag{10.13}$$

The scattering of a light ray by a single black hole can now be understood in terms of the possible values of the corresponding impact parameter. A light ray comes from infinity and approaches the black hole with impact parameter b and whose direction makes an angle ϕ with a reference axis. If $b < b_m$, the light ray falls into the black hole and disappears. On the other hand, for this light ray not to escape back to infinity, it is necessary that its impact parameter b be such that the scattering angle α is at least π. Thus we impose $b > b_{\text{esc}}$, where $\alpha(b_{\text{esc}}) = \pi$. Using (10.6), this means that

$$b_{\text{esc}} = b_m \left[\exp\left(\frac{\bar{c} - \pi}{\bar{a}} \right) + 1 \right] \approx 2.67332. \tag{10.14}$$

As a result, for a light ray deflected by a single black hole not to escape to infinity or to collide with a black hole, the impact parameter must belong to the narrow interval $b_m < b < b_{\text{esc}}$.

10.3. The Scattering Map

When working with a system of two black holes in orbit of each other, the map has no exact solution of field equations (unlike the case of a single black hole treated in the previous section). But the nonlinear interaction between their gravitational fields can be neglected if the distance D between the black holes is much higher than their Schwarzschild radius $R = 1$.

In this case, we consider the deflection of light from each black hole in a separate way using the expressions previously found for the Schwarzschild metric. In other words, in the light scattering by a given black hole, the effect of the other black hole is neglected. Similar approximations for the Schwarzschild metric are discussed by de Moura [8]. In particular, they may not hold if we consider the scattering of massive test particles.

The axial symmetry axis is the line connecting the two black holes. Assuming that the light rays have zero angular momentum in this direction, as a result, the light rays are constrained to move in the plane containing the two black holes. Figure 10.1 shows the basic geometry involved in the light scattering by the system. Coming from infinity, a light ray approaches the first black hole with impact parameter b, the direction makes an angle ϕ with the axial symmetry line. So that this light ray does not escape back to infinity it is necessary that $b < b_{\text{esc}}$, where b_{esc} is given by (10.14). As we do not want the light ray to fall to the first black hole as soon as they meet, we define the impact parameter as $b > b_m$. As the light ray is not deflected to infinity by the first black hole, make it goes to the other black hole and is again deflected. Again, if not deflected to infinity it returns to the vicinity of the first black hole, *ad infinitum*.

For convenience, we define discrete variables (b_n, ϕ_n), the impact parameter and escape angle, in reference to the axis symmetry line

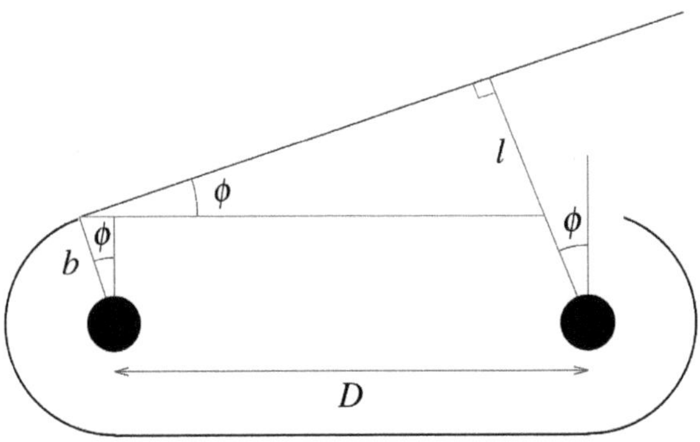

Fig. 10.1. Schematic figure for the trajectory of a light ray in the two black hole system.

in the neighborhood of the nth scattering. Odd (even) values of n correspond to the second (first) black hole. Using such definition, the differential equations for light scattering reduce to a discrete map,

$$b_{n+1} = b_n + D\phi_n, \tag{10.15}$$

$$\phi_{n+1} = \pi + \phi_n - \alpha(b_{n+1}) \pmod{2\pi}, \tag{10.16}$$

which is called scattering map by Moura and Letelier [8]. Some sign conventions are essential: positive values of b imply that the light ray goes from black hole 1 to 2 (from "right" to "left" in Fig. 10.1), and negative values otherwise; whereas positive values of ϕ correspond to counterclockwise rotations.

The Jacobian matrix of this transformation is given by

$$\mathbf{J} = \begin{pmatrix} 1 & D\cos\phi_n \\ -\alpha'(b_{n+1}) & 1 - D\alpha'(b_{n+1})\cos\phi_n \end{pmatrix}, \tag{10.17}$$

whose determinant is equal to the unity, hence the scattering map (10.15)–(10.16) is a twist and area-preserving mapping, corresponding to the following continuous-time Hamiltonian:

$$H(b, \phi, n) = \pi b - \int^b db'\alpha(b') + D\delta_1(n)\cos\phi, \tag{10.18}$$

where

$$\delta_1(n) = \sum_{m=-\infty}^{\infty} \delta(n-m) = 1 + 2\sum_{q=1}^{\infty} \cos(2\pi q n), \tag{10.19}$$

is a periodic delta function("Dirac comb").

Fixed points of the scattering map are $\phi^*_{1,2} = 0, \pi$ and $b^*_{1,2} = b_{\mathrm{esc}}$. The eigenvalues of the Jacobian matrix (10.17) at these points are

$$\lambda_{1,2} = \frac{\zeta}{2} \pm i\sqrt{1 - \left(\frac{\zeta}{2}\right)^2}, \tag{10.20}$$

where

$$\zeta = 2 - D\alpha'(b_{n+1})\cos\phi_n \tag{10.21}$$

is the trace of Jacobian matrix.

These fixed points are stable provided that $|\zeta| < 2$. It turns out that the point $(b_{\mathrm{esc}}, 0)$ is stable if $0 < D\alpha'(b_{\mathrm{esc}}) < 4$, whereas the other fixed point $(b_{\rm esc}, \pi)$ is stable if $-4 < D\alpha'(b_{\rm esc}) < 0$. In both cases, the light ray trajectory is such that $b_{n+1} = -b_n$. Considering $D > 0$ and $\alpha'(b_{\rm esc})) < 0$ the fixed point at $\phi = 0$ is always unstable (a hyperbolic saddle). If, further, we have that $|\alpha'(b_{\rm esc})| < (4/D)$, the other fixed point at $\phi = \pi$ is an elliptic center.

The Hamiltonian (10.18) is generally nonintegrable for $D \neq 0$. However, if the light deflection by one black hole has to be independent on the existence of the another black hole, then we must assume that D is typically a large number. In such an instance, both fixed points are unstable and we expect a sizeable area-filling chaotic orbit in the phase space (b, ϕ).

10.4. Escape Basins

The system proposed of two black holes is an example of open dynamical system, which is a system where trajectories (light rays) eventually escape from a given phase space region. If there are different ways by which trajectories can escape, then it is an applicable problem to identify the sets of initial conditions that causes trajectories to escape through a given exit. This array is called the escape basin, which corresponds to that exit [4]. In the case of two or more exits, we can identify the boundary that divides those escape basins, termed escape basin boundary. It is long known that conservative dynamical systems with chaotic dynamics have fractal escape basins and fractal escape basin boundaries [1].

With the desire to obtain the plot of the escape basins corresponding to the scattering map (10.15)–(10.16), we designated a set of initial conditions (b_0, ϕ_0) and iterated them to encounter to which basin they belong. We isolate the phase space region $\Omega = \{0 \leq \phi_0 \leq 2\pi, b_m < b_0 < b_{\rm esc}\}$ in a extensive number of points and iterate the map (10.15)–(10.16) for each of these initial conditions, saving the final outcome for each point. Succeeding a number of map iterations, depending on its initial conditions, a light ray may fall into one black hole (**A**), into the other black hole (**B**), or escapes in the direction of infinity (**C**). All outcomes can be treated as exits since we stop iterating the map once a light ray falls into a black hole. Accordingly, we denote the corresponding escape basins to be $\mathcal{B}(\mathbf{A})$, $\mathcal{B}(\mathbf{B})$ and $\mathcal{B}(\mathbf{C})$.

To do this analysis, we use a range of values for D from $D = 5$ until $D = 20$, which is a value sufficient to ensure that each black hole deflects light rays in an independent fashion. An orbit falls into a black hole whenever $b_n < b_m$ for a given escape time $n = \bar{n} < 10^4$. If $\bar{n}$ is even (odd) we know that the light ray falls into black hole **A** (**B**) and the corresponding initial condition is painted red (blue). If the orbit goes to infinity ($b_n > b_{\rm esc}$) for a given $n = \bar{n} < 10^4$ the corresponding initial condition is painted green. The red, blue and green regions are thus numerical approximation for the exit basins $\mathcal{B}(\mathbf{A})$, $\mathcal{B}(\mathbf{B})$ and $\mathcal{B}(\mathbf{C})$, respectively. There is a measured set of unstable periodic orbits which never escape and, since $D < \infty$, we cannot rule out orbits within very tiny periodic islands which do not escape at all, but whose effect in the exit basins would be negligible.

Since the labeling of black holes **A** and **B** is immaterial, their escape basins would be symmetric, i.e., they must have the same size. On the other hand, it is expected that the dominant basin is that of infinity (**C**). But the analyses require further magnifications as sequences of exits cannot be observed. This characterization is also possible by defining a function $g(b)$ such that $g(b) = 1$, if the orbit falls into black hole **A**, $g(b) = -1$, if it falls into **B**, and $g(b) = 0$ if the orbit escapes to infinity [8].

In order to investigate the escape basin boundary, we analyze two regions in the vicinity of the points with $\phi_0 = 0$ and $\phi_0 = \pi$. For both regions, we divide the impact parameter interval $b_m < b < b_{\rm esc}$ into 10^6 points. The corresponding escape basins are plotted in Figs. 10.2(a) and 10.2(d), respectively, as a horizontal bar with green, red and blue stripes. We also plotted the corresponding function $g(b)$ below the bars. In Figs. 10.2(b) and 10.2(e), we show magnifications of two intervals of Figs. 10.2(a) and 10.2(d), respectively, and Figs. 10.2(c) and 10.2(f) are further magnifications. These zooms clearly show that there are regions for which there are pieces of the three escape basins in all scales. This self-similarity is a signature of the fractality of the basins as well as of its basin boundary.

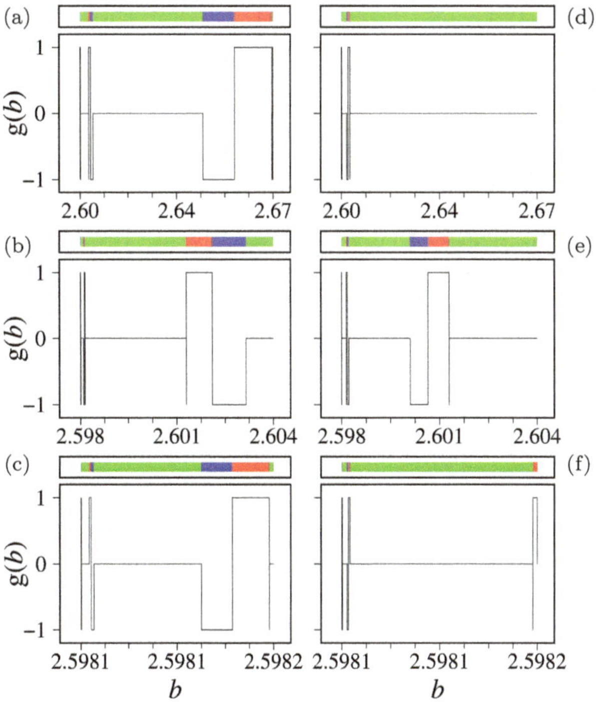

Fig. 10.2. For $D = 12,5$, the horizontal bars represent the escape basins as a function of the impact parameter b for $\phi_0 = 0$ in (a), with magnifications in (b) and (c); and $\phi_0 = \pi$ in (d), with magnifications in (e) and (f). We plot, below the horizontal bars, the corresponding values of the function $g(b)$ (see text for details).

A cursory inspection of Fig. 10.2 suggests that the escape basin boundary is self-similar and has a fractal structure that depends of b. Actually, the escape basin boundary cannot be completely fractal, since there are smooth parts of it corresponding to regions in which the function $g(b)$ has the same value for different ranges of b.

Since some regions of different escape basins are intertwined in arbitrarily fine scales, it is extremely difficult, if not impossible, to predict the final outcome of a light ray, given its initial condition being always known up to a given uncertainty. One of the observable consequences of the existence of fractal structures in phase space is final-state sensitivity, i.e., small uncertainties in the initial conditions may lead to large uncertainties with respect to the future behavior of the system [14].

10.5. Basin Entropies

We quantify the degree of uncertainty due to the fractality of the escape basin boundary by the determination of the so-called basin entropy and basin boundary entropy, a tool recently developed by Miguel Sanjuán and collaborators [5, 10]. Let us suppose that an open Hamiltonian system, like the scattering map, has N_E exits, by which typical trajectories can escape, with the corresponding escape basins. We cover the phase space region with a grid of N boxes of size ε, and consider the probability p_{ij} for the trajectory beginning to the jth box to escape through the ith exit, where $i = 1, 2, \ldots, N_E$ and $j = 1, 2, \ldots, m_i$, where $m_i \in [1, N_E]$ is the number of exits available to trajectories originating from the ith box.

On supposing that the trajectories beginning in a box are statistically independent, Daza *et al.* defined the entropy of the ith box as

$$S_i = -\sum_{j=1}^{m_i} p_{ij} \log(p_{ij}), \qquad (10.22)$$

in such a way that the entropy of the grid is obtained by summing over all N boxes, $S = \sum_{i=1}^{N} S_i$. The basin entropy is defined as $S_b = S/N$.

The basin entropy quantifies the degree of the uncertainty associated with the escape basin, and varies from 0, in the case of a unique exit, to $\log N_E$, if the escape basins are so completely intertwined that we may consider them as equiprobable. If we restrict the computation of the basin entropy to those N_b boxes covering the escape basin boundaries, then we have the basin boundary entropy $S_{bb} = S/N_b$, which quantifies the uncertainty related only to the boundary.

In order to compute the basin entropy and basin boundary entropy of the escape basin of Fig. 10.2, we analyze the regions

$\phi_0 = 0$ and $\phi_0 = \pi$, separately. For both regions, we consider boxes with 4, 5, 9, 10 initial conditions per box. For each initial condition, we computed a maximum number of 10^4 iterations of the map and exclude from the statistics those initial conditions leading to orbits that do not escape during this maximum iteration time. Then, we compute the probabilities (10.23) of obtain the number of points into each box n_A, n_B and n_C corresponding to the exits **A**, **B** and **C**, respectively.

$$p_A = \frac{n_A}{n_A + n_B + n_C}, \quad p_B = \frac{n_B}{n_A + n_B + n_C}, \quad p_C = \frac{n_C}{n_A + n_B + n_C}, \tag{10.23}$$

and the entropy for each box is

$$S = -p_A \log p_A - p_B \log p_B - p_C \log p_C. \tag{10.24}$$

The entropy related to the each region of the escape basin is obtained by summing the contribution of each box $S = \sum_{i=1}^{N} S_i$. Thus, the basin entropy results by normalizing this value $S_b = S/N$. In limit situations, if we have a single exit, the corresponding probability is equal to the unity ($p_A = 1$), hence $S_b = 0$, i.e., no uncertainty at all. The opposite situation consists in completely randomized basins with N equiprobable escapes, which result in $S_b = \log N$ as the upper bound of the basin entropy.

We obtain the so-called boundary basin entropy $S_{bb} = S/N_b$ limiting the computation of the basin entropy only to the boxes containing points of the boundary (N_b). Thus, the basin boundary entropy S_{bb} measures the complexity of the basin boundary. In the case of two basins with smooth boundaries, the number of boxes in the boundary is negligible for the computation of the boundary basin entropy S_{bb}, since there are many more boxes with just one basin. Thus, the maximum possible value of S_{bb} that a smooth boundary can have is $\ln 2$, which would a pathological case where every box in the boundary contains equal proportions. Therefore, if $S_{bb} > \ln 2$, the basin boundary is said to be fractal.

We plot in Fig. 10.3 the values of the basin entropy S_b and the basin boundary entropy S_{bb} of the escape basin considering the two regions (a) $\phi = 0$ and (b) $\phi = \pi$. The variable parameter is the

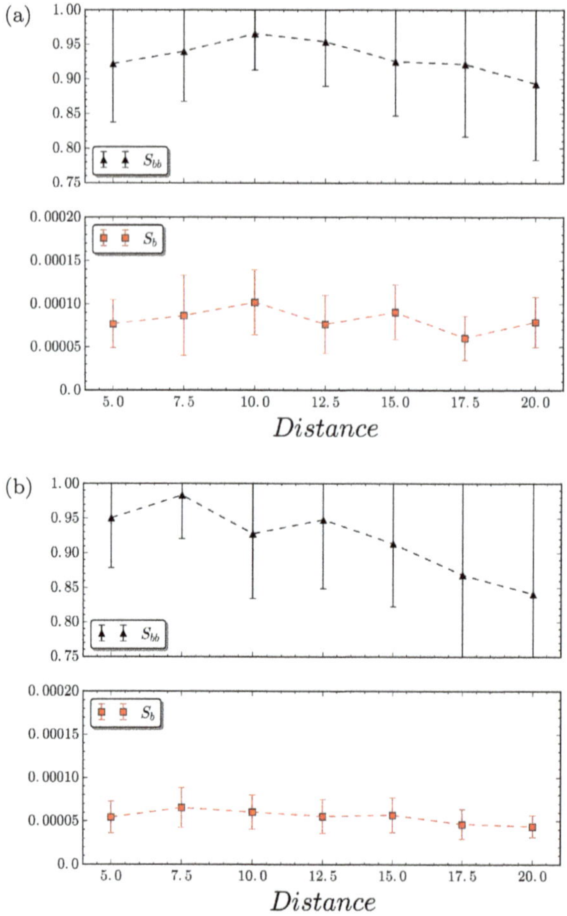

Fig. 10.3. Basin entropy S_b and basin boundary entropy S_{bb} as a function of distance between the black holes D for (a) $\phi = 0$ and (b) $\phi = \pi$.

distance between the black holes D. For the two cases of ϕ the value of basin entropy S_b is low, in fact most boxes contains initial conditions that escape through a single exit. However, the entropy of the boundary of basin S_{bb} obeys the inequality $S_{bb} > \ln 2$, i.e., the basin boundary is fractal.

As a general trend, the degree of complexity of the basin structure and its boundary decreases with D. We can observe in (a) that the entropies increase as D goes from 5 to 10 and suffer a decrease afterwards. The same happens in (b) but the maximum entropy value occurs in $D = 7.5$ and decreases for higher values of D. The decrease of entropy means that the basins become progressively less mixed and involved as the distance between the black holes increases.

10.6. Conclusion

An outstanding example of open Hamiltonian systems is the scattering of light rays by a binary black hole system. But as an open system, the chaotic motion is typically transient, the destiny of an incident light ray is either to escape to infinity or fall into one of the black holes. In these cases, we seek to obtain the respective exit basin, or the set of initial conditions that attend to a given outcome. The exit basins and their common boundary are fractal structures, essentially due to the presence of a nonattractive invariant set called chaotic saddle.

We considered the light ray scattering by a binary black hole system from the angle of an open Hamiltonian systems, focusing on the fractal structures present in the chaotic dynamics. The equations of general relativity for a light ray in the gravitational field of a binary Schwarzschild black hole system were integrated approximately to gather a discrete-time map, which exhibits chaotic dynamics for a wide range of its parameters, with the most important being the distance between the black holes. The chaotic dynamics here are transient, though, for the light rays can either escape to infinity or fall into one of the black holes.

The basin entropy, which is basically the information entropy related to the probability of going to a given basin, has been found to vary according to the distance used to represent the exit basins. The basin boundary entropy only takes into account the intervals containing basin boundary.

This work can be viewed as a connection to the research on black hole shadows as we can see in [9, 15], the shadow is a direct result of gravitational lensing as the black holes block some of the light rays that converge to it. As we know, the study of black hole shadows is important on the visualization and classification of black holes.

The shadow that it produces is an exit basin, so in this work we can use the exit basin we found to see the black hole shadow if necessary.

In summary, we have studied fractal structures that appear in the chaotic motion of a light ray under the gravitational field of two Schwarzschild black holes, a problem which is classically nonintegrable and has a nonattractive chaotic invariant set responsible for chaotic transient dynamics. Such fractal structures are responsible for various signatures of the so-called chaotic scattering, which is a phenomenon ubiquitous in nonintegrable open Hamiltonian systems.

Acknowledgments

This work has been supported by grants from the Brazilian Government Agencies CAPES and CNPq. One of us (R.L.V.) would like to thank Professor Miguel Angel Sanjuán for the several fruitful discussions on physics, nonlinear dynamics, and many other subjects.

References

[1] J. Aguirre, R. L. Viana and M. A. F. Sanjuán, Fractal structures in nonlinear dynamics, *Rev. Mod. Phys.* **81** (2009) 333–386; doi: 10.1103/RevModPhys.81.333.
[2] S. Weinberg, *Gravitation and Cosmology: Principles and Applications of the General Theory of Relativity* (Wiley, 1972).
[3] A. Péntek, Z. Toroczkai, T. Tél, C. Grebogi and J. A. Yorke, Fractal boundaries in open hydrodynamical flows: Signatures of chaotic saddles, *Phys. Rev. E* **51** (1995) 4076–4088; doi: 10.1103/PhysRevE.51.4076.
[4] S. Bleher, C. Grebogi, E. Ott and R. Brown, Fractal boundaries for exit in hamiltonian dynamics, *Phys. Rev. A* **38** (1988) 930–938; doi: 10.1103/PhysRevA.38.930.
[5] A. Daza, J. O. Shipley, S. R. Dolan and M. A. F. Sanjuán, Wada structures in a binary black hole system, *Phys. Rev. D* **98** (2018) 084050; doi: 10.1103/PhysRevD.98.084050.
[6] S. D. Majumdar, A class of exact solutions of Einstein's field equations, *Phys. Rev.* **72** (1947) 390–398; doi: 10.1103/PhysRev.72.390.
[7] A. Papapetrou, A static solution of the equations of the gravitational field for an arbitrary charge-distribution, *Proc. Royal Irish Acad. Sect. A: Math. Phys. Sci.* **51** (1945) 191–204.

[8] A. P. S. de Moura and P. S. Letelier, Scattering map for two black holes, *Phys. Rev. E* **62** (2000) 4784–4791; doi: 10.1103/PhysRevE.62.4784.

[9] P. V. P. Cunha and C. A. R. Herdeiro, Shadows and strong gravitational lensing: A brief review, *Gen. Relativ. Grav.* **50**(4) (2018) 42; doi: 10.1007/s10714-018-2361-9.

[10] A. Daza, A. Wagemakers, B. Georgeot, D. Guéry-Odelin and M. A. F. Sanjuán, Basin entropy: A new tool to analyze uncertainty in dynamical systems, *Sci. Rep.* **6**(1) (2016) 31416; doi: 10.1038/srep31416.

[11] A. Daza, B. Georgeot, D. Guéry-Odelin, A. Wagemakers and M. A. F. Sanjuán, Chaotic dynamics and fractal structures in experiments with cold atoms, *Phys. Rev. A* **95** (2017) 013629; doi: 10.1103/PhysRevA.95.013629.

[12] S. Chandresekhar, *The Mathematical Theory of Black Holes* (Oxford University Press, 1983).

[13] V. Bozza, Gravitational lensing in the strong field limit, *Phys. Rev. D* **66**, (2002) 103001; doi: 10.1103/PhysRevD.66.103001.

[14] S. W. McDonald, C. Grebogi, E. Ott and J. A. Yorke, Fractal basin boundaries, *Physica D* **17**(2) (1985) 125–153; doi: doi.org/10.1016/0167-2789(85)90001-6.

[15] A. Daza, J. O. Shipley, S. R. Dolan and M. A. F. Sanjuán, Wada structures in a binary black hole system, *Phys. Rev. D* **98** (2018) 084050; doi: 10.1103/PhysRevD.98.084050.

https://doi.org/10.1142/9789811221903_0011

Chapter 11

Characterizing Fractal Basin of Attraction in Planar Switched Systems

Yongxiang Zhang

School of Mathematics and Statistics, Qingdao University, Qingdao 266071, P. R. China
zyx9701@126.com; zyx9701@qdu.edu.cn

Based upon the auxiliary dynamical system method, we introduce some analytical tools to characterize the properties of fractal basin boundaries for planar switched systems with time-dependent switching. Under some generic assumptions, some sufficient conditions guaranteeing Wada basin boundaries are presented for the periodic switching signals. Other characterizing methods are based on the view point of limit sets and prime ends. By constructing the auxiliary dynamical system related to the switched systems, the fractal basin boundaries of planar switched systems can be proved if every diverging path in the basin of associated auxiliary system has the entire basin boundary as its limit set. Fractal basin boundary is also verified if every prime end that is defined in the basin of associated auxiliary system is a prime end of type 3 and all other prime ends are of type 1. Bifurcations of fractal basin boundary are investigated by analyzing what types of prime ends in the basin are involved. The basin boundary of switched system is also described by the indecomposable continuum. We verify numerically the existence of Wada basin boundaries in a switched Hénon map.

11.1. Introduction

Switched systems comprise an important class of dynamical systems that consist of a finite number of subsystems and a switching rule between these subsystems. These subsystems are usually described by a collection of indexed differential or difference equations [1].

The primary motivation for studying such switched systems comes partly from the numerous examples that are dependent on various environmental factors [2]. Another source of motivation comes from the rapidly developing area of switching control. The past decade has witnessed an enormous interest in this type of systems and such systems are common across a diverse range of areas of application, such as mechanical system, automotive industry, power systems, aircraft and traffic control, switching power converters, and many other fields [3, 4]. In examining these works and the broader literature, the stability analysis, stabilization and controllability have been dominant themes in switched systems research [5–7]. In contrast, the complex and novel dynamics of switched systems have received relatively little attention, even though such dynamical characteristics can provide valuable design criteria and useful performance measures for system properties.

In this chapter, we focus on fractal basin dynamics of the switched systems, which is one of the goals of nonlinear dynamics. The problem of determining the basin of attraction (or region of attraction; stability region) and boundary for a dynamical system is an important one in many applications, such as controlled systems [8, 9], physics [10], ecology [11] etc. The basin of attraction is of fundamental importance to be able to determine which attractor a specified initial condition goes to. One of the goals of dynamical systems is to describe the structure and properties of basins and their boundaries. For smooth dynamical systems, classical algebraic methods of investigation usually rely on Lyapunov function and Lasalle's extension of Lyapunov's theory [12, 13]. Methods based on a complete topological characterization of the stability boundary via the stable manifolds have been investigated [14, 15]. These basin boundaries are some smooth stable manifolds. A basin boundary is fractal if it contains a transversal homoclinic point [16]. The fact that a basin boundary is fractal has important practical consequences, which arbitrarily fine-scaled structure of fractal basin boundaries implies considerable sensitivity to small errors in initial conditions [17, 18]. Wada basin boundaries are not only fractal and the topological properties have been investigated by Kennedy and Yorke [17]. A basin boundary satisfies the Wada property if every open neighborhood of any point on the boundary has a nonempty intersection with at least three different basins. Wada basin boundaries have attracted a growing interest in recent years [18] and several applications to plasma physics [19], chaotic scattering [20–22], black hole system [23], different mechanical systems [24, 25], ecological systems [26], delayed systems [27] and quasiperiodically forced systems [28, 29] have been proposed. In the literature dealing with Wada basin boundaries, it is usually assumed that the dynamical system arises from a differentiable process (for two-dimensional diffeomorphisms). Though some numerical methods have been proposed [30–34], theories are still lacking for their analysis of the basin boundaries in switched systems. The classical theorems for fractal or Wada basin boundaries require the maps and inverse maps which are differentiable. These theorems are not straightforward for the application to switched systems. Thus, it is an interesting question to explore the theories and methods on fractal or Wada basin boundaries in switched systems.

The first purpose of this chapter is to describe fractal Wada basin boundaries to the generally discrete-time switched systems with time-dependent switching [35]. We construct an auxiliary dynamical system (ADS) related to the switched systems and address the relationship between the basin boundaries of switched systems and the boundaries of ADS. We argue that the basin boundaries of the ADS provide an evidence of the Wada basin boundaries for switched systems. Under some generic assumptions, we find conditions to satisfy the Wada property for some periodic switching signals. In particular, we provide some criteria of generalized basin cell analysis with which the Wada basin boundaries of switched systems can be determined. These methods will turn out to be computable and easily applicable to some examples of switched systems (the switched Hénon map is regarded as the modeling paradigm [36]). The practical consequences of encountering in switched systems are obviously very serious although the underlying equation is strictly speaking deterministic. This type of topological complexity of Wada basin boundaries provides insight into the qualitative unpredictability and uncertain dynamics for the switched systems.

The second purpose of this chapter is to introduce some analytical tools to characterize the properties of fractal basin boundaries for planar switched systems (with time-dependent switching) [37]. In smooth dynamical systems, some topological methods have been developed for the complete characterization basins and their boundaries. Usually, these basin boundaries are related to the manifolds of hyperbolic periodic points and basin boundaries are smooth [14, 15]. When fractal basin boundaries occur, indecomposable continua are useful for topological characterization of basin boundaries [38–41]. We construct an auxiliary dynamical system and find conditions to characterize fractal basin boundaries of planar switched systems from the view point of limit sets and prime ends. By constructing the auxiliary dynamical system, the fractal basin boundaries of planar switched systems can be proved if every diverging path in the basin of associated auxiliary system has the entire basin boundary as its limit set. Fractal property is also verified if every prime end that is defined in the basin of associated auxiliary system is a prime end of type 3 and all other prime ends are of type 1. Bifurcations of fractal basin boundary are investigated by analyzing what types of prime ends in the basin are involved. The basin boundary of switched system is also described by the indecomposable continuum.

11.2. Switched Systems and Auxiliary Dynamical Systems

To define a switched system, we need the notion of *a switching signal* [1, 2]. This is a piecewise constant function $\sigma : [0, +\infty) \to \Sigma$, where Σ is some index set, $\Sigma = \{1, 2, \ldots, J\}$. We assume for concreteness that σ is continuous from the right everywhere. An example of such a switching signal for the case $\Sigma = \{1, 2\}$ is depicted in Fig. 11.1(a). Thus, a continuous-time switched system (with time-dependent switching) can be described by the equation

$$\dot{x} = f_{\sigma(t)}(x), \tag{11.1}$$

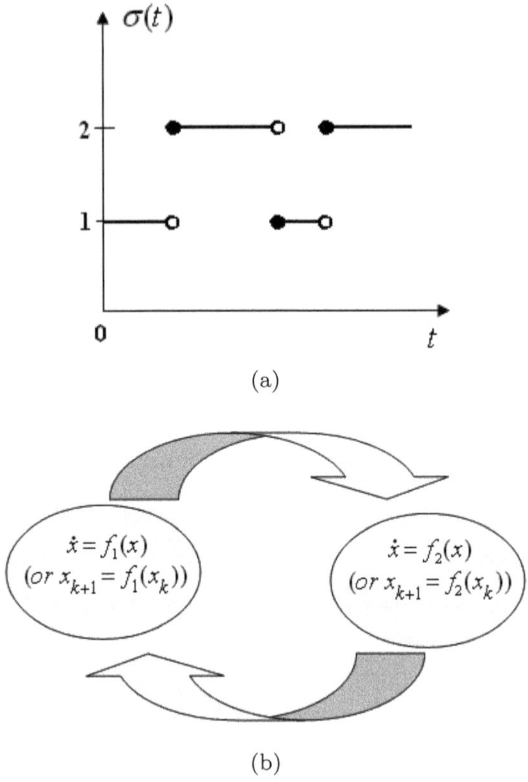

Fig. 11.1. (a) A switching signal; (b) a switched system model such as gene-regulator dynamics.

where the state $x \in R^n$ and $\sigma(\cdot)$, is a switching signal. Each f_i is globally Lipschitz continuous. The i's are picked in such a way that there are finite switched in finite time.

Similarly, we can represent a discrete-time switched system as a collection of difference equations

$$x_{k+1} = f_\sigma(x_k), \tag{11.2}$$

where $x_k \in R^n$. Here, we add the assumption that each f_i is globally Lipschitz continuous.

Many manmade systems and some physical systems can be modeled using such a framework. For example, a gene-regulatory network [42] consists of a set of interacting genes, each of which produces a protein through a process known as gene expression. Figure. 11.1(b) shows the switched system model of gene-regulatory dynamics and the switching signal is determined by the concentration of regulatory protein crossing a threshold. Figure. 11.1(b) also shows a discrete-time switched system, e.g., a switched Hénon map with a period-2 switching signal

$$(x_{n+1}, y_{n+1}) = \begin{cases} f_1(x_n, y_n) = (a_1 - x_n^2 + b_1 y_n, x_n) & \text{when } n \text{ is odd,} \\ f_2(x_n, y_n) = (a_2 - x_n^2 + b_2 y_n, x_n) & \text{when } n \text{ is even.} \end{cases} \tag{11.3}$$

For more examples, the readers can see some books on the switched systems [1, 3]. Though the switched systems provide an attractive paradigm for modeling a variety of practical situations, the analysis of such systems is far from straightforward. To fix the ideas clearly, we introduce some basic concepts. One can associate with a system (11.2) the following sequence, indexed by an initial state x_0, S_{x_0}: $(i_0, t_0), (i_1, t_1), \ldots (i_N, t_N), \ldots$. The switching sequence, along with (11.2), completely describes the orbit of the system according to the following rule: (i_k, t_k) means that the system evolves according to $x_{k+1} = f_{i_k}(x_k)$ for $t_k \leq t < t_{k+1}$. Throughout, we assume that the switching sequence is minimal in the sense that $i_j \neq i_{j+1}, j \in Z_+$. We can take projections of this sequence onto its first and second coordinates, yielding the sequence of indexed $\pi_1 : i_0, i_1, \ldots, i_N, \ldots$ and the sequence of switching times $\pi_2 : t_0, t_1, \ldots, t_N, \ldots$, respectively.

We also define the auxiliary dynamical system if σ is a periodic switching signal with the switching period $T = t_N - t_0$

$$x_{n+1} = f_{i_N}^{t_N - t_{N-1}} \circ f_{i_{N-1}}^{t_{N-1} - t_{N-2}} \circ \cdots \circ f_1{}^{t_1 - t_0} = F_\sigma(x_n). \tag{11.4}$$

11.3. Basic Concepts and Preliminaries

In this section, for clarity, we give some basic concepts including basin of attraction, stable manifold, unstable manifold, accessible point, Wada basin and Wada basin boundary, basin cell and prime ends. These concepts and terms play an important role in our understanding of the main theorems in this chapter.

11.3.1. *Stable manifold and unstable manifold*

Throughout this chapter, B is a connected, simply-connected open set in the plane R^2, M is either the plane or any other smooth two-dimensional manifold with boundary. Let F be a dissipative C^1-diffeomorphism from M to itself, that is, both F and its inverse are differentiable and the derivatives are continuous. Given a point x, the point z is in the ω-limit set of the orbit x, if there exists a sequence $\{t_n\}$, with $t_n \to \infty$, such that $F^{t_n} \to z$. We assume that B is invariant under F (i.e., $F(B) = B$). For a set B, write $\partial B = \overline{B} - B$ for the boundary of B, where $\overline{B}$ is the closure of the set B. A basin is the set of points x whose orbit $\{F^n(x)\}_{n\geq 0}$ converges to a specified compact invariant set such as an attractor. For an invariant set B, write $\partial B = \overline{B} - B$ for the boundary of B, where $\overline{B}$ is the closure of the set B. Obviously, the boundary ∂B of B is unstable when B is a basin of attraction. Let $B_1, B_2, \ldots, B_n$ be n disjoint basins. These basin boundaries will be denoted by $\partial B_1, \partial B_2, \ldots, \partial B_n$.

The point q is called a periodic point of period k if $F^k(q) = q$ and $F^i(q) \neq q$ for $1 \leq i < k$. A periodic point q is hyperbolic if the Jacobian matrix $DF(q)$ has no eigenvalues with absolute values 1. A periodic point q of period k is an attractor (respectively, saddle) if the eigenvalues e_1 and e_2 of $DF^k(q)$ satisfy $|e_1| < 1$ and $|e_2| < 1$ (respectively, $|e_1| < 1$ and $|e_2| > 1$). The compact sets we study have a finite number of components. For a periodic point q of periodic k, the stable manifold $W^s(q)$ and unstable manifold $W^u(q)$ are defined to be the following sets:

$$W^s(q) = \{x \in M | F^{kn}(x) \to F^{kn}(q), n \to \infty\},$$

$$W^u(q) = \{x \in M | F^{-kn}(x) \to F^{-kn}(q), n \to \infty\}.$$

The orbit of q is the set of k distinct forward images of q.

11.3.2. *Wada basin boundary and basin cell*

A point x on a basin boundary is a Wada point if every open neighborhood of x has a nonempty intersection with at least

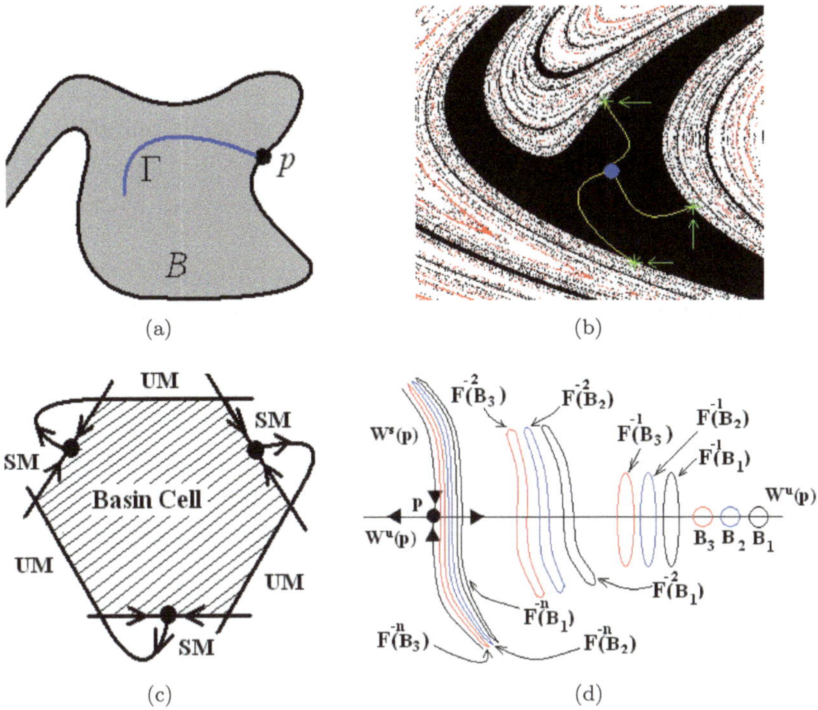

Fig. 11.2. (a) A accessible point p from the B. (b) Three green "$*$" are accessible from the black basin, because each of them is the first boundary point that a curve hits starting from the interior of the basin. (c) A schematic view for a basin cell, which is a trapping region generated by the stable manifold (SM) and unstable manifold **(UM)** of a period-3 orbit. (d) The λ-lemma when $W^u(p)$ intersects the basin $B_i (i = 1, 2, 3)$.

three basins. The boundary ∂B of basin B is called Wada basin boundary if all points on B are Wada points. Frequently, accessible periodic points play a crucial role in describing phenomena of basin boundaries. A point $p \in \partial B$ is accessible from B if a curve Γ can be drawn, starting in B so that p is the first boundary point of B (the intersection of the curve with the boundary), see Fig. 11.2(a). Three accessible points (three green "$*$") from the black basin are shown in Fig. 11.2(b). We also want to make use of the notions "trapping region" and "basin cell" [43, 44]. We say a compact region is a simply-connected, compact set with nonempty interior which consists of finitely many connected components. A compact region

Q is a trapping region if $F(Q) \subset Q$ and $F(Q) \neq Q$. If Q is a trapping region, then the basin of Q is the set of points which eventually map into the interior of Q. Each trapping region is constructed so that its boundary consists of pieces of stable and unstable manifolds of a periodic orbit that also lies on the boundary of the trapping region. Such trapping regions are called basin cells (see Fig. 11.2(c) for a schematic view of the six-sided basin cell).

For the invertible maps or smooth flows, Kennedy, Nusse and Yorke investigated the mechanism for Wada basin boundaries by the rigorously mathematical work [17, 43, 44]. In order to understand the Wada properties, Kennedy and Yorke introduce an important proposition [17], which states that: *let the sets $B_1, \ldots, B_k$ contained in R^2 be any open invariant sets. Let q be a point not in $B_i (1 \leq i \leq k)$ whose unstable manifold $W^u(q)$ intersects all the sets $B_1, \ldots, B_k$, then each x in the stable manifold $W^s(q)$ is in the boundary ∂B_k for each i.* This proposition is very useful to investigate the basin boundaries. Obviously, these sets are easily seen to be satisfied when $B_i (1 \leq i \leq k)$ is a basin of attraction. The proof is essentially the λ-lemma [45]. Since the unstable manifold of q intersects all the sets $B_1, \ldots, B_k$, we can choose some disks D_k in $B_k \cap W^u(q)$. Then $F^{-n}(D_k)$ gets arbitrarily close to each point of $W^s(q)$ as n increase, and therefore arbitrary close to each point of $\overline{W^s(q)}$. For arbitrary $\varepsilon > 0$, let $x \in W^s(q)$ and let $B_\varepsilon(x)$ denote the open ball centered at x with radius ε. There exists a positive integer n such that $(F^{-1})^n(D_k) \cap B_\varepsilon(x)$ is nonempty. This implies $B_\varepsilon(x)$ contains some points of the basin $B_i (1 \leq i \leq k)$. Since x on $W^s(q)$ and $\varepsilon > 0$ are arbitrarily chosen, then stable manifold $W^s(q)$ is contained in the boundary B_i for each i. This can be understood by referring to Fig. 11.2(d), where there are three coexisting basins. Suppose that B_1, B_2 and B_3 are small disks that belong to different basins. If we calculate the preimages of B_1, B_2 and B_3, that is their images when the time goes backward, we find that they successively approach the stable manifold of p. In other words, the preimages of B_1, B_2 and B_3 become exponentially stretched in the direction of the stable manifold and get arbitrarily close to it. In the limit, all points on the stable manifold of the periodic orbit p are boundary points, that is, there are points that belong to all three basins arbitrarily close to them. Therefore, all points on the stable manifold of p are Wada points.

11.3.3. *Prime ends and chains of regions*

Carathéodory introduced the notions of ends and prime ends for a simply connected region B, who classified into four kinds [41]. Carathéodory's Principle Theorem is expressed as follows: If F maps the unit disk conformally and one-to-one onto the region B, it induces a one-to-one mapping between the points of the unit circle and the prime ends of B. In other words, the prime ends are defined building blocks that provide a substituted for the points in the boundary ∂B so that the closure $\bar{B}$ of B can be homeomorphic to a closed disk. With Carathéodory's theorem, we can learn about the dynamics of F on the boundary of B by studying the corresponding dynamics on the circle. Thus, the foundation for the study of boundaries of simply-connected domains in the plane was laid. In order to understand the following theorems, we give and explain a brief account of definitions concerning prime ends and other principal elements. A simple arc c in B with two different endpoints on boundary ∂B and no other points on ∂B is called a crosscut of B, see Fig. 11.3, c_n is a crosscut and the diameter of c_n tends to 0 as $n \to \infty$. Here, let $\{c_n\}$ be a sequence of disjoint crosscuts such that c_n separates c_{n+1} from c_{n-1}. Therefore, there exists a corresponding sequence $\{B_{c_n}\}$ of subdomains of B such that B_{c_n} contains c_{n+1} except for its endpoints. The nested

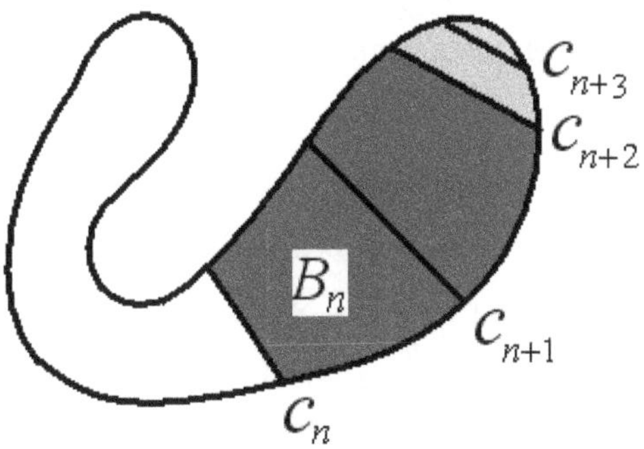

Fig. 11.3. Sequences of crosscuts $\{c_n\}$ and subdomains defining a prime.

infinite sequence $B_{c_1} \supset B_{c_2} \supset B_{c_3} \supset \cdots$ is called a chain. Two chains $\{B_{c_n}\}$ and $\{B_{c'n}\}$ are equivalent if every B_{c_n} contains some $B_{c'_n}$ and conversely, every $B_{c'_n}$ contains some B_{c_n}. A prime end is an equivalence class of chains. The set $Im(\gamma) = \cap \bar{B}_{c_n}$ is called the impression of the prime end γ and it is independent of the chain chosen. The impression $Im(\gamma)$ is a compact connected subset of ∂B and it is either single point or a continuum. The impression of a prime end is trivial (degenerate) if it consists of a single point. A point p_{prin} in $Im(\gamma)$ is called a principal point of the prime end γ if there exists a sequence $\{B_{c_n}\}_{n=1}^{\infty}$ of crosscuts such that $\{B_{c_n}\}_{n=1}^{\infty}$ converges to p_{prin} (p_{prin} is the only limit point). The set of all such points is called principal sets of the prime end γ and it is denoted by $\Pr(\gamma)$, which is compact, connected and nonempty. A prime end is said to be accessible if its' principal set consists of a single point. It is obvious that $\Pr(\gamma) \subset Im(\gamma)$. Every figure in Figs. 11.4(a)– 11.4(c) represents an open, simply-connect sets and segment AB is the impression of a prime end. In Fig. 11.4(a), P is the only principal point and it is the only accessible point. In Fig. 11.4(b), segment P_1P_2 is the principal set of AB. In Fig. 11.4(c), each point of AB is a principal point, in other words, segment AB is the impression of a prime end and the principal set of a prime end. Figure. 11.4(d) gives an example of a region whose boundary has a nontrivial prime end. Consider the simply-connect domain B around the origin and the boundary is obtained by adjoining to the unit circle (the radial segments $S_\theta = \{re^{2\pi i\theta} | 0.1 \leq r \leq 1\}$ for $\theta = 1/n, n \in Z$, and $\theta = 0$). There exists a single prime end whose impression is all of S_0 and the impression of all other prime ends is trivial.

11.3.4. *Alternate definitions of prime ends, extensions and the classification*

An alternate definition of prime ends is based on equivalence classes of paths to accessible boundary points. Two boundary paths Γ_1 and Γ_2 (two homeomorphic images $\eta_i = \eta_i(t)(i = 1, 2)$ of the open unit interval) in B are equivalence if they both approach the same boundary point $\eta_i(11.1)$ as $t \to 1$ and if, for each $\varepsilon > 0$, every

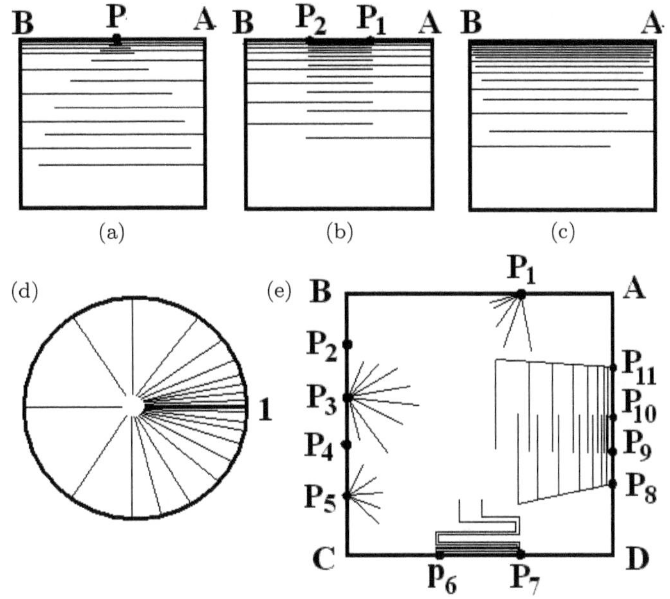

Fig. 11.4. Some artificial examples for the types of prime ends. (a) Segment AB is the impression of a prime end, P is the only principal point (the only accessible point). (b) Segment AB is the impression of a prime end and segment P_1P_2 is the principal set of AB. (c) Each point of AB is a principal point (segment AB is the impression of a prime end and the principal set of a prime end). (d) An example for nontrivial prime end. (e) An example for four types of prime ends.

point $\eta_1(t)(t > t_\varepsilon)$ can be joined to Γ_2 by a path (in B) and has diameter less than ε. Prime ends are equivalence classes of paths to accessible boundary points. The concept of prime ends has been extended to domains of infinite connectivity and to abstract spaces (e.g., topological space). When B is an unbounded set, a path Γ in B is called diverging path if $d_B(\Gamma(0), \Gamma(t))$ tends to infinity as t tends to 1, where $d_B(\cdots)$ is a path metric in B (its definition [46]). To each prime end γ, the impression $Im(\gamma)$ can be thought of as the union of all limit sets of half-open paths $\Gamma([0,1))$ converging to γ. The principal set $\Pr(\gamma)$ can be thought of as the intersection of all limit sets of half-open paths $\Gamma([0,1))$ converging to γ.

The prime end γ is called of type 1 if it impression consists of a single (principal) point (it is B-accessible); γ is of type 2 if its principal set is a single point and its impression is a nondegenerate continuum; γ is of type 3 if its principal set and its impression coincide and its principal set is a nondegenerate continuum ($\Pr(\gamma) = Im(\gamma)$ is a nondegenerate continuum); and γ is of type 4 if its principal set is a nondegenerate continuum and its complement in its impression contains a nondegenerate continuum ($\Pr(\gamma)$ and $Im(\gamma)\setminus\Pr(\gamma)$ are nondegenerate continuum. Figure 11.4(e) is very useful to understand the classification of the prime ends. In Fig. 11.4(e), each impression on the upper edge AB of square is of type 1. The point P_1 is the impression of infinitely many prime ends. There are two impressions of type 2 on the side BC, namely the segments P_2P_3 (with P_2 as principal point) and P_3P_5 (with P_4 as principal point). The sets consisting of P_3 and P_5, respectively, are the impressions of infinitely many prime ends of type 1. On the side CD, if there exists adjoining infinitely many the slits which approach the segment P_6P_7, the segment P_6P_7 is an impression of infinitely many prime ends of type 3. On the side DA and the adjoining slits, all impressions are of type 1, except that the segment P_8P_{11} is an impression of type 4. The points on the closed segment P_9P_{10} are principal points and the points on the remainder of the closed segment P_8P_{11} are nondegenerate continuum. The sets consisting of P_8 and P_{11}, respectively, are impressions of type 1.

11.4. Characterizing Fractal (or Wada) Basin Boundaries

In this section, we will discuss the fractal basin boundaries of switch system in two parts. In the first part, we present some sufficient conditions in order to verify the existence of Wada basin boundaries from the view point of manifolds of dynamical systems. In the second part, we investigate the fractal basin boundaries of switch systems from limit sets and prime ends in topology.

11.4.1. *Methods from manifolds of dynamical systems*

The main focus of this section is to present some sufficient conditions guaranteeing the existence of Wada basin boundaries for switched systems. Before discussing the Wada basin boundaries in the switched systems, we need to give some assumptions and lemmas. The point p is called a periodic point of period m if $F_\sigma^m(p) = p$ and $F_\sigma^i(p) \neq p$ for $1 \leq i < m$. For a periodic point p of period m, the stable manifold $W_{F_\sigma}^s(p)$ and unstable manifold $W_{F_\sigma}^u(p)$ are defined to be the following sets: $W_{F_\sigma}^s(p) = \{x \in M | F_\sigma^{mn}(x) \to F_\sigma^{mn}(p), n \to \infty\}$, $W_{F_\sigma}^u(p) = \{x \in M | F_\sigma^{-mn}(x) \to F_\sigma^{-mn}(p), n \to \infty\}$. Let $B_1^{F_\sigma}$, $B_2^{F_\sigma}, \ldots, B_n^{F_\sigma}$ be n disjoint basins for system (11.4). These basin boundaries will be denoted by $\partial B_1^{F_\sigma}$, $\partial B_2^{F_\sigma}, \ldots, \partial B_n^{F_\sigma}$ for system (11.4). In order verify the Wada property of system (11.4), we provide the following conditions that are similar to the case in the literature [43].

Throughout the text, let $F_\sigma{:}M \to M$ satisfy the following three conditions:

(a) Every accessible basin boundary point is on the stable manifold of some hyperbolic periodic point.
(b) For a given switching signal σ, there exist at least three disjoint basins for system (11.2) and N of them will be denoted by $B_1^{F_\sigma}$, $B_2^{F_\sigma}, \ldots, B_N^{F_\sigma}$, where $N \geq 3$.
(c) For a given switching signal σ, there exists a compact set $K \subset M$ such that K contains all points whose trajectories are bounded under forward and backward iterations of F_σ and if $x \in K$, then for every positive integer n, $F_\sigma(x) \notin K$.

The auxiliary dynamical system (11.4) plays a crucial role in describing phenomena of Wada basin boundaries for switched systems. Assumption (a) is established by Corollary 7.6′ in Ref. [47], assuming the periodic points are hyperbolic and each basin has at least one accessible periodic orbit. It can be checked by direct computation of the eigenvalues of the corresponding Jacobian matrix of $DF_\sigma^k(q)$. Assumption (a) can also be verified by the basin cell method [43]. Assumption (b) is a generic property for the Wada basin boundaries, and we can construct a switching signal that creates some new coexisting attractors, which is a necessary condition for Wada basin boundaries to exist. Assumption (c) is called lock-out property [43]. We might regard K as a large neighborhood of either an attracting set or a repelling set or a set that behaves like a saddle. This property implies any trajectory that leaves K cannot reenter K.

We will use the auxiliary dynamical system tool to prove that the switched systems (11.2) have Wada basin boundaries for some given switching signals.

Theorem 1. *If there exists a switching signal σ with the switching period T in the system (11.2), their basins satisfy the following assumptions. Let q_{F_σ} be F_σ a $B_1^{F_\sigma}$-accessible hyperbolic periodic point, and assume that the unstable manifold $W_{F_\sigma}^u(q_{F_\sigma})$ intersects $B_k^{F_\sigma}$ for some k, where $1 \leq k \leq N$. If the orbit of q_{F_σ} is the only $B_1^{F_\sigma}$-accessible hyperbolic periodic orbit, then ∂B_1^σ is a Wada basin boundary with respect to B_k^σ $(1 \leq k \leq N)$. If there are several $B_1^{F_\sigma}$-accessible hyperbolic periodic orbits (for F_σ), and if the unstable manifold of each of them intersects every $B_k^{F_\sigma}$ $(1 \leq k \leq N)$, then ∂B_1^σ is a Wada basin boundary with respect to B_k^σ $(1 \leq k \leq N)$.*

Proof. F_σ is a special type of dissipative diffeomorphism (composite diffeomorphism) because f_σ are the family of dissipative diffeomorphisms on the plane. Recall that there is a connection between f_σ and F_σ on the definition of periodic points. Note that there exists a simple one-to-one correspondence among the orbits of (11.2) and (11.4) because the period of an orbit of (11.2) is T-period of the corresponding orbit of (11.4). The orbits of system (11.4) contain in the orbits of system (11.2) and the attractors of system (11.4) contain in the attractors of system (11.2). The basin structures are quite the same although they have different attractors. Since the orbit of q is the only $B_1^{F_\sigma}$-accessible hyperbolic periodic orbit and assumption (c) holds, Corollary 7.6′ in Ref. [47] shows that all accessible points are on the stable manifold $W_{F_\sigma}^s(q_{F_\sigma})$. Since the accessible points are dense in the boundary, then every point on $W_{F_\sigma}^s(q_{F_\sigma})$ is a Wada point and the closure of $W_{F_\sigma}^s(q_{F_\sigma})$ equals $\partial B_1^{F_\sigma}$, so $\partial B_1^\sigma = \partial B_1^{F_\sigma}$ is a Wada basin boundary with respect to B_k^σ $(1 \leq k \leq N)$. Assume there are several $B_1^{F_\sigma}$-accessible hyperbolic periodic orbits $q_1^{F_\sigma}$, $q_2^{F_\sigma}, \ldots, q_j^{F_\sigma} \ldots$ (for F_σ), and if the unstable manifold of each of them intersects every $B_k^{F_\sigma}$ $(1 \leq k \leq N)$, then the closure of these hyperbolic periodic orbits $q_1^{F_\sigma}, q_2^{F_\sigma}, \ldots, q_j^{F_\sigma}$ contain in $\partial B_1^{F_\sigma}$ by λ-lemma. Since the system is dissipative [43], the periods of all accessible periodic orbits must be the same. Then every accessible point on $\partial B_1^{F_\sigma}$ is in the stable manifold of these periodic points. (Let F be a diffeomorphism of the plane R^2. Assume the periodic points of F in ∂B are hyperbolic. If there exists an accessible periodic point of minimum period q, then every accessible point in ∂B either is a periodic point of minimum period q or is in the stable manifold of such a periodic point. [47]) Then $\partial B_1^\sigma = \partial B_1^{F_\sigma}$ is a Wada basin boundary with respect to B_k^σ.

We remark that the basin have partially Wada basin boundaries [18] if there exist other accessible periodic orbits (their unstable manifold intersect only two basins). For a switching signal σ, assume that for every $j(1 \leq j \leq N)$, the unstable manifold of each ${B_j}^{F_\sigma}$-accessible hyperbolic periodic point intersects each of the basins $B_k^{F_\sigma}(1 \leq k \leq N)$. Then ∂B_1^σ is a Wada basin boundary and all of the basins have the same boundary, that is, $\partial B_1^\sigma = \partial B_k^\sigma$ for every $1 \leq k \leq N$. □

Theorem 2. *If there exists a switching signal σ with the switching period T in the system* (11.1), *their basins satisfy the following assumptions. Let q_{F_σ} be F_σ a hyperbolic periodic orbit that generates a basin cell $C_1^{F_\sigma}$. Let $B_k^{F_\sigma}$ $(1 \leq k \leq N)$ be N disjoint basins of $F_\sigma(N \geq 3)$, where $B_1^{F_\sigma}$ is the basin of $C_1^{F_\sigma}$. Assume that the unstable manifold $W_{F_\sigma}^u(q_{F_\sigma})$ intersects every $B_k^{F_\sigma}$ $(1 \leq k \leq N)$, then ∂B_1^σ is a Wada basin boundary with respect to B_k^σ $(1 \leq k \leq N)$.*

Proof. Similarly, we can characterize $\partial B_k^\sigma = \partial B_k^{F_\sigma}$. By the proposition in Ref. [43], (*Let P be a hyperbolic periodic orbit which generates a basin cell C_P. Let B_P be the basin of the cell C_P. Then every B_P-accessible point is on the stable manifold $W^s(P)$ of P*) the period orbit q_{F_σ} is the only periodic orbit on the boundary of $B_1^{F_\sigma}$ which is $B_1^{F_\sigma}$-accessible. Since the unstable manifold $W_{F_\sigma}^u(q_{F_\sigma})$ intersects every $B_k^{F_\sigma}$ $(1 \leq k \leq N)$, Theorem 1 implies that $\partial B_1^{F_\sigma}$ is a Wada basin boundary. Therefore, $\partial B_1^\sigma = \partial B_1^{F_\sigma}$ is a Wada basin boundary with respect to B_k^σ.

Obviously, Theorem 2 can also be extended by another statement. For a switching signal σ, let $q_{F_\sigma}(k)$ be a hyperbolic period-m_k orbit, where $1 \leq k \leq N (N \geq 3)$. Assume that for every $k (1 \leq k \leq N)$, (a) the periodic orbit $q_{F_\sigma}(k)$ generates a basin cell $C_k^{F_\sigma}$, and (b) the unstable manifold $W^u_{F_\sigma}(q_{F_\sigma}(k))$ intersects each basin $B_j^{F_\sigma} (1 \leq j \leq N)$. Then, for every j $(1 \leq j \leq N)$, the boundary ∂B_j^σ is a Wada basin boundary and $\partial B_j^\sigma = \partial B_1^\sigma$. □

We will present two examples with Wada basin boundaries in switched systems. The Hénon map is the most famous two-dimensional map. Due to its simplicity, it has become a benchmark system and has been extensively studied because of its generality, the complexity and richness of its dynamics. Let us consider two basin dynamics A_1 and A_2 defined by the two-dimensional Hénon maps:

$$A_1 : f_1(x_n, y_n) = (x_{n+1}, y_{n+1}) = (a_1 - x_n^2 + b_1 y_n, x_n),$$
$$A_2 : f_2(x_n, y_n) = (x_{n+1}, y_{n+1}) = (a_2 - x_n^2 + b_2 y_n, x_n).$$

We define the alternated dynamics A_1A_2 (or A_2A_1) (a period-2 switched signal as follows:

$$A_1A_2 : (x_{n+1}, y_{n+1}) = \begin{cases} f_1(x_n, y_n) = (a_1 - x_n^2 + b_1 y_n, x_n) & \text{when } n \text{ is odd,} \\ f_2(x_n, y_n) = (a_2 - x_n^2 + b_2 y_n, x_n) & \text{when } n \text{ is even,} \end{cases}$$

$$A_2A_1 : (x_{n+1}, y_{n+1}) = \begin{cases} f_1(x_n, y_n) = (a_1 - x_n^2 + b_1 y_n, x_n) & \text{when } n \text{ is even,} \\ f_2(x_n, y_n) = (a_2 - x_n^2 + b_2 y_n, x_n) & \text{when } n \text{ is odd.} \end{cases}$$

In order to study the alternated global dynamics A_1A_2 (or A_2A_1), we will give the auxiliary dynamics $A_1 \circ A_2 : (x_{n+1}, y_{n+1}) = f_2 \circ f_1(x_n, y_n)$ (or $A_2 \circ A_1 : (x_{n+1}, y_{n+1}) = f_1 \circ f_2(x_n, y_n)$). We now use this auxiliary dynamics approach to formulate our results for the switched dynamics A_1A_2 (or A_2A_1).

Corollary 1. *Let f_1 and f_2 are two orientation-reversing diffeomorphisms of the plane. Let σ_1 be a period-2 switching signal. Assume the* *hyperbolic fixed point* $q_{\sigma 1}$ *of* $f_2 \circ f_1$ *(or* $f_1 \circ f_2$*) are isolated in* $B_1^{F_{\sigma 1}}$ *and the unstable manifold* $W^u_{f_2 \circ f_1}(q_{\sigma 1})$ *(or* $W^u_{f_1 \circ f_2}(q_{\sigma 1})$*) intersects* $B_k^{F_{\sigma 1}}$ *for some* k*, where* $1 \le k \le N$*. Then* $\partial B_1^{\sigma 1}$ *is a Wada basin boundary with respect to* $B_k^{\sigma 1}$ $(1 \le k \le N)$*.*

Corollary 2. *If there exists a switching signal* σ_2 *with period-2 in the system* A_1A_2 *(or* A_2A_1*), their basins satisfy the following assumptions. Let* $q_{\sigma 2}$ *be* $f_2 \circ f_1$ *(or* $f_1 \circ f_2$*) a hyperbolic periodic orbit that generates a basin cell* $C_1^{F_{\sigma 2}}$*. Let* $B_k^{F_{\sigma 2}}$ $(1 \le k \le N)$ *be* N *disjoint basins of* $f_2 \circ f_1$ *(or* $f_1 \circ f_2$*)* $(N \ge 3)$*, where* $B_k^{F_{\sigma 2}}$ *is the basin of* $C_1^{F_{\sigma 2}}$*. Assume that the unstable manifold* $W^u_{f_2 \circ f_1}(q_{\sigma 2})$ *(or* $W^u_{f_1 \circ f_2}$ $(q_{\sigma 2})$*) intersects every* $B_k^{F_{\sigma 2}}$ $(1 \le k \le N)$*, then* $\partial B_1^{\sigma 2}$ *is a Wada basin boundary with respect to* $B_k^{\sigma 2}$ $(1 \le k \le N)$*.*

In particular, the fixed point of $f_2 \circ f_1$ (or$f_1 \circ f_2$) can be easily checked. We provide a simple criterion on the creation the Wada basin boundaries for the special switching signal in the following. An orientation-reversing homeomorphism of S^1 must have fixed points. The map $f_2 \circ f_1$ is orientation reversing and has no periodic orbits of minimum period greater than 1. Suppose f_1 and f_2 have a periodic orbit of minimum period k, $k \ge 3$, then $f_2 \circ f_1$ has a periodic orbit of minimum period $k/2$, if k is even, or of minimum period k, if k is odd.

The purpose of this section is to describe the basin dynamics obtained by combination of the basin dynamics A_1 and A_2. We describe some numerical experiments giving some evidences of Wada basin boundaries by a switched strategy A_1A_2 (or A_2A_1). For the first example, we choose the parameters $a_1 = 0.55$, $b_1 = 0.7$ and $a_2 = 0.6$, $b_2 = 0.9$. The basin of infinity is the set of all points (x, y) such that $|f^n(x, y)| \to \infty$ as $n \to \infty$. It is easily verified that A_1 and A_2 are orientation reversing. Figure 11.5(a) shows two basins for Hénon maps A_1. The black area is the basin of a period-2 attractor ($s_1 \approx (0.8446, -0.5446)$, $s_2 \approx (-0.5446, 0.8446)$ and the white area is the basin of the attractor at infinity. The basin boundaries are the stable manifold of the fixed point $p_1 \approx (-0.9066, -0.9066)$ (see the arrow for its position marked green "*") on the basin boundary that

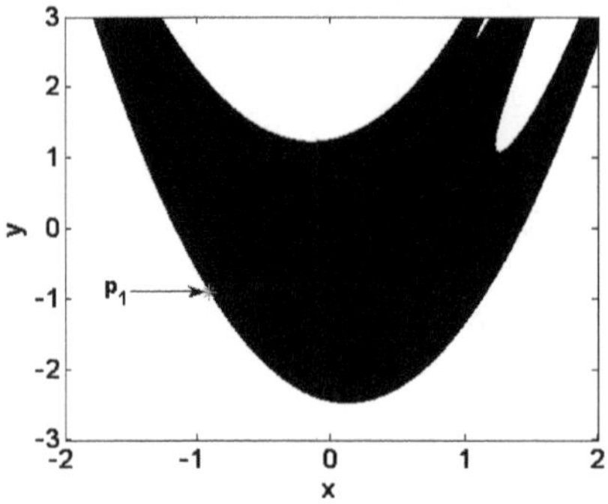

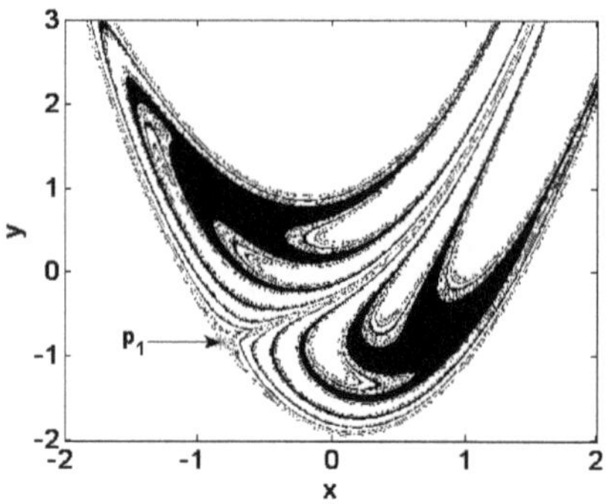

(a) (b)

Fig. 11.5. (a) There are two basins for the Hénon map $f_1(x, y) = (0.55 - x^2 + 0.7\,y, x)$. The black area is the basin of a period-2 attractor and the white area is the basin of the attractor at infinity. The basin boundaries are the stable manifold of the fixed point p_1 (see the arrow for its position marked green "$*$"), which are some smooth curves. (b) There are two basins for the Hénon map $f_2(x, y) = (0.6 - x^2 + 0.9\,y, x)$. The black area is the basin of a period-2 attractor and the white area is the basin of the attractor at infinity. The fractal basin boundaries are the stable manifold of the fixed point.

is accessible from two basins, which is a smooth curve. Figure 11.5(b) shows two basins for Hénon maps A_2. The black area is the basin of a period-2 attractor ($s_1 \approx (0.8197, -0.7197)$, $s_2 \approx (-0.7197, 0.8197)$) and the white area is the basin of the attractor at infinity. The basin boundaries are fractal because there are horseshoes in the basin boundary. However, the system A_1 and A_2 do not exhibit the Wada property.

For a given switching signal $\sigma_1 : A_1A_2$, four different attractors can coexist in the switched Hénon maps A_1A_2. Figure 11.6(a) shows four basins for four different attractors, which is a necessary condition for Wada basin boundary to exist. The black area is the basin of a period-2 attractor ($s_1 \approx (0.8096, -0.5542)$, $s_2 \approx (-0.5542, 0.8096)$). The blue area is the basin of another period-2 attractor ($s_1 \approx (0.8745, -0.7159)$, $s_2 \approx (-0.7159, 0.8745)$). The red area is the basin of a period-6 attractor ($s_1 \approx (-0.4061, 0.8185)$, $s_2 \approx (1.1718, -0.4061)$, $s_3 \approx (-1.1073, 1.1718)$, $s_4 \approx (0.4284, -1.1073)$, $s_5 \approx (-0.4087, 0.4284)$, $s_6 \approx (0.8185, -0.4087)$) and the white area is the basin of the attractor at infinity. It is noted that

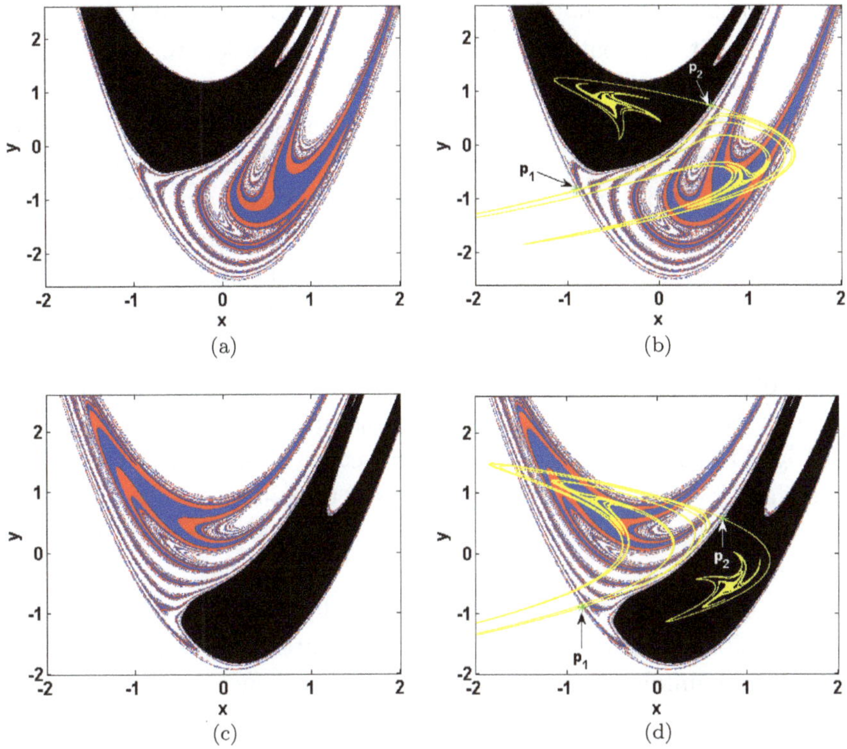

Fig. 11.6. (a) There are four basins for the switched Hénon map $A_1A_2(a_1 = 0.55, b_1 = 0.7$ and $a_2 = 0.6, b_2 = 0.9)$. The black area is the basin of a period-2 attractor, the blue area is the basin of another period-2 attractor, the red area is the basin of a period-6 attractor and the white area is the basin of the attractor at infinity. (b) The figure shows the unstable manifold of the fixed point p_1 and p_2 on the basin boundary for the auxiliary dynamics $A_1 \circ A_2$. p_1 is accessible from the basin of the infinity attractor and p_2 is accessible from the basin of a period-2 attractor (black area). The manifold of p_1 intersects all four basins and the manifold of p_2 intersects three basins. The black and white basin are two Wada basin boundaries. (c) There are four basins for the switched Hénon map $A_2A_1(a_1 = 0.55, b_1 = 0.7$ and $a_2 = 0.6, b_2 = 0.9)$. The black area is the basin of a period-2 attractor, the blue area is the basin of another period-2 attractor, the red area is the basin of a period-6 attractor and the white area is the basin of the attractor at infinity. (d) The figure shows the unstable manifold of the fixed point p_1 and p_2 on the basin boundary for the auxiliary dynamics $A_2 \circ A_1$. p_1 is accessible from the basin of the infinity attractor and p_2 is accessible from the basin of a period-2 attractor (black area). The manifold of p_1 intersects all four basins and the manifold of p_2 intersects three basins. The black and white basins have two Wada basin boundaries.

some attractors are not in their own basins which is different from the case for dissipative diffeomorphism. In order to understand the global structure on the basin boundary, we introduced the auxiliary dynamical system $A_1 \circ A_2$. The system $A_1 \circ A_2$ has four fixed points: $p_1 \approx (-0.8939, -0.8303)$, $p_2 \approx (0.5736, 0.7366)$, $p_3 \approx (-0.5542, 0.8096)$, $p_4 \approx (0.8745, -0.7159)$. We obtain a fixed point saddle p_1 on the basin boundary which is accessible from the white basin (whose attractor is at infinity). We also obtain another fixed point saddle p_2 on the basin boundary which is a accessible from the black basin (whose attractor is p_3). We find that the unstable manifold of p_1 intersects three basins and the unstable manifold of p_2 intersects each of the four basins; this unstable manifold is shown in Fig. 11.6(b) (the yellow curves). Since p_1 is isolated in the white basin and p_2 is isolated in the black basin, Corollary 1 shows the black basin boundaries and the white basin boundaries are two Wada basin boundaries.

We obtain a period-3 saddle S_3 on the basin boundary which is accessible from the blue basin (whose attractor is p_4) that the unstable manifold of S_3 intersects only two basins: the blue and the red basins, can also easily be seen from Fig. 11.6(b).When we find periodic orbits accessible from the red basin (whose attractor is a period-3 orbit, s_2, s_4 and s_6), the resulting periodic saddle is again S_3. So the S_3 is accessible from both the blue and red basins. Therefore, the basin boundaries of the red and blue basins are not Wada basin boundaries.

Similarly, for a given switching signal σ_2: A_2A_1, four different attractors can coexist in the switched Hénon maps A_2A_1 (with the same attractors for A_1A_2). However, their basin structure is quite different. Figure 11.6(c) shows their basins for four different attractors. The basins of period-2, another period-2, period-6, and infinity are shown by black, blue, red and white, respectively. The system $A_2 \circ A_1$ has four fixed points: ${p'}_1 \approx (0.7366, 0.5736)$, ${p'}_2 \approx (-0.8303, -0.8939)$, ${p'}_3 \approx (0.8096, -0.5542)$, ${p'}_4 \approx (-0.7159, 0.8745)$, and we obtain a fixed point saddle ${p'}_1$ on the basin boundary which is accessible from the white basin (whose attractor is at infinity). We also obtain another fixed point saddle ${p'}_2$ on the basin boundary which is a accessible from the black basin (whose attractor is p'_3). We find that the unstable manifold of p'_1 intersects three basins and the unstable manifold of p'_2 intersects each of the four basins (shown in Fig. 11.6(d)). Corollary 1 shows the black basin boundaries and the white basin boundaries are two Wada basin boundaries. We obtain a period-3 saddle S'_3 on the basin boundary which is accessible from the blue basin (whose attractor is p'_4) that the unstable manifold of S'_3 intersects only two basins:the blue and the red basins, can also easily be seen from Fig. 11.6(d). When we find periodic orbits accessible from the red basin (whose attractor is a period-3 orbit, s_1, s_3 and s_5), the resulting periodic saddle is again S'_3. So the S'_3 is accessible from both the blue and red basins. Therefore, the basin boundaries of the red and blue basins are not Wada basin boundaries.

For the second example, we choose the parameters, $a_1 = 0.66$, $b_1 = 0.52$ and $a_2 = 0.62$, $b_2 = 0.9$. Again, there are two attractors for each dynamical system (A_1 or A_2). The similar basin boundaries can be observed, which A_1 has a smooth basin boundary and A_2 has a fractal basin boundary (non-Wada). The system $A_1$and A_2 do not exhibit the Wada property. Figure 11.7(a) shows three basins for the switched Hénon maps A_1A_2. The blue area is the basin of a period-2 attractor, the red area is the basin of a period-4 attractor and the white area is the basin of the attractor at infinity. We consider the auxiliary dynamical system $A_1 \circ A_2$. The three points which are $p_1 \approx (1.1602, -0.6712)$, $p_2 \approx (0.5929, -0.1350)$ and $p_3 \approx (1.1008, -0.2297)$ constitute a period-3 orbit P for A_1A_2. It is easily numerically verified that P is a hyperbolic period-3 orbit, where the eigenvalues of $D(f_2 \circ f_1)^3$ at $p_n (1 \leq n \leq 3)$ are approximately 4.7183 and 0.0217. Assume that parts of the stable and unstable manifolds of $p_n (1 \leq n \leq 3)$ have been computed. The basin cell (the black region in Fig. 11.7 (b)) has been chosen so that the sum of the lengths of the stable and unstable segments of the six manifolds is minimal. Figure 11.7(b) shows three saddles on the basin boundary (marked by green "+") and one branch of the unstable manifold (marked by black curves) of each of the points of the saddles P. Computing the other branch of the unstable manifold

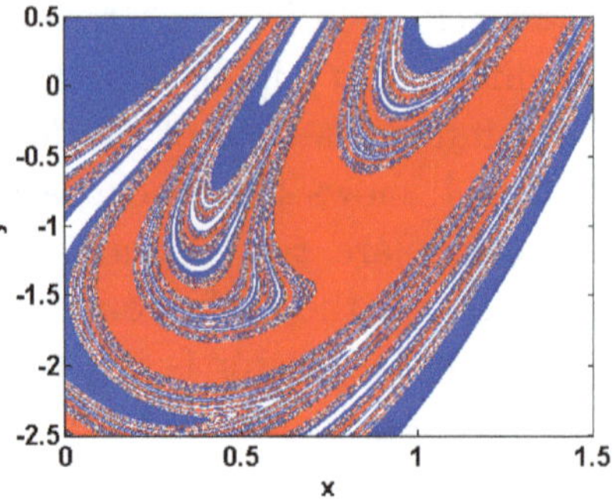

(a)

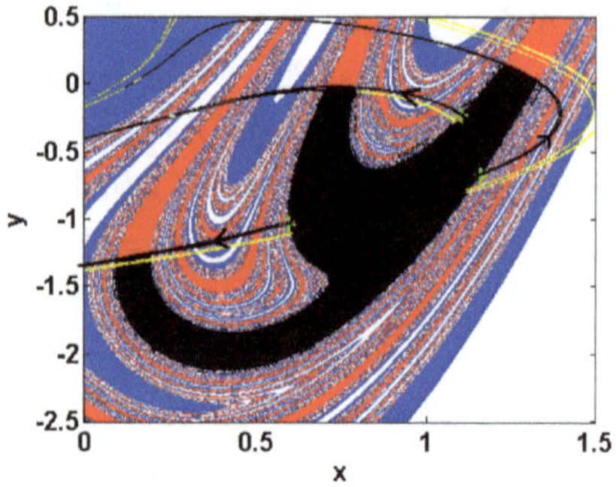

(b)

Fig. 11.7. (a) There are three basins for the switched Hénon map A_1A_2 ($a_1 = 0.66$, $b_1 = 0.52$ and $a_2 = 0.62$, $b_2 = 0.9$). The blue area is the basin of another period-2 attractor, the red area is the basin of a period-4 attractor and the white area is the basin of the attractor at infinity. (b) The black area is the basin cell for the period-3 saddles (marked with green "+") of the auxiliary dynamics $A_1 \circ A_2$: $f_2 \circ f_1$. The arrows on the black curves indicate whether it is a part of the unstable manifold of the corresponding point of p_1. The boundary of the red basin is a Wada basin boundary.

of each of the periodic points of the orbit P, we observe that each goes into the cell and approaches the attractors. Corollary 2 shows that the red basin has a Wada basin boundary.

11.4.2. *Methods from limit sets and prime ends in topology*

Let $F_\sigma : M \to M$ be an orientation-preserving C^1-diffeomorphism for which (11.1) there exists a connected, simply-connected, trapping region Q whose basin $B_Q^{F_\sigma} \subset M$ has the properties $B_Q^{F_\sigma}$ is d_B-unbounded, $\mathrm{int}(\overline{B}_Q^{F_\sigma})$ is connected and $\overline{B}_Q^{F_\sigma} \neq M$, $\overline{B}_Q^{F_\sigma} \neq Q$, and (11.2) there exists exactly one $B_Q^{F_\sigma}$-accessible saddle-hyperbolic period-m orbit in $\partial\overline{B}_Q^{F_\sigma}$.

For the diffeomorphism F_σ, Nusse and Yorke utilize prime theory as a tool and exploit the relation between prime ends and homoclinic tangencies. They also analyze the types of prime ends for basins having a basin cell. Figures 11.8(a) and 11.8(b) give two schematic views for the relation between homoclinic tangencies and prime ends,

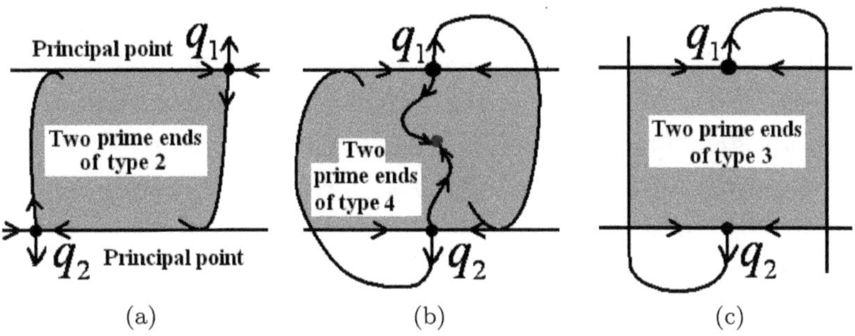

Fig. 11.8. Schematic views for the homoclinic tangencies, basin cells and prime ends.

which q_1 and q_2 are two B^{F_σ}-accessible saddle-hyperbolic periodic orbits of period 2. In Fig. 11.8(a), the inner unstable manifold (in B^{F_σ}) of $q_1(q_2)$ has a tangency with its stable manifold if and only if B^{F_σ} has exactly two prime ends of type 2 (q_1 and q_2 are two principal points). In Fig. 11.8b, the inner unstable manifold of q_1 and q_2 lies wholly in B^{F_σ}. The outer unstable manifold of q_1 (or q_2) crosses a stable branch of $q_1(q_2)$ and has a tangency (in $\overline{B}^{F_\sigma}$) with another stable branch of q_2 (or q_1) and has no crossings with that branch if and only if B^{F_σ} has two prime ends of type 4 and all other prime ends of B^{F_σ} are prime ends of type 1. In Fig. 11.8(a), the impression of the type 2 prime end equals the closure of a stable branch of q_1 and q_2. In Fig. 11.8(b), the impression of the type 4 prime end equals the closure of the stable manifold of q_1 and q_2. Its principal set is the closure of a stable branch of q_1 and q_2. Figure 11.8(c) shows a schematic view for prime ends and a basin cell formed by the stable and unstable manifolds of an accessible period-2 orbit (q_1 and q_2). B^{F_σ} has a basin cell if and only if B^{F_σ} has two type 3 prime ends and all other prime ends are type 1 prime ends.

We extend some results of smooth systems [41, 48] to switched system (11.2) and describe the fractal basin boundary of switched systems.

Theorem 3. *If there exists a switching signal σ with the switching period T in the system* (11.1), *their basins satisfy the following* *assumptions. Let q_{F_σ} be F_σ exactly one $B_Q^{F_\sigma}$-accessible hyperbolic periodic point. The orbit q_{F_σ} has a homoclinic point, and the stable manifold of q_{F_σ} has no tangency with the unstable manifold of q_{F_σ}. There are diverging paths in $B_Q^{F_\sigma}$ and for every diverging path Γ in $B_Q^{F_\sigma}$, the limit set of Γ equals the boundary of $B_Q^{F_\sigma}$. Then, the basin boundary of B_Q^{σ} is fractal.*

Proof. Recall that there is a connection between f_σ and F_σ on the definition of periodic points. Note that there exists a simple one-to-one correspondence among the orbits of (11.2) and (11.4) because the period of an orbit of (11.2) is T-period of the corresponding orbit of (11.4). The orbits of system (11.4) contain in the orbits of system (11.2) and the attractors of system (11.4) contain in the attractors of system (11.2). The structures of basin of attraction are quite the same although they have different attractors [36]. The theorem in Ref. [46] described the relationship between basin cells and limit sets of diverging paths for the diffeomorphism. Assume B is the basin of attraction and P is exactly one B-accessible saddle-hyperbolic periodic orbit of diffeomorphism F. It is shown that if the orbit P has a homoclinic point and the manifolds of P have no tangency, then a basin B has a basin cell if and only if the limit set of every diverging path in the basin equals the basin boundary. Since there are diverging paths in $B_Q^{F_\sigma}$ and for every diverging path Γ in $B_Q^{F_\sigma}$, the limit set of Γ equals the boundary of $B_Q^{F_\sigma}$, then $B_Q^{F_\sigma}$ has a basin cell. When the basin cell of $B_Q^{F_\sigma}$ exist, fractal basin boundary of $B_Q^{F_\sigma}$ can be characterized. Therefore, $\partial B_Q^{\sigma} = \partial B_Q^{F_\sigma}$ is fractal. □

Theorem 4. *If there exists a switching signal σ with the switching period T in the system* (11.2), *their basins satisfy the following assumptions. Let q_{F_σ} be F_σ exactly one $B_Q^{F_\sigma}$-accessible hyperbolic period-m orbit. If $B_Q^{F_\sigma}$ has m prime ends of type* 3, *and all other prime ends of $B_Q^{F_\sigma}$ are prime ends type* 1, *then the basin boundary of B_Q^{σ} is fractal.*

Proof. In [41], Theorem BC-PE3 describes a relationship between the basin cells and prime ends of type 3. It is shown that a basin has a basin cell if and only if the basin has m prime ends of type 3, and all other prime ends are prime ends of type 1 (see Fig. 11.8(c) for $m = 2$). Therefore, the basin $B_Q^{F_\sigma}$ has a basin cell and there must be a transversal homoclinic point. Thus, $\partial B_Q^{F_\sigma}$ is a fractal basin boundary. Similarly, we can characterize $\partial B_Q^{\sigma} = \partial B_Q^{F_\sigma}$ according to Theorem 1. Therefore, ∂B_Q^{σ} is also a fractal basin boundary. □

Theorem 5. *If there exists a switching signal σ with the switching period T in the system (11.2), their basins satisfy the following assumptions. Let $F_\sigma(\mu, x) : M \to M$ be one-parameter family of orientation-preserving C^1-diffeomorphism, such that for $a \leq \mu \leq b (a, b \in R)$ the map satisfies the following conditions:*

(a) *There exists a connected, simply-connected, trapping region $Q(\mu)$ whose basin $B_Q^{F_\sigma}(\mu) \subset M$ has the properties $B_Q^{F_\sigma}(\mu)$ is d_B-unbounded, $\mathrm{int}\overline{B}_Q^{F_\sigma}(\mu)$ is connected and simply connected, and $\overline{B}_Q^{F_\sigma}(\mu) \neq M$.*

(b) *There exists a $B_Q^{F_\sigma}(\mu)$-accessible saddle-hyperbolic period-m orbit $q_{F_\sigma}(\mu)$ in $\partial\overline{B}_Q^{F_\sigma}(\mu)$.*

(c) *For $a < \mu < b$, the basin $B_Q^{F_\sigma}(\mu)$ has m prime ends of type 3, and all other prime ends of $B_Q^{F_\sigma}(\mu)$ are prime ends of type 1. For $\mu \in \{a, b\}$, the basin $B_Q^{F_\sigma}(\mu)$ has either a prime end of type 2 or a prime end of type 4.*

Then, the basin boundary $\partial B_Q^{\sigma}(\mu)$ can jump suddenly at $\mu = a$ or $\mu = b$ (such changes are called basin boundary metamorphoses [49], e.g., fractal basin boundaries are created or destroyed).

The proof of Theorem 5 is based on the theorem on the creation and destruction of basin cells [41]. For every basin B_μ and every B_μ-accessible periodic orbit P_μ (saddle-hyperbolic) of one-parameter family of orientation-preserving C^1-diffeomorphisms F_μ ($a \leq \mu \leq b$), if P_μ generates a basin cell for $a < \mu < b$ and P_μ does not generate a basin cell for $\mu \in \{a, b\}$, then at $\mu = a$, either F_μ has a saddle-node bifurcation at P_μ, or the basin has either a prime end of type 2 or a prime end of type 4. For the auxiliary dynamical system (11.4), the condition (c) implies the creation of basin cells ($a < \mu < b$) according to the theorem in [41] on basin cells and prime ends of type 3. The basin cells disappear at $\mu = a$ or $\mu = b$. The metamorphoses of basin cells can imply metamorphoses of the basin boundary $\partial B_Q^{F_\sigma}$. The proof is similar to the proof of Theorem 3 ($\partial B_Q^\sigma = \partial B_Q^{F_\sigma}$). Thus, the basin boundary (∂B_Q^σ) metamorphoses can occur at $\mu = a$ or $\mu = b$.

In order to understand Theorem 5, we illustrate our ideas by the Fig. 11.8. If the basin has either a prime end of type 2 or a prime end of type 4 in the auxiliary dynamical system, the homoclinic tangencies of period orbits in the basin boundary occur. If the basin has either a prime end of type 3 and all other prime ends are type 1, the basin has a basin cell. Basin cells play an important role to verify fractal basin boundary for the diffeomorphism. When basin cells (its boundary contains a transversal homoclinic point) exist, fractal basin boundaries can be characterized robustly. Thus, $\mu = a$ (or $\mu = b$) is a basin boundary bifurcation point from the viewpoint of dynamical system.

A result is related to the theory of indecomposable continua for the basin boundaries of switched systems. A continuum is a compact, connected metric space. A continuum C is said to be indecomposable if C is not the union of two proper subcontinua of C. The following theorem provides verifiable conditions guaranteeing that the basin boundary is an indecomposable continuum.

Theorem 6. *If there exists a switching signal σ with the switching period T in the system* (11.2), *their basins satisfy the following assumptions. Let Q_{F_σ} be F_σ a hyperbolic periodic orbit with period-m. The unstable manifold of Q_{F_σ} intersects each $B_k^{F_\sigma}$, where $1 \leq k \leq n (n \geq 3)$. If the basin $B_Q^{F_\sigma}$ has m prime ends of type* 3, *and all other prime ends are prime ends of type* 1, *then the basin boundary of B_Q^σ is an indecomposable continuum.*

Proof. Because the basin $B_Q^{F_\sigma}$ has m prime ends of type 3, and all other prime ends are prime ends of type 1, then the basin $B_Q^{F_\sigma}$ has a basin cell formed by the stable and unstable manifold of period orbits. Since the unstable manifold of Q_{F_σ} intersects each $B_k^{F_\sigma}$, where

$1 \leq k \leq n(n \geq 3)$, Theorem 4 implies that $\partial B_Q^{F_\sigma}$ is a Wada basin boundary [43]. The periodic orbit Q_{F_σ} is the only periodic orbit on the boundary of $B_C^{F_\sigma}$, which is $B_C^{F_\sigma}$-accessible. Let q_{F_σ} be any of the periodic points of the orbit Q_{F_σ}. The definition of basin cell together with the λ-Lemma (see, e.g., [18]) implies that there exists a transversal homoclinic point q_{F_σ}. We now have: (a) $\overline{\partial B_Q^{F_\sigma}}$ is compact, since M is compact by assumption; (b) the periodic point q_{F_σ} admits a transversal homoclinic point; and (c) the unstable manifold of q_{F_σ} contains a one-dimensional disk that does not intersect the stable manifold of that point, since Q_{F_σ} is a $B_C^{F_\sigma}$-accessible periodic orbit. The result due to Barge [50] applied on the map F_σ^{-1} (all three conditions are satisfied) now implies that $\partial B_Q^{F_\sigma}$ is an indecomposable continuum. Similarly, we can characterize $\partial B_Q^\sigma = \partial B_Q^{F_\sigma}$. Therefore, ∂B_Q^σ is also an indecomposable continuum. □

11.5. Conclusions

In this chapter, we constructed the auxiliary dynamical systems according to the switched systems. We presented some conditions on Wada basin boundaries of the switched systems by the auxiliary dynamical system method. It is showed how to design the time-dependent switching signals inducing the Wada basin boundary. The fractal basin boundary is also described from the view point of limit sets and prime ends. In this type of switched system, the fractal basin boundary is verified by the limit set of diverging path in the basin of auxiliary system. We have shown that the basin of switched system has fractal boundary if every prime end (in the basin of auxiliary system) is a prime end of type 3 and all other prime ends are of type 1. We have given some conditions on the basin bifurcations of switched systems from the view point of prime ends. In particular, we also discussed the relationship between the indecomposable continua and prime ends for the switched system in order to understand the fractal basin boundaries of switched systems. Very little work has been done on the fractal basin boundaries related to the switched systems. The topological idea has been introduced into switched systems, as enriching the existing methods on the global analysis of nonsmooth systems.

One of the main consequences of fractal or Wada basin boundaries is the existence of a high number of patterns and the difficulty of predicting to which of them a given initial condition leads [18]. Furthermore, the Wada basin boundaries in the phase space may lead to another unpredictability of parameter space [51]. The sensitive dependence of asymptotic attractors on both initial conditions and parameters can occur in the switched systems. The implications of these results for engineering systems are cautionary. It may be of interest for an engineer to predict which of these finial states will exist for a given initial condition. When the switching gives rise to the Wada basin boundaries, the existence of this property suggests something more troublesome. Therefore, the existence of the Wada properties should be considered when building an engineer device. This Wada basin dynamics phenomenon may be not easy to encounter in practical application of engineering. However, there exist the possibilities that actual switched systems work near the critical value of parameters due to change of parameters. From an engineering point of view, it is easy to design these parameters because the practicing engineer can select some metabolic parameters in which stable responses can be expected to occur. On the other hand, if the inherent parameters in the systems have been given or limited, other parameters can be optimized by analysis of Wada basin dynamics so that the system does not work near the critical parameters. It is still an interesting question whether switching systems have complex fractal basins of attraction structure like smooth systems [52–54].

Acknowledgment

This work was supported by the National Natural Science Foundation of China (No. 11002092 and 11572205).

References

[1] D. Liberzon, *Switching in System and Control*, (Birkhäuser, Cambridge, MA, (2003).

[2] D. Liberzon and A. S. Morse, Basic problems in stability and design of switched systems, *IEEE Control Syst. Mag.* **19**(5) (1999) 59–70.

[3] A. C. J. Luo, *Discrete and Switching Dynamical Systems*, (L&HC and HEP, 2012).

[4] R. Shorten, F. Wirth, O. Mason and K. Wulff, Stability criteria for switched and hybrid systems, *SIAM Rev.* **49**(4) (2007) 545–592.

[5] J. P. Hespanha, Uniform stability of switched linear systems: extensions of LaSalle's invariance principle, *IEEE Trans. Automat. Control*, **49**(4) (2004) 470–482.

[6] D. W. Ding and G. H. Yang, H-infinity static output feedback control for discrete-time switched linear systems with average dwell time, *IET Control Theory Appl.* **4**(3) (2010) 381–390.

[7] D. W. Ding and G. H. Yang, Finite frequency H-infinity filtering for uncertain discrete-time switched linear systems, *Prog. Nat. Sci.* **19**(11) (2009) 1625–1633.

[8] D. Limon, T. Alamo and E. F. Camacho, Enlarging the domain of attraction of MPC controllers, *Automatica* **41**(4) (2005) 629–635.

[9] J. Lee and H. D. Chiang, Theory of stability regions for a class of nonhyperbolic dynamical systems and it's application to constraint satisfaction problems, *IEEE Trans. Circuits Syst. I* **49**(2) (2002) 196–209.

[10] Y. C. Lai and T. Tél, Fractal basin boundaries, *Appl. Math. Sci.* **173** (2011) 147–185.

[11] M. Scheffer, S. Carpenter, J. A. Foley, C. Folke and B. Walker, Catastrophic shifts in ecosystems, *Nature* **413**(10) (2001) 591–596.

[12] G. Chesi, Estimating the domain of attraction via union of continuous families of Lyapunov estimates, *Syst. Control Lett.* **56**(4) (2007) 326–333.

[13] W. Tan and A. Packard, Stability region analysis using polynomial and composite polynomial Lyapunov functions and sum of squares programming, *IEEE Trans. Automat. Control* **53**(2) (2008) 565–571.

[14] H. D. Chiang, M. W. Hirsch and F. F. Wu, Stability regions of nonlinear autonomous dynamical systems, *IEEE Trans. Automat. Control* **33**(1) (1988) 16–27.

[15] J. Zaborszky, G. Huang, B. Zheng and T. Leung, On the phase portrait of a class of large nonlinear dynamic systems such as the power system, *IEEE Trans. Automat. control* **33**(1) (1988) 4–15.

[16] S. W. Mcdonald, C. Grebogi, E. Ott and J. A. Yorke, Fractal basin boundaries, *Physica D* **17**(2) (1985) 125–153.

[17] J. Kennedy and J. A. Yorke, Basin of Wada. *Physica D* **51**(1–3) (1991) 213–225.

[18] J. Aguirre, R. L. Viana and M. A. F Sanjuán, Fractal structures in nonlinear dynamics, *Rev. Mod. Phys.* **81**(1) (2009) 333–386.

[19] R. L. Viana, E. C. Da Silva, T. Kroetz, I. L. Caldas, M. Roberto and M. A. F. Sanjuán, Fractal structures in nonlinear plasma physics, *Phil. Trans. R. Soc. A* **369**(1935) (2011) 371–395.

[20] D. Sweet, E. Ott and J. A.Yorke, Topology in chaotic scattering, *Nature* **399**(6734) (1999) 315–316.

[21] J. M. Seoane and M. A. F. Sanjuán, New developments in classical chaotic scattering, *Rep. Prog. Phys.* **76**(1) (2013) 016001.

[22] J. Aguirre and M. A. F. Sanjuán, Limit of small exits in open Hamiltonian systems, *Phys. Rev. E* **67**(5) (2003) 056201.

[23] A. Daza, J. O. Shipley, S. R. Dolan and M. A. F. Sanjuán, Wada structures in a binary black hole system, *Phys. Rev. D* **98**(8) (2018) 084050.

[24] F. Bellido and J. B. Ramirez-Malo, Periodic and chaotic dynamics of a sliding driven oscillator with dry friction, *Int. J. Nonlinear Mech.* **41**(6–7) (2006) 860–871.

[25] Y. Zhang, G. Luo, Q. Cao and M. Lin, Wada basin dynamics of a shallow arch oscillator with more than 20 coexisting low-period periodic attractors, *Int. J. Nonlinear Mech.* **58**(1) (2014) 151–161.

[26] J. Vandermeer, Wada basins and qualitative unpredictability in ecological models: a graphical interpretation, *Ecol. Model.* **176**(1–2) (2004) 65–74.

[27] A. Daza, A. Wagemakers and M. A. F. Sanjuán, Wada property in systems with delay, *Commun. Nonlinear Sci. Numer. Simulat.* **43**(2) (2017) 220–226.

[28] Y. Zhang, Strange nonchaotic attractors with Wada basins, *Physica D* **259**(9) (2013) 26–36.

[29] Y. Zhang, Wada basins of strange nonchaotic attractors in a quasiperiodically forced system, *Phys. Lett. A* **377**(18) (2013) 1269–1273.

[30] A. Daza, A. Wagemakers, and M. A. F. Sanjuán, Ascertaining when a basin is Wada: the merging method, *Sci. Rep.* **8**(1) (2018) 9954.

[31] A. Wagemakers, A. Daza and M. A. F. Sanjuán, The saddle-straddle method to test for Wada basins, *Commun. Nonlinear Sci. Numer. Simulat.* **84**(5) (2020) 105167.

[32] A. Daza, A. Wagemakers, M. A. F. Sanjuán and J. A. Yorke, Testing for basins of Wada, *Sci. Rep.* **5**(1) (2015) 16579.

[33] A. Daza, A. Wagemakers, B. Georgeot, D. Guéry-Odelin and M. A. F. Sanjuán, Basin entropy: a new tool to analyze uncertainty in dynamical systems, *Sci. Rep.* **6**(1) (2016) 31416.

[34] Y. Zhang and H. Zhang, Wada basin boundaries in switched systems, *Nonlinear Dyn.* **76**(1) (2014) 281–288.

[35] Y. Zhang, Switching-induced Wada basin boundaries in the Hénon map, *Nonlinear Dyn.* **73**(4) (2013) 2221–2229.
[36] Y. Zhang, characterizing fractal basin boundaries for planar switched systems, *Fractals* **25**(3) (2017) 1750031.
[37] M. A. F. Sanjuán, J. Kennedy, E. Ott and J. A. Yorke, Indecomposable continua and the characterization of strange sets in nonlinear dynamics, *Phys. Rev. Lett.* **78**(10) (1997) 1892–1895.
[38] M. A. F. Sanjuán, J. Kennedy, C. Grebogi and J. A. Yorke, Indecomposable continua in dynamical systems with noise: Fluid flow past an array of cylinders, *Chaos* **7**(1) (1997) 125–138.
[39] J. Kennedy, M. A. F. Sanjuán, J. A. Yorke and C. Grebogi, The topology of fluid flow past a sequence of cylinders, *Topol. Appl.* **94**(1–3) (1999) 207–242.
[40] H. E. Nusse and J. A. Yorke, Bifurcation of basins of attraction from the view point of prime ends, *Topol. Appl.* **154**(13) (2007) 2567–2579.
[41] P. Baldi and G. Hatfield, *DNA Microarrays and Gene Expression*, (Cambridge University Press, Cambridge, 2002).
[42] H. E. Nusse and J. A. Yorke, Wada basin boundaries and basin cells, *Physica D* **90**(3) (1996) 242–261.
[43] H. E. Nusse and J. A. Yorke, Basin of attraction, *Science* **271**(5254) (1996) 1376–1380.
[44] J. Guckenheimer and P. Holmes, *Nonlinear Oscillations, Dynamical Systems, and Bifurcation of Vector Fields* (Springer, New York, 1983).
[45] H. E. Nusse and J. A. Yorke, Characterizing the basins with the most entangled boundaries, *Ergod. Theory Dynam. Sys.* **23**(9) (2003) 895–906.
[46] K. Alligood and J. A. Yorke, Accessible saddles on fractal basin boundaries, *Ergod. Theory Dynam. Sys.* **12**(3) (1992) 377–400.
[47] H. E. Nusse and J. A. Yorke, The structure of basins of attraction and their trapping regions, *Ergod. Theory Dynam. Sys.* **17**(2) (1997) 463–481.
[48] C. Grebogi, E. Ott and J. A. Yorke, Basin boundary metamorphoses: changes in accessible boundary orbits, *Physica D* **2**(1–3) (1987) 243–262.
[49] M. Barge. Homoclinic intersections and indecomposability, *Proc. Amer. Math. Soc.* **101**(3) (1987) 541–544.
[50] Y. Zhang and G. Luo, Unpredictability of the Wada property in the parameter plane, *Phys. Letts. A* **376**(45) (2012) 3060–3066.
[51] Y. Zhang, H. Zhang and W. Gao, Multiple Wada basins with common boundaries in nonlinear driven oscillators, *Nonlinear Dyn.* **79**(4) (2015) 2667–2674.

[52] Y. Zhang, X. Xie and G. Luo, Multiple nested basin boundaries in nonlinear driven oscillators, *Commun. Nonlinear Sci. Numer. Simulat.* **44**(4) (2017) 220–228.
[53] Y. Zhang and L. F. Lu, Basin boundaries with nested structure in a shallow arch oscillator, *Nonlinear Dyn.* **77**(4) (2014) 1121–1132.
[54] F. Zhang and Y. Zhang, Wada fractal basins of attraction in a zero-stiffness vibration isolation system, *Fractals* **28**(4) (2020) 2050023.

https://doi.org/10.1142/9789811221903_0012

Chapter 12

Shadowing, Errors, and Exact Orbits of Quadratic Maps

Jason A. C. Gallas
Instituto de Altos Estudos da Paraíba, Rua Silvino Lopes 419-2502, 58039-190 João Pessoa, Brazil
Complexity Sciences Center, 9225 Collins Avenue Suite 1208, Surfside FL 33154, USA
Max-Planck Institute for the Physics of Complex Systems, Nöthnitzer Str. 38, 01187 Dresden, Germany

We consider two problems in dynamical systems. First, shadowing, namely the impact of small numerical errors in calculations of orbital points of the two-dimensional Hénon map is investigated. The unique map inverse allows one to access numerical errors by performing, under size-controlled representation of real numbers, a "there-and-back game", namely, iterating the map arbitrarily forward and, from there, backtracking using the unique inverse. Numerical errors are determined by measuring the distance between the points of departure and return. Second, we report exact analytical results for the quadratic map in the fully chaotic partition generating limit. In particular, exact expressions for periodic orbits and for preperiodic points allow one to extract precise coordinates of points belonging to periodic orbits. Such coordinates provide valuable checkpoints against which to gauge propagation errors for some classes of numerical calculations. The simple systems considered here provide fruitful workhorses to investigate numerical noise and error propagation when performing large number of iterates.

12.1. Introduction

The overwhelming majority of models that are interesting from a practical or fundamental point of view turns out to be hard to deal with numerically. This is so because it is quite difficult to numerically compute approximate solutions of an initial value problem which remains close to true solutions over a long time interval. To begin with, what should be understood by a "true solution" of a problem that cannot be integrated exactly? This question is relevant because virtually every equation of motion does not allow a solution to be represented in closed form and, therefore, to mimic the ad-hoc and very narrow class of problems which admit exact solutions and which are traditionally found in introductory textbooks of calculus and classical dynamics. Systems allowing exact analytical solutions are the exception, not the rule.

The task of solving typical equations of motion becomes generally impossible if one fixes the precision of the numerical representation of real numbers and demands solutions to remain valid over arbitrarily long time intervals. Numerical solutions normally involve hundred of thousands or even millions of iterates. To numerically solve equations of motion means to iterate discrete maps, either stepping forward a numerical integrator, or iterating a map assumed to represent the dynamics. In this context, it becomes natural to inquire about the exact meaning and significance of *numerical solutions* of equations of motion, and for how long such numerical approximations remain valid. But, how to make sense of this question when "true" solutions are unknown and impossible to obtain? Against what can we gauge numerically obtained solutions? It is one thing for a given dynamical system to have true orbits, that may behave in complicated or even chaotic ways, and another for such orbits to be computable, or observable in real life.

To address numerical computability problems, mathematicians came up with a number of concepts and claims centered around the notion of *shadowing*, all arising from a shadowing theorem for Anosov diffeomorphisms [1]. Anosov and Bowen showed that in systems which are *uniformly hyperbolic*, approximate numerical (i.e., "noisy") orbits will stay "close", or shadow, a true orbit for arbitrarily long times. Note the use of "a true", not "the true" orbit. Regrettably, a key issue in this context is that the shadowing theorem does not apply to real-life systems, which tend to be never hyperbolic let alone uniformly hyperbolic. Thus, strictly, one is still left in the dark regarding true mathematical orbits, having to rely on numerical experiments and rules of thumb derived from them which, nevertheless, use to emerge dressed as theorems, with dubious significance for applications. The rigorous study of shadowability has been a surprisingly difficult and technically sophisticated mathematical playground which so far seems to have only been thoroughly carried out in the case of one-dimensional maps, like maps of intervals or the circle, or the two-dimensional Hénon map [2, 3]. At present, despite the immense literature available, it seems fair to say that shadowability is still a subject where problems vastly outnumber significant results.

The purpose of this paper is, first, to describe a simple numerical experiment profiting from the fact that the Hénon map has a unique inverse. As it is typical for any map having a unique inverse, one can in principle iterate the map forward any number of times and, using the inverse, return to the original point of departure. Here, we check how this "there-and-back" excursion fares numerically. Generically, dynamical systems with unique inverse provide a wide class of models to study round-off and propagation errors numerically using simple examples. Second, we provide a survey of some recent results and examples of a general methodology to obtain exact expressions for periodic orbits of the quadratic map in the partition generating limit, as well as the coordinates of their orbital points. Knowledge of the precise location of orbital points is a mean of obtaining exact markers, reference points in phase space, that help to quantify the impact of unavoidable round-off and noise errors that parasitize numerical solutions of equations of motion. Curiously, while the literature on shadowability is already immense, apparently in all this literature there are no explicit discussions of the impact of shadowability on real-life applications. Of course, as for Lorenz's butterfly effect, it is not difficult to grasp the *conceptual* interest in shadowability. But an example of a practical situation where shadowability really makes a detectable difference seems to be still missing.

Before proceeding, we draw attention to the notion of *noninvertible* map. It is common to say that a dynamical system is noninvertible if the backward time evolution is either undefined or multivalued. In the present paper, the term *unique inverse* is used to clearly indicate situations where no branch ambiguity arises when taking the inverse. Of course, in general the number of inverses can vary from point to point or be undefined. Systems with multivalued inverse pose no intrinsic difficulty, just much more labor.

12.2. The There-and-Back Game with the Hénon Map

As mentioned, maps with a unique inverse are good examples to learn about difficulties and round-off errors in numerically computed orbits.

Consider a standard textbook-example of a map with a unique inverse, the Hénon map [4–6]

$$\tilde{x} = a - x^2 + by, \tag{12.1}$$

$$\tilde{y} = x. \tag{12.2}$$

When $b \neq 0$, the unique inverse is

$$x = \tilde{y}, \tag{12.3}$$

$$y = (\tilde{x} - a + \tilde{y}^2)/b. \tag{12.4}$$

Clearly, the inverse means that, at least formally, one can forward iterate the map any arbitrary number of times and, from there, precisely return to the initial point of departure. But, does this also hold for numerical work? If not, after how many forward iterates can one return within a prescribed radius of the initial point? In other words, how far can we move from the initial point and safely come back? To investigate this question we use `Maple`, a computer algebra software which allows fixing the number of significant digits in computations with real numbers.

Table 12.1 illustrates typical results obtained when iterating the Hénon map forward and backward 30 times while fixing number of digits in the computations. As parameters, we select Hénon's classical values $(a, b) = (1.4, 0.3)$, and as initial point in phase space, we arbitrarily chose the point $(x_0, y_0) = (0.2, 0.7)$. This point lies relatively close to the chaotic attractor. To reduce the table size,

Table 12.1. Differences observed in the arrival distance from an arbitrarily selected initial point $(x_0, y_0) = (0.2, 0.7)$, after performing n forward followed by n inverse iterates of the Hénon map. The topmost set of data was obtained with 10 digits arithmetics, the middle set with 15 digits, and the bottom set with 30 digits. Here, $(a, b) = (1.4, 0.3)$. Manifestly, the huge distances shown have no practical use, but are just indicative of the numerical capabilities of the computer algebra system used.

Iterate	x	y	Distance from departure point
5	0.2000000000	0.7000000000	0
10	0.2014125000	0.7023640733	0.2753×10^{-2}
15	−2.801580008	25.58914483	25.06948
20	0.2120×10^{24}	0.1498×10^{42}	0.1498315×10^{42}
25	0.4649×10^{604}	0.7205×10^{1208}	0.7205×10^{1208}
30	0.1048×10^{75725}	$0.3663 \times 10^{151449}$	$0.3663 \times 10^{151449}$
5	0.200000000000000	0.700000000000000	0
10	0.200000005703233	0.700000009518633	0.1109×10^{-7}
15	0.200018165229800	0.700030318666333	0.3534×10^{-4}
20	0.0815040284379333	0.548873019665267	0.1920
25	873.988260542437	2546234.16352894	2546233.61
30	0.2165×10^{574}	0.1563×10^{1148}	0.1563×10^{1148}
5	0.200000000000000	0.700000000000000	0
10	0.200000000000000	0.700000000000000	0.1028×10^{-22}
15	0.200000000000000	0.700000000000000	0.1162×10^{-19}
20	0.199999999999999	0.699999999999999	0.3407×10^{-15}
25	0.200000000001711	0.700000000002855	0.3329×10^{-11}
30	0.199999988783928	0.699999981280499	0.2182×10^{-7}

iterates are recorded *modulo 5*. The top set of data in Table 12.1 was obtained with 10 digits arithmetics, the middle set with 15 digits, and the bottom set with 30 digits.

As seen from Table 12.1, moving five iterates forward and then backward always allow one to safely return to the original departure point, independently of using 10, 15 or 30 digits arithmetic. However, the situation starts to change considerably after moving 10 iterates forth and back. Ten iterates are barely able to bring one back to the point of departure with 10 digits arithmetics. Double precision computations normally use 15–16 digits and, therefore, allow safe excursions of about 10 iterates. However, 15 digit arithmetics fails spectacularly after some 20 iterates or so. Even when using 30 digits, which require the ability of performing arithmetic with dedicated software, does not allow very far excursions. One may suspect the divisions by b in Eq. (12.4) to cause loss of precision. By performing similar experiments with $b = \pm 1$, it is not difficult to realize this not to be the case.

Out of curiosity, we decided to compare the distribution of stability regions between the direct Hénon map, Eqs. (12.1)–(12.2), and the corresponding inverse map, Eqs. (12.3)–(12.4). To this end, phase diagrams were computed in the control parameter space as described in the literature [7, 8]. Figure 12.1 illustrates the results obtained for representative parameter windows. As seen from this figure, the formal similarity of both maps is reflected in the relatively isomorphic distribution of the stability phases, despite the strong

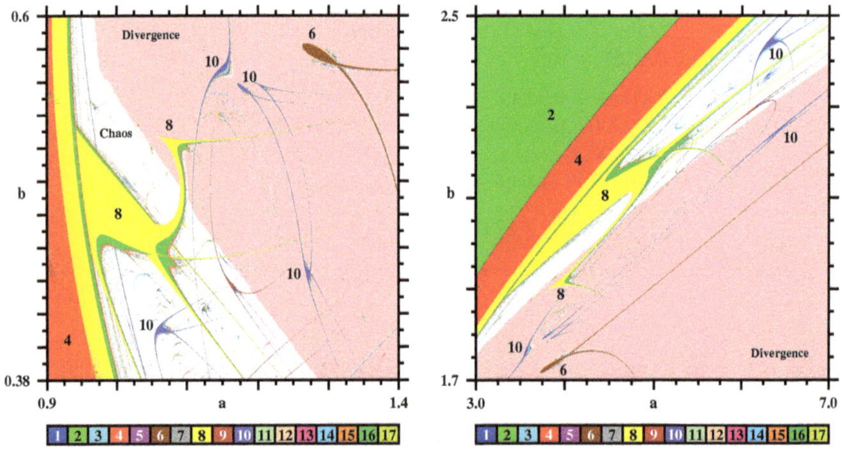

(a) Stability under forward iteration (b) Stability under backward iteration

Fig. 12.1. The future and the past: Topological similarity between shape and distribution of stability phases in the control parameter space of (a) the Hénon map, and (b) the inverse Hénon map. Colors/shadings denote periods mod 17, as defined by the colorbar. White denotes chaos. Numbers represent the period of the main domains of major islands and are added to facilitate visual comparisons. Divergent (unbounded) orbits constitute roughly half of both windows. Resolution of each panel: 1200×1200 parameter points.

shear present in Fig. 12.1(b) and the fact the window lies well outside the familiar stripe $-1 \leq b \leq 1$ typical of forward iteration of the Hénon map [7]. Thus, to iterate forward or backward in time a map which has a unique inverse seems to produce quite similar distribution of stable periodic motions.

Before moving on, it is interesting to observe that, with the possible exception of the weak interaction [9], all known laws of physics do not distinguish between future and past, although observed phenomena are irreversible. It is an important and difficult open problem to find a theoretical explanation for the observed asymmetry in the direction of time, *time's arrow* in the words of Eddington [10].

12.3. Exact Coordinates of Orbital Points

Now, we consider the determination of exact orbital points in the so-called $a = 2$ *partition generating limit* [4–6] of the quadratic or, equivalently, logistic map

$$x_{t+1} \equiv f(x_t) = a - x_t^2, \quad t = 0, 1, 2, \ldots. \tag{12.5}$$

For systems like this, with algebraic equations of motion, it is possible to use the standard *elementary symmetric functions* [11] to express individual periodic orbits as functions of the orbital points, which are roots of the periodic orbits. Using the so-called *Vieta's formulas*, the elementary symmetric functions may be expressed in a general manner by means of the coefficients of the orbital equation of motion, without the equation itself being solved. Of particular interest in the present context is an observation by John Wallis (1616–1703), in *De Algebra Tractatus, Historicus et Practicus*, that "the coefficient of the second term [of any polynomial], reckoning downward from the highest, is the aggregate of all the roots...".

Accordingly, a fruitful quantity to express individual orbits of polynomial dynamical systems is the scalar which represents the sum σ of the orbital points. With σ, all possible orbits can be extracted from a single σ-parameterized polynomial $\psi_k(x)$, called the *orbital carrier* [12, 13]. For each period k, the corresponding values of the scalar σ are obtained as roots of an auxiliary polynomial $\mathbb{S}_k(\sigma)$, whose degree informs the total number of k-periodic orbits that exist in the system. As it is not difficult to realize, the study of the polynomial doublets $\psi_k(x)$ and $\mathbb{S}_k(\sigma)$ shifts the traditional study of *numerically approximate orbital points* to a new level, namely to the study of σ-parameterized *algebraically exact equations of motion.* Every periodic orbit can be parameterized by the sum σ of orbital points, and the method to do it is effective [13].

A few examples help to understand what this is all about, and how it works. Consider the polynomial doublets $\psi_k(x)$ and $\mathbb{S}_k(\sigma)$ which encode all possible orbits of periods 4, 5 and 6 in the partition generating limit of the quadratic map. Explicitly, for period 4 the doublet is [13]

$$\begin{aligned}\psi_4(x) &= x^4 - \sigma x^3 + \frac{1}{2}(\sigma^2 + \sigma - 8)x^2 - \frac{1}{6}(\sigma^3 + 3\sigma^2 - 20\sigma + 2)x \\ &\quad + \frac{1}{24}(\sigma - 3)(\sigma^3 + 9\sigma^2 - 2\sigma - 16), \qquad (12.6)\end{aligned}$$

$$\mathbb{S}_4(\sigma) = (\sigma + 1)(\sigma^2 - \sigma - 4). \qquad (12.7)$$

The degree of $\mathbb{S}_4(\sigma)$ tells us that there are three period-4 orbits, one with integer coefficients and two with quadratic numbers as coefficients.

For period 5, the analogous doublet encoding all possible orbits is

$$\begin{aligned}\psi_5(x) &= (360\sigma^2 - 360\sigma - 240)x^5 - 120\sigma(3\sigma^2 - 3\sigma - 2)x^4 \\ &\quad + 60(\sigma^2 + \sigma - 10)(3\sigma^2 - 3\sigma - 2)x^3 \\ &\quad - (60\sigma^5 + 90\sigma^4 - 1800\sigma^3 + 1710\sigma^2 + 1860\sigma - 1200)x^2 \\ &\quad + (15\sigma^6 + 45\sigma^5 - 735\sigma^4 + 375\sigma^3 \\ &\quad + 3480\sigma^2 - 2700\sigma - 1200)x \\ &\quad - 3\sigma^7 - 12\sigma^6 + 192\sigma^5 + 30\sigma^4 - 2061\sigma^3 \\ &\quad + 1446\sigma^2 + 4248\sigma - 3600, \qquad (12.8)\end{aligned}$$

$$\mathbb{S}_5(\sigma) = (\sigma - 1)(\sigma^2 + \sigma - 8)(\sigma^3 - \sigma^2 - 10\sigma + 8). \qquad (12.9)$$

The total number of period-5 orbits is six, the degree of $\mathbb{S}_5(\sigma)$. The orbit corresponding to the choice $\sigma = 1$ has integer coefficients, two orbits have quadratic numbers as coefficients, and three orbits have coefficients given by algebraic numbers of degree 3.

Abbreviating $\varphi \equiv 3\sigma^3 - 7\sigma^2 - 13\sigma + 13$, for period 6 the doublet encoding all possible orbits is

$$\begin{aligned}
\psi_6(x) = {} & 160\varphi^2\sigma^2(x^6 - \sigma x^5) + 80\sigma^2(\sigma+4)(\sigma-3)\varphi^2 x^4 \\
& - 40\sigma\varphi\big(2\sigma^7 + \sigma^6 - 86\sigma^5 + 126\sigma^4 \\
& + 358\sigma^3 - 343\sigma^2 - 50\,\sigma + 56\big)x^3 \\
& + 20\sigma^2\varphi\big(\sigma^7 + 3\sigma^6 - 70\sigma^5 + 48\sigma^4 + 679\sigma^3 \\
& - 683\sigma^2 - 1218\sigma + 1048\big)x^2 \\
& - 4\sigma\varphi\big(\sigma^9 + 6\sigma^8 - 91\sigma^7 - 78\sigma^6 + 1693\sigma^5 \\
& - 976\sigma^4 - 6911\sigma^3 + 5496\sigma^2 \\
& + 2508\sigma - 2128\big)x + 2\sigma^{14} + 14\sigma^{13} \\
& - 247\sigma^{12} - 268\sigma^{11} + 7984\sigma^{10} \\
& - 8072\sigma^9 - 80966\sigma^8 + 157668\sigma^7 \\
& + 184938\sigma^6 - 530694\sigma^5 + 88965\sigma^4 \\
& + 373032\sigma^3 - 197156\sigma^2 - 13440\sigma + 15680,
\end{aligned} \tag{12.10}$$

$$\begin{aligned}
\mathbb{S}_6(\sigma) = {} & (\sigma+1)(\sigma-1)(\sigma^3 - 21\sigma + 28) \\
& \times(\sigma^4 + \sigma^3 - 24\sigma^2 - 4\sigma + 16).
\end{aligned} \tag{12.11}$$

Now, the total number of orbits is nine, the degree of $\mathbb{S}_6(\sigma)$. Two orbits have integer coefficients, three have coefficients given by algebraic numbers of degree 3, and four have coefficients given by algebraic numbers of degree 4. All these exact informations regarding the algebraic nature of the orbits are not possible to extract from numerical approximations of the orbital equations of motion, independently of the number of digits used for their computation.

As mentioned, when substituted into $\psi_k(x)$, each individual root of $\mathbb{S}_k(\sigma) = 0$ "projects" $\psi_k(x)$ into the specific orbital equation corresponding to σ. Normally, the polynomials $\mathbb{S}_k(\sigma)$ factor over the integers: factors with degree $\partial_k > 1$ correspond to orbital *clusters*, namely to irreducible polynomial aggregates commingling together a total of ∂_k orbits. Linear factors, multiple or not, correspond to nonclustered *single* orbits of period k. For instance, substituting $\sigma = -1$ into $\psi_4(x)$ we obtain the nonclustered orbit $o_{4,1}(x)$, while for $\sigma = (1-\sqrt{17})/2$ and $\sigma = (1+\sqrt{17})/2$, roots of the quadratic factor in $\mathbb{S}_4(\sigma)$, we get the orbits $o_{4,2}(x)$ and $o_{4,3}(x)$, respectively:

$$o_{4,1}(x) = x^4 + x^3 - 4x^2 - 4x + 1, \tag{12.12}$$

$$\begin{aligned} o_{4,2}(x) = x^4 &- \frac{1}{2}(1-\sqrt{17})x^3 - \frac{1}{2}(3+\sqrt{17})x^2 \\ &- (2+\sqrt{17})x - 1, \end{aligned} \tag{12.13}$$

$$\begin{aligned} o_{4,3}(x) = x^4 &- \frac{1}{2}(1+\sqrt{17})x^3 - \frac{1}{2}(3-\sqrt{17})x^2 \\ &- (2-\sqrt{17})x - 1. \end{aligned} \tag{12.14}$$

Note that the last two orbits have more complicated algebraic coefficients enforced by the σ values underlying them. When multiplied together, $o_{4,2}(x)$ and $o_{4,3}(x)$ produce an orbital cluster, or aggregate:

$$\begin{aligned} c_{4,1}(x) &= o_{4,2}(x) \times o_{4,3}(x) \\ &= x^8 - x^7 - 7x^6 + 6x^5 + 15x^4 - 10x^3 - 10x^2 + 4x + 1, \end{aligned} \tag{12.15}$$

a cluster obtainable directly by eliminating σ between $\psi_4(x)$ and $\sigma^2 - \sigma - 4$.

Note that the product of $o_{4,2}(x)$ and $o_{4,3}(x)$, which individually have algebraic coefficients, resulted in a cluster with integer coefficients, a generic characteristic of conjugated orbits with algebraic coefficients. Technically, $o_{4,2}(x)$ and $o_{4,3}(x)$ are defined by *relative* [14] quartic equations of motion, because their coefficients are algebraic numbers, not integers. Manifestly, $c_{4,1}(x)$ decomposes over the field $\mathbb{Q}(\sqrt{17})$. Furthermore, note that $o_{4,1}(x)$ provides an exact representation of the orbit and has simple integer coefficients. In sharp contrast, the numerical representation of $o_{4,2}(x)$ and $o_{4,3}(x)$ will have necessarily approximate numerical coefficients. Thus, the symmetries that are clearly visible between Eqs. (12.13) and (12.14) are totally obliterated when considering numerically approximated "projections" of the orbits. This unambiguous dichotomic distinction between algebraically exact and numerically approximate orbits remains valid for other periods and displays clearly the powerful insight obtained by working with exact equations of motion.

Doublets like Eqs. (12.6)–(12.7), (12.8)–(12.9), and (12.10)–(12.11) may be determined systematically, in principle for any arbitrary period. Doublets for arbitrary values of a of the quadratic map and arbitrary (a, b) of the Hénon map are available [15, 16].

Similarly, eliminating σ between $\psi_5(x)$ and, successively, $\sigma - 1$, $\sigma^2 + \sigma - 8$, and $\sigma^3 - \sigma^2 - 10\sigma + 8$, we get, apart from multiplicative constants used to eliminate denominators in $\psi_5(x)$, the following orbits and orbital clusters:

$$
\begin{aligned}
o_{5,1}(x) &= x^5 - x^4 - 4x^3 + 3x^2 + 3\,x - 1, \\
c_{5,1}(x) &= x^{10} + x^9 - 10x^8 - 10x^7 + 34x^6 + 34x^5 - 43x^4 \\
&\quad - 43x^3 + 12x^2 + 12x + 1, \\
c_{5,2}(x) &= x^{15} - x^{14} - 14x^{13} + 13x^{12} + 78x^{11} - 66x^{10} - 220x^9 + 165x^8 \\
&\quad + 330x^7 - 210x^6 - 252x^5 + 126x^4 + 84x^3 - 28x^2 - 8x + 1.
\end{aligned}
$$

The clusters factor into quintics over $\mathbb{Q}(\sqrt{33})$ and $\mathbb{Q}\big(\sqrt[3]{-62+95\sqrt{-3}}\big)$, respectively, thereby providing explicit analytic expressions for the remaining five period-5 orbits. Such clustered orbits involve relative [14] quintic equations, namely quintics whose coefficients are algebraic numbers (not integers), and which cannot be represented exactly in numerical computations.

For period-6, there are nine possible orbits encoded simultaneously by $\psi_6(x)$ and $\mathbb{S}_6(\sigma)$. Apart from multiplicative constants used to eliminate denominators in $\psi_6(x)$, by selecting $\sigma = 1$ and $\sigma = -1$ we get the orbits with integer coefficients and discriminants $\Delta_{6,1}$ and $\Delta_{6,2}$:

$$
\begin{aligned}
o_{6,1}(x) &= x^6 - x^5 - 5x^4 + 4x^3 + 6x^2 - 3x - 1, \\
\Delta_{6,1} &= 371293 = 13^5, \\
o_{6,2}(x) &= x^6 + x^5 - 6x^4 - 6x^3 + 8x^2 + 8x + 1, \\
\Delta_{6,2} &= 453789 = 3^3 \cdot 7^5.
\end{aligned}
$$

It is interesting to observe that $o_{6,1}(x)$ and $o_{6,2}(x)$ imply a contrived twist in the current understanding of polynomial interdependence. Individually, they are obtained as two seemingly independent "projections" which, however, have the carrier $\psi_6(x)$ as their common origin. Therefore, rather than being *de facto* independent orbits, they are in a certain sense "conjugated" orbits. Furthermore, since all nine period-6 orbits arise from the same carrier $\psi_6(x)$, the orbits $o_{6,1}(x)$ and $o_{6,2}(x)$ are also conjugated to all other seven orbits. This illustrates the existence of a complex and quite subtle *carrier-conjugation*, i.e., an arithmetical interdependence lurking among all period-k orbits, which is rather different from the field isomorphisms familiar from Galois theory of equations [22]. For arbitrary periods k, the orbital carriers $\psi_k(x)$ allow all their carrier-conjugated orbits to have coefficients belonging to very distinct number fields, a remarkable twist in the concept of conjugation which is not compatible with the standard canons of number field theory.

As before, for roots of the cubic and quartic factors in Eq. (12.11), the resulting coefficients in Eq. (12.10) are more complicated algebraic numbers, not integers. When all orbits arising from the same σ-factor are multiplied together, one obtains a cluster, a polynomial aggregate with integer coefficients and degree $\partial = mk$, multiple of the period k, where $m > 1$ is an integer. For instance,

$$\begin{aligned} c_{6,1}(x) &= x^{18} - 18x^{16} + x^{15} + 135x^{14} - 15x^{13} - 546x^{12} + 90x^{11} \\ &\quad + 1287x^{10} - 276x^9 - 1782x^8 + 459x^7 + 1385x^6 - 405x^5 \\ &\quad - 534x^4 + 170x^3 + 72x^2 - 24x + 1, \end{aligned} \tag{12.16}$$

$$\begin{aligned} c_{6,2}(x) &= x^{24} + x^{23} - 24x^{22} - 23x^{21} + 252x^{20} \\ &\quad + 229x^{19} - 1521x^{18} - 1292x^{17} \\ &\quad + 5832x^{16} + 4540x^{15} - 14822x^{14} - 10282x^{13} + 25284x^{12} \\ &\quad + 15001x^{11} - 28667x^{10} - 13653x^9 + 20886x^8 + 7168x^7 \\ &\quad - 9126x^6 - 1802x^5 + 2085x^4 + 101x^3 - 180x^2 + 12x + 1. \end{aligned} \tag{12.17}$$

Table 12.2. The nine period-6 orbits $o_{6,j}$ of the map $x_{t+1} = 2 - x_t^2$. Here, $\sigma_{6,j} = \sum x_j$ is the sum of the orbital points. The triad and quartet of $\sigma_{6,j}$ values are roots of the cubic and quartic factors in Eq. (12.11), respectively.

Orbit	x_1	$\sigma_{6,j}$
$o_{6,1}$	−1.770912051306	1
$o_{6,2}$	−1.911145611572	−1
$o_{6,3}$	−1.990061550730	−5.142457360
$o_{6,4}$	−1.756443146740	1.491252188
$o_{6,5}$	−0.912421314706	3.651205171
$o_{6,6}$	−1.990663269435	−5.287613777
$o_{6,7}$	−1.916491658218	−0.902246984
$o_{6,8}$	−1.559348708126	0.756484903
$o_{6,9}$	−0.971966826485	4.433375858

Table 12.2 collects numerical approximations for one orbital point for all nine period-6 orbits, together with the corresponding sums $\sigma_{6,j}$. All other orbital points may be easily obtained from x_1 by iterating the map $x_{t+1} = 2 - x_t^2$.

As for the remaining polynomials, the orbit $o_{6,1}(x)$ factors into a pair of cubics over $\mathbb{Q}(\sqrt{13})$. The cluster $c_{6,2}(x)$ factors into two equations of degree 9 over $\mathbb{Q}(\sqrt{21})$. However, such 9-degree polynomials mix roots of two distinct orbits, since their degree is not a multiple of six. For the cluster $c_{6,1}(x)$, the proper six cubics are obtained over $\mathbb{Q}(\sqrt[3]{\alpha})$, where $\alpha = -154 + 42\sqrt{-3} - 18\sqrt{-7} + 30\sqrt{21}$. For the cluster $c_{6,2}(x)$, eight cubics are obtained over $\mathbb{Q}(\sqrt{\beta})$, where $\beta = 65 - 13\sqrt{5} + 15\sqrt{15} - 3\sqrt{65}$. These factorizations provide explicit and exact solutions for all period-6 orbits and their orbital points, which are roots of the cubics. Manifestly, the factors of $\mathbb{S}_k(\sigma)$ reveal how orbits are distributed into clusters and single orbits, if any.

We remark that, while the derivation of the carriers $\psi_k(x)$ becomes increasingly more laborious as the period k grows, knowledge of approximate numerical values of the orbital points suffices to

Table 12.3. The eighteen period-7 orbits, characterized by one orbital point and the sum $\sigma_{7,j}$ of all points. The remaining orbital points follow by iterating $x_{t+1} = 2 - x_t^2$.

Orbit	x_1	$\sigma_{7,j}$
$o_{7,1}$	−1.9786867361502203955847071	−2.8882360088434114464952734
$o_{7,2}$	−1.8108964749862932314435851	−0.6150716258115650649303224
$o_{7,3}$	−1.0418806809758605723525628	4.5033076346549765114255959
$o_{7,4}$	−1.9976281104816460923503281	−7.2481321962698823504243586
$o_{7,5}$	−1.8848761656674288384473805	−1.2290602270284331761618286
$o_{7,6}$	−1.9409835883148106464466346	−0.77490800237038950156013044
$o_{7,7}$	−1.6122889835107255448466855	1.8441318528399982410921511
$o_{7,8}$	−1.7197459866836201484847537	2.7448245616149058389987627
$o_{7,9}$	−1.2029816300037407895687593	3.6631440112138009480554039
$o_{7,10}$	−1.9975528324285225652564052	−7.1754350638396879312221350
$o_{7,11}$	−1.9780114089762614447581210	−2.9359943127153307953060572
$o_{7,12}$	−1.8812582592076877545063517	−1.6023708208031626612700681
$o_{7,13}$	−1.9391197295964931429508120	−0.52557288374288388601167975
$o_{7,14}$	−1.8049930381548525612803578	0.019650748046206826205286560
$o_{7,15}$	−1.6004083969600341010740502	2.1979580475223842977306042
$o_{7,16}$	−1.7110701448170319282769643	2.3647347938971127665283378
$o_{7,17}$	−1.1795694263410389611326784	3.2907783266632442601391821
$o_{7,18}$	−1.0142477277395461818441842	5.3662511649721171232065293

disclose exact expressions for the polynomials $\mathbb{S}_k(\sigma)$ and its factors, thereby revealing the algebraic character of all orbits encoded by the carriers.

For instance, although the exact analytic expression for $\psi_7(x)$ remains yet to be obtained, Table 12.3 displays data from which one easily finds exact expression and factors for $\mathbb{S}_7(\sigma) = s_3(\sigma)s_6(\sigma)s_{18}(\sigma)$, namely

$$s_3(\sigma) = \prod_{j=1}^{3}(\sigma - \sigma_{7,j}) = \sigma^3 - \sigma^2 - 14\sigma - 8,$$

$$s_6(\sigma) = \prod_{j=4}^{9}(\sigma - \sigma_{7,j}) = \sigma^6 + \sigma^5 - 39\sigma^4$$

$$+ 63\sigma^3 + 110\sigma^2 - 136\sigma - 128,$$

$$\begin{aligned} s_9(\sigma) = \prod_{j=10}^{18}(\sigma - \sigma_{7,j}) &= \sigma^9 - \sigma^8 - 56\sigma^7 \\ &+ 118\sigma^6 + 573\sigma^5 - 1249\sigma^4 \\ &- 1582\sigma^3 + 2700\sigma^2 + 1576\sigma - 32. \end{aligned}$$

As it is not difficult to recognize, analogously as before, the individual factors composing $\mathbb{S}_k(\sigma)$ reveal important informations about the number fields underlying conjugate orbits. For instance, $s_3(\sigma)$, $s_6(\sigma)$ and $s_{18}(\sigma)$ define the algebraic nature of the number fields composing the conjugate clusters entangled in $\psi_7(x)$ which, in their turn, define all individual period-7 orbits as well as the algebraic nature of their orbital points. First, for period 7, there are no orbits with integer coefficients. Second, there are three clusters defining period-7 orbits. Third, the orbits forming these three clusters are relative polynomials of degree 7, with coefficients defined by algebraic numbers of degrees 3, 6, and 18. These predictions should be validated as soon as $\psi_7(x)$ is explicitly determined. Manifestly, similar predictions may be obtained for all periods $k > 7$. There are 30 period-8 orbits, 56 of period 9, 99 of period 10, and so on [17, 18]. Explicit expressions for $\mathbb{S}_8(\sigma)$, $\mathbb{S}_9(\sigma)$, $\mathbb{S}_{10}(\sigma)$, as well as their composing factors, are already available [19].

12.4. Orbital Inheritance: Solving High Degree Equations

Another possibility to obtain exact analytical representation for periodic orbital points, that may act as reference points to investigate numerical errors, is to use "inherited" orbits [13, 19–21]. A simple example allows one to grasp easily what inheritance means. To this end, consider the nonlinear transformation $x^3 - 3x$ applied to $o_{6,2}(x)$, namely [13]

$$c_{6,1}(x) = o_{6,2}(x^3 - 3x). \tag{12.18}$$

This identity shows that, as soon as the roots z_i of the orbit $o_{6,2}(x)$ are determined analytically in an exact manner, namely

$$o_{6,2}(x) = \left(x^3 + \frac{1}{2}(1-\sqrt{21})x^2 - \frac{1}{2}(1+\sqrt{21})x + \frac{1}{2}(5+\sqrt{21})\right) \times\left(x^3 + \frac{1}{2}(1+\sqrt{21})x^2 - \frac{1}{2}(1-\sqrt{21})x + \frac{1}{2}(5-\sqrt{21})\right), \tag{12.19}$$

three new orbits follow from them by solving the six cubics

$$x^3 - 3x - z_i = 0, \quad i = 1, 2, \ldots, 6, \tag{12.20}$$

which, obviously, may be also determined exactly. Thus, these eighteen roots provide exact analytical representations in terms of radicals for all orbital points composing the 18th-degree cluster $c_{6,1}(x)$ of Eq. (12.16). We remark that exact explicit solutions for nontrivial polynomials of degrees 5 and higher are increasingly more difficult and rare to find [22]. In particular, we are not aware of any previous explicit exact solution of a degree-18 polynomial.

Analogous results concerning orbital inheritance for periods $k \leq 12$ are available in the literature [13].

12.5. Preperiodic Points as Access to Exact Orbital Points

In the partition generating limit it is easy to obtain preperiodic points which, when used as initial conditions to iterate the map $x_{t+1} = 2 - x_t^2$, grant access to the exact coordinates of orbital points of periodic orbits [19]. Such useful preperiodic points are roots of an infinite family of polynomials $Q_\ell(x)$, which may be easily generated by a recurrence relation, Eq. (12.21) below. These preperiodic points were previously used [19] to extract specific orbital equations embedded in polynomial clusters with degree exceeding one billion and, therefore, totally out of reach by ordinary brute-force polynomial factorization. But we now find that preperiodic points can be also used to generate systematically, one by one, *explicit coordinates of periodic points for all orbits* of the quadratic map, Eq. (12.5).

The $Q_\ell(x)$ polynomials are obtained as irreducible factors composing an auxiliary family of polynomials, $T_\ell(x)$, generated recursively. Starting from two given initial *seed functions*, $T_0(x)$ and $T_1(x)$, all subsequent auxiliary $T_\ell(x)$ are obtained from the recurrence [23]

$$T_\ell(x) = xT_{\ell-1}(x) - T_{\ell-2}(x), \quad \ell = 2, 3, 4, \ldots. \tag{12.21}$$

For our present purpose, we fix $T_0(x) = 2$ and $T_1(x) = x$. Instead of the above recurrence, one may also obtain $T_\ell(x)$ directly, with no need of knowing every $T_{\ell'}(x)$ with $\ell' < \ell$, using Pincherle's relation [24]

$$T_\ell(x) = \left(\frac{x - \sqrt{x^2-4}}{2}\right)^\ell + \left(\frac{x + \sqrt{x^2-4}}{2}\right)^\ell, \quad \ell = 0, 1, 2, \ldots. \tag{12.22}$$

Clearly, $T_1(x) = Q_1(x) = x$, and $T_2(x) = Q_2(x) = x^2 - 2$. For $\ell > 2$, the polynomials $T_\ell(x)$ are reducible over the integers, products of *cyclotomic-like* irreducible factors $Q_\ell(x)$, except for $\ell = 2^n$, $n = 1, 2, 3, \ldots$ when $T_\ell(x) = Q_\ell(x)$, which are irreducible. Every new $T_\ell(x)$ generated by Eq. (12.21) contributes a new irreducible factor $Q_\ell(x)$, new in the sense of not appearing for any index ℓ' smaller than ℓ. Thus, the next first few polynomials are

$$T_3(x) = Q_1(x)Q_3(x), \quad T_4(x) = Q_4(x), \quad T_5(x) = Q_1(x)Q_5(x),$$
$$T_6(x) = Q_2(x)Q_6(x), \quad T_7(x) = Q_1(x)Q_7(x), \quad T_8(x) = Q_8(x),$$

where

$$Q_3(x) = x^2 - 3, \quad Q_4(x) = x^4 - 4x^2 + 2 \quad Q_5(x) = x^4 - 5x^2 + 5,$$
$$Q_6(x) = x^4 - 4x^2 + 1, \quad Q_7(x) = x^6 - 7x^4 + 14x^2 - 7,$$

and $Q_8(x) = x^8 - 8x^6 + 20x^4 - 16x^2 + 2$. The first twenty $Q_\ell(x)$ are listed in Table 1 of an open access paper [19] while the first 100 $Q_\ell(x)$ are shown in Table A.1 in the appendix below. A key observation is that the *irreducible* $Q_\ell(x)$ are the building blocks of the *reducible* auxiliary $T_\ell(x)$. Using the roots of $Q_\ell(x)$ as starting conditions to iterate the quadratic map, Eq. (12.5), one finds that after a finite preperiodic start, i.e., a certain number of nonrepeating iterates, the iteration lands on a cycle of k distinct points that repeat forever.

12.6. Conclusions and Outlook

We are used to think that the most common classes of models of dynamical systems involve either *discrete-time* maps or *continuous-time* differential equations. This simplistic separation ignores the fact that, when it comes to find *explicit solutions* for equations of motion, apart from a measure-zero set of equations discussed in textbooks of calculus and dynamics, the solution of differential equations are normally obtained after representing the differential equations by approximate time-discrete maps which, unfortunately, depend on the integration step or steps of the specific procedure adopted. Thus, to find explicit solutions of differential equations means to artfully select a range for the integration step where solutions remain stable in the sense that they do not change *significantly* under slight changes of the integration step. As it is known, the finiteness of the integration step is an unavoidable source of error in the numerical solution of differential equations. Even in rare cases when formulas for error estimates are available, such estimates are virtually never addressed in real-life applications. In sharp contrast, maps can yield the "correct" temporal evolution free from the aforementioned errors of numerical methods, provided that the maps faithfully represent the dynamics, and that calculations are performed under strict control of the numerical precision. Thus, in one way or another, modulo a measure-zero set of analytically solvable problems, to find explicit solutions for equations of motion means, in fact, to iterate maps or, equivalently, to assume that the *dynamics is inevitably governed by discrete-time variables.*

For the large class of maps with unique inverse, we believe the there-and-back game summarized in Table 12.1 to provide a viable tool to gauge numerical precision of calculations. For the complementary class of systems with multiple inverses the there-and-back game should of course be still possible, albeit much more complicated to implement. We argued that one should seek exact analytical solutions of simple problems and use them as reference marks in phase-space to test if numerical work is being done with enough precision. Of course, in general we will still remain in the dark as usual, having to resort to the standard rules of thumb. Be it as it may, we believe that physical systems with unique inverses should be explored under the new light discussed here, in particular to understand the interconnections between stability diagrams like the ones in Fig. 12.1.

From a theoretical point of view, we argued that orbital carriers, imply the startling and subtle notion of carrier-conjugates, i.e., an unexpected arithmetical interdependence lurking among all period-k orbits, which is distinct from the familiar field isomorphisms from Galois theory of equations. For any arbitrary period k, the orbital carrier $\psi_k(x)$ allows all its carrier conjugated orbits to have coefficients belonging to very distinct number fields, a remarkable twist in the concept of conjugation, not compatible with the standard canons of number field theory. Orbital carriers certainly exist for any dynamical system governed by algebraic equations of motion, a wide class of systems, although it may not be feasible to obtain them explicitly for arbitrary systems. Fortunately, as illustrated above, the quadratic map of Eq. (12.5) allows one to explore carriers explicitly.

Appendix

Table A.1 lists the first 100 irreducible polynomials $Q_\ell(x)$ as factors of the corresponding $T_\ell(x)$. Manifestly, the *irreducible* $Q_\ell(x)$ are the building blocks of the *reducible* auxiliary $T_\ell(x)$. The polynomials $Q_\ell(x)$ divide themselves naturally into two groups, according to the odd or even character of ℓ. Further, $Q_1(x) = x$ is a factor of $T_\ell(x)$ for odd values of ℓ, while $Q_{2^\ell}(x)$ is a factor of $T_{2^\ell}(x)$, for $\ell = 1, 2, 3, \ldots$. Additionally, the power $p \equiv 2^\ell$ divides the indices of all factors composing $T_p(x)$.

The regularity observed in Table A.1 between the base factors of ℓ and the subindexes of $Q_\ell(x)$ reveals how to generate the decomposition indices of $T_\ell(x)$ automatically. For example, for odd values of $\ell = b_1^{p_1} b_2^{p_2}$ the following Maple driver generates the list of

Table A.1. The first 100 factors $Q_\ell(x)$ of $T_\ell(x)$.

i Odd			*i* Even		
ℓ	Factors	$T_\ell(x)$	ℓ	Factors	$T_\ell(x)$
1	1	Q_1	2	2	Q_2
3	3	Q_1Q_3	4	2^2	Q_4
5	5	Q_1Q_5	6	$2\cdot 3$	Q_2Q_6
7	7	Q_1Q_7	8	2^3	Q_8
9	3^2	$Q_1Q_3Q_9$	10	$2\cdot 5$	Q_2Q_{10}
11	11	Q_1Q_{11}	12	$2^2\cdot 3$	Q_4Q_{12}
13	13	Q_1Q_{13}	14	$2\cdot 7$	Q_2Q_{14}
15	$3\cdot 5$	$Q_1Q_3Q_5Q_{15}$	16	2^4	Q_{16}
17	17	Q_1Q_{17}	18	$2\cdot 3^2$	$Q_2Q_6Q_{18}$
19	19	Q_1Q_{19}	20	$2^2\cdot 5$	Q_4Q_{20}
21	$3\cdot 7$	$Q_1Q_3Q_7Q_{21}$	22	$2\cdot 11$	Q_2Q_{22}
23	23	Q_1Q_{23}	24	$2^3\cdot 3$	Q_8Q_{24}
25	5^2	$Q_1Q_5Q_{25}$	26	$2\cdot 13$	Q_2Q_{26}
27	3^3	$Q_1Q_3Q_9Q_{27}$	28	$2^2\cdot 7$	Q_4Q_{28}
29	29	Q_1Q_{29}	30	$2\cdot 3\cdot 5$	$Q_2Q_6Q_{10}Q_{30}$
31	31	Q_1Q_{31}	32	2^5	Q_{32}
33	$3\cdot 11$	$Q_1Q_3Q_{11}Q_{33}$	34	$2\cdot 17$	Q_2Q_{34}
35	$5\cdot 7$	$Q_1Q_5Q_7Q_{35}$	36	$2^2\cdot 3^2$	$Q_4Q_{12}Q_{36}$
37	37	Q_1Q_{37}	38	$2\cdot 19$	Q_2Q_{38}
39	$3\cdot 13$	$Q_1Q_3Q_{13}Q_{39}$	40	$2^3\cdot 5$	Q_8Q_{40}
41	41	Q_1Q_{41}	42	$2\cdot 3\cdot 7$	$Q_2Q_6Q_{14}Q_{42}$
43	43	Q_1Q_{43}	44	$2^2\cdot 11$	Q_4Q_{44}
45	$3^2\cdot 5$	$Q_1Q_3Q_5Q_9Q_{15}Q_{45}$	46	$2\cdot 23$	Q_2Q_{46}
47	47	Q_1Q_{47}	48	$2^4\cdot 3$	$Q_{16}Q_{48}$
49	7^2	$Q_1Q_7Q_{49}$	50	$2\cdot 5^2$	$Q_2Q_{10}Q_{50}$
51	$3\cdot 17$	$Q_1Q_3Q_{17}Q_{51}$	52	$2^2\cdot 13$	Q_4Q_{52}
53	53	Q_1Q_{53}	54	$2\cdot 3^3$	$Q_2Q_6Q_{18}Q_{54}$
55	$5\cdot 11$	$Q_1Q_5Q_{11}Q_{55}$	56	$2^3\cdot 7$	Q_8Q_{56}
57	$3\cdot 19$	$Q_1Q_3Q_{19}Q_{57}$	58	$2\cdot 29$	Q_2Q_{58}
59	59	Q_1Q_{59}	60	$2^2\cdot 3\cdot 5$	$Q_4Q_{12}Q_{20}Q_{60}$
61	61	Q_1Q_{61}	62	$2\cdot 31$	Q_2Q_{62}
63	$3^2\cdot 7$	$Q_1Q_3Q_7Q_9Q_{21}Q_{63}$	64	2^6	Q_{64}
65	$5\cdot 13$	$Q_1Q_5Q_{13}Q_{65}$	66	$2\cdot 3\cdot 11$	$Q_2Q_6Q_{22}Q_{66}$
67	67	Q_1Q_{67}	68	$2^2\cdot 17$	Q_4Q_{68}
69	$3\cdot 23$	$Q_1Q_3Q_{23}Q_{69}$	70	$2\cdot 5\cdot 7$	$Q_2Q_{10}Q_{14}Q_{70}$
71	71	Q_1Q_{71}	72	$2^3\cdot 3^2$	$Q_8Q_{24}Q_{72}$
73	73	Q_1Q_{73}	74	$2\cdot 37$	Q_2Q_{74}

(*Continued*)

Table A.1. (*Continued*)

i Odd			i Even		
ℓ	Factors	$T_\ell(x)$	ℓ	Factors	$T_\ell(x)$
75	$3 \cdot 5^2$	$Q_1Q_3Q_5Q_{15}Q_{25}Q_{75}$	76	$2^2 \cdot 19$	Q_4Q_{76}
77	$7 \cdot 11$	$Q_1Q_7Q_{11}Q_{77}$	78	$2 \cdot 3 \cdot 13$	$Q_2Q_6Q_{26}Q_{78}$
79	79	Q_1Q_{79}	80	$2^4 \cdot 5$	$Q_{16}Q_{80}$
81	3^4	$Q_1Q_3Q_9Q_{27}Q_{81}$	82	$2 \cdot 41$	Q_2Q_{82}
83	83	Q_1Q_{83}	84	$2^2 \cdot 3 \cdot 7$	$Q_4Q_{12}Q_{28}Q_{84}$
85	$5 \cdot 17$	$Q_1Q_5Q_{17}Q_{85}$	86	$2 \cdot 43$	Q_2Q_{86}
87	$3 \cdot 29$	$Q_1Q_3Q_{29}Q_{87}$	88	$2^3 \cdot 11$	Q_8Q_{88}
89	89	Q_1Q_{89}	90	$2 \cdot 3^2 \cdot 5$	$Q_2Q_6Q_{10}Q_{18}Q_{30}Q_{90}$
91	$7 \cdot 13$	$Q_1Q_7Q_{13}Q_{91}$	92	$2^2 \cdot 23$	Q_4Q_{92}
93	$3 \cdot 31$	$Q_1Q_3Q_{31}Q_{93}$	94	$2 \cdot 47$	Q_2Q_{94}
95	$5 \cdot 19$	$Q_1Q_5Q_{19}Q_{95}$	96	$2^5 \cdot 3$	$Q_{32}Q_{96}$
97	97	Q_1Q_{97}	98	$2 \cdot 7^2$	$Q_2Q_{14}Q_{98}$
99	$3^2 \cdot 11$	$Q_1Q_3Q_9Q_{11}Q_{33}Q_{99}$	100	$2^2 \cdot 5^2$	$Q_4Q_{20}Q_{100}$

$Q_\ell(x)$ indices which appear in the factorization $T_\ell(x)$:

```
with(ListTools):
b1 := 3:     b2 := 5:
p1 := 3:     p2 := 2:        L := [NULL]:
for ijk from 0 to p1 do
for lmn from 0 to p2 do
   aux:= b1^ijk * b2^lmn:
   L  := [op(L),aux]:   od: od:
print( b1^p1 * b2^p2, sort(MakeUnique(L)) );
```

This driver may be easily adapted to deal with more complicated values of ℓ. For higher odd values of ℓ we determined explicitly the decompositions:

$$225 = 3^2 5^2 : \quad T_{225} = Q_1Q_3Q_5Q_9Q_{15}Q_{25}Q_{45}Q_{75}Q_{225},$$
$$441 = 3^2 7^2 : \quad T_{441} = Q_1Q_3Q_7Q_9Q_{21}Q_{49}Q_{63}Q_{147}Q_{441}.$$

However, for $675 = 3^3 \cdot 5^2$ the software and hardware available to us could not generate $T_{675}(x)$. The driver above was then used to obtain the following illustrative lists of decomposition indices:

$$675 = 3^3 5^2 : \quad [1, 3, 5, 9, 15, 25, 27, 45, 75, 135, 225, 675],$$
$$729 = 3^2 9^2 : \quad [1, 3, 9, 27, 81, 243, 729],$$
$$1089 = 3^2 11^2 : \quad [1, 3, 9, 11, 33, 99, 121, 363, 1089].$$

Table A.1 leads us to conjecture an interesting theoretical possibility about the $Q_\ell(x)$ polynomials, which share remarkable similarities with the *cyclotomic polynomials* [25] which may be generated by the quadratic map for $a = 0$. As it is known, every cyclotomic field is an Abelian extension of the rational numbers $\mathbb{Q}$. In this context, an important discovery is the so-called Kronecker–Weber theorem, stating that every finite Abelian extension of $\mathbb{Q}$ can be generated by roots of unity, i.e., Abelian extensions are contained within some cyclotomic field. In other words, every algebraic integer whose Galois group is Abelian can be expressed as a sum of roots of unity with rational coefficients. For details see, e.g., [26]. The study of the partition generating limit of the quadratic map $x_{t+1} = a - x_t^2$ seems to lend hope that for $a = 2$ the map may also share an analogous correspondence with Abelian equations as the one embodied in the Kronecker–Weber theorem [27] when $a = 0$. The dynamics for other values of a, when real and complex orbits coexist, is totally open to investigation. Integer and rational values of a are first good candidates to learn how bifurcation cascades unfold arithmetically. In particular, the case $a = 1$, say, offers the possibility of learning about the interplay of infinite cascades of coexisting orbits governed by towers of real and complex algebraic quantities, a new and totally unexplored world. The investigation of these problems will certainly provide useful insight regarding the intricate algebraic nesting induced by the dynamics Clearly, these open problems deserve to be investigated.

Acknowledgments

This work was started during a visit to the Max-Planck Institute for the Physics of Complex Systems, Dresden, gratefully supported by an Advanced Study Group on *Forecasting with Lyapunov vectors*. The author was partially supported by CNPq, Brazil, grant 304719/2015-3.

References

[1] K. Palmer, *Shadowing in Dynamical Systems, Theory and Applications* (Springer, Dordrecht, 2000).

[2] A. Carvalho, M. Lyubich and M. Martens, Renormalization in the Hénon family, I: Universality but non-rigidity (2005); https://arxiv.org/pdf/math/0508477.pdf.

[3] M. Lyubich and M. Martens, Renormalization in the Hénon family, II: The heteroclinic web (2008); https://arxiv.org/pdf/0804.0780.pdf.

[4] J. Argyris, G. Faust, M. Haase and R. Friedrich, *An Exploration of Dynamical Systems and Chaos*, 2nd edn. (Springer, Berlin, 2015).

[5] S. Strogatz, *Nonlinear Dynamics and Chaos*, 2nd edn. (CRC Press, Boca Raton, 2015).

[6] M. Cencini, F. Cecconi and A. Vulpiani, *Chaos — From Simple Models to Complex Systems* (World Scientific, Singapore, 2010).

[7] J. A. C. Gallas, Structure of the parameter space of the Hénon map, *Phys. Rev. Lett.* **70** (1993) 2714–2717.

[8] R. E. Francke, T. Pöschel and J. A. C. Gallas, Zig-zag networks of self-excited periodic oscillations in a tunnel diode and a fiber-ring laser, *Phys. Rev. E* **87** (2013) 042907.

[9] J. P. Lees *et al.* (The BABAR Collaboration), Observation of time-reversal violation in the B^0 meson system, *Phys. Rev. Lett.* **109** (2012) 211801.

[10] A. S. Eddington, *The Nature of the Physical World* (Macmillan, London, 1928).

[11] H. G. Funkhouser, A short account of the history of symmetric functions of roots of equations, *Amer. Math. Monthly* **37** (1930) 357–365.

[12] J. A. C. Gallas, Lasers, stability, and numbers, *Phys. Scripta* **94** (2019) 014003.

[13] J. A. C. Gallas, Orbital carriers and inheritance in discrete-time quadratic dynamics, *Int. J. Mod. Phys. C* **31** (2020) 2050100.

[14] J. Sommer, *Vorlesungen über Zahlentheorie* (Teubner, Leipzig, 1907).

[15] A. Endler and J. A. C. Gallas, Arithmetical signatures of the dynamics of the Hénon map, *Phys. Rev. E* **65** (2002) 036231.

[16] A. Endler and J. A. C. Gallas, Existence and characterization of stable ghost orbits in the Hénon map, *Physica A* **344** (2004) 491–497.

[17] J. A. C. Gallas, Counting orbits in conjugacy classes of the Hénon Hamiltonian repeller, *Phys. Lett. A* **360** (2007) 512–514.

[18] O. J. Brison and J. A. C. Gallas, What is the effective impact of the explosive orbital growth in discrete-time one-dimensional polynomial dynamical systems? *Physica A* **410** (2014) 313–318.

[19] J. A. C. Gallas, Preperiodicity and systematic extraction of periodic orbits of the quadratic map, *Int. J. Mod. Phys. C* **31** (2020).

[20] J. A. C. Gallas, Nonlinear dependencies between sets of periodic orbits, *Europhys. Lett.* **47** (1999) 649–655.

[21] J. A. C. Gallas, Infinite hierarchies of nonlinearly dependent periodic orbits, *Phys. Rev. E* **63** (2000) 016216.

[22] J. P. Tignol, *Galois Theory of Algebraic Equations* (World Scientific, Singapore, 2002).

[23] J. A. C. Gallas, Equivalence among orbital equations of polynomial maps, *Int. J. Mod. Phys. C* **29** (2018) 1850082.

[24] S. Pincherle, L'iterazione completa di $x^2 - 2$, *Realle Accad. dei Lincei, Rend. della Classe Sci. Fisiche, Mat. Naturali (Roma), Ser. 5* **29**(1) (1920) 329–333.

[25] L. C. Washington, *Introduction to Cyclotomic Fields* (Springer, New York, 1982).

[26] H. M. Edwards, Roots of solvable polynomials of prime degree, *Expo. Math.* **32** (2014) 79–91.

[27] N. Schappacher, On the history of Hilbert's twelfth problem, a comedy of errors, in *Matériaux pour l'Histoire des Mathématiques au XX*e *Siècle.* Actes du colloque à la mémoire de Jean Dieudonné. Nice, 1996, Séminaires et Congrès 3 (Societé Mathématique de France, Paris, 1996), pp. 243–273.

Part III

Recent Trends in Applications of Chaotic and Complex Systems

https://doi.org/10.1142/9789811221903_0013

Chapter 13

Electromechanical Models of Micro- and Nanoresonators

A. V. Lukin*, O. V. Privalova*, I. À. Popov*,
D. Yu. Skubov* and L. V. Shtukin*,†

**SPbPU, Saint-Petersburg, Russia*
†IPME RAS, Saint-Petersburg, Russia

This work is devoted to construction and numerical investigation of the dynamic of electromechanical models of micro-and nanoresonators. The main purpose of this resonators is detected the presents of adherent micro-or nanoparticles and measurement its mass. In this work, the periodical oscillations of mono and many layers nanocapacitors, from which nanoresonator situated in electric field, is constructed are studied. It is significant to note that the time of overcharge of the nanocapacitor is much less than the period of excited oscillations. This fact reduces the excitement of the nonlinear oscillations, which are yielded difficulty in accuracy of measurement and gives the possibility to apply asymptotic methods in numerical investigation. Therefore, the previous numerical and physical experiment is proved to be necessary.

13.1. Introduction

The new technologies and materials discovered in recent time enable to development the principal new micro-and nanoelectromechanical systems (MEMS, NEMS), particularly, micro-and nanoresonators. One of possible application of the graphene nanoresonators is using of these devices as detectors of nanoparticle mass, adherent on graphene layer. In result of adherence of the particle (biological cell or even molecules) to elastic graphene surface its resonance frequency is changed. This effect gives us the possibility to determine the mass of nanoparticle.

Now we have the great number of works devoted to study of nanoresonators based on graphene [1–4]. The main problem in creating of nanoresonator is an increasing of the sensitivity to nanomass of this particle. This sensitivity determines the alteration of eigenflexural oscillations frequency at setting a nanoparticle on a nanoresonator. At low Q-factor of resonators (order 100) it will be occur, that this alteration comparable with width of resonance diapason. This fact does not give us the possibility to obtain the accurate data of particle mass in case if its mass represents the value, which is order 1% from mass of graphene layer of resonator, and compels to use new models and methods.

Now the certain articles consider resonators with a few graphene layers [5–8]. This paragraph is a prolongation of works [9] and is devoted to investigation of possible increasing the precisions of measurement of resonators frequency with low Q-factor.

13.2. Monolayer Nanoresonator

The scheme of monolayer resonator is showed in Fig. 13.1.

The monolayer nanoresonator consists from graphene layer clamped on isolated rigid supports. Under graphene layer on some distance the plate conductive surface is situated. In a space between graphene layer and conductive surface with help of EMF the electric field is generated. The characteristic sizes of graphene resonator: the length of layer span — $500 \div 1000\,\text{nm}$, the width of

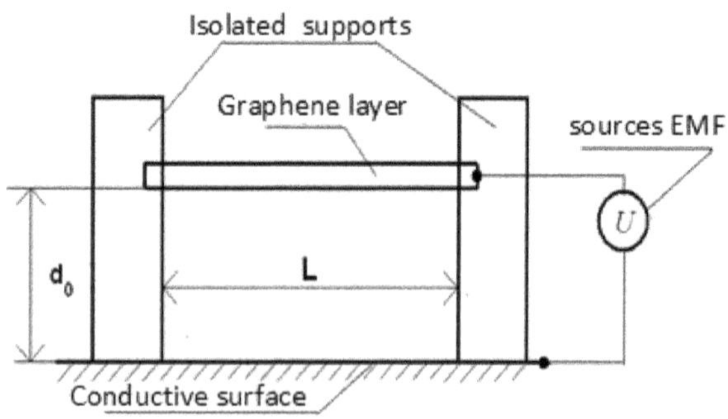

Fig. 13.1. Monolayer nanoresonator.

layer — $10 \div 20\,\text{nm}$, the gap between graphene layer and conductive surface also $10 \div 20\,\text{nm}$.

Let us consider the process of motion of this system taking into account the electrical forces. The different mechanical model of graphene layer may be considered — the tension string, the beam with flexural stiffness and simultaneously under the influence of axial tension, elastic membrane, thin tensile plate, etc. These models may be both linear and nonlinear, including geometrical and physical nonlinearity. In common case, the equations of motions may be written in a view

$$L(w) + R\ddot{w} = F_e(w,t). \tag{13.1}$$

Here w is the flexure of graphene layer, $L(w)$ the differential operator of elasticity, R the operator of inertia, $F_e(w,t)$ the attracted force, induced by assistance of electrical field.

By the long-term exposure of constant EMF, the system becomes in equilibrium position, which may be founded from the system of Eq. (13.1) with absence of inertial force

$$L(w_0) = F_e(w_0). \tag{13.2}$$

Equilibrium position does not coincide with zero flexure of graphene layer, because the load induced by assistance of electrical field do not equal zero. At increasing of flexure of layer the electrical force grows unlimited then the elastic force stays limited. Thus, the equilibrium positions, depending from EMF, may be either two or nonexistent at all. In extreme case, we have a single equilibrium. From character of acting forced it is a clear that equilibrium may have two positions. Lesser flexure is stable, and other with large flexure — unstable.

Let us consider the oscillations of graphene layer close to stable equilibrium at action of AC voltage. In this case, the solution of Eq. (13.1) searches as deflection from equilibrium position $v = w - w_0$. Then, Eq. (13.1) taking in account inertial force has a form

$$L(w_0 + v) + R\ddot{v} = F_e(w_0 + v). \tag{13.3}$$

For obtaining the equations of graphene layer motions at regimes closed to resonance let us use Galerkin method. The solution represents as series of space coordinate function V_k with coefficients x_k depended from time

$$v = x_1(t)V_1 + x_2(t)V_2 + \cdots . \tag{13.4}$$

As coordinate functions, the continuum of eigenfunction without taking into account their nonlinear elastic properties and action of electric field is considered. The eigenfunctions satisfy to equations

$$L_0 V_k - {\lambda_k}^2 R V_k = 0, \quad k = 1, 2, \ldots \tag{13.5}$$

where $L_0(V)$ is the linearized operator $L(V)$, ${\lambda_k}^2$ is the eigenfrequency of free oscillations increasing in series.

Only one component from series makes sense to keep near of resonance, which corresponds to resonance frequency λ_i. As usual, it may be λ_1:

$$v = x(t)V_1. \tag{13.6}$$

Then, using the Galerkin procedure to Eq. (13.3), taking into account (13.6), gives the equation of oscillations close to the resonance regime:

$$m\ddot{x} + \beta\dot{x} + P(x) - \frac{1}{2}\frac{Q^2}{C^2}\frac{dC}{dx} = 0. \tag{13.7}$$

In Eq. (13.7), the motion of overcharge of capacitor can be added. This capacitor is the development of graphene layer and plane conductive surface. The capacitance depends on the distance between armatures $d_0 - x$, where d_0 is the initial gap of condenser and x is the value of deformation of elastic graphene layer. The equation of electric circuit is

$$\dot{Q} + \frac{1}{RC(x)}Q = I(t). \tag{13.8}$$

Here, P are elastic forces, $Q(t)$ the value of capacitor charge, $C(x)$ the capacity of capacitor, R the internal resistance of EMF source and $I(t)$ the current in electric circuit.

Equations (13.7) and (13.8) are described as the motion of electromechanical system (Fig. 13.2), which consists of oscillation system with one degree of freedom and electric circuit consists of in-series EMF, resistor and capacitor. The interaction between mechanical system and electric circuit determines the electric field in capacitor, which creates the mechanical force affected on the oscillated mass. In the same time, the displacement of the mass determines the change of capacity.

In case if, EMF is a source with harmonic voltage $U(t) = U_0 \cos \omega t$ and at the small time capacitor discharging in compare with period of voltage, i.e., at fulfillment of condition $RC_0 \ll \frac{2\pi}{\omega}$, then we may neglect the first summand of Eq. (13.8). When charge of capacitor in time will be changed by the following form:

$$Q = C(x)U_0 \cos \omega t, \tag{13.9}$$

the electromechanical model of resonator, which is analogy of nanoresonator, is showed in Fig. 13.2.

By substituting the last expression (13.9) into Eq. (13.7), we obtain the equation of graphene layer oscillations in regime close to resonance

$$m\ddot{x} + \beta\dot{x} + P(x) - \frac{1}{2}C_0{U_0}^2\frac{d_0}{(d_0 - x)^2} = 0. \tag{13.10}$$

For the dynamic analysis, let us transform Eq. (13.10) into dimensionless form, which includes nondimensional variables: time $\tau = \lambda t$, flexure $\xi = \frac{x}{d_0}$, frequency $\Omega = \frac{\omega}{\lambda}$, where λ is the eigenfrequency of linearized oscillation system without action of electric field.

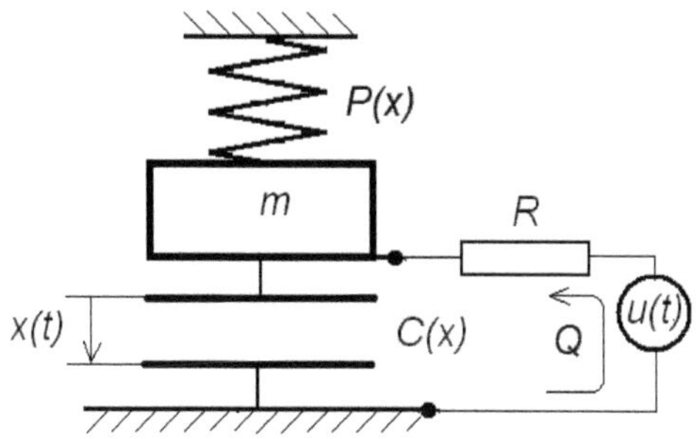

Fig. 13.2. Electromechanical model of monolayer nanoresonator.

Equation (13.10) represents the following form:

$$\ddot{\xi} + 2n\dot{\xi} + P(\xi) - b^2 \frac{1 + \cos 2\Omega\tau}{(1-\xi)^2} = 0, \tag{13.11}$$

where parameter $b^2 = \frac{C_0 U_0^2}{2m\lambda^2 d_0^2}$ is a ratio of amplitudes of electric energy and kinetic energy of oscillations with frequency λ and amplitude is equal to the gap of capacitor. Further, we consider only the case $\xi < 1$, i.e., maximum of flexure deformation of the graphene layer do not exceed the initial value of the gap.

Equation (13.11) consists two nonlinear components. The second component connects with nonlinear influence of electrical force, acted at the presence of electric field and indefinitely increasing at approach of graphene layer flexure to the value of initial gap. Really, this force plays the role of negative stiffness. First, nonlinear component $P(\xi)$ connects with nonlinear elastic of graphene layers at its transversal deformation. Graphene layers by the method of its production fix so that its edges do not make travel either transversal or longitudinal directions. The stiffness of graphene layer on the expansion is so large that equivalent Young's modulus for graphene is five times more than by the constructed steels. At flexure of graphene layer, the longitudinal tension will be increased owing to its stretching in longitudinal direction that stipulates the assistant for its transversal stiffness. This is result of many physical experiments.

As the next analog to the monolayer nanoresonator, let us consider the static deformation of string, previously tension, the force of tension of which T_0. String clamped, so that the displacement of its edges in transversal and longitudinal directions are not possible, and in the middle it is loaded by transversal force P (Fig.13.3).

Total value of tension is summed from initial and supplemental component owing to stretching of string at its transversal loading:

$$T = T_0 + T_* \varepsilon, \varepsilon = \frac{ds - dx}{dx} = \sqrt{1 + (tg\alpha)^2} - 1. \tag{13.12}$$

Substituting the last expression into conditions of equilibrium and using demonstrable geometrical correlation between flexure w and angle α, we obtain connection between longitudinal force P and

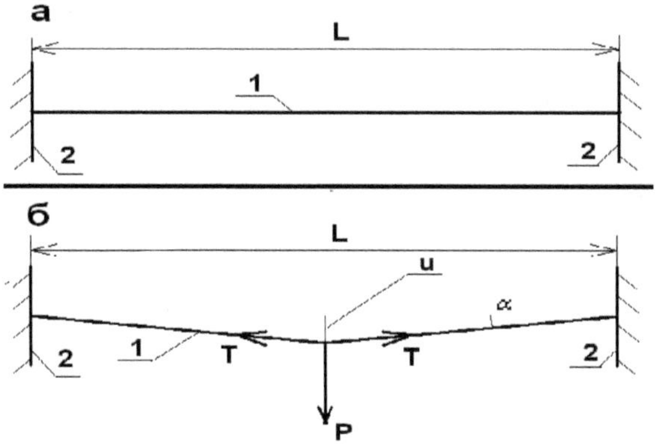

Fig. 13.3. Transversal deformation of string.

flexure

$$P = 4\frac{w}{L}\left(T_0 + T\left(\sqrt{1+\left(2\frac{w}{L}\right)^2} - 1\right)\right) \Big/ \sqrt{1+\left(2\frac{w}{L}\right)^2}. \quad (13.13)$$

The linearized approximation (13.13) gives the classical expression for transversal stiffness, so-called "inextensible" string, while at the same time it can be more correctly believed to be infinite tension. The next approximation gives cubic relation:

$$\begin{aligned} P &= 4T_0\frac{w}{L} + (T_* - T_0)\left(2\frac{w}{L}\right)^3 \\ &= 4T_0\left(\frac{w}{L} + \frac{2(T_* - T_0)}{T_0}\left(\frac{w}{L}\right)^3\right) = c(\xi + \gamma\xi^3). \quad (13.14) \end{aligned}$$

The influence of cubic term that is more strong than initial tension T_0 is smaller and larger than longitudinal stiffness of string T_* is greater. Let us introduce dimensionless variable $\xi = \frac{w}{L}$ and time $\tau = \lambda t$, where λ is the eigenfrequency of small oscillation of string. Taking into account (13.14), let us overwrite Eq. (13.11) in dimensionless form:

$$\ddot{\xi} + 2n\dot{\xi} + \xi + \gamma\xi^3 - \dot{b}^2\frac{1+\cos 2\Omega\tau}{(1-\xi)^2} = 0. \quad (13.15)$$

For Eq. (13.15), it is interesting to obtain the resonance curve — the dependence of amplitude of oscillations in steady-state regime from exciting frequency. Amplitude–frequency characteristic for Eq. (13.15) may be obtained by harmonic-balanced method. However, the nonlinear algebraic equations, obtained by harmonic-balanced method, do not solve analytically. Therefore, for Eq. (13.15) numerical experiment with scanning of exciting frequency was fulfilled.

At first, the Cauchy task is solved for exciting frequency close on resonance frequency at zero initial data. Then, after the steady-state was determined, the Cauchy task is solved again for nearest value of frequency, and initial data were taken from solution for previous value of frequency in time of installation of steady-state regime. This procedure returns at many values of frequency. Responsible resonance curves are shown in Figs. 13.4, and 13.5.

Here resonance curves that have characteristic breaking conditions either by elasticity nonlinear characteristic (at small initial tension) or by nonlinear negative electric stiffness (at large initial tension and large amplitude of oscillations) are demonstrated. Resonance curves with characteristic, which has breaking, observe in real physical experiments; in particular, these results are demonstrated in [9].

In laboratory of "Mechanical and Processes of Control" department, the demonstrational device, which enables to observe "breaking" resonance curves, was produced. As an oscillation system, a segment of steel wire clamped by edge was used with possible of

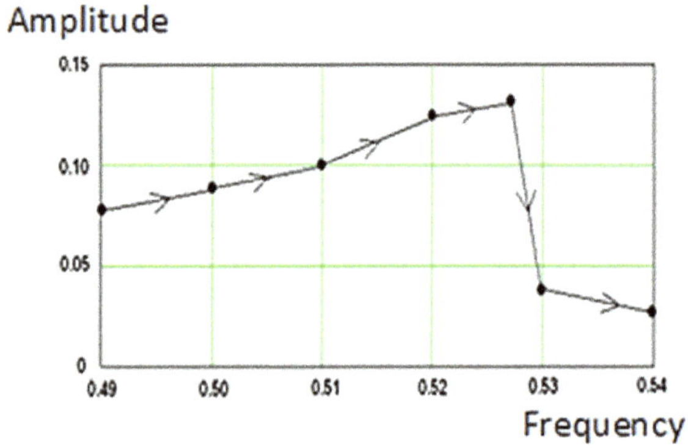

Fig. 13.4. Resonance curve at small initial tension ($\gamma = 10$).

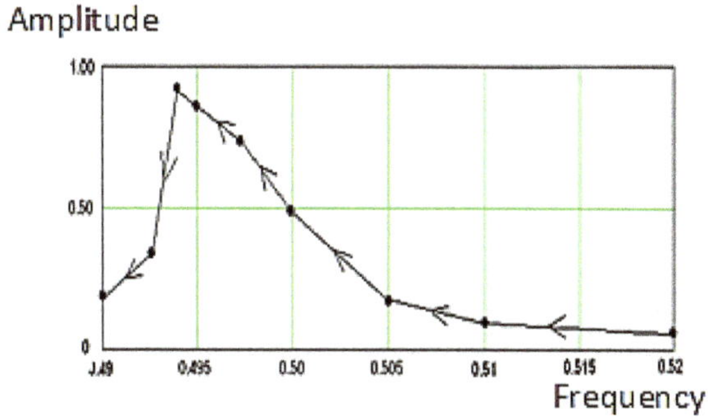

Fig. 13.5. Resonance curve at large initial tension ($\gamma = 0.1$).

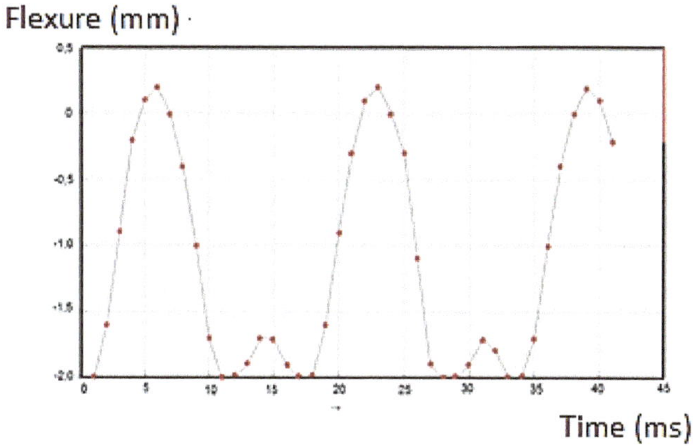

Fig. 13.6. The oscillogram of oscillation of wire at exciting frequency 62 Hz.

previous tension. The oscillation of wire is excited by magnetic system, which consists steel magnetic conductor (spring), permanent magnet and coil with alternating current. The coil feeds from generator of standard signals with possible adjustment of amplitude and frequency. The dependence of magnetic force, acted to wire, from values of wire flexure is analogy to the same dependence of electric force, acted in graphene. Namely, this force rises monotonous at increasing of flexure. The equation, described the oscillation of wire in this experiment, is analogy to Eq. (13.11). In Figs. 13.6 and 13.7, the examples of oscillogram and resonance curves, obtained on the experimental device, are shown.

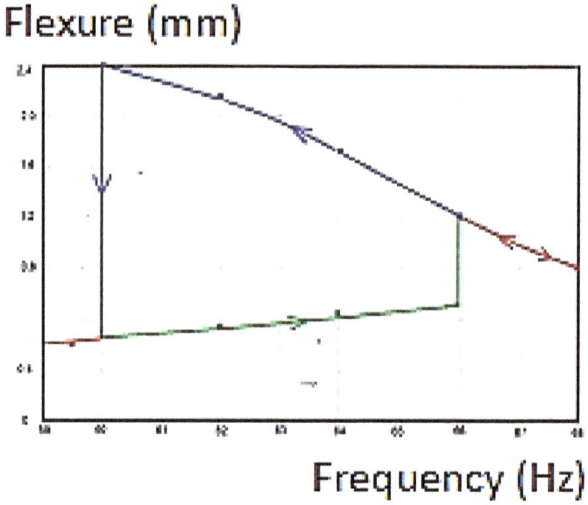

Fig. 13.7. The resonance curve.

The direction of scanning of frequency at execution of experiment is shown in Fig. 13.7. On the resonance curve, characteristic breaking at scanning of top-down exciting frequency is visible.

13.3. Differential Resonator

The scheme of differential resonator that consists two graphene layers was proposed in [20, 21]. Scheme of this resonator is shown in Fig. 13.8. Differential resonator consists of two parallel graphene layers: basic and additional. Additional layer is situated under basic layer and represents the plate conductive surface. The source of permanent EMF (basic) is aimed for creation of force interaction between basic and additional layers, and the source EMF (additional) is proposed for assign the initial condition.

The principal of work of differential resonator is as follows. The aim of one cycle of measuring is that voltage from additional source in a form of shot impulse excites combined free oscillations of graphene layers. Partial eigenoscillations of each layer (without connection) are very close and connection between them is soft in comparison with the natural elastic of each layer. Thus, this system has two

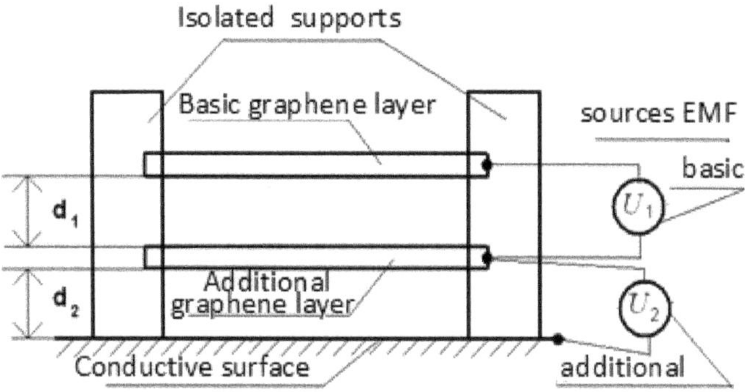

Fig. 13.8. The scheme of differential resonator.

graphene layers closely approximated one from another and close to partial frequencies. Thus, free oscillations of this system will have the character of beating. The envelope of this process is determined by detection. The characteristic envelope frequency is equal to the half of the difference of partial frequencies and much less than the partial frequency of every layer. At adherence of particle on top layer the partial frequency of this layer decreases, and characteristic frequency is also changed, then the small alteration of partial frequency gives the large alteration of envelope characteristic frequency. After some period, the free oscillations are damped and cycle measuring may be repeated.

In laboratory of "Mechanical and Process of Control" department, as well as at first case demonstrational device for observing the beating process was prepared. This device consists of the two parallel steel cantilever beams.

13.4. Parametrical Resonator

The new scheme of graphene resonator, in which the oscillation is created only by parametric action, is proposed. The scheme of this resonator is shown in Fig. 13.9.

Parametrical resonator consists of one graphene layer, which is enclosed by two conductive surfaces. This combination is formed by the two capacitors with one common armature — graphene layer.

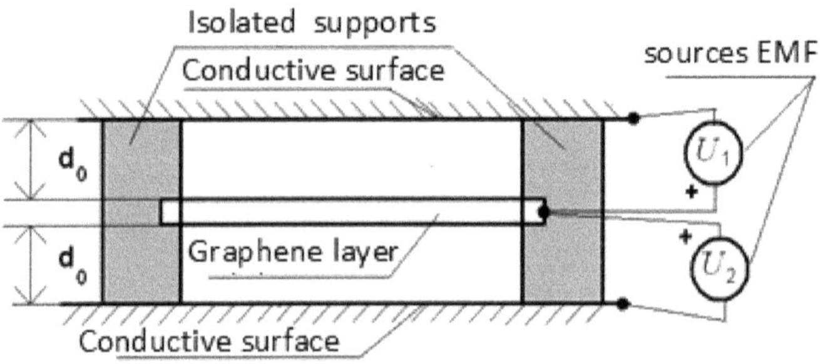

Fig. 13.9. Parametrical resonator.

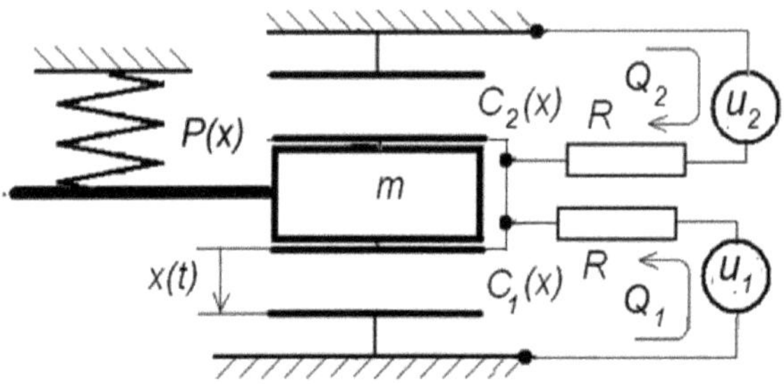

Fig. 13.10. Electromechanical models of parametrical resonator.

The sources of AC EMF with synchronous frequency connect to each of the capacitor. Macromodel analogy to parametrical resonator, which was showed in Fig. 13.9, is shown in Fig. 13.10

This model includes two capacitors so that the decreasing of distance between armatures of first gives increasing the distance in second. Analogically, equation of motion Eq. (13.10) for model of parametrical resonator may be written as

$$m\ddot{x} + \beta\dot{x} + cx - \frac{Q_1^2}{2C_1}\frac{dC_1}{dx} - \frac{Q_2^2}{2C_1}\frac{dC_2}{dx} = 0. \tag{13.16}$$

The capacities of capacitors are changed at displacement of central mass x (that analogy flexure of graphene layer) and have a view

$$C_1 = C_0\frac{d_0}{d_0 - x}, \quad C_2 = C_0\frac{d_0}{d_0 + x},$$

where the same designations are input.

Equation (13.16) with account relation between period of overcharging of capacitor circuit and period of external sources $RC_0 \ll \frac{2\pi}{\omega}$ may be written in dimensionless form:

$$\ddot{\xi} + 2n\dot{\xi} + \xi - \frac{b^2\xi}{(1-\xi^2)^2}(1 + \cos 2\Omega\tau) = 0. \tag{13.17}$$

The initial position of equilibrium of this resonator coincides with strainless position of graphene layer $\xi = 0$. At absent of external voltages, this position is stable. However, at existence of external voltage, this position began unstable. The linearized equation (13.17) has the following form:

$$\ddot{\xi} + 2n\dot{\xi} + (1 - b^2)\left(1 + \frac{b^2}{1-b^2}\cos 2\Omega\tau\right)\xi = 0. \tag{13.18}$$

This is Mathieu equation, in which for parameter b^2 both coefficient of pulsation and eigenfrequency are dependent. The border of unstable field is shown in Fig. 13.11.

For unstable equilibrium position, the diapason of frequency between left and right branches is defined. If this diapason became narrow, then Q-factor of oscillation system is less. That parametrical resonance is different from usual. This fact may be used for increasing the accuracy of measuring the resonance frequency at low Q-factor.

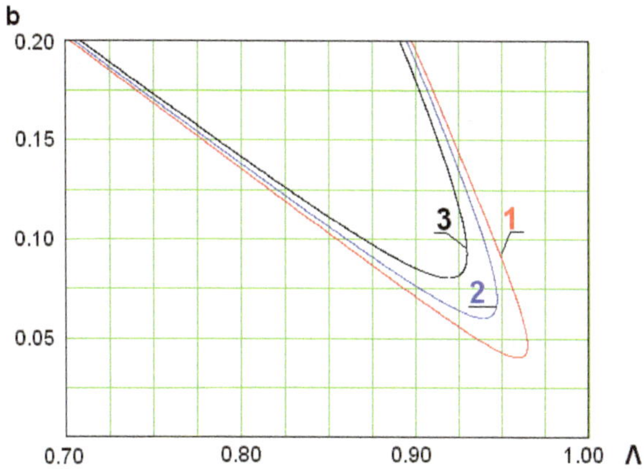

Fig. 13.11. The borders of unstable field, $\Lambda = \sqrt{1-b^2}$: 1 — Q-factor 100, 2 — 70, 3 — 50.

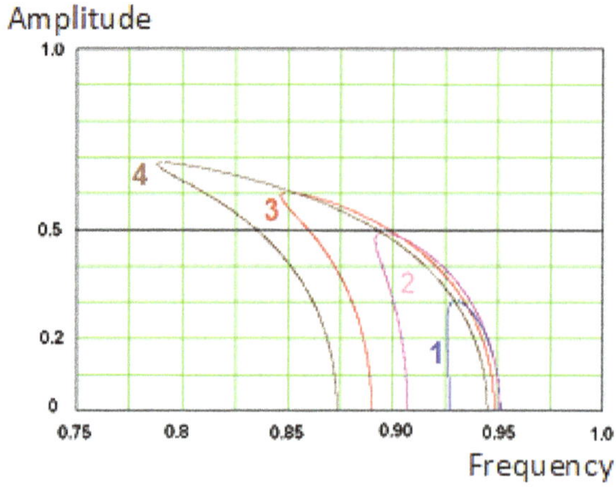

Fig. 13.12. Amplitude of stationary regime.

At parametric oscillations closed on equilibrium position, the amplitude is limited by the nonlinear of the oscillations system. Amplitude of stationary regime may be found by harmonic-balanced method, i.e., to search solution of Eq. (13.17) in a view of harmonic oscillations with frequency Ω. In Fig. 13.12, the dependence of amplitude of stationary regime from frequency of external voltage at different values of parameter b^2 is shown.

13.5. Parametrical Nanoresonator with Magnetic Exciting of Oscillation

In this section, the equilibrium and oscillation nearly situated graphene layers with harmonic currents excited parametrical resonance oscillations are considered. These parametrical oscillations are used for determine the mass of adherent nanoparticle. The task of this type is a prolongation of our works [11–13], in which the oscillations of the different types of nanoelectromechanical system (NEMS) excite by alternative electric field. In this section, parametrical oscillations excited by alternative current conducted through two closely spaced graphene layers are considered. The determination of nanoparticle mass, adherent to one of parallel graphene layers, consists in determination of shifting the zone of parametrical resonance.

The main problem at creating a resonator, as detector of adherent particle, is an increasing of the sensitivity to nanomass of this nanoparticle. The explanation of advantage of using the parametrical resonance at adherent of nanoparticle on graphene layer at low Q-factor (order ~100) was done in previous section.

In this section the method of the determination is a prolongation of works [20–22]. In these works the change of the spectrum of mechanical oscillations of different types NEMS, excited by adherence of nanoparticle, is determined by the alteration of amplitude–frequency characteristic in a region of parametrical or forced resonance.

In this section, our work propose the new scheme of resonator consisted of two graphene layers. The scheme of installation of graphene layers is analogy that it was fulfilled in differential resonator [22]. In difference from differential resonator, the graphene layers are used as conductors on which alternating current is conducted. Hence, around conductive layers, the alternating magnetic field is excited. In every layer, the alternating in time magnetic force is arise and their mechanical oscillations are excited. It is shown that this action may get excitation of parametrical resonance. As it known, the width of zone of parametrical resonance is narrowed at decreasing of Q-factor of electromechanical system. This condition may be useful for increasing the accuracy of measuring the frequency of resonator at its low Q-factor.

The scheme of this resonator is shown in Fig. 13.13

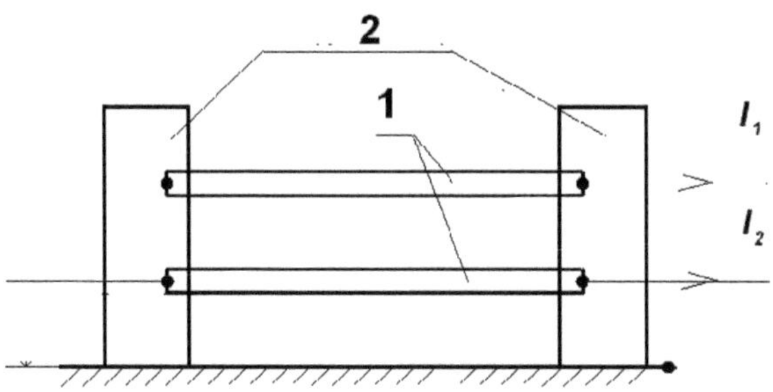

Fig. 13.13. The scheme of resonator: 1 — graphene layers, 2 — supports.

Resonator consists from two graphene layers situated on some distance one from another. At every layer, the current is conducted as shown in Fig. 13.13.

The analogous experiment was fulfilled in laboratory of "Mechanical and Process of Control" department. In this experiment, two tension steel wires with alternating current are used.

The forces-distributed interaction between conductors with currents I_1, I_2 in case of its even directivity attraction is determined by relation

$$\mathbf{q} = \frac{\mu_0 I_1 I_2}{2\pi} \frac{\mathbf{r}}{r^2}, \tag{13.19}$$

where $\boldsymbol{r}$ the radius vector, connecting the points of deformable conductors, initially lies on common normally [10].

First, let us fulfill the solution of task of static deformation initially parallel conductors at action of distributed attraction force. The boundary problem of symmetric deformation relatively dimensionless distance between strings $1 - v$, where $v = 2v_1/h$ is summary deflection of strings from nondeformable positions, h the initial gap, has the following form [9]:

$$v'' + \frac{\lambda^2}{1 - v} = 0, \quad v(0) = v(1) = 0. \tag{13.20}$$

Here $\lambda^2 = \frac{\mu_0 I^2}{2\pi}$. The small free oscillations of tension string (filament) relatively nondeformable position are described by equation

$$\frac{T}{l^2} v'' - \rho \ddot{v} = 0. \tag{13.21}$$

with boundary conditions $v(0, t) = v(1, t) = 0$. The next designations are inputs: $\rho = \rho_V S$ — linear density of string, ρ_V — volume density of nanostring material, S — square of transversal profile, and touch $(\,)'$ designates the derivative on dimensionless coordinate. The frequency of transverse oscillations on first eigenform is given by $\overline{\omega} = \pi\sqrt{\frac{T}{\rho l^2}}$.

Let us get estimate the value of current, at which the static deformation by attraction of graphene layers visually used parameters of graphene nanoresonator: length $l = 1000\,\text{nm}$, width $= \delta = 10\,\text{nm}$, thickness $b = 0.3\,\text{nm}$, resonance frequency $\overline{\omega} = 30\,\text{MHz}$ and Q-factor $\sim$100. From formula for frequency of free oscillation, we find the tension force

$$T = \rho_V S\, l^2\overline{\omega}^2/\pi^2 = 1.5 \times 10^{-12}\,\text{N}.$$

Using the expression for parameter $\overline{\omega}$, and believed that its value is order 1, for gap $h \sim 10\,\text{nm}$, we obtain value of current strength $I_0 = \frac{h\overline{\omega}}{l}\sqrt{\frac{\pi T}{\mu_0}} = 40\,\text{mcA}$.

Introducing nondimensional time $\tau = \frac{\overline{\omega}}{\pi}t$, we transform the equation of free oscillation (13.21) into the form:

$$v'' - \ddot{v} = 0 \tag{13.22}$$

with boundary condition $v(0,\tau) = v(1,\tau) = 0$, () the difference by nondimensional time. The eigenforms and frequency of free oscillations for tension string are given by

$$V_k(s) = C_k \sin \pi k s,\ \Omega_k = \pi k\,,\ k = 1, 2, \ldots.$$
$$s = 0, 1. \tag{13.23}$$

At normalization $\int_0^1 V_k V_j ds = \delta_{kj}$, $C_k = \sqrt{2}$.

Let us consider the dynamical task, introduced into Eq. (13.20) inertial component and believed that currents are harmonic with amplitude I_0 ($I(\tau) = I_0 \sin \omega\tau$, $\omega = \frac{\pi\tilde{\omega}}{\overline{\omega}}$ $\tilde{\omega}$ is the frequency of alternating current). Boundary problem takes the following form:

$$v'' + \frac{\lambda^2(1 - \cos 2\omega\tau)}{2(1 - v)} - \ddot{v} = 0,$$
$$v(0,\tau) = v(1,\tau) = 0. \tag{13.24}$$

For simplification in future, let us consider the small oscillation relatively static equilibrium position, which was founded taking into account the nonlinearity of deformation (for example, at (13.15)). The complete dynamical flexure we search as sum of static and alternating component is given by $v(s,\tau) = v_0(s) + \tilde{v}(s,\tau)$. Then, the equation of oscillations relatively dynamical flexure takes the following form:

$$\tilde{v}'' + \frac{\lambda^2 \tilde{v}}{(1 - v_0)(1 - v_0 - \tilde{v})} - \frac{\lambda^2 \cos 2\omega\tau}{(1 - v_0 - \tilde{v})} - \ddot{\tilde{v}} = 0. \qquad (13.25)$$

The boundary conditions are homogenous $\tilde{v}(0, \tau) = \tilde{v}(1, \tau) = 0$. The solution is fulfilled in a diapason of parameter value $\lambda < \lambda_*$, where value λ_* conforms to parameter, responsible to static point of branching. Static solution has two equilibrium positions: stable and unstable.

Let us investigate small free oscillations (without consideration the component, clearly depended from damping), relatively stable static equilibrium form taking in account negative stiffness, excited by electrical force at deformable equilibrium form of graphene layer. The motion of this type without dissipation is described by linearized Eq. (13.25):

$$\zeta'' + \frac{\lambda^2}{(1 - v_0(s, \lambda))^2} \zeta - \ddot{\zeta} = 0. \qquad (13.26)$$

This is a linear equation with alternative coordinate-wise positive coefficient

$$a(s, \lambda) = \frac{\lambda^2}{(1 - v_0(s, \lambda))^2}.$$

Thus, we obtain the boundary problem: the equation

$$\zeta'' + a(s, \lambda)\zeta - \ddot{\zeta} = 0 \qquad (13.27)$$

and boundary condition: $\zeta(0, \tau) = \zeta(1, \tau) = 0$.

In the first place, it is necessary to find the static flexure using relation (13.27) without inertial term, both for stable and unstable forms $v_0(s, \lambda)$, solved them as a system of implicit equations.

By the frequency and form of small string oscillations as eigenfunctions taking into account attraction represented in the form of the time harmonic function of boundary problem (13.28):

$\zeta = Z(s)\sin(\Lambda\tau + \alpha)$, we receive the space boundary problem:

$$Z'' + a(s,\lambda)Z + \Lambda^2 Z = 0,$$
$$Z(0) = Z(1) = 0. \tag{13.28}$$

The symmetric forms of oscillations, taking into account attraction represented in the form of series of odd forms of string oscillations without external action, determined by relation (13.23):

$$Z(s) = \sum_{i=1,3,\ldots}^{\mathrm{n}} \beta_i V_i(s). \tag{13.29}$$

Projective conditions for coefficients of mode expansion β_{i} have the following form:

$$\int_0^1 (Z'' + a(s,\lambda)Z + \Lambda^2 Z)V_i(s)ds = 0, \quad i = 1, 3, \ldots. \tag{13.30}$$

After transformation of system (13.30) taking into account (13.23) we receive a system of linear homogenous algebraic equations relatively of coefficients β_{i}:

$$\beta_{\mathrm{i}}(\Lambda^2 - \Omega_i^2) + \sum_{k=1,3,\ldots}^{n} \beta_k a_{ki} = 0, \; i = 1, \ldots, \tag{13.31}$$

where coefficient $a_{ki} = \int_0^1 a(s,\lambda)V_i(s)V_k(s)ds$. By equating to zero the determinant of system (13.31), we find the frequencies Λ_i, $i = 1, 3, \ldots$ where every value coincides with manifold of coefficients ${\beta_k}^{(i)}$, $k = 1, 3, \ldots$, and, accordingly, to their modes of oscillations $Z_i(s)$.

Keeping in expansion (13.23) only the first form, we may find approximate alteration of first eigenfrequency, which corresponds to static attraction of strings:

$$\Lambda_1^2 = \pi^2 - \int_0^1 a(s,\lambda){V_1}^2(s)ds. \tag{13.32}$$

As the function $a(s,\lambda) = \frac{\lambda^2}{(1-v_0(s,\lambda))^2}$ is positive on the segment $[0,1]$ on coordinate s, so integrand also is positive — integral nonnegative, and, accordingly, the frequency of free oscillations taking into account attraction decreases. Hence, the decreasing is droningly to come up to zero at value of parameter $\lambda = \lambda_*$, responsible to point of branching of static boundary problem. Namely, in this case, variational equation for nonlinear boundary problem (13.28) has nonzero eigenfunctions coincide with solution of boundary problem (13.20) at $\Lambda = 0$. The dependence of first eigenfrequency, which taking in account attraction at increasing of parameter λ, i.e., considering of increasing of current amplitude, is shown in Fig. 13.14.

Here Λ_1 the relative frequency, i.e., divided on the first eigenfrequency of string without taking into account their attraction in nondimensional time, is equal to π.

Let us transform Eq. (13.25), leave alone in it only linear relatively $\tilde{v}$ component,

$$\tilde{v}'' + \frac{\lambda^2(1-\cos 2\omega\tau)\tilde{v}}{(1-v_0)^2} - \frac{\lambda^2 \cos 2\omega\tau}{(1-v_0)} - \ddot{\tilde{v}} = 0. \tag{13.33}$$

The boundary conditions are homogenous $\tilde{v}(0,\tau) = \tilde{v}(1,\tau) = 0$. Initial condition for deformed string conforms to stable static form at given value of parameter λ. Thus, we may suppose that $\tilde{v}(s,0) = 0$, $\dot{\tilde{v}}(s,0) = 0$.

In case of small oscillations, we consider approximate solution of boundary problem (13.33) in the form $\tilde{v} = Z_1(s)u(\tau)$, where Z_1(s) is the first mode shape taking into account attraction of strings.

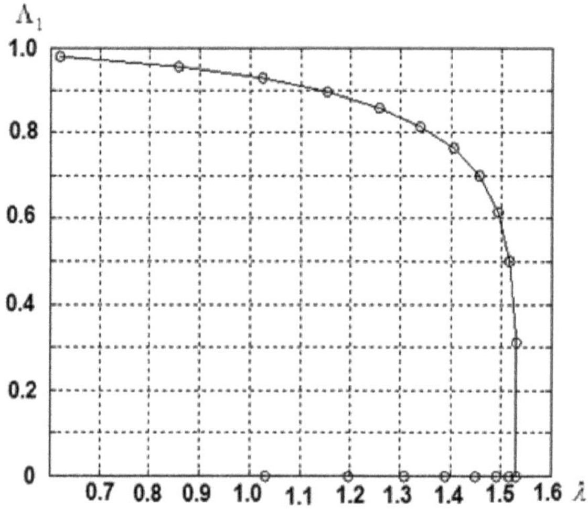

Fig. 13.14.

Projecting equation on this form, we have differential equation in time:

$$u\int_0^1 Z_1''Z_1ds + u(1-\cos 2\omega\tau)\int_0^1 \frac{\lambda^2 {Z_1}^2}{(1-v_0)^2}ds - \ddot{u}\int_0^1 {Z_1}^2 ds$$

$$= \cos\, 2\omega\tau \int_0^1 \frac{\lambda^2 Z_1}{(1-v_0)}ds. \tag{13.34}$$

Let us designate the integrals, included into (13.34), as

$$\int_0^1 \frac{\lambda^2 {V_1}^2}{(1-v_0)^2}ds = p(\lambda), \quad \int_0^1 \frac{\lambda^2 V_1}{(1-v_0)}ds = f(\lambda).$$

Using equation for mode Z_1 (13.34), we obtain

$$\ddot{u} + (\Lambda_1^2 + p(\lambda)\cos 2\omega\tau)u = -f(\lambda)\cos 2\omega\tau. \tag{13.35}$$

Equation (13.35) is a linear nonautonomous differential equation with periodic coefficient and periodic right part. Homogenous part of Eq. (13.35) taking into consideration the dissipation component presents in the following form:

$$\ddot{u} + 2n\dot{u} + (a(\lambda) + p(\lambda)\cos\ 2\omega\tau)u = 0, \tag{13.36}$$

where $a(\lambda) = \Lambda_1^2$. Equation (13.36) has a form of Mathieu equation with constant exciting coefficient. Hence, both the frequency of free oscillations and coefficient of exciting depend from parameter λ, determined by current strength in conductors. Parametrical resonance takes place at next relations of frequencies:$\omega = \omega_0, \omega_0/2, \omega_0/3, \ldots$, where $\omega_0 = \Lambda_1$. The right part of equation (13.36) has harmonic component with frequency 2ω therefore at $2\omega = \omega_0$ the resonance of forced oscillation, the frequency of which coincides with second parametric resonance, takes place. At $\lambda \to 0$, $v_0 \to 0$, $a(\lambda) = \Lambda_1 \to \pi$, $p(\lambda) \to 0$, consequently, nondimensional frequency of excitement current of parametrical oscillation tends to π.

In the first approximation periodic solution coincided with border of first (main) parametric resonance searches by harmonic-balanced method with frequency equal to half of frequency of external action

$$u = A\cos\omega\tau + B\sin\omega\tau. \tag{13.37}$$

In result, we receive two linear algebraic equations relatively coefficients A and B that give the equation for frequencies appropriated for borders of main parametrical resonance:

$$(a(\lambda) - \omega^2)A + 0.5\,p(\lambda)A + 2n\omega B = 0,$$
$$(a(\lambda) - \omega^2)B - 0.5\,p(\lambda)B - 2n\omega A = 0. \qquad (13.38)$$

The dependence of the zone of parametrical resonance from λ in mentioned approximation is determined by equal to zero the determinant of system (13.38) (Fig. 13.15).

The main idea of determination of nanoparticle mass, adherent to graphene layer, consists in alteration of parametrical resonance zone taking into account the change of eigenfrequency and mode of free flexure oscillations. For simplification in further we assume that significant alteration has place only with first symmetrical form and, thereafter, with first eigenfrequency, what may be, for example, at adherent of nanoparticle near string middle. Hence, both coefficients $p(\lambda)$, $f(\lambda)$ in Eqs. (13.35), (13.36) and eigenfrequency Λ_1 are changed. The aim of analytical represent of approximated zone of parametrical resonance consists in estimation of conformity spectral properties of nanostring alteration at adherent of the nanoparticle with alteration of excitement frequency (frequency of current).

Let us consider forced and parametrical nonlinear oscillations of string taking into account damping, described Eq. (13.34) in the assumption of viscous damping for considerable nanoscales. As an early space distribution let us search in form $\tilde{v} = Z_1(s)u(\tau)$, taken

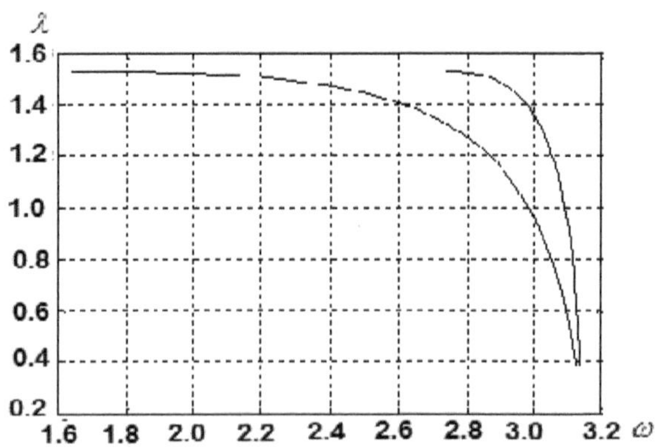

Fig. 13.15. The zone of parametric resonance.

only first mode of free oscillations, which was found only keeping the first mode of free oscillation taking into account attraction of the strings. Projective equation on the first mode of free oscillations, considered nonlinear components, is written in the following view:

$$\varphi(u)(\ddot{u}+2n\dot{u})+(b(\lambda)\Lambda_1^2+\lambda^2 e(\lambda))u^2-\Lambda_1^2 c(\lambda)u$$
$$=\lambda^2 d(\lambda)\cos(2\omega\tau). \tag{13.39}$$

Here, next designations are input:

$$\varphi(u,\lambda)=b(\lambda)u-c(\lambda),$$
$$b(\lambda)=\int_0^1(1-v_0(s,\lambda))Z_1{}^3(s)ds,$$
$$c(\lambda)=\int_0^1(1-v_0(s,\lambda))^2 Z_1^2(s)ds, \tag{13.40}$$
$$d(\lambda)=\int_0^1(1-v_0(s,\lambda))Z_1(s)ds,$$
$$e(\lambda)=\int_0^1\frac{Z_1^3(s)}{(1-v_0(s,\lambda))}ds.$$

Coefficient $\varphi(u)$ at highest derivations in Eq. (13.39) depends from unknown u and changing in time may have the return point, where $\varphi(u)$ is equal to 0.

The parameters, corresponded to contact of strings at their oscillations, are determined the equality to zero the function $1-v$. Substitution $v=v_0+\tilde{v}$, where $\tilde{v}=Z_1(s)u(\tau)$, gives possibility to estimate the critical value of coefficient $u(\tau)$, at which strings may touch. Cleary, $u_{\text{crit}}=\min\left(\frac{1-v_0}{Z_1}\right)=\frac{1-\max(v_0)}{\sqrt{2}}=0$, where $\max(v_0)$ corresponds to flexure in a middle, coefficient $\sqrt{2}$ at Z_1 inputs from condition of normalization. Thus, it is needed to check, how the critical value of amplitude of oscillation corresponds of string touch, and critical point in dependence $\varphi(u,\lambda)$ from parameter λ correlate.

From Fig.13.16, we can see that at every values of parameters the level of return point is situated over the level of critical value, i.e., addition condition has a form $u<u_{\text{crit}}$. By indicating the value, it is necessary to check the calculation the motion of dynamical deformation at solution of Eq. (13.39).

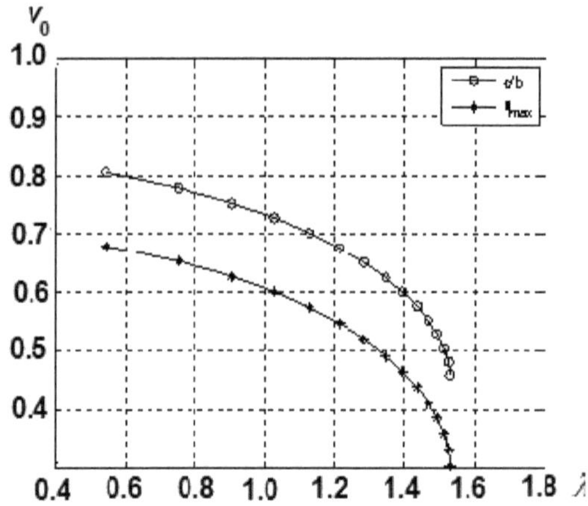

Fig. 13.16. The dependence of critical value of maximal flexure v_{cr} from λ.

By the same meaning, the main interest represents the behavior of system at the frequencies near to main parametrical resonance and, consequently, at value of current frequency near to free oscillations frequency. Let us stay on the determination of amplitude–frequency characteristic of parametrical resonance, believed that $\omega \sim \Lambda_1$.

Attempts to receive the stationary regimen in zone of parametrical resonance by harmonic balance method give the system algebraic nonlinear equations, the analytic solution of which looks impossible. Therefore, the numerical experiment with nonlinear Eq. (13.39) was fulfilled as near so inside zone of parametrical resonance. In time of realization of numerical experiment, the Cauchy task solved firstly nearly of right border of parametrical resonance zone at zero initial condition. At achievement of steady-state regime, the Cauchy task solved again for slightly smaller value of frequency, but at initial condition, taken from solution in some moment of time. This procedure repeated at many values of frequencies until the left border of parametric zone is reached. In Figs. 13.17–13.20, the examples of oscillograms, obtained in result of fulfillment of the numerical experiments, are shown.

From these oscillograms, it is seen that out of zone of parametrical resonance amplitude of oscillation extremely small and oscillations

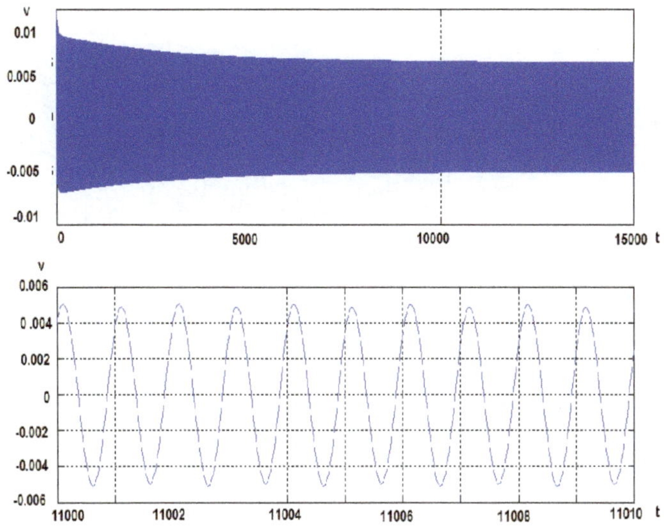

Fig. 13.17. Oscillogram of oscillations at $\omega = 3.1190$.

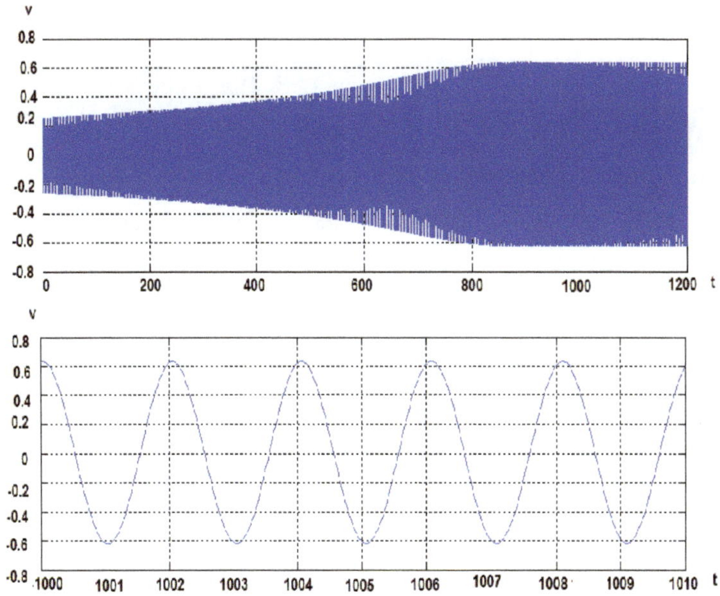

Fig. 13.18. Oscillogram of oscillation at $\omega = 3.1160$.

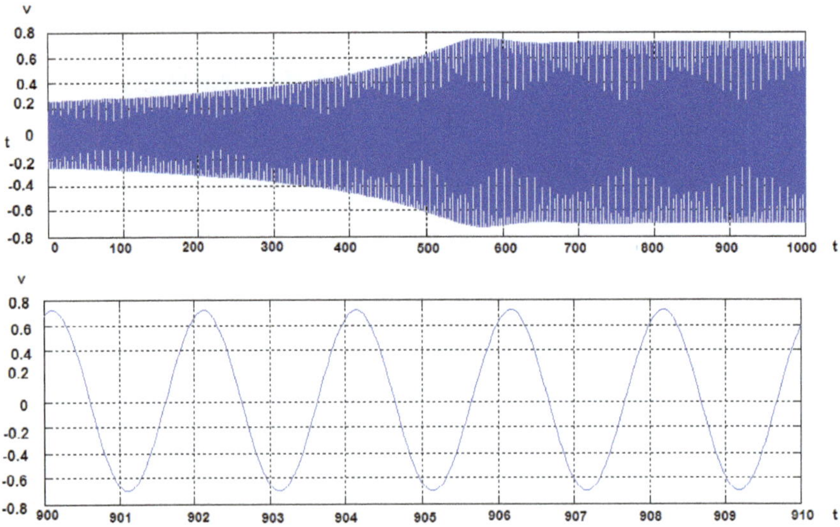

Fig. 13.19. Oscillagram of oscillation at $\omega = 3.1130$.

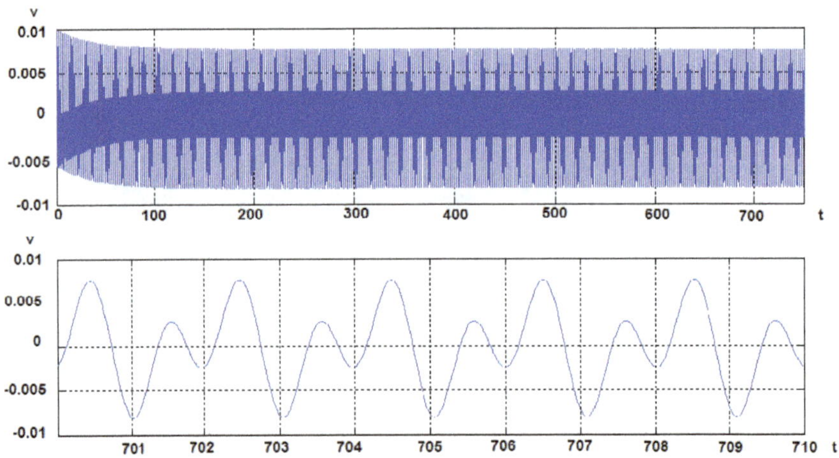

Fig. 13.20. Oscillogram of oscillation at $\omega = 3.1129$.

have the second harmonic of excitement frequency. Inside of parametrical resonance zone, the amplitude receives the value is order 0.7 of initial distance between graphene layers and consist only first harmonic of excitement frequency.

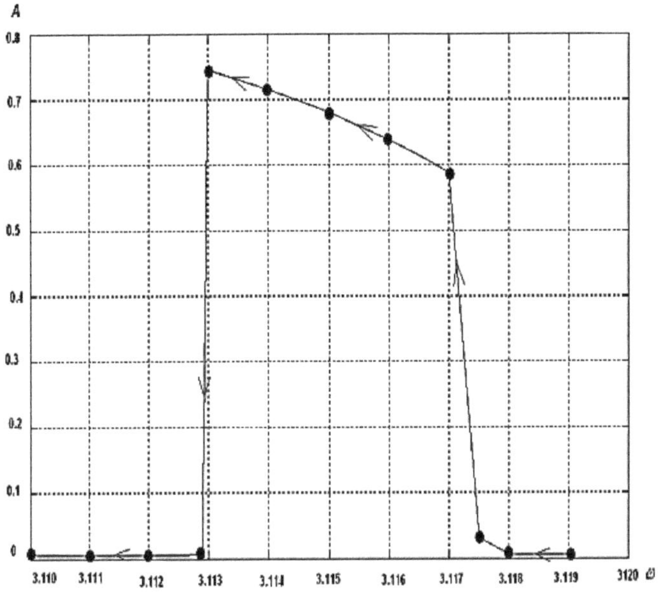

Fig. 13.21. The resonance curve in a zone of parametric resonance.

After results of numerical experiment, the resonance curve was built, which is shown in Fig. 13.21.

In Fig. 13.21, the points designate the values of steady-state amplitudes of oscillations at appropriated values of frequency. Arrowheads show the direction of frequency alteration at transition from one regime to another.

The zone of parametrical resonance obtains the enough narrow. The resonance curve turns out nonsymmetrical relatively to medium-sized frequency band (zone of resonance). Moreover, we can observe the breaking on the resonance curve at scanner of excited frequency top-down. To the small (relatively with zone of resonance) alteration of frequency corresponds abrupt decreasing of amplitude of steady-state regimes.

Thus, the new scheme of graphene nanoresonator was proposed. Investigation of possible regimes of this resonator work is shown that in this device excitement of parametric oscillation with large amplitude is possible. The distribution of amplitudes on frequency diapason has resonance character and resonance curve has characteristic breaking.

Obtained results may be useful at using of this construction resonator as detector of mass of nanoparticle adherent on the one of its layers. For determination of size of adherent particle, it may be used the value of frequency breaking, which more accuracy to measuring then the frequency of maximum on the resonance curve.

13.6. Self-oscillation Regime of Nanoresonator

In overwhelming majority of experimental works on study the graphene resonator [1–9], so as in works, where the various mechanical models of nanoresonators are considered [10, 11], it is proposed to determine the eigenfrequency of resonator by the method of building the amplitude-frequency characteristic (resonance curve). It means, in essence, the next. On the resonator, the forced periodical action with some frequency is realized. Further, the amplitude of steady-state oscillations of resonator is measured. Then the frequency of action is changed and again the amplitude of steady-state oscillation is measured. This procedure is realized in frequency diapason inside of which hypothetically the eigenfrequency of resonator is situated. Then the dependence of amplitude of steady-state oscillations from frequency of action is built. The frequency corresponding to maximum value of amplitude assumes as required eigenfrequency. The preference of this method is simplicity of experiment fulfillment.

Let us note the disadvantage of this method. First, the fulfillment of experiment demands the significant expenditures of time. The measuring must be carried out for large values of frequencies. Second, the determination of maximum values of amplitude may have the large inaccuracy. Namely, nearly of resonance the amplitude deeply changes at small alteration of frequency, which makes difficult the searching of this extremum. Using nonlinear resonance curves, which have abrupt breaking, significantly simplify the task of searching resonance frequency. Then we have another unpleasant factor — the frequency of breaking strongly depends on amplitude of action.

Oscillating system, in which self-oscillations possible, will not have this enumerated imperfection. Appearance of self-oscillations regime is possible at availability of positive feedback between oscillations system and the source of excitement of oscillations [22].

The important degree of self-oscillations regime is self-regulation on the resonance frequency at slow (in comparison with period of oscillation) alteration of parameters of oscillations system. For example, at adherent of nanoparticle on the graphene layer, the mass, taken part in oscillation process, is changed that stipulate to alteration of self-oscillation frequency.

The proposed scheme of self-oscillation resonator is shown in Fig. 13.22.

Self-oscillation resonator consists from amplifier, which must have positive feedback connection. Feedback circuit properly has graphene resonator and vibration transducer of graphene layer.

Output current of amplifier conducts by graphene layer. Graphene layer itself is situated in magnetic field. Induction vector is normal to

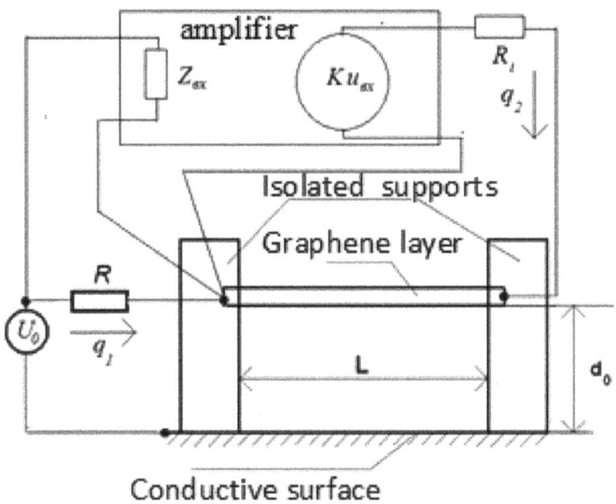

Fig. 13.22. The scheme of self-oscillation resonator.

current direction and direction of layer flexure. Thus, the magnetic force acts on graphene layer and excites its flexure.

Graphene layer and conductive surface under layer are same and are early produce capacitor. To armature of capacitor, we apply the source of constant EMF U_0. At alteration of flexure of graphene layer, the capacity is changed that excites overcharge of capacitor. The current of overcharge depends on velocity of flexure alteration. Voltage applied to resistor R is proportionately to current of overcharge and input on entry of amplifier, thereby feedback of amplifier close. This feedback may be positive or negative, which is determined by magnetic field direction.

Let us use mechanical model of graphene layer at its transversal oscillation represented in [21]. This model represents itself electromechanical system, consisted from oscillation mechanical system with one degree of freedom and electrical circuit from in-series source EMF, resistor and capacitor, presented in Fig. 13.23. Interference of mechanical system and electric circuit stipulates due to exciting of mechanical force acted on the oscillation mass and displacement of mass excites the change of capacitance of capacitor. In a case of current transmission on graphene layer and at the present of magnetic field, the magnetic force will be done on the mass. The value of this force is proportional to value of current $\dot{q}_2$, magnetic induction B and length of conductor (graphene layer) L.

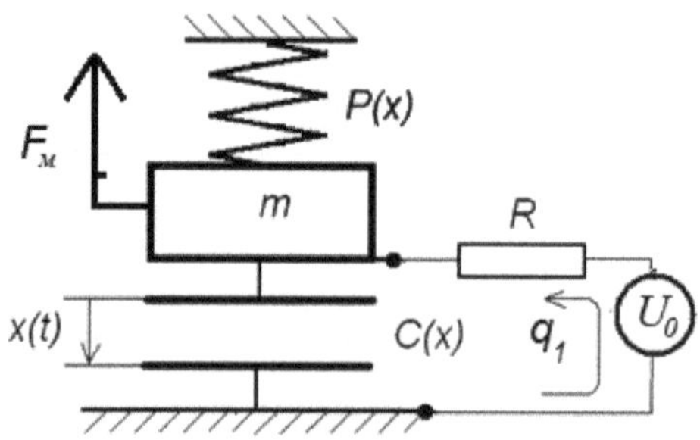

Fig. 13.23. Electromechanical model of self-oscillating resonator.

The equation of motion has the same form that in [21] with addition force connected with availability of magnetic field (last term in Eq. (13.41))

$$m\ddot{x} + b\dot{x} + P(x) - \frac{1}{2}\frac{q_1^2}{C} + BL\dot{q}_2 = 0. \tag{13.41}$$

For two electric contours adequacy with scheme showed in Fig. 13.22, let us write Kirchhoff equations

$$U_0 - R\dot{q}_1 - \frac{q_1}{C} = 0, \quad F(u_{\text{BX}}) - R_i\dot{q}_2 = 0, \tag{13.42}$$

where $F(u_{\text{BX}})$ is the characteristic of amplifier. To this equation, it is next to add

$$u_{\text{BX}} = R\dot{q}_1. \tag{13.43}$$

For convenience instead of charges in contours, let us input the new unknowns. In particular, it is more simply to input the voltage on condenser v, which is expressed from the charge

$$v = \frac{q_1}{C}. \tag{13.44}$$

Then the first from equation of system (13.42) takes a form

$$R(C\dot{v} + \dot{C}v) + v = U_0, \tag{13.45}$$

and output current of amplifier also may be expressed through voltage on the condenser v capacitance C and their derivatives

$$\dot{q}_2 = F\left(\frac{R}{R_i}(C\dot{v} + \dot{C}v)\right). \tag{13.46}$$

Finally, capacitance of capacitor may be expressed through flexure of graphene layer

$$C = C_0\frac{d_0}{(d_0 - x)}, \tag{13.47}$$

where C_0 is the capacitance of condenser at nondeformable graphene layer.

Then, we obtain the system of equations, described the oscillation process in considerable electromechanical system, which is written in form

$$m\ddot{x} + b\dot{x} + P(x) - \frac{1}{2}C_0 v_0^2 \frac{d_0}{(d_0 - x)^2}$$
$$-BLF\left(\frac{R}{R_i}C_0\left(\frac{d_0}{d_0 - x}\dot{v} + \frac{d_0}{(d_0 - x)^2}v\dot{x}\right)\right) = 0,$$
$$RC_0\left(\frac{d_0}{d_0 - x}\dot{v} + \frac{d_0}{(d_0 - x)^2}v\dot{x}\right) + v = U_0. \quad (13.48)$$

The characteristic of amplifier may be difference. Let us believe that in common case the characteristic of amplifier has a form

$$F(u) = V_0 \frac{2}{\pi} \text{arctg}\left(\frac{\pi}{2}K\frac{u}{V_0}\right). \quad (13.49)$$

Here, K is the coefficient of increasing at small input signals and V_0 the voltage of limited signals on the exit of amplifier. This characteristic allows to taking into account almost linear sector at small input signals, smooth nonlinear at large input signals and limitation of output signal under some level, depended from feeding of amplifier.

The voltage V_0 must choose the following consideration. At appearance of self-oscillating regime, the amplitude of output signal from amplifier will be closed to V_0. In steady-state self-oscillation regime, it is necessary that amplitude of graphene layer will be enough large, but not be exceed the value of initial gap. From this, at first, it is necessary to solve the task about resonance forced oscillations of graphene layer under action of magnetic interference and to choose the value of output current of amplifier so that amplitude of graphene layer in resonance regime will be specified value in order 0.3–0.5 from the value from initial gap. Without loss of generality, we may choose $V_0 = U_0$, and necessary of output current provided by the value of resistor R_i.

Equations (13.48) may be written in nondimensional form:

$$\xi'' + 2\nu\xi' + \xi + \gamma\xi^3 - \beta^2 \frac{\eta}{(1-\xi)^2}$$
$$- \alpha F \left(\delta \left(\frac{1}{1-\xi}\eta' + \frac{1}{(1-\xi)^2}\eta\xi' \right) \right) = 0,$$
$$\delta \left(\frac{1}{1-\xi}\eta' + \frac{1}{(1-\xi)^2}\eta\xi' \right) + \eta = 1, \qquad (13.50)$$

where nondimensional variables $\xi = \frac{x}{d_0}$ is the dimensionless flexure, $\eta = \frac{v}{U_0}$ the dimensionless voltage on the capacitor, $\tau = \lambda t$ the nondimensional time, λ the eigenfrequency of graphene resonator. Then, in elastic characteristic $P(\xi)$ taking into account only linear part, $()'$ is the derivative on nondimension time.

In Eq. (5.50), three dimensionless multipliers are inputted, which have concrete physical sense. First, $\frac{1}{2}\frac{C_0 {U_0}^2}{m{d_0}^2\lambda^2} = \beta^2$, the relation of energy of electrical field to maximum of kinetic energy of oscillation at amplitude equal to full gap. This parameter was met early at consideration of resonator oscillations [20, 21]. At that estimation, it corresponds to value of order 0,01–0,1.

The second multiplier

$$\frac{\frac{U_0}{R_i}BL}{m\lambda^2 d_0} = \frac{\frac{U_0}{R_i}BL}{cd_0} = \alpha$$

the relation of magnetic force at current, created by source with output voltage U_0 and internal resistor R_i and elastic force, corresponds to flexure of graphene layer on the value of initial gap. Estimation of magnetic force has been discussed earlier. Multipliers' α-value is inversed Q-factor of oscillations system.

The third multiplier $RC_0\lambda = 2\pi\frac{RC_0}{T} = \delta$ is the relation of time constant of capacitor charging and the period of eigenfrequency. At the large value of this multiplier, the condenser does not have time to overcharge by one period of graphene oscillations. At too small value

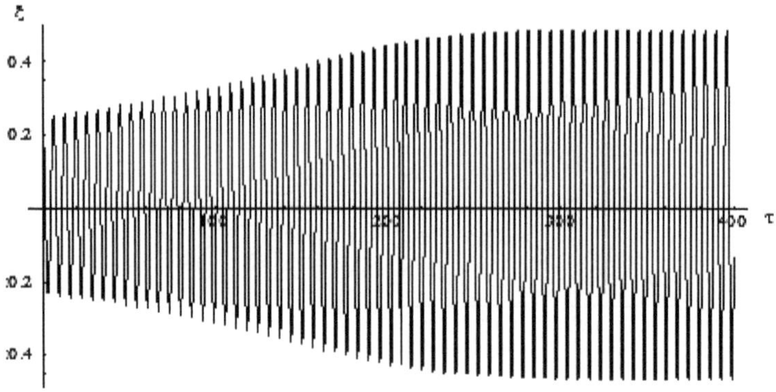

Fig. 13.24. Oscillogram of self-oscillating process.

of this multiplier, the voltage on the resistor R will be also small. Estimation of this parameter also is of order 0,01–0,1.

Finally, in Eq. (13.50) we have more than one dimensionless multiplier — coefficients of increasing K. Choosing of this coefficient gives the possibility of excitation of oscillations.

Excitation of self-oscillations is possible then natural damping may be compensated by arrival of energy from external source, from Eq. (13.50), we have two components, connecting velocity ξ', and they have unlike signs. First, from its viscous friction, corresponded energy absorption, and second is a force, creating by interference of current and magnetic field, corresponded for feeding of energy. By choosing the amplifier coefficients, we obtain the positive balance of energy and as result — excitation of oscillations.

The fulfilled numerical experiment with Eq. (13.50) is shown that, at corresponded of choosing the amplifier coefficient K and limited voltage V_0, it may obtain excitation of oscillation with output on the steady-state regime. In Fig. 13.24, oscillogram obtained by numerical integration of the system (13.50) is shown.

At this oscillogram, we see the process of excitation and next limitation of amplitude. At adherent of a nanoparticle on graphene layer the change of effective mass is occurred. By Eq. (13.50), taking into account the alteration of coefficient at highest derivative, we obtain

$$(1+\mu)\xi'' + 2\nu\xi' + \xi + \gamma\xi^3 - \beta^2\frac{\eta}{(1-\xi)^2}$$
$$-\alpha F\left(\delta\left(\frac{1}{1-\xi}\eta' + \frac{1}{(1-\xi)^2}\eta\xi'\right)\right) = 0,$$
$$\delta\left(\frac{1}{1-\xi}\eta' + \frac{1}{(1-\xi)^2}\eta\xi'\right) + \eta = 1, \tag{13.51}$$

where μ is the relative increasing of effected mass.

In Fig. 13.25, the oscillograms of steady-state self-oscillation at various values of additional mass are shown.

$$1 - \mu = 0, \quad 2 - \mu = 0.1, \quad 3 - \mu = 0.2.$$

The increasing of the mass of adherent particle is extended to the period of steady-state self-oscillation and, consequently, the frequency decrease.

Separately let us consider the case of small time constant of time of capacitor charging. Then by second equation of (13.50) we may neglect the speed of alteration, that is, the capacitor voltage and in

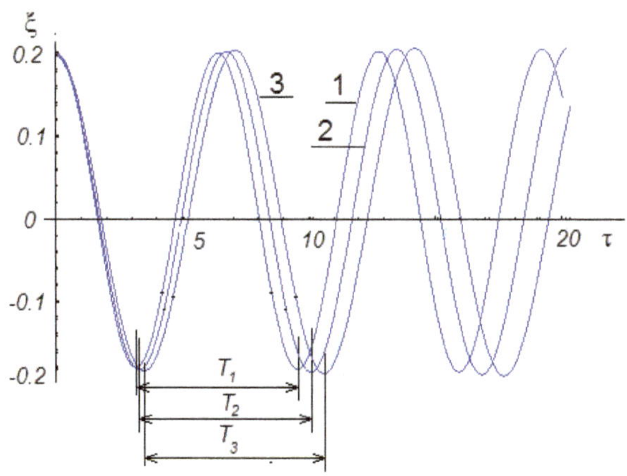

Fig. 13.25. Steady-state self-oscillations at different values of adherent particle.

an explicit form we obtain the voltage on the capacitor.

$$\eta = \frac{1}{1 + \frac{\xi'}{(1-\xi)^2}}. \tag{13.52}$$

Then, substituing (13.51) into the first equation of (13.50) we obtain the following equation:

$$\xi'' + 2\nu\xi' + \xi + \gamma\xi^3 - \beta^2 \frac{1}{(1-\xi)^2} - \alpha F\left(\delta\left(\frac{1}{(1-\xi)^2 + \xi'}\xi'\right)\right) = 0. \tag{13.53}$$

For obtaining the representation about possibility of limit cycle from Eq. (13.53), let us consider phase portrait of free oscillation of graphene layer without magnetic field. Corresponding equation of free oscillation has the following form:

$$\xi'' + \xi + \gamma\xi^3 - \beta^2 \frac{1}{(1-\xi)^2} = 0. \tag{13.54}$$

In Fig. 13.26, phase portrait, described by Eq. (13.54), is shown.

Evidently, this system has two equilibrium positions: stable (center) and unstable (saddle). Equation (13.54) has energy integral

$$\frac{1}{2}(\xi')^2 + \frac{1}{2}\xi^2 + \frac{\gamma}{4}\xi^4 - \frac{\beta^2}{1-\xi} = H, \tag{13.55}$$

which allows to find H, corresponded to homoclinical separatrix. For its determination, it is necessary to substitute phase coordinate of

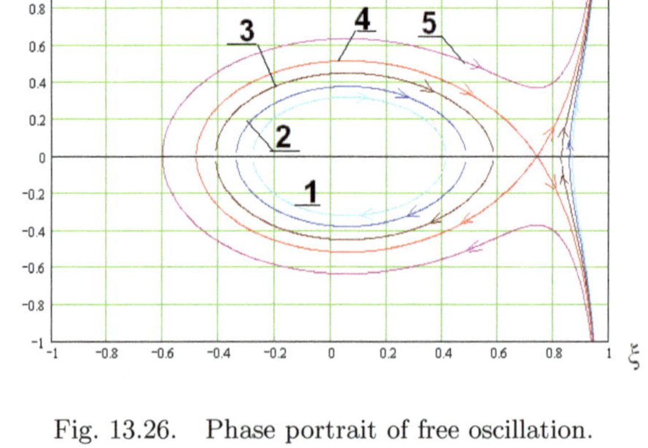

Fig. 13.26. Phase portrait of free oscillation.

critical point (saddle). Then, it is possible to find intersection of separatrix with abscissa axes $0, \xi$, and solve the equation

$$\frac{1}{2}\xi^2 + \frac{\gamma}{4}\xi^4 - \frac{\beta^2}{1-\xi} = H. \tag{13.56}$$

Let us consider the motion nearly stable position at small amplitudes, i.e., believing that flexure ξ is small comparable with 1 (gap in resonator). Then Eq. (13.53) may linearize relatively by ξ

$$\xi'' + 2\nu\xi' + \xi - \beta^2 - \alpha F(\delta\xi') = 0. \tag{13.57}$$

Keeping in amplifier characteristic only linear and cubic component, we obtain the following equation:

$$\xi'' + (2\nu - \alpha\delta K + \varepsilon(\xi')^2)\xi' + \xi - \beta^2 - \alpha F(\delta\xi') = 0. \tag{13.58}$$

It is known that this equation has single limit cycle. Approximate estimation of its amplitude a is given by

$$a = \sqrt{\frac{3}{4}\frac{\varepsilon}{\alpha\delta K - 2\nu}}. \tag{13.59}$$

Existence of limit cycle is only possible at fulfilling the condition $\alpha\delta K > 2\nu$, i.e., at condition that inputting energy from source at small amplitudes is more than loosing energy for the account damping, and at large amplitude inversely smaller. Estimation of its amplitude gives the possibility to show that limit cycle belongs to the area, which lies inside separatrix, determined by relation (13.56). The stable of limit cycle is determined by the sign of its characteristic ratio [13]

$$h = \frac{1}{T}\int_0^T \left(-2\nu + \alpha\delta K - 3\varepsilon y^2 + \alpha\delta\frac{\partial F}{\partial y}\right) d\tau. \tag{13.60}$$

Here, integral takes closed phase trajectory, corresponded limit cycle for periodic function $y(\tau) = \xi'(\tau)$ with period T. Therefore, we obtain its expression

$$h = (\alpha\delta K - 2\nu) - \frac{1}{T}\int_0^T \left(3\varepsilon y^2 - \alpha\delta\frac{\partial F}{\partial y}\right) d\tau, \tag{13.61}$$

which determines its stability in case then $h < 0$.

Thus, the scheme of self-oscillation nanoresonator on graphene nanolayer is proposed.

13.7. "Pull-In" Effect for Micro-Rod with Tensile Load

In this section the effect of "pull-in" instability is considered for rod with tensile load. "Pull-in" instability, significant nonlinear effect, connected with disappearance of equilibrium forms of elastic part of sensitive or acting mechanism, which in our task is elastic part of double-sided embedded micro-rod. Its parameters: length l, fracture stiffness EI, tensile load P, ρF the density of unit of length. That is, more it is proposed that flexure stiffness much less than stretching stiffness. The model of this mechanism is shown in Fig. 13.27.

Dynamical equation of rod taking into account tensile load $P > 0$ has the following form:

$$EIy^{IY} - Py'' + \rho F\ddot{y} = 0, \tag{13.62}$$

where y is the flexure, $y' = \frac{\partial y}{\partial x}, \dot{y} = \frac{\partial y}{\partial t}$, $x \in [0, l]$, $P = ES\frac{\Delta l}{l}$ the longitude load. Boundary problem, from Eq. (13.62) consists condition of double-sided embedded:

$$y(0) = y(l) = 0, \quad y'(0) = y'(l) = 0. \tag{13.63}$$

Scale factor is eigenfrequency of longitude oscillations $\Omega_0 = \frac{1}{l}\sqrt{\frac{P}{\rho F}}$. Let us input nondimensional variables $s = \frac{x}{l}, \tau = \Omega_0 t$ and

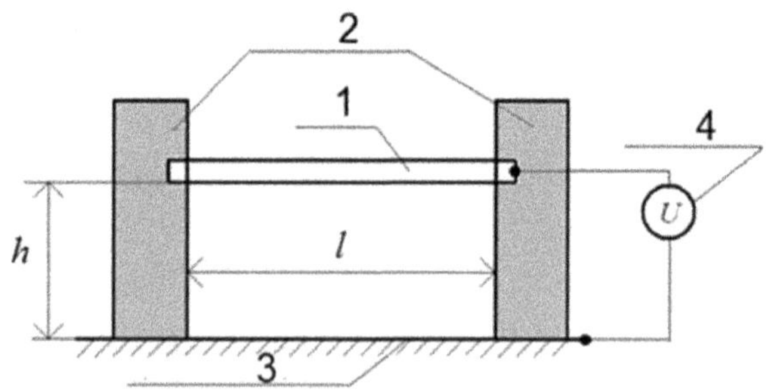

Fig. 13.27. The investigated model: 1 — elastic element; 2 — isolated supports; 3 — fixed electrode; 4 — EMF source.

dimensionless flexure $w = y/h$ (h-value of gap). Equation (13.62) may be rewritten in the following form:

$$\varepsilon\alpha w^{IY} - w'' + \ddot{w} = 0, \tag{13.64}$$

where nondimensional small parameters $\varepsilon\alpha = \frac{EI}{Pl^2}$ and $\varepsilon = \frac{h}{l}$ (small parameter), and parameter $\alpha = \frac{EI}{Phl}$ is proposed order unit.

First, let us consider free oscillations of tensile rod. The solution of (13.64) is searched in the form $w = W(s)e^{i\Omega\tau}$, here Ω is the eigenfrequency. We obtain homogenous differential equation:

$$\varepsilon\alpha W^{IY} - W'' - \Omega^2 W = 0. \tag{13.65}$$

By $W = Ce^{\lambda s}$, we obtain nondimensional characteristic biquadratic equation:

$$\varepsilon\alpha\lambda^4 - \lambda^2 - \Omega^2 = 0, \tag{13.66}$$

which has two roots:

$$\lambda_{1,2}^2 = \frac{1}{2\varepsilon\alpha} \pm \frac{1}{2\varepsilon\alpha}\sqrt{1 + 4\varepsilon\alpha\Omega^2}. \tag{13.67}$$

These roots may be separated in two groups: positive and negative. By designations

$$\nu = \sqrt{\frac{1}{2\varepsilon\alpha}(\sqrt{1 + 4\varepsilon\alpha\Omega^2} - 1)}, \quad \kappa = \sqrt{\frac{1}{2\varepsilon\alpha}(\sqrt{1 + 4\varepsilon\alpha\Omega^2} + 1)},$$

the solution of Eq. (13.65) may be written in the following form:

$$W_i = A\sin\nu s + B\cos\nu s + C\,\mathrm{sh}\,\kappa s + D\,\mathrm{ch}\,\kappa s \tag{13.68}$$

Using boundary conditions, we obtain the following homogenous linear algebraic system:

$$\begin{gathered} B + D = 0, \quad \nu A + \kappa C = 0, \\ A\sin\nu + B\cos\nu + C\,\mathrm{sh}\,\kappa + D\,\mathrm{ch}\,\kappa = 0, \\ A\cos\nu - \nu B\sin\nu + \kappa C\,\mathrm{ch}\,\kappa + \kappa D\,\mathrm{sh}\,\kappa = 0. \end{gathered} \tag{13.69}$$

By determination in zero, the determinant system of Eq. (13.69) leads to characteristic equation:

$$2\nu\kappa(\cos\nu\,\mathrm{ch}\,\kappa - 1) + (\nu^2 - \kappa^2)\sin\nu\mathrm{sh}\kappa = 0, \tag{13.70}$$

where $\nu\kappa = \Omega\sqrt{\frac{1}{\varepsilon\alpha}}$, $\nu^2 - \kappa^2 = -\frac{1}{\varepsilon\alpha}$.

From (13.70), we find eigenfrequency $\Omega_i, i = 1, 2, \ldots$. The mode of free oscillations W_i, corresponded i eigenfrequency, has the following form:

$$W_i(s) = A_i\left(\sin \nu_i s - G_i \cos \nu_i s - \frac{\nu_i}{\kappa_i}\operatorname{sh} \kappa_i s + G_i \operatorname{ch} \kappa_i s\right), \quad (13.71)$$

here

$$G_i = \frac{\sin \nu_i - \frac{\nu_i}{\kappa_i}\operatorname{sh} \kappa_i}{\cos \nu_i - \operatorname{ch}\kappa_i}$$

Coefficients A_i are determined from condition of orthonormalization:

$$\int_0^1 W_i(s)W_k(s)ds = \delta_{ik}.$$

The results of calculation of eigenfrequencies in dependence from parameter $\varepsilon\alpha$ are given in the following table:

$\varepsilon\alpha$	Ω_1	Ω_2	Ω_3
0.01	4.10	9.12	15.60
0.1	7.78	20.65	39.50
0.2	10.60	28.44	54.98

In Fig. 13.28, the two first modes are shown.

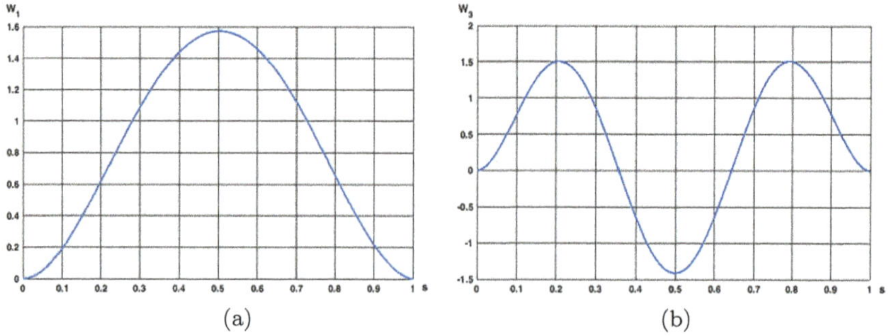

Fig. 13.28. Two lowest symmetrical modes.

Electrostatic forces may be obtained from expressions of electrostatic potential. Distributions of potential $\Psi(x, z)$ and ponderomotive forces of attraction are obtained from solution of Laplace equation

$$\Delta\Psi = 0. \tag{13.72}$$

Further, only symmetrical case of deformation of conductive layer may be considered. Upper conductive layer has electric potential V, Low layer and environ isolators have zero potential believed that side faces separate from conductive layer by thin-isolated layers. The border condition in this case has next form:

$$\begin{aligned} &x = 0 \quad \Psi = 0, \quad x = \frac{1}{2}, \quad \frac{\partial\Psi}{\partial x} = 0, \\ &z = h \quad \Psi = V, \quad z = 0, \quad \Psi = 0. \end{aligned} \tag{13.73}$$

In dimensionless coordinates $\xi = 2x/l, \zeta = z/h$, the nondimension expression for potential $\psi = \Psi/V$ satisfies to (13.72).

The calculation of electrostatic field carries out, taking into account boundary effect, the ratio from value of gap h to the length of layer l, $\varepsilon = \frac{h}{l}$ is the small parameter. Using the method of variables separation of homogenous equation without boundary condition $\psi_0 = \zeta$ and inputted the boundary coordinates $\eta = \xi/\varepsilon$, we obtain the boundary problem relatively $\psi = \tilde{\psi} + \zeta$

$$\begin{aligned} &\varepsilon^2 \frac{\partial^2\tilde{\psi}}{\partial\xi^2} + \frac{\partial^2\tilde{\psi}}{\partial\zeta^2} = 0 \quad \text{or} \\ &\frac{\partial^2\tilde{\psi}}{\partial\eta^2} + \frac{\partial^2\tilde{\psi}}{\partial\zeta^2} = 0 \end{aligned} \tag{13.74}$$

with boundary conditions:

$$\begin{aligned} &\xi = 0(\eta = 0), \quad \tilde{\psi} = -\zeta, \quad \xi = 1, \frac{\partial\tilde{\psi}}{\partial\tau} = 0, \\ &\zeta = 0, \quad \tilde{\psi} = 0; \quad \zeta = 1, \quad \tilde{\psi} = 0. \end{aligned} \tag{13.75}$$

Using symmetric of electroelastic system, all relations may be written only for left part of system ($\xi = 2s$, $0 \leq s \leq 0.5$, $0 \leq \xi \leq 1$). The solution of boundary problem is searched by method of boundary layer $\tilde{\psi} = \tilde{\psi}_0(\eta, \xi) + \varepsilon\tilde{\psi}_1(\eta, \xi, \varepsilon)$. In a case of nondeformable plate layer, the expression of electrostatic potential has the following form:

$$\psi = \zeta + \frac{2}{\pi}\sum_{n=1}^{\infty}\frac{(-1)^n}{n}\exp\left(-\frac{\pi n\xi}{\varepsilon}\right)\sin\pi n\zeta. \tag{13.76}$$

Taking into account the deviation of upper layer of "capacitor" $w(s)$, electric potential takes the following expression:

$$\psi = \frac{\zeta}{u(\xi)} + \frac{2}{\pi}\sum_{n=1}^{\infty}\frac{(-1)^n}{n}\exp\left(-\frac{\pi n\xi}{\varepsilon}\right)\sin\pi n\zeta,$$
$$u(\xi) = 1 - w(\xi). \tag{13.77}$$

The distribution of potential ψ is shown in Fig. 13.29.

Distribution of attraction force, acted on the normal to upper layer, may be found from the following expression:

$$f(s) = \left[\varepsilon^2\left(\frac{\partial\psi}{\partial\xi}\right)^2 + \left(\frac{\partial\psi}{\partial\zeta}\right)^2\right]_\Gamma, \ \zeta_\Gamma = u(s) = 1 - w(s). \tag{13.78}$$

Differentiating potential with respect to ξ, we obtain

$$\frac{\partial\psi}{\partial\xi} = \frac{\zeta w'}{2(1-w)^2} - \frac{2}{\varepsilon}\sum_{n=1}^{\infty}(-1)^n\exp\left(-\frac{2\pi ns}{\varepsilon}\right)\sin\pi n\zeta \tag{13.79}$$

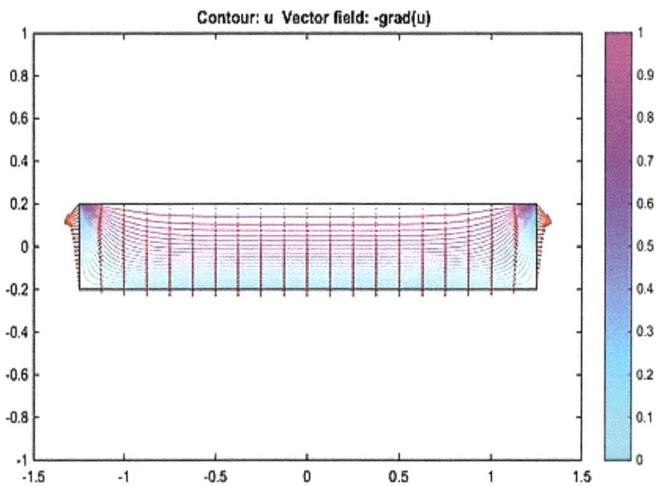

Fig. 13.29. Distribution of electrostatic potential and electric tension ($\varepsilon = 0.1$).

Here and in what follows, we neglect the component contained ε^2 in expression (13.78). Also, we neglect the alteration of normal to bent layer, which gives $\zeta = 1$. Thus, the electric force, acted on the normal to deformable elastic layer may be calculated, using expression

$$f(s) = \left(\frac{\partial\psi}{\partial\zeta}\right)^2, \quad \frac{\partial\psi}{\partial\zeta} = \frac{1}{1-w} + S, \tag{13.80}$$

where

$$S = 2\sum_{n=1}^{\infty} \exp\left(-\frac{2\pi n s}{\varepsilon}\right). \tag{13.81}$$

In Fig. 13.30, the graphic of function S, responsible edge effect, is shown.

Thus, we see fast fading of function, which is corresponded edge effect. Static task of electroelastic is described by equation of rod flexure under action electric force:

$$\varepsilon\alpha w^{IV} - w'' = \gamma^2 f(s, w), \tag{13.82}$$

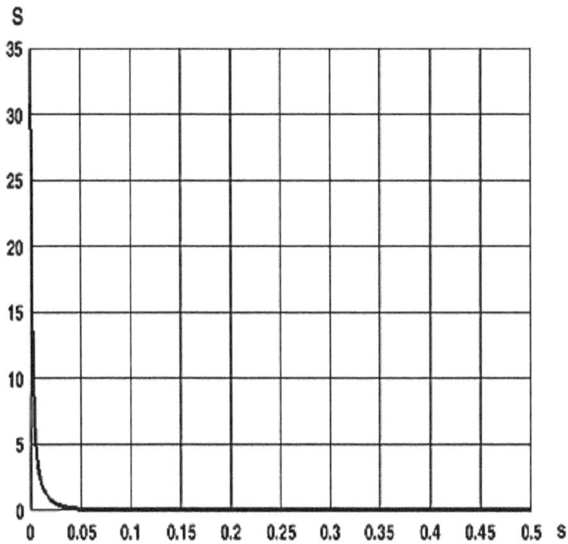

Fig. 13.30. Description of edge effect.

where $\gamma^2 = \frac{\epsilon_0 V^2 l^2}{2Ph^3}$ (ϵ_0 is the dielectric permittivity) and homogenous boundary condition:

$$w(0) = 0, w'(0) = 0, w(1) = 0, w'(1) = 0.$$

For solution of this problem, the Newton–Kantorovich method [9] is used. Corresponding operational equation may be written in the following form:

$$L(w) = 0. \tag{13.83}$$

Here

$$L(w) = \varepsilon\alpha w^{IV} - w'' - \gamma^2 f. \tag{13.84}$$

Let us find $L'(w)$ (Frechet derivative):

$$L'(w) = \varepsilon\alpha\frac{d^4}{ds^4} - \frac{d^2}{ds^2} - \gamma^2\frac{\partial f}{\partial w}. \tag{13.85}$$

The Taylor series $L(w)$ relatively $w = w_0$ in a linear approximation has the following form:

$$L(w) = L(w_0) + L'(w_0)(w - w_0) = 0. \tag{13.86}$$

Let us determine step of approximation:

$$\Delta_n = w_{n+1} - w_n, \quad n = 0, 1, 2, \dots,$$

and use iteration procedure

$$L'(w_n)\Delta_n + L(w_n) = 0, \tag{13.87}$$

or

$$\varepsilon\alpha{\Delta_n}^{IV} - {\Delta_n}'' - \gamma^2 r_n \Delta_n + \varphi_n = 0, \tag{13.88}$$

Here

$$\varphi_n = \varepsilon\alpha{w_n}^{IV} - {w_n}'' - \gamma^2\left(\frac{\partial\psi}{\partial\zeta}\right)^2\Bigg|_{w=w_n},$$

$$r_n = 2\frac{1}{(1-w_n)^2}\left(\frac{\partial\psi}{\partial\zeta}\right)\Bigg|_{w=w_n}.$$

If we know $w_n(s)$ — n approximation of unknown function (initial approximation is assigned), then, after solution of Eq. (13.87), we receive next approximation

$$w_{n+1} = w_n + \Delta_n. \tag{13.89}$$

By solution of equation (13.88), we search the sum of free oscillations mode (13.71):

$$\Delta_n(s) = \sum_{k=1}^{K} \beta_k W_k(s). \tag{13.90}$$

Summation is realized taking into account only symmetrical modes. For determination coefficients β_k, the Galerkin method is used. Projection equations have the following form:

$$\int_0^{\frac{1}{2}} (\varepsilon\alpha\Delta_n^{IV} - \Delta_n'' - \gamma^2 r_n \Delta_n + \varphi_n).w_k(s)ds = 0. \tag{13.91}$$

Using expansion procedure (13.90), we receive nonlinear system for coefficients:

$$\Omega_k^2 \beta_k - 2\gamma^2 \sum_{r=1}^{K} R_{kr}\beta_r = 2\gamma^2 C_k - 2\Omega_k^2 B_k, \tag{13.92}$$

where

$$\begin{aligned} R_{kr} &= \int_0^{1/2} r_n W_k W_r ds, \\ B_k &= \int_0^{1/2} W_k w_n ds, \\ C_k &= \int_0^{1/2} W_k \left(\frac{\partial \psi}{\partial \zeta}\right)^2 \Bigg|_{w=w_n} ds. \end{aligned} \tag{13.93}$$

By solving system (13.92), i.e., found unknown coefficients β_k from system (13.92) and, consequently, Δ_n, the next approximation $w_{n+1} = w_n + \Delta_n$ may be found.

The dependence $w_{\max}$ from value of parameter γ for the case, $\varepsilon\alpha = 0.1$ and $\varepsilon\alpha = 0.2$ (taking into account edge effect) is shown in Fig. 13.31. The dependence $w = w(s)$ at $\varepsilon = 0.1, \gamma = 1.5$ on the left part of rod is shown in Fig. 13.32. Calculation is fulfilled taking into account 12 symmetric forms.

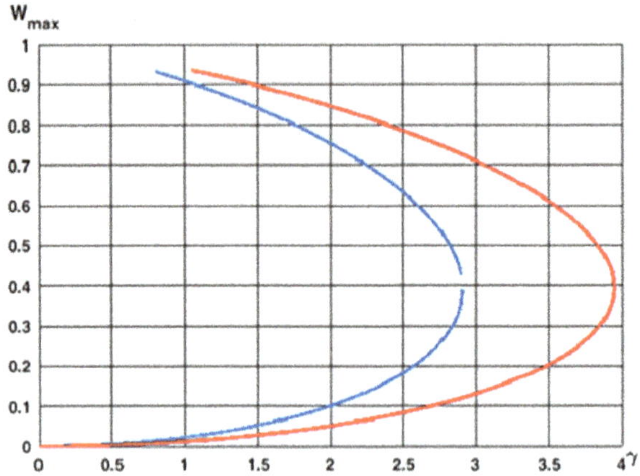

Fig. 13.31. Result of calculation of branch curve by iteration method ($\varepsilon = 0.1$).

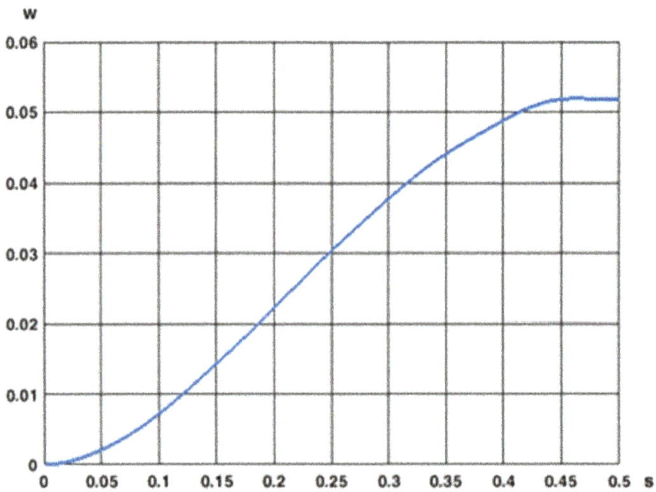

Fig. 13.32. The form of half of rod under action of electric force.

Internal curve corresponds to the case $\alpha = 1$, external $\alpha = 2.$, i.e., the increasing of stiffness shifts the branch curve to right.

For solution of this task without edge effect, it is sufficient in expression (13.80) for $\frac{\partial \psi}{\partial \zeta}$ function $S = 0$ is believed, i.e., solving the equation:

$$\varepsilon\alpha w^{IV} - w'' = \gamma^2 \frac{1}{(1-w)^2}. \tag{13.94}$$

Comparing this solution with previous solution and taking into account the edge effect may show that in our model this effect does not give large influence on branch diagram.

The task of flexure of tensile string without flexure stiffness under action of electrostatic force (without first component in Eq. (13.94)) has analytic solution. In this case, boundary problem has the following form:

$$w'' + \gamma^2 \frac{1}{(1-w)^2} = 0,$$
$$w(0) = w(1) = 0. \tag{13.95}$$

The solution of this task fulfills with taking into account symmetric of flexure relatively of the string middle $w'(\frac{1}{2}) = 0$. In this case, we have integral relation

$$\frac{1}{2}(w')^2 + \frac{\gamma^2}{1-w} = C, \tag{13.96}$$

where constant C is determined from condition of symmetry as $C = \frac{\gamma^2}{1-w_m}$, w_m is the maximum of flexure in a middle of string. Using the boundary condition, we receive the implicit equation relatively w_m:

$$\int_0^{w_m} \sqrt{\frac{1-w}{w_m - w}} dw = \gamma\sqrt{\frac{1}{2(1-w_m)}}. \tag{13.97}$$

Then, we obtain algebraic equation

$$\sqrt{w_m} + (1-w_m) ln \sqrt{\frac{1+\sqrt{w_m}}{1-\sqrt{w_m}}} = \gamma\sqrt{\frac{1}{2(1-w_m)}}. \tag{13.98}$$

This equation may be rewritten as

$$\gamma = \sqrt{2(1-w_m)}\sqrt{w_m} + \sqrt{2}\cdot(1-w_m)^{3/2} ln \sqrt{\frac{1+\sqrt{w_m}}{1-\sqrt{w_m}}}. \tag{13.99}$$

Resultant dependence $w_m(\gamma)$ (13.99) is shown in Fig.13.33.

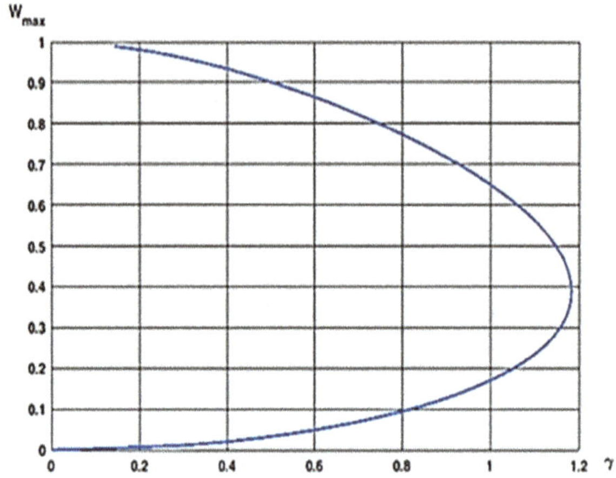

Fig. 13.33. Analytic diagram of branching.

With increasing of parameter γ from zero until point of branch, this equation has two solutions. To point of branching corresponded the touch of curve $\gamma(\omega_m)$ (13.99) in a point $w_m \sim 0.4$, which corresponded to $\gamma_* \sim 1.2$. Behind the point of bifurcation, this equation does not have solution. Evidently, at $\gamma < \gamma_*$, the solution with smaller amplitude is stable and distinguished from solution with greater amplitudes, which take place at equal value of electric potential.

Thus, in our chapter, some electromechanical models of nanoresonators were constructed. It is shown that the nanoresonators with parametrical effect have the greatest advantage taking into account the largest accuracy of measurement of the mass of sticking nanoparticle at low Q-factor.

Acknowledgments

The study was supported under grant RFBR 20-01-00537.

References

[1] C. Chen *et al.*, Performance of monolayer graphene, *Nature Nanotechnol.* **4** (2009) 861–867.

[2] X. Q. Kitipornchai and S. K. M. Liew, Resonance analysis of multi-layered graphene sheets used as nanoscale resonators, *Nanotechnology.* **16** (2005) 2086–2091.
[3] L. Y. Z. Xu and Q. Zheng, The interlayer shear effect on grapheme multilayer resonators, *J. Mech. Phys. Solids* **59** (2011) 1613–1622.
[4] T. Natsuki, J. Shi and X. Q.-Q. Ni, Vibration analysis of nanomechanical mass sensors using double-layered grapheme sheets resonators, *J. Appl. Phys.* **114** (2013).
[5] A. M. van der Zande *et al.*, Large-scale arrays of single-layer graphene resonators, *Nano Lett* **10** (2010) 4869–4873.
[6] C. Chen and J. Hone, Graphene nanoelectromechanical systems, *Proc. IEEE* **101** (7) 1766–1779.
[7] N. F. Morozov, I. E. Berinsky, D. A. Indeitsev, O. V. Privalova, D. Yu. Skubov and L. V. Shtukin, Breaking of oscillations of graphene layer as a method of determination of its spectrum characteristic. RAS **456**(5) (2014) 537–540.
[8] L. V. Shtukin, I. E. Berinsky, D. A. Indeitsev, N. F. Morozov and D. Yu. Skubov, Electromechanical models of nano-resonators, *Phys. Meso-mech.* **20**(6) (2017) 5–13.
[9] D. Yu. Skubov and K. Sh. Khodzhaev, *Non-linear Electromechanics*, (Springer, Berlin, 2008), 395 pp.

https://doi.org/10.1142/9789811221903_0014

Chapter 14

The Retina as a Dynamical System

Bruno Cessac

Universite Côte d'Azur, Inria, France
Biovision team and Neuromod Institute,
2004 route des Lucioles, 06901 Sophia Antipolis, France
bruno.cessac@inria.fr

Considering the retina as a high-dimensional, nonautonomous, dynamical system, layered and structured, with nonstationary and spatially inhomogeneous entries (visual scenes), we present several examples where dynamical systems, bifurcations, and ergodic theory provide useful insights on retinal behavior and dynamics.

14.1. Introduction

The retina is the entrance to the visual system [1].[1] The development of new technologies and experimental methodologies (MEA, 2-photon, genetic engineering, pharmacology) and the resulting experiments have made it possible to show that the retina is not a mere camera, transforming the flow of photons coming from a visual scene in sequences of action potentials interpretable by the visual cortex. It appears, in contrast, that the specific and hierarchical structure of the retina allows it to pre-process visual information, at different scales, in order to reduce redundancy and increase the speed, efficiency and reliability of visual responses [2]. This is particularly salient in the processing of motion, which is the essence of what our visual system receives as input: there are no static objects since our body, our head, our eyes are constantly on the move. In particular, it has been shown experimentally that the retina can detect a moving

[1] https://webvision.med.utah.edu/.

object on a background having a distinct motion, anticipate the trajectory of a moving object [2] or, even, respond to changes in its trajectory [3]. Such processing are obviously present in the visual cortex, but retinal pre-processing could make visual processing more efficient, for example, by compensating for the delay — mainly due to photo transduction — between the capture of the photons emitted by an object by the photo-receptors and the response of the brain [4].

The retina is thus a fascinating object and its study opens the door to better understand how our brain processes information (here, visual information). In this chapter, we adopt a specific, modeler, point of view: The retina is a dynamical system. This might look, on the one hand, quite restrictive, and, on the other hand, not really original. Restrictive because, focusing on a dynamical system perspective, we are somewhat constrained by the structure of equations and the limited tools to analyze them; we are therefore bound to specific problems and questions which necessarily provide a limited view of the retina's behavior. Not very original because, for sure, the retina is made of neurons, synapses, neurotransmitters, ionic channels, ruled by the laws of physics, written in terms of differential equations. So, yes, retina dynamics can be studied in the realm of dynamical systems theory, but the same is done since long with the brain and cortical neuronal networks models [5]. What's new with the retina?

So, the scope of this chapter is to convince the reader that the retina is a specific organ, with specific neurons and circuits, rather different from their cortical counterparts, leading to specific questions — requiring specific techniques — that can be addressed in the realm of dynamical systems theory. More precisely, the retina is a high-dimensional, *nonautonomous* dynamical system, layered and structured, with nonstationary and spatially inhomogeneous entries (visual scenes). It is therefore necessary to go, as much as we can, beyond the comfortable assumption of stationarity.

Specific local retinal circuits made of specific neurons, connected with specific nonlinear synapses, efficiently process the local visual information. Most of the involved neurons — except ganglion cells, the output of the retina — are *not spiking*, in strong contrast with cortical neurons. These local circuits are connected together, spanning the whole retina in a regular tiling, providing parallel channels of information processing [6–8] that can hardly (from a dynamical system perspective) be considered as independent. Thus, we have to consider, still from a dynamical system perspective, which type of spatio-temporal correlations can be induced by the conjunction of intrinsic neurons dynamics, their interactions by synapses, and their stimulation by a nonstationary drive. This question is particularly salient when the stimulation comes from an object moving along a deterministic trajectory entangling space and time.

In addition, the retina behavior can be affected by specific changes in its structure or physiology. During the early stage of its development, the retina of vertebrates undergo retinal waves, disappearing a few days after birth, which are essential to structure the visual system and make it efficient. Later in its life, the retina can be affected by specific pathologies, impairing vision. Finally, pharmacology can affect the retinal response to visual stimulation in a controlled way. These changes and their impact on retina's dynamics can be addressed, too, from the point of view of dynamical systems and bifurcations theory.

This chapter presents three examples along these lines. After a brief introduction on retina's structure and dynamics (Sec. 14.2), we present, in Sec. 14.3, a dynamical system study of retinal waves, published in [9, 10]. This is an interesting example of multi spatio-temporal scales dynamics where the conjunction of nonlinearities and closeness to a bifurcation point, combined with a nonlinear coupling on a regular lattice structure, can produce a highly disordered spatio-temporal dynamics, where the impact of variations of physiological parameters (development, pharmacology) and its consequence on waves dynamics can be addressed. In Sec. 14.4, we consider the effect of lateral amacrine cells connectivity on the response to a moving object, and especially, the anticipation of its trajectory [11]. Finally, in Sec. 14.5, we provide a conceptual and operational framework to study the nonstationary spike correlations induced by a time-dependent stimulus in a neuronal network. Although the related work was initially not specific to the retina [12–16] it has resulted in algorithmic tools and a software used to analyze retinal data [17–19] that we shall not discuss here though, focusing on the mathematical and conceptual consequences of these studies. The last section tries and generalizes the point of view considering the retina as a dynamical system to more general perspectives.

This chapter mostly present results, published elsewhere, done in collaboration with Ignacio Ampuero, Rodrigo Cofré, Lionel Gil, Dora Karvouniari, Olivier Marre, Serge Picaud, Selma Souihel and I would like to warmly acknowledge them.

14.2. Brief Overview of the Retina

14.2.1. *Structure*

The organization of the retina can be found, e.g., in [1] or on the website https://webvision.med.utah.edu/book/part-i-foundations/simple-anatomy-of-the-retina/. It is structured in neuronal and synaptic layers with five neuronal types: photo-receptors, rod and cones (P), horizontal cells (HCells), bipolar cells (BCells), amacrine cells (ACells), ganglion cells (GCells), to which are added glial cells (Mueller's cells). A feedforward connectivity can be distinguished, the P-B-G path which leads from the photo transduction to the spike trains emitted by the GCells towards the cortex. There is also a lateral connectivity due to Hcells, at the origin of the Center-Surround structure of the receptive fields, and the Acells whose role is still poorly understood and which are one of the main objects of this chapter.

We can roughly subdivide the retina into two blocks. The first, ending at the OPL (Outer Plexiform Layer), comprises the P, HCells, BCells and the related synapses. The "input" of this block is the flow of photons emitted by the outside world and picked up by the photo-receptors. The "output" is the set of membrane potentials of Bcells, temporally modulated by visual input. More precisely, the voltage of each BCells integrates, spatially and temporally, the local visual information of the photo-receptors which are connected to it, with a lateral modulation due to the HCells. Each BCells is thus sensitive to specific local characteristics of the visual scene, defining its Receptive Field (RF). Mathematically, the receptive field is characterized by the convolution between a spatio-temporal kernel and the stimulus (see Sec. 14.4.1). Nonlinear effects such as gain control must also be taken into account for a realistic characterization of the BCells response.

The second block is the Inner Plexiform Layer (IPL), comprising the ACells and GCells and the afferent synapses. Its "input" is the output of the OPL, and its output, the trains of action potentials emitted by the GCells, which transmit the flow of information from the retina to the thalamus and the cortex via their axons which constitute the optic nerve. ACells are difficult to study experimentally because hardly accessible from electrophysiology measurements. There are also a large number of cell subtypes in the ACells class (around 40) of which only a small number have duly identified functions. It is however recognized that they play an essential role in the treatment of motion [20–22].

14.2.2. *Modelling*

Neurons in the retina have the same biophysics as their cortical counterparts. However, they operate under different regimes. Thus, with the exception of the Gcells, the retinal neurons do not emit action potentials. Their activity and interactions therefore take place through graded (continuous) membrane potentials as opposed to the sharp peak of an action potential. Furthermore, there is no long-term synaptic plasticity in the retina. Finally, if cortical processing units, such as cortical columns, involve thousands of neurons — making the mathematical description in terms of mean field and neural masses relevant — in the retina the role of these units is played by elementary circuits involving a very small number of neurons, of the order of unity.

To summarize, the retina is an organ able to perform quite complex tasks in (pre-)processing visual information, with a very low energy consumption, using analog computing where most neurons do not spike (except GCells). These astonishing capacities are in particular due to specific, local circuits, involving a small number of neurons with specific synapses.

As a consequence, models start from the same biophysical principles as cortical neurons, but the resulting dynamics is different. The mathematical studies can target, as well, different scopes.

14.3. The Multiscale Dynamics of the Retina: Retinal Waves

14.3.1. *Context*

During the early development in the visual system of many vertebrate species [23–31], bursts of activity, called *retinal waves*, occur spontaneously and contribute to the visual system organization [23, 26, 29, 30]. They are characterized by localized groups of neurons becoming simultaneously active, initiated at random points and propagating at speeds ranging from 100 μm/s (mouse, [31, 32]) up to 400 μm/s (chick, [25]), with randomly moving boundaries [30, 33]. Wave activity begins in the early development, long before the retina is responsive to light, and it stops a few weeks after birth. It emerges due to several biophysical mechanisms, which changes during development, dividing retinal waves into three stages (I, II, III) [29]. Each stage, mostly studied in mammals, is characterized by a certain type of network interaction: gap junctions (electric synapses) for stage I; cholinergic transmission for stage II; and glutamatergic transmission for stage III.

It has been shown that this intermittent spatio-temporal activity is an inherent property of specific retinal cells and their network interactions [34]. More precisely, the generation of waves at any stage requires three conditions [30, 35]. (i) A source of depolarization for wave initiation. Given that there is no external input (e.g., from visual stimulation in the early retina), there must be some intrinsic mechanism by which neurons become active. (ii) A network of excitatory interactions for propagation. Once some neurons become spontaneously active, how do they excite neighbouring neurons.

(iii) A source of inhibition that limits the spatial extent of waves and dictates the minimum interval between them.

In this chapter, we focus on stage II retinal waves where a wide literature exists on the experimental [23–32] and modeling side [24, 36–40]. The study presented here relies on the works [9, 10, 41].

During this period, the cells involved are called Starbust Amacrine Cells (SACs) (so-called because of their dendritic tree shape which make them look like the explosion of a star). They interact via the neurotransmitter Acetylcholine (Ach) with nicotinic receptors. Zheng *et al.* [34] have shown that SACs emit spontaneous intrinsic calcium bursts, which stop after a few hundreds of ms. When bursting, a SAC emits Ach, which can in turn excite SACs in its neighborhood, thereby triggering a wave. Stage II waves are therefore spatio-temporal phenomena resulting from the coupling of local bursters (SACs), spatially coupled by Acetylcholine. In addition, the SAC's bursting generates a strong after hyperpolarization current (sAHP) which, eventually, drives the SAC to rest, i.e., the bursting phase stops, and induces a long refractory period. In a network, the propagation of a wave generates therefore sAHP refractory zones preventing the propagation of a new wave for a period of order a few seconds [34]. The sAHP plays thus a prominent role in shaping waves periodicity and boundaries.

From this brief introduction, one may believe that retinal waves show relatively constant characteristics. This is not the case. In contrast, they show strong variability. The sizes and durations are distributed according a power law [39] and, within stage II, this distribution evolves. One also observes a wide variability in waves inter bursts, period or duration, across different species (e.g., rabbit vs. mouse) [30], but also a variability between individuals within a species. Finally, retinal waves are quite sensitive to pharmacological agents that can increase or decrease their activity.

In this section, we want to show how these strong variabilities in retinal waves can be explained through a unique mechanism, grounded on dynamical systems analysis and bifurcations theory.

14.3.2. *The dynamics of Starburst Amacrine Cells*

The details of the model are given in [9, 10, 41]. Here, we give the main equations and results. SACs are considered as points on a lattice, i.e., we do not consider their spatial structure.

14.3.2.1. *Fast potassium channels*

In this model, the fast repetitive firing of SACs is due to the competition between a depolarizing Ca^{2+} current and a fast potassium current. The membrane voltage $V(t)$ of a SAC obeys:

$$C_m \frac{dV}{dt} = I_L(V) + I_C(V) + I_K(V, N) + I_{\text{ext}} + \sigma\xi_t, \tag{14.1}$$

where C_m is the membrane capacitance, $I_L = -g_L(V - V_L)$ is the leak current, with g_L, leak conductance, and V_L, leak reversal potential. The calcium current reads:

$$I_C(V) = -g_C M_\infty(V)(V - V_C), \tag{14.2}$$

where $g_C M_\infty(V)$ is the voltage-dependent conductance of the Ca^{2+} channel, with $M_\infty(V) = \frac{1}{2}[1 + \tanh(\frac{V-V_1}{V_2})]$. The *fast* potassium current takes the form:

$$I_K(V, N) = -g_K N(V - V_K), \tag{14.3}$$

where the evolution of the voltage-gated K^+ channel gating variable $N(t)$ is given by:

$$\tau_N \frac{dN}{dt} = \Lambda(V)(N_\infty(V) - N), \tag{14.4}$$

with $\Lambda(V) = \cosh(\frac{V-V_3}{2V_4})$, and $N_\infty(V) = \frac{1}{2}[1 + \tanh(\frac{V-V_3}{V_4})]$, and where τ_N, the characteristic time of the activation variable N, is of order 5 ms. Although the form of Eq. (14.3), (14.4) is based on the Morris–Lecar model [42], the variable N is fast, in contrast to the original paper, calibrated in order to capture the frequency of the fast repetitive firing of SACs (around 20 Hz [34]).

The term $\sigma\xi_t$ accounts for noise in the dynamics, where ξ_t is a white noise and σ the noise intensity. The role of noise is further discussed below. For the moment, we take $\sigma = 0$.

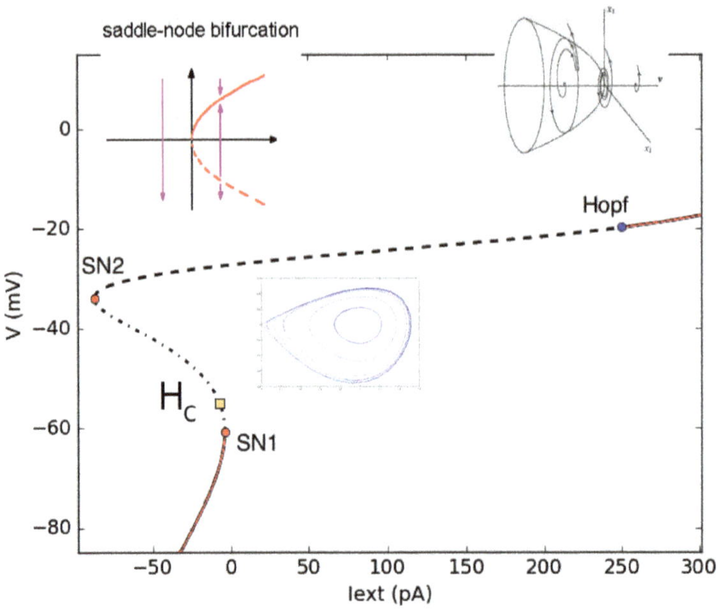

Fig. 14.1. **Bifurcation diagram of** (14.1) in the presence of a constant current (I_{ext}). SN stands for "saddle-node". Red full lines correspond to stable fixed points. Small dashed to a saddle-point. Long dashes correspond to a unstable focus. In the inset, we have illustrated the typical form of the corresponding bifurcations. See also Fig. 14.2.

For the moment too, I_{ext} is considered as a constant, tunable current. The parameters of the model have, mostly, been tuned from experimental papers (see [9] for details). For these parameters, we observe the following behavior when varying the current I_{ext} (Fig. 14.1). For low, negative current, the dynamical system admits a unique stable fixed point (red, full, branch on the left of the diagram). When I_{ext} increases from negative values, there is a saddle-node bifurcation (SN_2), at $I_{\text{ext}} = I_{\text{SN}_2}$, where this fixed point coexists with two unstable fixed points, i.e., an unstable focus (long dashed lines in Fig. 14.1) and a saddle (small dashed lines in Fig. 14.1). At the homoclinic bifurcation point (H_c) ($I_{\text{ext}} = I_{H_c}$), a stable limit cycle emerges with fast oscillations (of order a few ms). The stable fixed point coexists with an unstable focus and this stable limit cycle. At the saddle-node bifurcation SN_1 ($I_{\text{ext}} = I_{\text{SN}_1}$), the stable fixed point coalesces with the unstable focus, and only the stable limit cycle remains. Note therefore that the homoclinic bifurcation and the saddle-node SN_1 are very close together (less than 2 pA). For simplicity, the next description will do as if they were occurring at the same point although their separation plays an important role in retinal waves distribution of size and durations [10]. Finally, for a relatively high current ($I_{\text{ext}} = I_{\text{Hopf}} \sim 250\,\text{pA}$), the stable limit cycle disappears by Hopf bifurcation giving rise to a stable state.

Therefore, essentially, going from a negative to a positive current, one switches from a stable rest state with low voltage ($\sim -70\,\text{mV}$, close to the leak reversal potential V_L) to fast oscillations with high average voltage ($\sim -30\,\text{mV}$), whereas going from a positive to a negative current, one switches from fast oscillations to stable rest state.

14.3.2.2. *Slow after-hyperpolarization current*

In the fast oscillations regime, the average membrane potential has a sufficiently high value to trigger the entrance of calcium via voltage-gated calcium channels. This induces a cascade of reactions resulting in the opening of potassium channels and the production of a slow hyperpolarization current I_{sAHP}. This cascade involves the internal calcium concentration C, the fraction of saturated Calmodulin S and the fraction R of bounded terminals in the corresponding potassium channels. The corresponding equations are as follows [9]:

$$\begin{cases} \tau_C \dfrac{dC}{dt} = -\dfrac{\alpha_C}{H_X} C + C_0 + \delta_C I_C(V), \\ \tau_S \dfrac{dS}{dt} = \alpha_S C^4 (1 - S) - S, \\ \tau_R \dfrac{dR}{dt} = \alpha_R S (1 - R) - R, \end{cases} \tag{14.5}$$

with the characteristic times $\tau_C \sim 2\,\text{s}$ and $\tau_S, \tau_R \sim 40\,\text{s}$. Thus, the calcium dynamics is slow compared to the fast oscillations (timescale of order a few ms) while the variables R, S are slow compared to calcium dynamics. The other parameters are fixed from biophysics [41].

This cascade of reactions induces a sAHP current of the form

$$I_{\text{sAHP}}(V, R) = -g_{\text{sAHP}} R^4 (V - V_K) \equiv -G_S(V - V_K), \qquad (14.6)$$

where g_{sAHP} is the maximum sAHP conductance and $G_S = g_{\text{sAHP}} R^4$ is the effective conductance. It involves the probability, R, that one terminal of a slow potassium channel is bounded. This probability has the power four because four bound terminals are needed to open a Ca^{2+}-gated K^+ channel.

Since the time variation of G_S is extremely slow compared to the fast oscillations of (14.1) one can average over the fast term $V - V_K$ in (14.6), and consider I_{sAHP} as a slow varying parameter, corresponding to I_{ext} in the bifurcation diagram 14.1. (See also Fig. 14.2 for a bifurcation diagram dealing with the slow variables.)

So, the picture is the following. In the fast oscillations regime, V is high, thereby triggering the increase (in absolute value) of a slow negative (potassium) current I_{sAHP} eventually leading the system to cross the bifurcation point SN_1. Here, oscillations stop. Because, sAHP is slow there is some inertia where I_{sAHP} goes on decreasing, hyperpolarizing the neuron. In this stage, calcium entrance stops, so, eventually, the I_{sAHP} becomes less and negative and returns back to its rest value $I_{\text{sAHP}} = 0$. The corresponding periodic variation of the variables C, R, S is shown in Fig. 14.2 bottom, right.

Here, actually, two situations are possible. In the case depicted in Fig. 14.1, $I_{\text{SN}_1} > 0$, so when $I_{\text{ext}} = I_{\text{sAHP}} = 0$, fast oscillations restart and we are back to the initial situation. Thus, we have a simple mechanism explaining the period bursting of cells (alternation of fast oscillations and rest state) observed in experiments. We call this case spontaneous bursting (SB).

By a slight change of parameters, it is also possible to make the SN_1 and homoclinic bifurcation arising for a strictly positive value of I_{ext}. Then, $I_{\text{ext}} = I_{\text{sAHP}} = 0$ corresponds to the rest state on the stable branch left-wise to SN_1. If, however, the SN_1 bifurcation is close to $I_{\text{ext}} = I_{\text{sAHP}} = 0$ ($I_{\text{SN}_1} \lesssim 0$), then it is possible to restart oscillations with a bit of noise (the term $\sigma\xi_t$ in (14.1)). Then, the same process as above occurs, but, now, the sequence of bursts is not periodic any more. It occurs on random times. We speak of noise-induced bursting (NIB).

In the NIB regime the average time between two bursts depends strongly on the closeness of I_{SN_1} to 0. As shown in [9] this sensitivity provides a simple mechanism to explain the variability in inter burst intervals observed across and within species.

14.3.2.3. *Acetylcholine coupling*

In this stage of development SACs are interacting via the neuro-transmitter Acetylcholine (Ach) with nicotinic receptors. We consider from now a network of N SACs, with index $i = 1, \dots, N$, organized on a regular lattice (e.g., nearest neighbors or a connectivity that mimics the biological reality [10]) in one, or two dimension.

The concentration A_j (in nM) of Ach emitted by a pre-synaptic cell j, with voltage V_j, is given by the differential equation:

$$\frac{dA_j}{dt} = -\mu A_j + \beta_A T_A(V_j), \tag{14.7}$$

where μ is a degradation coefficient, β_A is the maximal production rate of A (in nMs^{-1}), and the production term:

$$T_A(V_j) = \frac{1}{1 + e^{-\kappa_A (V_j - V_A)}}, \tag{14.8}$$

where V_A is the Nernst potential associated to Ach. The parameters $\mu, \kappa_A, \beta_A, V_A$ have been fitted from experiments [10, 34]. They are uniform on the SACs population (independent of j).

The Ach released in the synaptic cleft from pre-synaptic neuron j to post-synaptic neuron i binds to two nicotinic receptors triggering the opening of specific ionic channels. This induces, in the post synaptic neuron i an excitatory current $-g_A(V_i - V_A)\frac{A_j^2}{\gamma_A + A_j^2}$. Let us denote $\mathcal{B}_i$ the neighborhood of i, that is, the pre-synaptic neurons connected to i by cholinergic coupling. The total cholinergic current injected in neuron i is

$$I_{A_i} = -g_A(V_i - V_A) \sum_{j \in \mathcal{B}_i} \frac{A_j^2}{\gamma_A + A_j^2} \equiv -G_A(V_i - V_A), \tag{14.9}$$

where g_A is the maximal Ach conductance and $G_A = g_A \sum_{j \in \mathcal{B}_i} \frac{A_j^2}{\gamma_A + A_j^2}$ the effective conductance. The evolution of G_A is slow compared to the fast oscillations ($\sim$ a few seconds).

Thus, the Ach current induces a nonlinear coupling between cells.

14.3.3. *Retinal waves propagation*

We have now the three ingredients to trigger, propagate and control waves: fast bursting, mutual excitation and slow hyperpolarization. Let us see how these mechanism collaborate or compete to produce the wide variability in retinal waves dynamics.

14.3.3.1. *Toward a generic mechanism controlling waves*

In general, thus, a SAC is subject to two competing current: the (extremely slow) hyperpolarizing current sAHP (14.6) and the (slow) depolarizing cholinergic current (14.9). One easy way to summarize the competition between these two slows and competing mechanism is to draw the equivalent of the bifurcation diagram Fig. 14.1 of (14.1) in a plane G_S, G_A. This way, G_S, G_A become two free parameters that we can vary independently.

The complete picture is represented in Fig. 14.2 (see legend for detail). On the top figure we see the bifurcation diagram of (14.1) in the plane G_S, G_A. The region of fast oscillations is region C. This is visible in the heat map of Fig. 14.2 (Bottom, Left), where the color represents the amplitude of oscillations (no oscillation in the rest state, corresponding to black regions). In the same figure, we have sketched a typical pathway showing the competition between sAHP and Ach current and the resulting dynamics. Finally, in Fig. 14.2 (Bottom, Right) we show the typical evolution of the variables C, R, S determining the sAHP conductance, in a cycle where bursting alternates with slow hyperpolarization. There is a remarkable line in the bifurcation diagram: the line where SN_1 (red) and H_c (blue) are superimposed. Although they are not identical, they are very close (see also Fig. 14.1) and are indistinguishable at this scale of G_S, G_A. As already explained when commenting Fig. 14.1 we will consider

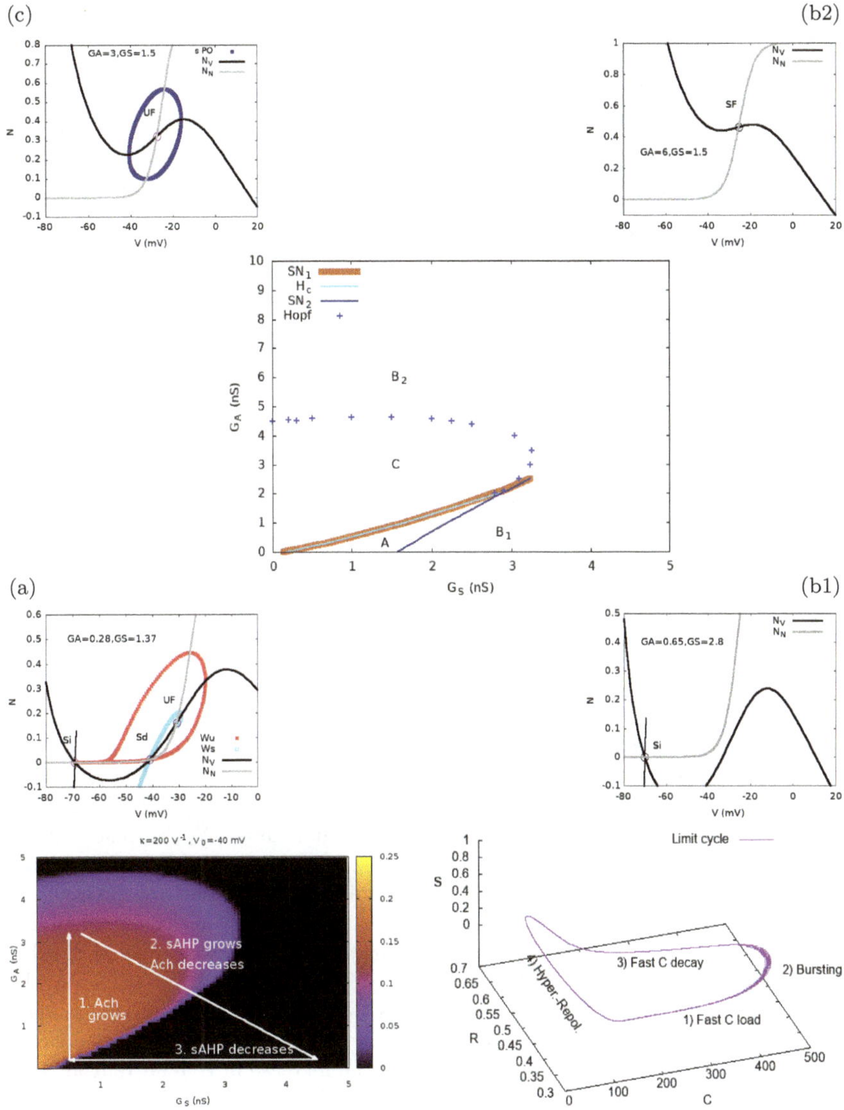

Fig. 14.2. (*Continued*)

that these two bifurcations occurs at the same point. It corresponds to the switch from stable rest state to fast oscillations.

When dealing with a network of SACs, one observes spatio-temporal variations of G_S, G_A, organized in regions of excitable cells, or refractory cells. These regions evolve in time. When an SAC starts to burst it recruits, via Ach coupling, excitable cells, which in turns can recruit other cells, generating a wave, propagating until the front meets a refractory zone with high sAHP, corresponding to cells which have recently undergone a wave. Then, what determines if the wave propagates to the next cells is a threshold condition directly related to the SN_1–H_c bifurcation. This results in a high spatio-temporal variability where waves propagate in an sAHP landscape, itself slowly evolving in time, and driven by waves dynamics.

There is some analogy with forest fires [39, 40], although the detailed dynamics of SACs wave is a bit more subtle as we now discuss.

Fig. 14.2. **The synthesized mechanism of waves dynamics. Top, Center.** Bifurcation diagram of the fast dynamics (14.1) in the plane G_S (effective sAHP conductance), G_A (effective Ach conductance). The panels (A,B_1,B_2,C) around the bifurcation map are the phase portrait with nullclines (N_V, in black, is the V nullcline, N_N, in gray, is the N nullcline), stable manifold of hyperbolic fixed points (W_s in blue, panel A), unstable manifold of the saddle-fixed points (W_u in red, panel A), stable periodic orbit (in blue, panel C). "S_d" means "Saddle", "SF" means "Stable Focus", "UF" means "Unstable Focus", "SPo" means "Stable Periodic orbit". **Bottom, Left.** Heat map of the voltage fluctuations in the G_S, G_A map. The colored region corresponds to region C in the bifurcation map. The white arrow sketches the typical evolution of G_S, G_A for a cell embedded in a wave. If the cell is at rest (region B_1 or A) the increase in Ach current leads it to region C where this cell bursts. It starts to produce sAHP so G_S increases. At some time G_A starts to decrease when neighboring SACs stop bursting. Although the increase in G_S and the decrease in G_A are not independent, we sketch here the compound effect along a straight line in the plane G_S, G_A. At some time, the cell leaves region C and returns to region B_1. Then, sAHP decreases and the cell returns to its initial state. **Bottom, Right.** Typical course of the variables C, R, S for a cell. "Hyper–Repol" means "Hyperpolarization–Repolarization".

14.3.3.2. *Chain of SACs*

We start from a one-dimensional chain of N SACs and we consider first the case ... $N = 2$. The first cell, C_1, starts to burst, and, if G_A is large enough, induces the second cell, C_2 to burst. Then, C_2 generates an Ach current which *prolongs* the burst of C_1 and increases the peak in Ach production for the two cells. Both cells are then mutually coupled and this mutual coupling reinforces their synchrony. Here, we have assumed that all cells have the same time constant and we only observe 1:1 locking, but, in a population of heterogeneous cells it is possible to exhibit $m : n$ synchronies in patterns corresponding to Arnold tongues [43, 44]. We will not develop on this situation here. The 2 SACs synchrony lasts until I_{sAHP} is large enough to stop bursting in both cells, but not necessarily at the same time.

From these considerations, it is possible to compute the time t_B when the cell C_2 starts to burst if the cell C_1 has started to burst at time 0 [41]. This time depends on model's parameters fixing the value of I_{SN_1} and Ach production. It also depends on the value of the refractory variable R at time t_B, illustrating, once again the delicate balance between I_{sAHP} (G_S) and I_A (G_A).

It is also possible to show that, in the NIB regime ($I_{\text{SN}_1} > 0$), there is a minimal value g_{A_c} for g_A, the maximal conductance, below which C_1 cannot excite C_2. This observation is fundamental to understand the propagation of wave in a one-dimensional chain of N SACs. In Fig. 14.3, we show the initiation and propagation of waves for increasing g_A. On the left, the entrance of a SAC in the bursting regime initiates a propagation which ends very soon. This is because, for this regime of parameters, the Ach current, even at its maximum is not sufficient to cross the SN barrier. Only noise affords it, with a low probability, depending on noise intensity and distance between g_A and g_{A_c}. On the opposite, on the right-most figure we see waves essentially propagating through the whole lattice. This is because the sAHP barrier is not sufficient to stop waves propagation. Therefore, there is an intermediate regime (Fig. 14.3, middle) where waves propagation has to cope with the sAHP landscape left by the previous waves, and where there is a wide spread in the distribution of waves (power laws).

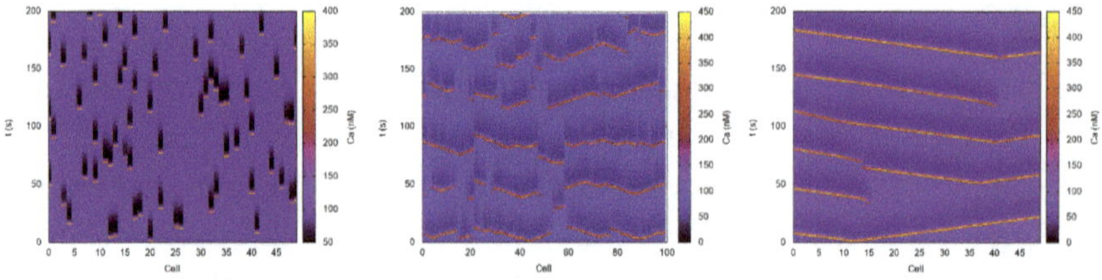

Fig. 14.3. Example of propagation of waves, near the g_A critical point g_{A_c}, in the presence of noise ($\sigma = 3\,\mathrm{pA\,ms}^{-\frac{1}{2}}$) for 50 cells. Left: $g_A = 0.04\,\mathrm{nS}$; Middle: $g_A = 0.07\,\mathrm{nS}$; Right: $g_A = 0.1\,\mathrm{nS}$. Red zones are active (bursting) zones; black zones are hyperpolarized (sAHP) zones; purple zones correspond to SACs at rest. In the middle figure, red zones (propagating waves) can be stopped by hyperpolarized regions.

One can show that there is a characteristic correlation length, corresponding to the average size of active regions and depending on the parameters g_{sAHP} and g_A [44]. In addition, near g_{A_c}, one observes power laws in the waves distribution (size and duration) and the correlation length increases [10]. At high g_A waves spread along the full lattice and the correlation length is the lattice size.

14.3.3.3. *Two-dimensional dynamics*

The situation in two dimensions is very similar. The main difference is the topology of wave fronts, which are points in one dimension and becomes complex lines in two dimensions. Therefore, it is quite more difficult to determine the wave shape, especially when two waves interact, rendering difficult a statistical analysis of waves statistics. Example of two-dimensional simulations (made by Lionel Gil) can be found at https://www.youtube.com/watch?v=shMR3NMCBDE.

A big difficulty in simulating the two-dimensional case is the large timescale separation between dynamics. To simulate the complete dynamics, one has to use an integration scheme with a very small step, lower than one millisecond, while the timescale of duration of sAHP regions last of order minutes. Therefore, it requires a very long time to simulate a large number of waves to get their statistics.

At this stage, it would certainly be more useful to derive, from this model, another model averaging over the fast dynamics to produce transport equations correctly accounting for waves initiation, propagation and stop. Some attempts in this direction has been made [24, 35, 37, 39, 40] but none of them really take into account the closeness to bifurcation, at the core of the wide variability in waves characteristics (see Sec. 14.3.4.1), where one would like to have, in addition, a scaling theory to describe waves statistics when approach the critical point. A potential candidate is a Kardhar–Parisi–Zhang-like equation for retinal waves taking into account the spatio-temporal variations of sAHP (in preparation and [45]).

14.3.4. *Consequences*

We would like now to shortly propose some potential consequences of this analysis.

14.3.4.1. *Explaining variability with a unique mechanism*

As announced in the introduction of this section the wide variability in waves distribution can be explained by the closeness to a saddle-node bifurcation point. The spatio-temporal variability in waves distribution for a given set of parameters have been explained throughout this section. Now, during development, and even in the stage II phase, the maximal Ach conductance g_A is known to evolve — Zheng *et al.* [46], have shown that the intensity of the Ach coupling is *monotonously decreasing* with time — thereby inducing strong variations in the waves evolution. We have also seen that the closeness to SN in the NIB regime leads to strong variations in the interburst interval. This could explain the variability observed across and within species [9]. Finally, one interest of our model is to take into account physiological parameters that can be modified upon pharmacology, inducing potential bifurcations. Some of them are studied in [9].

More generally, the variations in physiological parameters can be induced by: (i) development; (2) pharmacology; (3) pathologies. Indeed, during retinal diseases such as Retinitis Pigmentosa or Age Macular Degeneration, the structure and dynamics of the retina is impacted inducing, in addition to the loss of sight, spatio-temporal phenomena such as large scales oscillations that could be studied with the same type of analysis made here (local bifurcation and extension to network) although the involved cells are not SACs [47].

14.3.4.2. *Dynamical response to stimuli*

Waves distribution can exhibit power laws (in our model, close to the critical line between A and C in Fig. 14.2). This was first observed by M. Hennig *et al.*, in a paper combining experiments and modeling. Investigating the dynamics of stage II retinal waves, they show that the network of SACs is capable of operating at a transition point between purely local and global functional connectedness, corresponding to a percolation phase transition, where waves of activity — often referred to as "avalanches" — are distributed according to power laws (see Fig. 14.4 of [39]). They interpret this regime as an indication that early spontaneous activity in the developing retina is regulated according to the following principle; maximize randomness and variability in the resulting activity patterns. This remark is in complete agreement with the idea of dynamic range maximization [48] and could be of central importance for our understanding of the visual system. In fact, it suggests that, during its formation, the visual system could be driven by spontaneous events, namely the retinal waves, exhibiting the characteristics of a second-order phase transition. However, giving solid evidence that a second-order phase transition occur in this model, namely satisfies the relations between critical exponents predicted, e.g., by the renormalization group is a hard task [48–51], even numerically, because of the large range in timescales variations in the dynamics, requiring prohibitive simulations times (especially in two dimensions) to obtain a sufficiently large sample of waves. There might be more hope to develop a model where fast time scale is averaged out [40]. To our best knowledge, no renormalization group analysis has been yet in models of retinal waves.

It has been suggested [52, 53] that such the scale-free organization observed in many neural systems [53, 54] could foster information storage and transfer, improvement of the computational capabilities [55], information transmission [54–56], sensitivity to stimuli and enlargement of dynamic range [57–60]. In this spirit, retinal waves could certainly play a role in the organization of the visual system to a point with maximal efficacy in response to the wide spatio-temporal range of visual stimuli.

14.4. Response to Stimuli

Many branches of dynamical system theory, many dynamical system studies, deal with stationary (time translation invariant) situations. This is somewhat similar in the study of neuronal models where entries are mostly static (constant or white noise). Unfortunately, the retina (and thereby the visual cortex) receives constantly non-stationary entries. Indeed, *in vivo*, the retina has to cope with moving objects in a moving world.

These remarks open up the necessity to adapt or extend theoretical methods of dynamics analysis to nonstationary situations. To our best knowledge, such methods are at an early stage of development, and the reader should not expect to find anything exceptional on this topic in these pages. Nonetheless, a few interesting results can be obtained, enlightening the potentialities of the retina. In this section, we give an example of nonstationary response of the retina, in the context of retinal anticipation. We use the fact that bipolar and amacrine cells are not spiking to propose an approximation based on piecewise linear dynamics where nonstationarity can be handled. This allows us to predict a potentially new effect of lateral amacrine connectivity, fully studied in [11, 61]: the possibility that a stimulus, moving on a deterministic trajectory, triggers a wave of activity, *ahead* of the stimulus, thereby allowing the retina to anticipate motion, enhancing local nonlinear gain control mechanisms reported in [3, 4], and shortly introduced below.

14.4.1. *Convolution kernel*

The "first" stage of the retina, ending at the OPL, including photo-receptors, Hcells and Bcells is often represented by a kernel, integrating the processing of these different layers and characterizing the so-called Receptive Field (RF) of a Bcell. In this description, the voltage of a Bcell i is modulated, stimulus-driven, by the term:

$$V_{i_{\text{drive}}}(t) = [\mathcal{K}_{B_i} \overset{x,y,t}{*} \mathcal{S}](t) = \int_{x=-\infty}^{+\infty} \int_{y=-\infty}^{+\infty} \int_{s=-\infty}^{t}$$
$$\times \mathcal{K}(x - x_i, y - y_i, t - s)\, \mathcal{S}(x, y, s) dx\, dy\, ds, \quad (14.10)$$

where $\overset{x,y,t}{*}$ means space–time convolution.

On biological grounds, a Receptive Field (RF) is a region of the visual field (the physical space) in which stimulation alters the voltage of the Bcell. Mathematically, this is, typically, the product[2] $\mathcal{K}(x, y, t) = \mathcal{K}_S(x, y)\, \mathcal{K}_T(t)$. Here,

$$\mathcal{K}_S(x, y) = \frac{A_1}{2\pi\sqrt{\det C_1}}\, e^{-\frac{1}{2}\tilde{X}_i.C_1^{-1}.X_i} - \frac{A_2}{2\pi\sqrt{\det C_2}}\, e^{-\frac{1}{2}\tilde{X}_i.C_2^{-1}.X_i} \quad (14.11)$$

is the spatial part of the RF. This is a difference of (oriented) Gaussian that mimics the center-surround organization of Ref. [62], where $X_i = \begin{pmatrix} x - x_i \\ y - y_i \end{pmatrix}$, $\sim$ denotes the transpose, x_i and y_i are the coordinates of the receptive field center which coincide with the coordinates of the cell, C_1, C_2 are positive definite matrix whose main principal axis represents the preferred orientation of the RF. The two Gaussians of the RF are thus concentric. The temporal part of the RF, $\mathcal{K}_T$, can have different form, but, it decays sufficiently fast (e.g., exponentially) as time $t \to +\infty$. Typically, this is the difference of decaying exponentials.

In our model, the BCell voltage is the sum of the external drive (14.10) received by the BCell and of a post-synaptic potential P_{B_i} induced by connected ACells:

$$V_{B_i}(t) = V_{i_{\text{drive}}}(t) + P_{B_i}(t). \quad (14.12)$$

The form of P_{B_i} is given by Eq. (14.17) below. $P_{B_i}(t) = 0$ when no ACells are considered.

[2]More generally, spatio-temporal kernels are not separable in the product of a space- and a time-dependent function. The mathematical analysis described here actually does not require this separability but we choose this form for simplicity.

14.4.2. *Nonlinearities*

BCells have voltage threshold [4]:

$$\mathcal{N}_B(V_{B_i}) = \begin{cases} 0 & \text{if } V_{B_i} \leq \theta_B, \\ V_{B_i} - \theta_B & \text{else,} \end{cases} \tag{14.13}$$

and gain control, a desensitization when activated by a steady illumination [63], mediated by a rise in intracellular calcium Ca^{2+}, preventing thus prolonged activity [3, 64]. Gain control is associated to an activity variable A_{B_i} obeying the differential equation:

$$\frac{dA_{B_i}}{dt} = -\frac{A_{B_i}}{\tau_a} + h_B \mathcal{N}(V_{B_i}(t)). \tag{14.14}$$

The bipolar output to ACells and GCells is then characterized by a nonlinear response to its voltage variation, given by :

$$R_{B_i}\left(V_{B_i}, A_{B_i}\right) = \mathcal{N}_B\left(V_{B_i}\right) \mathcal{G}_B\left(A_{B_i}\right), \tag{14.15}$$

where

$$\mathcal{G}_B(A_{B_i}) = \begin{cases} 0 & \text{if } A_{B_i} \leq 0, \\ \dfrac{1}{1 + A_{B_i}^6}, & \text{else.} \end{cases} \tag{14.16}$$

Gain control plays an important rule in regulating neurons activity, as it adjusts the relationship between dramatically changing inputs and the limited range of amplitudes in the neuron's response of neurons [65–67]. It has been shown in [4] that gain control in BCells and GCells allows the retina to anticipate the trajectory of a moving object, by shifting its peak of activity before the peak obtained without gain control. Mathematically, this anticipation effect is obtained thanks to Eqs. (14.14)–(14.16). When a BCell is submitted to a visual stimulation its voltage V_{B_i} increases,[3] thereby increasing the activity A_{B_i}. When A_{B_i} becomes sufficiently large,

[3]For simplicity, we do not distinguish here between ON or OFF cells and on the nature of the stimulus.

$\mathcal{G}_B(A_{B_i})$ drops down, hence R_{B_i} drops down before the voltage V_{B_i} drops down, thereby advancing the peak of activity.

14.4.3. *Network*

We consider a regular lattice of BCells and a regular lattice of ACells, with the same number of cells. ACells are connected to BCells via a connectivity matrix W_B^A such that $W_{B_i}^{A_j} \leq 0$ is the inhibitory[4] strength of the synaptic connection from ACell A_j to BCell B_i with the convention that $W_{B_i}^{A_j} = 0$ if there is no connection from A_j to B_i. In general $W_{B_i}^{A_j} \neq W_{B_j}^{A_i}$. Likewise, BCells are connected to ACells via a connectivity matrix W_A^B where the synapses $W_{A_j}^{B_i} \geq 0$ (excitatory).

We consider synaptic responses modeled by exponential alpha profiles [68]. Thus, the PSP induced by ACells on BCell i is

$$P_{B_i}(t) = \sum_{j=1}^{N_B} W_{B_i}^{A_j} \int_{-\infty}^{t} \alpha_B(t-s)\, V_{A_j}(s)ds, \quad \alpha_B(t) = e^{-\frac{t}{\tau_B}} H(t), \tag{14.17}$$

where the Heaviside function H ensures causality.

Conversely, the BCell B_i connected to A_j induces, on this cell, a synaptic response characterized by a post-synaptic potential (PSP) $P_{A_j}(t)$. We consider here passive ACells so that their voltage $V_{A_j}(t)$ is equal to this PSP:

$$V_{A_j}(t) = \sum_{i=1}^{N_A} W_{A_j}^{B_i} \int_{-\infty}^{t} \alpha_A(t-s)\, R_{B_i}(s)ds, \tag{14.18}$$

with $\alpha_A(t) = e^{-\frac{t}{\tau_A}} H(t)$. Note that the voltage of the BCell is rectified and gain-controlled.

14.4.4. *Waves of activity induced by a stimulus*

The question we are interested in is the following. Consider a stimulus moving through the receptive field of BCells. How does ACells lateral

[4]Most ACells are inhibitory, glycinergic or GABAergic.

connectivity influence the response of BCells to that stimulus? To answer this question we rewrite the joint dynamics of BCells voltage, BCells activity, ACells voltage in the form of a $3N$-dimensional dynamical system. We use Greek indices $\alpha, \beta, \gamma = 1, \dots, 3N$ and define the state vector $\mathcal{X}$ as:

$$\vec{\mathcal{X}}_\alpha = \begin{cases} V_{B_i}, & \alpha = i, \\ V_{A_i}, & \alpha = N+i, \qquad i = 1, \dots, N. \\ A_i, & \alpha = 2N+i \end{cases}$$

Likewise, we define the stimulus vector $\vec{\mathcal{F}}_\alpha = F_{B_i}$, if $\alpha = 1, \dots, N$ and $\vec{\mathcal{F}}_\alpha = 0$ otherwise. F_{B_i} is a function of the drive (14.10). Then, the dynamical system describing the joint dynamics has the general form:

$$\frac{d\vec{\mathcal{X}}}{dt} = \mathcal{H}(\vec{\mathcal{X}}) + \vec{\mathcal{F}}(t). \tag{14.19}$$

where $\mathcal{H}(\vec{\mathcal{X}})$ is a nonlinear function, due to function $R_{B_i}(V_{B_i}, A_{B_i})$, featuring gain control and low-voltage threshold. This system is a nonlinear, nonautonomous, dynamical system.

We do not know about any general method to analyze this dynamics. However, the nonlinearity comes from the function(14.16), $\mathcal{G}_B(A_{B_i})$ which can be approximated by a piecewise linear function. The idea of using such a phase space decomposition with piecewise linear approximations has been used, in a different context by S. Coombes *et al.* [69] and in [12, 13, 70]. Then, there is a domain Ω of $\mathbb{R}^{3N}$, where $R_{B_i}(V_{B_i}, A_{B_i}) = V_{B_i}$ so that (14.19) is linear and can be written in the following form:

$$\frac{d\vec{\mathcal{X}}}{dt} = \mathcal{L}.\vec{\mathcal{X}} + \vec{\mathcal{F}}(t), \tag{14.20}$$

where $\mathcal{L}$ is a $3N \times 3N$ matrix, depending on the connectivity matrices W_B^A, W_A^B (see [11] for detail). Being in Ω corresponds to intermediate activity, where neither BCells gain control (14.16) nor low threshold (14.13) are active. One can then study this case and describe then what happens when trajectories of (14.19) get out of this domain, activating low-voltage threshold or gain control.

The general solution of (14.20) is

$$\vec{\mathcal{X}}(t) = \int_{t_0}^{t} e^{\mathcal{L}(t-s)}.\vec{\mathcal{F}}(s)\, ds. \tag{14.21}$$

The behavior of the solution (14.21) depends on the eigenvalues $\lambda_\beta, \beta = 1, \ldots, 3N$ of $\mathcal{L}$ and its eigenvectors, $\vec{\mathcal{P}}_\beta$, with entries $\mathcal{P}_{\alpha\beta}$. The matrix $\mathcal{P}$ transforms $\mathcal{L}$ in Jordan form ($\mathcal{L}$ is not diagonalizable when $h_B \neq 0$). Whatever the form of the connectivity matrices W_A^B, W_B^A the N last eigenvalues are always $\lambda_\beta = -\frac{1}{\tau_a}, \beta = 2N+1, \ldots, 3N$, where τ_a is the characteristic time of the BCells activity variable (14.14).

From (14.21), one can show the following general result (not depending on the specific form of W_A^B, W_B^A, they just need to be square matrices and to be diagonalizable):

$$\mathcal{X}_\alpha(t) = V_{\alpha_{\text{drive}}}(t) + \mathcal{E}^B_{B,\alpha}(t) + \mathcal{E}^B_{A,\alpha}(t) + \mathcal{E}^B_{a,\alpha}(t),$$
$$\alpha = 1, \ldots, 3N, \tag{14.22}$$

where the drive term (14.10) is extended here to $3N$-dimensions with $V_{\alpha_{\text{drive}}}(t) = 0$ if $\alpha > N$. It corresponds to the direct effect of the drive on BCells and this is the only nonvanishing term when $W_A^B = W_B^A = 0$. The other terms have the following definition and meaning. $\mathcal{E}^B_{B,\alpha}(t)$, corresponds to the *indirect* effect, via the ACells connectivity, of the BCells drive on BCells voltages (i.e., the drive excites BCell i which acts on BCell j via the ACells network); $\mathcal{E}^B_{A,\alpha}(t)$ corresponds to the effect of BCell drive on ACells voltages, and, finally $\mathcal{E}^B_{a,\alpha}(t)$ corresponds to the effect of the BCells drive on the dynamics of BCell activity variables (see [11] for the explicit form). All these terms involve the spectrum of $\mathcal{L}$ and its eigenvectors, making a connection between spectral properties and the network response to a spatio-temporal stimulus.

In particular, the eigenvalues of $\mathcal{L}$ can be real or complex, stable (real part < 0) or unstable, essentially depending on the connectivity matrices. In this linear response description, the network effect is the superimposition of stable or unstable eigenmodes that can interfere in a constructive way, amplifying the stimulus response, and generate a wave of activity propagating ahead of the stimulus [11]. Depending on the connectivity matrices (graph structure and intensity of weights) these eigenmodes can also interfere and weaken the response to stimuli. Depending on parameters too, the dynamics (14.19) can stay within the set Ω where the linear approximation (14.20) holds. However, the interesting regime is when dynamics gets out of Ω. It happens if some eigenmodes are unstable, but it can happen too when all modes stable, whenever the cumulative of the drive and network effect decreases the BCell voltage sufficiently low to activate the rectification term, or increases the BCell activity sufficiently to trigger gain control (14.16). In all cases getting out of Ω generates a mechanism saturating the instability, preventing divergence and thereby stabilizing the cell in a range where it is still able to respond to the stimuli. In particular, the activation of the gain control mechanism induces anticipation and this activation can be enhanced by the wave induced from the ACells network.

14.4.5. *Consequences*

It is known that the visual system is able to anticipate the motion of an object moving on a deterministic, regular, trajectory [71–74]. In the early visual cortex, an object moving across the visual field triggers a wave of activity ahead of motion, thanks to the cortical lateral connectivity [75–77]. In the retina, anticipation takes a different form. One observes a peak in the firing rate response of GCells to a moving object, occurring *before* the peak response to the same object when flashed. This effect can be explained by purely local mechanisms, at individual cells level, like gain control, here characterized by (14.16) [3, 4]. To our best knowledge, collective effects similar to the cortical ones — that is, a rise in the cell's activity before the object enters in its receptive field due to a wave of activity ahead of the moving object — have not been reported yet experimentally.

In contrast, our analysis predicts such an effect and emphasizes a potential role of amacrine lateral connectivity on anticipation (gap junctions may also play a role but we don't comment it here). The connectivity graph is involved in the spectrum of a propagation operator controlling the time evolution of the network response to a moving stimulus. In the linear approximation, we obtain a superimposition of several effects which can collaborate and enhance anticipation, or compete.

Our model was constructed with a simplification of ACells dynamics, which are simple linear integrators. Nonlinear mechanisms in the dynamics of these cells could enhance resonance-like effects in the network and, thereby, favor the propagation of a lateral wave of activity induced by a moving stimulus [11]. Thus, a more detailed analysis would require a closer investigation of BCells and ACells dynamics and connectivity, especially implying to define more specifically the type of ACell (AII, Starburst, A17, wide field, medium field, narrow field, etc.) and the type of functional circuit one wants to consider.

The retina is only the early stage of the visual system. Visual responses are then processed via the thalamus and the cortex. As exposed in the introduction, anticipation is also observed in V1 with a different modality than in the retina. How do these two effects combine? How does retinal anticipation impact cortical anticipation? To answer these questions at a computational level one would need to propose a model of the retino-thalamico-cortical pathway which, to the best of our knowledge, has never been done. This is a work in progress (S. Souihel, M. di Volo, S. Chemla, A. Destexhe, F. Chavane and B. Cessac, in preparation). See [61] for preliminary results.

14.5. The Structure of Correlations

The trajectory of an object — which is, in general, quite more complex than a moving bar with constant speed — across our visual field generally involves long-range correlations in space and in time. Local information about this motion is encoded by retinal ganglion cells. Decoders based on the firing rates of these cells can extract some of the motion features [78–84]. Yet, lateral connectivity in the retina — especially via amacrine cells connecting bipolar cells — plays a central role in motion processing (see, e.g., [2]). In addition, ganglion cells are directly connected through electric synapses and indirectly via ACells. What could be the role of this lateral connectivity in motion processing? Clearly, one may expect it to induce spatial and temporal correlations in spiking activity, as an echo, a trace, of the object's trajectory. These correlations cannot be read in the variations of firing rate; they also cannot be read in synchronous pairwise correlations as the propagation of information due to lateral connectivity necessarily involves *delays*. This example raises the question about what information can be extracted from spatio-temporal correlations in a network of connected neurons submitted to a transient stimulus. What is the effect of the stimulus on these correlations? How can one handle this information from data where one has to measure *transient* correlations?

Here, we address these questions in a context linking dynamics to statistics of orbits, that is, ergodic theory and statistical mechanics.

14.5.1. *The spontaneous activity of Ganglion cells*

GCells are spiking cells. Henceforth, they can be described either by a spiking neuronal model, like the Integrate and Fire [85], or by a firing rate model. We consider a dual approach where GCells spike according to a joint firing probability of the form:

$$\mathbb{P}_n[\omega(n) \,|\, \omega_{-\infty}^{n-1}] \equiv e^{\phi(n,\omega)}. \tag{14.23}$$

Here, time is binned with a discrete time interval $\delta > 0$ (typically a few ms). We consider a set of N GCells with index $k = 1, \ldots, N$. We are just interested in the "spike state" of GCells, defined by a variable $\omega_k(n) \in \{0, 1\}$: $\omega_k(n) = 1$ if GCell k spikes in the time bin n and $\omega_k(n) = 0$ otherwise. The vector of all spike states ("spike pattern") at time bin n is noted $\omega(n) = \big(\omega_k(n)\big)_{k=1}^{N}$. A "spike history" is a list of successive spike patterns from some time m to some time $n \geq m$, $\omega_m^n \equiv \omega(m), \omega(m+1), \ldots, \omega(n)$. Thus, $\omega_{-\infty}^{n-1}$ denotes the history prior to time n, where the past goes to $-\infty$. Therefore, $\mathbb{P}_n[\omega(n) \,|\, \omega_{-\infty}^{n-1}]$ is the conditional probability to observe the spike pattern n given the infinite history prior to n. The index n, in $\mathbb{P}_n[\,|\,]$, expresses that the transition probabilities depend explicitly on time (we do not assume stationarity). Note that, such transition probabilities, generalizing the notion of Markov chain to infinite memory, are rigorously defined provided they decay sufficiently fast (e.g., exponentially) with the history depth. Under suitable conditions[5] there is a unique invariant probability associated with this set of transition probabilities (generalizing the unique invariant probability of a Markov chain with a primitive transition matrix [87, 88]), called "Chain with complete connection" [86, 89] in probability theory, "g-measure" in ergodic theory [90], or "Gibbs measure". For the last terminology they have indeed a close connection to their statistical mechanics counterpart [91–93], if one assimilates the sequence of spike patterns as a left-sided infinite chain of sites, labeled by n, the time index, with states taking values in $\{0, 1\}^N$. These sites can then be assimilated to "spins" taking 2^N possible values. In Eq. (14.23), $\phi(n, \omega)$ is then naturally called a (normalized) Gibbs potential. It is normalized because it is the log of a probability so there is no need to divide by a partition function (see [16, 93] for the link between "thermodynamic" like Gibbs potentials and normalized potentials).

What is the link of this rather abstract paragraph and GCells in the retina? The idea is the following. It is experimentally easy to measure GCells collective activity (voltage, from Multi-Electrodes Arrays) and to extract spikes via spike sorting [94, 95]. From this information one attempts to construct a probability distribution that mimics the collective spikes statistics. This statistics is constrained by "hidden" information, typically the network of amacrine cells which correlates GCells by indirect interactions, or gap junctions which connect them directly. Of course, the applied stimulus plays a role and what the GCells receive is the processing of this stimulus

---

[5]We assume that $\mathbb{P}_n[\omega(n) \,|\, \omega_{-\infty}^{n-1}] > 0$ so that $\phi(n, \omega) = \log \mathbb{P}_n[\omega(n) \,|\, \omega_{-\infty}^{n-1}] > -\infty$. We note, for $n \in \mathbb{Z}$, $m \geq 0$, and r integer:

$$\omega \overset{m,n}{=} \omega', \text{ if } \omega(r) = \omega'(r), \ \forall r \in \{n - m, \dots, n\}.$$

For m be a positive integer, the m-variation of $\mathbb{P}_n[\omega(n) \,|\, \cdot]$ is:

$$\mathrm{var}_m[\mathbb{P}_n[\omega(n) \,|\, \cdot]] = \sup\{| \ \mathbb{P}_n[\omega(n) \,|\, \omega_{-\infty}^{n-1}] - \mathbb{P}_n[\omega(n) \mid \omega'^{n-1}_{-\infty}] \ |: \omega \overset{m,n}{=} \omega'\}.$$

A sufficient condition to ensure that this system of transition probabilities has a unique invariant measure compatible with it is that var_m decays exponentially fast with m [86].

through the whole retinal network above GCells (see Sec. 14.2). If one has only information about the spikes the most general form of statistics is (14.23). The probability to observe a spike pattern in a time bin n depends on the whole spike history of the GCells network. This history includes the stimulus influence. Of course, for specificity and tractability, one has to consider simplifications of this formalism, for example that successive times are independent. In this case, one obtains a class of models usually designed as Maximum Entropy Models (although Maximum Entropy Principle is quite more general than that [96]). A classical model in this setting is the so-called Ising model — where the potential ϕ takes the form of the energy in a spin-glass with binary spins — or variations (type-dependent local field, triplets or quadruplets interactions) [97–104], see [105] for a physicists-oriented review.

Another approach, considering causality and time correlations, consists of considering conditional probabilities of the form:

$$\mathbb{P}_n[\omega_k(n) = 1 \,|\, \omega_{-\infty}^{n-1}] = f\left(\sum_{j \in \mathcal{B}_k} \sum_{l \leq n-1} J_{kj}(l)\, \omega_j(l) + S_k(n)\right) \quad (14.24)$$

called Generalized Linear Model [106]. f is a nonlinear function. As in Sec. 14.3, $\mathcal{B}_k$ is a set of cells connected to neuron k, $S_k(n)$ is the stimulus viewed by neuron k, namely processed through the retina network. The $J_{kj}(l)$s corresponds to "effective" interactions. In a classical neuronal model, this would be the direct synaptic interactions between neurons, with a possible time integration reflected in the time dependence (see Sec. 14.5.2 for an example). Here, GCells are not directly connected so the $J_{kj}(l)$s should be considered as parameters that need to be fit to best approximate statistics.

MaxEnt and GLM are in fact closely related through Gibbs distributions as developed in [93, 96].

14.5.2. *An example: the discrete time integrate and fire model*

There are many models for neuronal dynamics both at the level of individual neurons and neuronal networks [5, 107, 108]. A canonical example of such a model, actually the first one proposed to the scientific community was introduced by Lapicque in 1907 [85], the Integrate and Fire model. Here, we consider a discrete time version of this model (and discuss the extension to continuous time at the end of the section). We consider N neurons with index $k = 1, \ldots, N$ and membrane potential V_k, depending on a discrete time n. We fix a variable θ called "spiking threshold" such that voltages dynamics obeys [12, 13]:

$$\begin{cases} V^k(n+1) = \gamma V^k(n) + \sum_{j \in \mathcal{B}(k)} W_{kj}\omega_j(n) \\ \qquad\qquad + I_k(n) + \sigma_B \xi_k(n), \quad \text{if } V^k(n) < \theta, \text{ Integrate phase;} \\ V^k(n+1) = 0 \text{ and } \omega_k(n) = 1, \quad \text{if } V^k(n) \geq \theta, \text{ Firing phase,} \end{cases} \tag{14.25}$$

where $0 < \gamma < 1$. If, at some, discrete time n, $V^k(n)$ exceeds the threshold θ the membrane potential is reset at time $n+1$ and a spike is recorded, at n for neuron k, i.e., $\omega_k(n) = 1$. We have assumed that synapses are instantaneous. Then, the synaptic input is $\sum_j W_{kj}\omega_j(n)$, that is a pre-synaptic neuron j acts on the post-synaptic neuron k whenever j spikes, $\omega_j(n) = 1$. $I_k(n)$ is the time discretization of the external stimulus influence (for example, a time discretization of the BCell response (14.10)). Note that it depends on n, hence dynamics (14.25) is nonstationary. $\xi_k(n)$ are independent standard Gaussian random variables representing the effect of noise.

It is easy to integrate Eq. (14.25) conditionally upon a fixed spike sequence ω. A trajectory $V \equiv \{ V_k(n), k = 1, \ldots, N, n \in \mathbb{Z} \}$ is *compatible*[6] with the spike sequence ω if $\chi(V_k(n) > \theta) = \omega_k(n)$, $\forall k = 1, \ldots, N, n \in \mathbb{Z}$, where $\chi(A)$ is the indicator function of the logical event A, $\chi(A) = 1$ if A is true, $\chi(A) = 0$ otherwise. We note $\tau_k(\omega, n) = \max\{ l, l < n \,|\, \omega_k(n) = 1 \}$ the last time before n where

[6]Compatibility has to do here with symbolic coding of voltage trajectories and Markov partitions, see [12] for detail.

neuron k has spiked, thus whose voltage was reset to 0. Then

$$V^k(n+1) = \sum_{j=1}^{N} W_{kj}\,\eta_{kj}(n,x) + \sum_{l=\tau_k(\omega,n)}^{n} \gamma^{n-l} I_k(l)$$
$$+\sigma_B \sum_{l=\tau_k(\omega,n)}^{n} \gamma^{n-l}\xi_k(l), \tag{14.26}$$

where

$$\eta_{kj}(n,\omega) = \sum_{l=\tau_k(\omega,n)}^{n} \gamma^{n-l}\omega_j(l) \tag{14.27}$$

integrates the influence of pre-synaptic neuron j on the time interval $[\tau_k(\omega,n)+1,n]$. Each spike emitted by this neuron, at time l in this time interval contributes with the weight γ^{n-l} and there is no contribution at times where this neuron did not spike, $\omega_j(l)=0$. The condition $\gamma<1$ somewhat defines the exponential decay in the spike history dependence via the characteristic time:

$$\tau_\gamma = -\left[\frac{1}{\log\gamma}\right]. \tag{14.28}$$

Likewise, $\sum_{l=\tau_k(\omega,n)}^{n}\gamma^{n-l}I_k(l)$ integrates the stimulus influence on neuron k and $\sigma_B\sum_{l=\tau_k(\omega,n)}^{n}\gamma^{n-l}\xi_k(l)$ is the integrated noise term. This is a Gaussian random variable, with mean zero and variance $\sigma(k,n,x)=\sigma_B^2\,\frac{1-\gamma^{2(n+1-\tau_k(\omega,n))}}{1-\gamma^2}$.

Thanks to the integrated Eq. (14.26) and because the integrated noise is Gaussian, it is now easy to compute the probability that neuron k spikes at time $n+1$ *given* the history ω, $\mathbb{P}_n[\omega_k(n)=1\,|\,\omega_{-\infty}^{n-1}]=\mathbb{P}_n[V_n^k\geq\theta\,|\,\omega_{-\infty}^{n-1}]$. This is

$$\mathbb{P}_n[\omega_k(n)=1\,|\,\omega_{-\infty}^{n-1}]$$
$$= f\left(\frac{\theta-\sum_{j=1}^{N}W_{kj}\eta_{kj}(n,\omega)-\sum_{l=\tau_k(\omega,n)}^{n}\gamma^{n-l}I_k(l)}{\sigma_B\sqrt{\frac{1-\gamma^{2(n+1-\tau_k(\omega,n))}}{1-\gamma^2}}}\right), \tag{14.29}$$

where $f(z)=\int_z^{+\infty}\frac{e^{-\frac{u^2}{2}}}{\sqrt{2\pi}}\,dz$.

As integrated noise variables for different ks are independent we have, for any spike pattern $\omega(n)$,

$$\mathbb{P}_n[\omega(n) \,|\, \omega_{-\infty}^{n-1}] = \prod_{k=1}^{N} (\omega_k(n)) \; \mathbb{P}_n[\omega_k(n) = 1 \,|\, \omega_{-\infty}^{n-1}] \\ + (1 - \omega_k(n)) \; \left(1 - \mathbb{P}_n[\omega_k(n) = 1 \,|\, \omega_{-\infty}^{n-1}]\right) \tag{14.30}$$

giving an explicit form to (14.23) and linking explicitly the dynamics (14.25) to statistics, with an explicit dependence in network parameters. Note, in particular, the form similar to the GLM model (14.24), although quite more general (as the $J_{ij}(l) = W_{ij}\gamma^{n-l}$ are here explicitly written).

Let us comment this result. First, (14.30) is the transition probability, of the form (14.23) where the normalized Gibbs potential ϕ can be explicitly written, in terms of the synaptic interactions and the parameter σ_B controlling the noise. Here, the memory depth, going back to $\tau_k(\omega, n)$, the last time in the past where neuron k was reset, depends on ω, defining a *variable length Markov chain.* In addition, $\tau_k(\omega, n)$, has a function of ω, is unbounded. One can find for all $r \in \{-\infty, n\}$ a sequence ω such that $\tau_k(\omega, n) = r$ (take the sequences where all $\omega_k(n) = 0$, $k = 1, \ldots, N$, $n > r$ and $\omega_k(n) = 1$ for at least one k). We have thus to deal with the extension of Markov chains, to chains with unbounded memory also called chains with complete connections [89]. The existence of a Gibbs distribution compatible with this chain is guarantee by the exponential decay of the memory controlled by $\gamma < 1$ [14]. Finally, in (14.30), the transition probabilities depend explicitly on time, because of the stimulus-dependent term. They are therefore nonstationary.

The form (14.24) can also be mathematically justified for a network of Integrate and Fire neurons with continuous time [15]. The main difficulty is that, in this case, the set of spikes in uncountable (because the spike is instantaneous in IF models, which is a biological, physical, and mathematical nonsense). Using the realistic fact that spikes have a duration, it is nevertheless possible to associate to the continuous time dynamics of the IF a countable set of spikes [70], for which one can establishes a generalization of Eq. (14.29), which, actually, allows to consider conductance-based IF [109] where the conductance depends on the spike history. The form (14.24) fails when there are fast synapses such as gap junctions and more general cases have to be considered [15], which encompass Maximum Entropy Models and GLMs in the formalism of Gibbs distributions [16, 96].

14.5.3. *Response to stimuli*

The main advantage of the general form (14.23) is that it handles nonstationarity, i.e., the dependence of the statistics in a spatio-temporal stimulus. More explicitly, a form like (14.23) shows that the statistics of spikes at time n depends on a delicate alchemy involving intrinsic neurons dynamics, their interactions, and the stimulus.

Obviously, the form (14.23) is too general to be useful unless additional assumptions are made. Such a natural assumption is to consider that the stimulus is "small" and additive, that is, that one can make a Taylor expansion of the potential as powers of the stimulus. In this case, it is possible to show that the response of the neuronal system to a stimulus, that is, the variations in spike statistics induced by the stimulation, can be described in terms of a linear response theory [110].

The main result is that the variation in the average of an observable f (i.e., a real function of spikes blocs) resulting from the application of a stimulus reads:

$$\delta\mu[f](n) \equiv \mu^S[f](n) - \mu^{(sp)}[f] = (K_f * S)(n), \tag{14.31}$$

where $\mu^{(sp)}$ is the Gibbs distribution in spontaneous activity (without stimulus), μ^S is the Gibbs distribution in evoked activity (with stimulus), $\mu^S[f](n)$ means the average of f with respect to μ^S, which depends on time (if the stimulus does), and $\mu^{(sp)}[f]$ means the average of f with respect to $\mu^{(sp)}$, which does not depends on time. This variation in average is given by a convolution of a kernel K_f, depending on f, and of the stimulus.

The form (14.31) is classical, looking very much like, e.g., the RF modeling (14.10). The detail of this simple canonical form hides in fact a high complexity. In particular, it depends on f and on the parameters determining dynamics (e.g., synaptic weights). In general, K_f can be expressed as a sum correlations between f and spikes events of the form $\omega_{k_1}(n_1)\ldots\omega_{k_r}(n_r)$, called monomials or order r. Monomials of order 1 have the form $\omega_{k_1}(n_1)$ and just tell us if neuron k_1 is spiking at time n_1, monomials of order 2 have the form $\omega_{k_1}(n_1)\omega_{k_2}(n_2)$ and tell us if neuron k_1 is spiking at time n_1 *and* neuron k_2 is spiking at time n_2, and so on. The correlations are computed with respect to the stationary measure $\mu^{(sp)}$ so that only the difference in times matters when computing correlations. In addition, the influence of monomials in the expansion decreases with their order, so that one can obtain a reasonable expression of the convolution kernel considering averages of order two monomials (time-dependent pairwise correlations).

This is therefore a result in the form of a fluctuation–dissipation "theorem" in statistical physics, with the difference that one considers time-dependent correlations. This formula has proven to give astonishingly good results when computing the response to a time-dependent stimulus in the model (14.25) [110].

14.5.4. *Consequences*

The result presented here introduces a general mathematical formalism allowing one to handle spike correlations as a result of a neuronal network activity in response to a stimulus. The most salient result of this work is that the difference of an observable average in response to an time-dependent external stimulus of weak amplitude can be computed from the knowledge of the spontaneous correlations, i.e., from the dynamics without the stimulus. This result is not surprising from a nonequilibrium statistical physics perspective (Kubo relations, fluctuation–dissipation relation [111, 112]). However, to the best of our knowledge, it has been established for the first time for nonstationary spiking neuronal networks in the recent paper [110]. The novelty of our approach is that it provides a consistent treatment of the expected perturbation of higher order correlations, going in this way, beyond the known linear perturbation of firing rates and instantaneous pairwise correlations; in particular, it extends to time-dependent correlations.

As we have shown, and as expected, the stimulus-response and dynamics are entangled in a complex manner. For example, the response of a neuron k to a stimulus applied on neuron j does not only depends on the synaptic weight W_{kj} but, in general, on all synaptic weights, because the dynamics create complex causality loops which build up the response of neuron k [16, 113, 114]. We formally obtained a linear response function in terms of the parameters of a spiking neuronal network model and the spike history of the network. Although a linear treatment may seem a strong simplification, our results suggest that already, in this case, the connectivity architecture should not be neglected. In the presence of stimuli, the whole architecture of synaptic connectivity, history and the dynamical properties of the networks are playing a role in the correlations through the perturbed potential. It follows from our analysis that this effective connectivity depends on stimulus, as already observed for the effective connectivity defined through the Ising model [115]; it depends also on synaptic weights, in a complex manner. It was observed in [114] that the existence of resonances in the power spectrum generates stimuli dependent graphs, e.g., controlled by the frequency of an harmonic stimulus. It would be interesting to check whether such a property holds here as well.

14.6. Discussion and Perspectives

Although the retina might be viewed as nothing more than a part of the brain, it shows up strong differences, which has to be thoroughly considered for a modeler interest in the retina dynamics. In this chapter, we have shown how specific behaviors of the retinal network can be considered in the context of dynamical systems theory. Yet, a large part of the modeling and of the mathematical machinery used for our models analysis are not really specific to the retina and could be addressed in the cortex as well.

In this section, we would like to consider questions, more specific to the retina structure and dynamics, that could be tackled in the realm of dynamical systems theory.

14.6.1. *Functional retinal circuits*

The feed-forward/lateral structure of the retina, the diversity of cells subtypes for a given type (e.g., around 40 amacrine cell subtypes), synaptic types offer a gigantic combinatorial of potential retinal circuits. There are however few functional circuits, duly identified, and characterized from the experimental point of view. Important examples are the ON, OFF, and ON–OFF pathways [21]. A central concept in visual system study is the notion of "feature detectors" where the visual system separates the visual image into components. From this perspective, highlights and shadows are two of the most fundamental features of images. In all vertebrate visual systems, the separation of highlights and shadows is performed by two separate pathways, starting at the retina: the ON and the OFF pathways. This organization allows to provide information to the ganglion cells concerning brighter than background stimuli (ON-center) or darker than background stimuli (OFF-center). More generally, the trajectory of an object (e.g., a bright bar) on a moving background necessarily involves these pathways performing local detection of highlights and shadows. These two pathways are, in fact, connected via AII amacrine cells and gap junctions, indirectly connecting ganglion cells thereby potentially creating spatio-temporal correlations between local information channels. This example outlines the functional role of specific retina circuits in processing visual information. Other pathways implied in differential motion would also be interesting to study, as discussed in Sec. 14.4.

These circuits are small in the sense that they involve a few, specific cells, with specific synapses. Here, thus, the details of the biophysical modeling play a distinguished role, especially the physiological parameters involved (see also Sec. 14.6.2). Yet, these circuits performing local information extraction from a visual scene, span the whole retina, resulting in a multi-scale, stimulus-driven dynamics, that can hardly be treated with classical mean-field methods used in the cortex [116–118], especially taking into account that most of the involved neurons are not spiking. A study in this spirit has been proposed in Sec. 14.3 dealing with retinal waves. Yet, the situation was still "simple" as we were considering one type of cells and synapses, without sensory entry. A more general, multi-scale, study of retina circuits like the ON–OFF would deserve considerably more efforts, but, is, perhaps, the necessary way to understand how retina efficiently encodes a visual scene.

14.6.2. *Changing physiological parameters*

The interest of a careful modeling of the retina by dynamical systems is also found in the impact of variations of parameters on local and global functioning of the retina. How do specific local characteristics of neurons, synapses, retinal circuits — which may be affected by physiological modifications (development, short-term plasticity, pharmacology, pathology) — impact the large-scale dynamics of the retina, in particular in the presence of spatio-temporal visual stimuli? Living systems, such as the retina, are somewhat robust to small variations in their physiological parameters ... up to some extent. One of the profound lessons to be drawn from the theory of bifurcations is that continuous, smooth, variations of parameters lead to smooth changes ... until bifurcations points are reached. There, changes are dramatic and drastic. The interest of studying retinal circuits, from the local scale to the whole retina scale, in a dynamical systems perspective, is precisely to anticipate potential bifurcations induced by physiological parameters variations. An example as been presented in Sec. 14.3. In the related paper [9], the effect of variations of parameters such as the conductance of fast potassium channels has lead us to conjecture a role played by specific channels, called Kv3, and the impact of pharmacological actions. In this perspective, dynamical systems modeling and bifurcations theory are theoretical complement to experimental investigations aiming at better understand the functioning of the retina in normal . . . or pathological conditions.

Let us indeed briefly deal with retinal pathologies. Visual impairment affects millions of people in the world. About 39 million were considered blind in 2014 [119]. Roughly 80% of visual impairment is preventable or curable, including operable conditions such as cataracts. However, retinal-degeneration disorders had no cure until recently [119]. Retinal degeneration is the deterioration of the retina caused by the progressive and eventual death of the cells of the retina. Important examples are Age Macular Degeneration (AMD) or Retinitis Pigmentosa.

In the recent years, a new hope of slowing or even reversing vision loss from retinal disorders has grown with technics such as gene therapy, cell transplants and retinal prostheses which have reached the level of clinical trials testing, mostly in treating rare congenital disorders. In parallel, several biotechnology firms have formed to take treatments through clinical testing. Institutes, like the Institut de la Vision in Paris, at the intersection between fundamental and clinical research, and industrial development are born. There are large avenues for thorough modeling of these diseases, the progressive degenerations of cells, the effect of therapies, using methods from dynamical systems theory. A promising example is the modeling of the spread and treatment of hyperoxia-induced photoreceptor degeneration in retinitis pigmentosa [120–122]. Such an approach could also benefit to the large development of retinal prostheses, technological devices used to stimulate specific, still responsive, cells in a degenerated retina, thereby (partially) restoring vision. For example, a modeling approach combined with dynamical system approach could be used to characterize the effect and the optimal regime of electric stimulation with electrodes, in a retina whose physiology, circuitry, and dynamics constantly evolve along the progression of the disease [123–125].

This emphasizes the universality of dynamical systems theory and opens up perspectives to better understand vision and its pathologies, and to help elaborating treatments.

Acknowledgments

This chapter mostly present results, published elsewhere, done in joint collaborations with I. Ampuero, R. Cofré, L. Gil, D. Karvouniari, O. Marre, S. Picaud, S. Souhiel and I would like to thank them. This work benefited from the support of the Ecole Doctorale des Sciences et Technologies de l'Information et de la Communication de Nice-Sophia-Antipolis (EDSTIC), the Agence Nationale de la

Recherche (ANR), project Trajectory https://anr.fr/Project-ANR-15-CE37-0011, and the Neuromod Institute of the University Côte d'Azur.

References

[1] J. Besharse and D. Bok, *The Retina and its Disorders* (Elsevier Science, 2011).
[2] T. Gollisch and M. Meister, Eye smarter than scientists believed: neural computations in circuits of the retina, *Neuron* **65**(2) (2010) 150–164.
[3] E. Y. Chen, O. Marre, C. Fisher, G. Schwartz, J. Levy, R. A. da Silviera and M. J. I. Berry, Alert response to motion onset in the retina, *J. Neurosci.* **33**(1) (2013) 120–132.
[4] M. Berry, I. Brivanlou, T. Jordan and M. Meister, Anticipation of moving stimuli by the retina, *Nature* **398**(6725) (1999) 334—338.
[5] G. B. Ermentrout and D. Terman, *Foundations of Mathematical Neuroscience. Interdisciplinary Applied Mathematics* (Springer, 2010).
[6] D. H. Hubel and T. N. Wiesel, Receptive fields of optic nerve fibres in the spider monkey, *J. Physiol.* **154** (1960) 572–580.
[7] H. Barlow, Possible principles underlying the transformation of sensory messages, *Sensory Commun.* (1961) pp. 217–234.
[8] B. Olshausen and D. Field, Sparse coding with an overcomplete basis set: A strategy employed by V1? *Vision Res.* **37** (1998) 3311–3325.
[9] D. Matzakou-Karvouniari, L. Gil, E. Orendorff, O. Marre, S. Picaud and B. Cessac, A biophysical model explains the oscillatory behaviour of immature starburst amacrine cells, *Sci. Rep.* **9** (2019) 1859.
[10] D. Matzakos-Karvouniari, B. Cessac and L. Gil, Noise driven broadening of the neural synchronisation transition in stage ii retinal waves, *Phys. Rev. E* (2019), submitted for publication.
[11] S. Souihel and B. Cessac, On the potential role of lateral connectivity in retinal anticipation, *J. Math. Neuro.* (2020), Under Revision.
[12] B. Cessac, A discrete time neural network model with spiking neurons. rigorous results on the spontaneous dynamics, *J. Math. Biol.* **56**(3) (2008) 311–345.
[13] B. Cessac, A discrete time neural network model with spiking neurons ii. dynamics with noise, *J. Math. Biol.* **62**(6) (2011) 863–900.
[14] B. Cessac, Statistics of spike trains in conductance-based neural networks: Rigorous results, *J. Math. Neurosci.* **1**(8) (2011) 1–42.
[15] R. Cofré and B. Cessac, Dynamics and spike trains statistics in conductance-based integrate-and-fire neural networks with chemical and electric synapses, *Chaos, Solitons Fractals* **50**(13) (2013) 3.

[16] R. Cofré and B. Cessac, Exact computation of the maximum-entropy potential of spiking neural-network models, *Phys. Rev. E.* **89** (2014) (052117).

[17] B. Cessac and A. Palacios. Spike train statistics from empirical facts to theory: the case of the retina, in *Modeling in Computational Biology and Biomedicine: A Multidisciplinary Endeavor*, eds. F. Cazals and P. Kornprobst, Lectures Notes in Mathematical and Computational Biology (LNMCB) (Springer, 2012), pp. 261–302.

[18] J. Vasquez, O. Marre, A. Palacios, M. Berry and B. Cessac, Gibbs distribution analysis of temporal correlation structure on multicell spike trains from retina ganglion cells, *J. Physiol. Paris.* **106**(3–4) (2012) 120–127.

[19] B. Cessac, P. Kornprobst, S. Kraria, H. Nasser, D. Pamplona, G. Portelli and T. Viéville, PRANAS: A new platform for retinal analysis and simulation, *Front. Neuroinform.* **11** (2017) 49; http://journal.frontiersin.org/article/10.3389/fninf.2017.00049.

[20] S. Baccus and M. Meister, Fast and slow contrast adaptation in retinal circuitry, *Neuron.* **36**(5) (2002) 909–919.

[21] R. Nelson and H. Kolb, On and off pathways in the vertebrate retina and visual system, *Visual Neurosci.* **1** (2004) 260–278.

[22] J. Johnston and L. Lagnado, General features of the retinal connectome determine the computation of motion anticipation, *Elife* (2015).

[23] R. O. L. Wong, M. Meister and C. J. Shatz, Transient period of correlated bursting activity during development of the mammalian retina, *Neuron* **11**(5) (1993) 923–938.

[24] M. B. Feller, D. A. Butts, H. L. Aaron, D. S. Rokhsar and C. J. Shatz, Dynamic processes shape spatiotemporal properties of retinal waves, *Neuron.* **19**(2) (1997) 293–306; http://www.sciencedirect.com/science/article/pii/S089662730080940X.

[25] E. Sernagor, S. Eglen and M. O'Donovan, Differential effects of acetylcholine and glutamate blockade on the spa- tiotemporal dynamics of retinal waves, *J. Neurosci.* **20** (2000) RC56.

[26] S. I. Firth, C.-T. Wang and M. B. Feller, Retinal waves: mechanisms and function in visual system development, *Cell Calcium.* **37**(5) (2005) 425–432; http://www.sciencedirect.com/science/article/pii/S0143416005000278. Calcium in the function of the nervous system: New implications.

[27] D. K. Warland, A. D. Huberman and L. M. Chalupa, Dynamics of spontaneous activity in the fetal macaque retina during development of retinogeniculate pathways, *J. Neurosci.* **26**(19) (2006) 5190–5197.

[28] E. Sernagor and N. Grzywacz, Spontaneous activity in developing turtle retinal ganglion cells: pharmacological studies, *J. Neurosci.* **19**(10) (1999) 3874–3887.

[29] E. Sernagor and M. Hennig, *Retinal Waves: Underlying Cellular Mechanisms and Theoretical Considerations* (Elsevier, 2012), Chapter 1.

[30] K. J. Ford and M. B. Feller, Assembly and disassembly of a retinal cholinergic network, *Visual Neurosci.* **29** (2012) 61–71.

[31] A. Maccione, M. H. Hennig, M. Gandolfo, O. Muthmann, J. Coppenhagen, S. J. Eglen, L. Berdondini and E. Sernagor, Following the ontogeny of retinal waves: pan-retinal recordings of population dynamics in the neonatal mouse, *J. Physiol.* **592**(7) (2014) 1545–1563.

[32] J. H. Singer, R. R. Mirotznik and M. B. Feller, Potentiation of L-type calcium channels reveals nonsynaptic mechanisms that correlate spontaneous activity in the developing mammalian retina, *J. Neurosci.* **21**(21) (2001) 8514–8522.

[33] K. J. Ford, A. L. Félix and M. B. Feller, Cellular mechanisms underlying spatiotemporal features of cholinergic retinal waves, *J. Neurosci.* **32**(3) (2012) 850–863.

[34] J. Zheng, S. Lee and Z. J. Zhou, A transient network of intrinsically bursting starburst cells underlies the generation of retinal waves, *Nat. Neurosci.* **9**(3) (2006) 363–371.

[35] J. Gjorgjieva and S. J. Eglen, Modeling developmental patterns of spontaneous activity, *Current Opinion Neurobiol.* **21**(5) (2011) 679–684; http://www.sciencedirect.com/science/article/pii/S0959438811000833. Networks, circuits and computation.

[36] D. A. Butts, M. B. Feller, C. J. Shatz and D. S. Rokhsar, Retinal waves are governed by collective network properties, *J. Neurosci.* **19**(9) (1999) 3580–3593.

[37] K. B. Godfrey and N. Swindale, Retinal wave behavior through activity-dependent refractory periods, *PLoS Comput Biol.* **3**(11) (2007) e245.

[38] K. B. Godfrey and S. J. Eglen, Theoretical models of spontaneous activity generation and propagation in the developing retina, *Mol. BioSyst.* **5** (2009) 1527–1535.

[39] M. Hennig, C. Adams, D. Willshaw and E. Sernagor, Early-stage waves in the retinal network emerge close to a critical state transition between local and global functional connectivity, *J. Neurosci.* (2009).

[40] B. Lansdell, K. Ford and J. Kutz, A reaction-diffusion model of cholinergic retinal waves, *PLoS Comput. Biol.* **10**(12) (2014) 1–14.

[41] D. Karvouniari. Retinal waves: theory, numerics, experiments, Ph.D. thesis, EDSTIC (2018).

[42] C. Morris and H. Lecar, Voltage oscillations in the barnacle giant muscle fiber, *Biophys J.* **35**(1) (1981) 193–213; http://www.biomedsearch.com/nih/Voltage-oscillations-in-barnacle-giant/7260316.html.

[43] V. I. Arnold, *Geometrical Methods in the Theory of Ordinary Differential Equations* (Springer, 1983).
[44] D. Karvouniari, L. Gil, O. Marre, S. Picaud and B. Cessac. Mathematical and experimental studies on retinal waves, in *2nd Int. Conf. Mathematical Neuroscience (ICMNS)*, Juan les Pins, France (May, 2016).
[45] B. Cessac. The retina: a fascinating object of study for a physicist, in *UCA Complex Days* (2018).
[46] J.-J. Zheng, S. Lee and Z. Zhou, A developmental switch in the excitability and function of the starburst network in the mammalian retina, *Neuron* **44**(5) (2004) 851–864.
[47] H. Choi, L. Zhang, M. S. Cembrowski, C. F. Sabottke, A. L. Markowitz, D. A. Butts, W. L. Kath, J. H. Singer and H. Riecke, Intrinsic bursting of aii amacrine cells underlies oscillations in the RD1 mouse retina, *J. Neurophysiol.* (2014).
[48] T. L. Ribeiro and M. Copelli, Deterministic excitable media under poisson drive: Power law responses, spiral waves, and dynamic range, *Phys. Rev. E* **77** (2008) 051911.
[49] J. Sethna, Hysteresis, avalanches, and disorder induced critical scaling: A renormalization group approach, *Phys. Rev. B* **53** (1995).
[50] S. Ma, *Modern Theory Of Critical Phenomena* (Routledge, 2001).
[51] D. Volchenkov, P. Blanchard and B. Cessac, Quantum field theory renormalization group approach to self-organized critical models: The case of random boundaries., *Int. J. Mod. Phys. B* **16**(08) (2002) 1171–1204.
[52] C. Haldeman and J. Beggs, Critical branching captures activity in living neural networks and maximizes the number of metastable states, *Phys. Rev. Lett.* **94** (2005) 058101.
[53] W. L. Shew and D. Plenz, The functional benefits of criticality in the cortex, *The Neuroscientist* **19**(1) (2013) 88–100.
[54] J. M. Beggs and D. Plenz, Neuronal avalanches in neocortical circuits, *J. Neurosci.* **23**(35) (2003) 11167–11177.
[55] N. Bertschinger and T. Natschläger, Real-time computation at the edge of chaos in recurrent neural networks, *Neural Comput.* **16**(7) (2004) 1413–1436.
[56] W. L. Shew, H. Yang, S. Yu, R. Roy and D. Plenz, Information capacity and transmission are maximized in balanced cortical networks with neuronal avalanches, *J. Neurosci.* **31**(1) (2011) 55–63.
[57] O. Kinouchi and M. Copelli, Optimal dynamical range of excitable networks at criticality, *Nature Phys.* **2** (2006) 348–351.
[58] W. L. Shew, H. Yang, T. Petermann, R. Roy and D. Plenz, Neuronal avalanches imply maximum dynamic range in cortical networks at criticality, *J. Neurosci.* **29**(49) (2009) 15595–15600.

[59] S. H. Gautam, T. T. Hoang, K. McClanahan, S. K. Grady and W. L. Shew, Maximizing sensory dynamic range by tuning the cortical state to criticality, *Plos Comput. Biol.* **11**(12) (2015) 1–15.

[60] M. Girardi-Schappo, G. S. Bortolotto, J. J. Gonsalves, L. T. Pinto and M. H. R. Tragtenberg, Griffiths phase and long-range correlations in a biologically motivated visual cortex model, *Sci. Rep.* **6** (2016) 29561.

[61] S. Souihel. Generic and specific computational principles for visual anticipation of motion trajectories, Ph.D. Thesis, Université Nice Côte d'Azur, EDSTIC (2019).

[62] D. Hubel and T. Wiesel, Receptive fields, binocular interaction and functional architecture in the cat visual cortex., *J Physiol.* **160** (1962) 106–154.

[63] Y. Yu and T. Sing Lee, Adaptive contrast gain control and information maximization, *Neurocomput.* **65–66** (2005) 111–116. http://www.sciencedirect.com/science/article/pii/S0925231204004503. Computational Neuroscience: Trends in Research 2005.

[64] J. Snellman, T. Kaur, Y. Shen and S. Nawy, Regulation of on bipolar cell activity, *Progress Retinal Eye Res.* **27**(4) (2008) 450–463.

[65] R. M. Shapley and J. D. Victor, How the contrast gain control modifies the frequency responses of cat retinal ganglion cells, *J. Physiol.* **318**(1) (1981) 161–179.

[66] N. Priebe and D. Ferster, A new mechanism for neuronal gain control (or how the gain in brains has mainly been explained), *Neuron* **35**(4) (2002) 602–604.

[67] S. Jarvis, K. Nikolic and S. R. Schultz, Neuronal gain modulability is determined by dendritic morphology: A computational optogenetic study, *Plos Comput. Biol.* **14**(3) (2018) 1–21.

[68] A. Destexhe, Z. Mainen and T. Sejnowski, Synthesis of models for excitable membranes, synaptic transmission and neuromodulation using a common kinetic formalism, *J. Comput. Neurosci.* **1**(3) (1994) 195–230.

[69] S. Coombes, Y. M. Lai, M. Şayli and R. Thul, Networks of piecewise linear neural mass models, *European J. Appl. Math.* **29**(5) (2018) 869–890. doi: 10.1017/S0956792518000050.

[70] B. Cessac and T. Viéville, On dynamics of integrate-and-fire neural networks with adaptive conductances, *Front. Neurosci.* **2**(2) (2008). http://www.frontiersin.org/computational_neuroscience/10.3389/neuro.10/002.2008/abstract.

[71] R. L. D. Valois and K. K. D. Valois, Vernier acuity with stationary moving gabors, *Vision Res.* **31**(9) (1991) 1619–1626. http://www.sciencedirect.com/science/article/pii/004269899190138U.

[72] R. Nijhawan, Motion extrapolation in catching, *Nature* **370** (1994) 256–257.
[73] M. V. C. Baldo and S. A. Klein, Extrapolation or attention shift?, *Nature.* **378** (1995) 565–566.
[74] R. Nijhawan, Visual decomposition of colour through motion extrapolation, *Nature.* **386** (1997) 66–69.
[75] G. Benvenuti, S. Chemla, A. Boonman, L. Perrinet, G. S. Masson and F. Chavane, Anticipatory responses along motion trajectories in awake monkey area v1, preprint (2020); https://www.biorxiv.org/content/early/2020/03/29/2020.03.26.010017.
[76] M. Subramaniyan, A. S. Ecker, S. S. Patel, R. J. Cotton, M. Bethge, X. Pitkow, P. Berens and A. S. Tolias, Faster processing of moving compared with flashed bars in awake macaque v1 provides a neural correlate of the flash lag illusion, *J. Neurophysiol.* (2018).
[77] D. Jancke, W. Erlaghen, G. Schöner and H. Dinse, Shorter latencies for motion trajectories than for flashes in population responses of primary visual cortex, *J. Physiol.* **556** (2004) 971–982.
[78] S. E. Palmer, O. Marre, M. J. Berry and W. Bialek, Predictive information in a sensory population, *Proc. Natl. Acad. Sci. USA* **112**(22) (2015) 6908–6913.
[79] J. Salisbury and S. Palmer, Optimal prediction in the retina and natural motion statistics, *J. Statist. Phys.* **162** (2016).
[80] S. Deny, U. Ferrari, E. Macé, P. Yger, R. Caplette, S. Picaud, G. Tkačik, and O. Marre, Multiplexed computations in retinal ganglion cells of a single type, *Nature Commun.* (2017).
[81] A. J. Sederberg, J. N. MacLean and S. E. Palmer, Learning to make external sensory stimulus predictions using internal correlations in populations of neurons, *Proc. Natl. Acad. Sci. USA* **115**(5) (2018) 1105–1110.
[82] M. Srinivasan, S. Laughlin and A. Dubs, Predictive coding: A fresh view of inhibition in the retina, *Proc. Royal Soc. London. Ser. B, Biol. Sci.* **216**(1205) (1982) 427–459.
[83] T. Hosoya, S. A. Baccus and M. Meister, Dynamic predictive coding by the retina, *Nature* **436** (2005) 71–77.
[84] D. B. Kastner and S. A. Baccus, Spatial segregation of adaptation and predictive sensitization in retinal ganglion cells, *Neuron* **79**(3) (2013) 541–554; http://www.sciencedirect.com/science/article/pii/S0896627313005278.
[85] L. Lapicque, Recherches quantitatifs sur l'excitation des nerfs traitee comme une polarisation, *J. Physiol. Paris* **9** (1907) 620–635.
[86] R. Fernandez and G. Maillard, Chains with complete connections: General theory, uniqueness, loss of memory and mixing properties, *J. Statist. Phys.* **118**(3–4) (2005) 555–588.

[87] F. R. Gantmacher, *The Theory of Matrices* (AMS Chelsea Publishing 1998).
[88] E. Seneta, *Non-negative Matrices and Markov Chains* (Springer, 2006).
[89] O. Onicescu and G. Mihoc, Sur les chaînes statistiques, *C. R. Acad. Sci. Paris.* **200** (1935) 511–512.
[90] M. Keane, Strongly mixing g-measures, *Publ. Math. Informat. Rennes* (1), (1970–1971) 132–154.
[91] H.-O. Georgii, *Gibbs Measures and Phase Transitions* (De Gruyter, 1988).
[92] A. LeNy, Introduction to (generalized) gibbs measures, *Ensaios Mat.* **15** (2008) 1–126.
[93] R. Cofré, C. Maldonado and B. Cessac, Thermodynamic formalism in neuronal dynamics and spike train statistics, *Entropy.* **22**(1330) (2020).
[94] P. Yger, G. L. Spampinato, E. Esposito, B. Lefebvre, S. Deny, C. Gardella, M. Stimberg, F. Jetter, G. Zeck, S. Picaud, J. Duebel and O. Marre, A spike sorting toolbox for up to thousands of electrodes validated with ground truth recordings *in vitro* and *in vivo*, *eLife.* **7** (2018) e34518.
[95] A. P. Buccino, C. L. Hurwitz, J. Magland, S. Garcia, J. H. Siegle, R. Hurwitz, and M. H. Hennig, Spikeinterface, a unified framework for spike sorting, Preprint (2019).
[96] B. Cessac and R. Cofré, Spike train statistics and Gibbs distributions, *J. Physiol. Paris.* **107**(5) (2013) 360–368. Special issue: Neural Coding and Natural Image Statistics.
[97] E. Schneidman, M. Berry, R. Segev and W. Bialek, Weak pairwise correlations imply strongly correlated network states in a neural population, *Nature* **440**(7087) (2006) 1007–1012.
[98] J. Shlens, G. Field, J. Gauthier, M. Grivich, D. Petrusca, A. Sher, A. Litke, and E. Chichilnisky, The structure of multi-neuron firing patterns in primate retina, *J. Neurosci.* **26**(32) (2006) 8254.
[99] G. Tkacik, J. S. Prentice, V. Balasubramanian and E. Schneidman, Optimal population coding by noisy spiking neurons, *Proc. Natl. Acad. Sci.* **107**(32) (2010) 14419–14424.
[100] E. Ganmor, R. Segev and E. Schneidman, The architecture of functional interaction networks in the retina, *J. Neurosci.* **31**(8) (2011) 3044–3054.
[101] E. Ganmor, R. Segev and E. Schneidman, Sparse low-order interaction network underlies a highly correlated and learnable neural population code, *Proc. Natl. Acad. Sci. USA* **108**(23) (2011) 9679–9684.

[102] G. Tkacik, O. Marre, T. Mora, D. Amodei, M. J. B. II and W. Bialek, The simplest maximum entropy model for collective behavior in a neural network, *J. Statist. Mech. Theory Exp.* **2013**(03) (2013) P03011.

[103] E. Granot-Atedgi, G. Tkačik, R. Segev and E. Schneidman, Stimulus-dependent maximum entropy models of neural population codes, *PLos Comput. Biol.* **9**(3) (2013) 1–14.

[104] T.-A. Nghiem, B. Telenczuk, O. Marre, A. Destexhe and U. Ferrari, Maximum-entropy models reveal the excitatory and inhibitory correlation structures in cortical neuronal activity, *Phys. Rev. E* **98** (2018) 012402.

[105] C. Gardella, O. Marre and T. Mora, Modeling the correlated activity of neural populations: A review, *Neural Comput.* **31**(2) (2019) 233–269.

[106] J. W. Pillow, J. Shlens, L. Paninski, A. Sher, A. M. Litke, E. Chichilnisky, and E. P. Simoncelli, Spatio-temporal correlations and visual signalling in a complete neuronal population, *Nature* **454**(7207) (2008) 995–999.

[107] P. Dayan and L. Abbott, *Theoretical Neuroscience: Computational and Mathematical Modeling of Neural Systems* (MIT Press, 2001).

[108] W. Gerstner and W. M. Kistler, Mathematical formulations of hebbian learning., *Biol. Cybernet.* **87** (2002) 404–415.

[109] M. Rudolph and A. Destexhe, Analytical integrate and fire neuron models with conductance-based dynamics for event driven simulation strategies, *Neural Comput.* **18** (2006) 2146–2210; http://www.mitpressjournals.org/doi/abs/10.1162/neco.2006.18.9.2146.

[110] B. Cessac, I. Ampuero and R. Cofre. Linear response for spiking neuronal networks with unbounded memory (2020), under revision.

[111] R. Kubo, The fluctuation-dissipation theorem, *Rep. Progress in Phys.* **29**(1) (1966) 255–284.

[112] D. Ruelle, Smooth dynamics and new theoretical ideas in nonequilibrium statistical mechanics., *J. Statist. Phys.* **95** 393–468 (1999).

[113] B. Cessac and J. Sepulchre, Stable resonances and signal propagation in a chaotic network of coupled units, *Phys. Rev. E* **70** (2004) 056111.

[114] B. Cessac and J. Sepulchre, Transmitting a signal by amplitude modulation in a chaotic network, *Chaos* **16** (2006) 013104.

[115] S. Cocco, S. Leibler and R. Monasson, Neuronal couplings between retinal ganglion cells inferred by efficient inverse statistical physics methods, *Proc. Natl. Acad. Sci. USA* **106**(33) (2009) 14058–14062.

[116] O. Faugeras, J. Touboul and B. Cessac, A constructive mean field analysis of multi population neural networks with random synaptic weights and stochastic inputs, *Front. Comput. Neurosci.* **3**(1) (2009); http://www.frontiersin.org/computational_neuroscience/10.3389/neuro.10/001.2009/abstract.

[117] J. Schuecker, S. Goedeke, D. Dahmen and M. Helias, Functional methods for disordered neural networks, Preprint (2016).

[118] M. Helias and D. Dahmen, *Statistical Field Theory for Neural Networks*, Lecture Notes in Physics, vol. 970, (Springer, 2020).

[119] C. Lok, Curing blindness: Vision quest, *Nature* **513** (2014) 160.

[120] E. Camacho *et al.*, Quantifying the metabolic contribution to photoreceptor death in retinitis pigmentosa via a mathematical model, *Theor. Biol.* **408** (2016) 75–87.

[121] P. A. Roberts, E. A. Gaffney, P. J. Luthert, A. J. Foss and H. M. Byrne, Mathematical models of retinitis pigmentosa: The oxygen toxicity hypothesis, *J. Theoret. Biol.* **425** 53–71 (2017); http://www.sciencedirect.com/science/article/pii/S002251931730214X.

[122] P. A. Roberts, E. A. Gaffney, J. P. Whiteley, P. J. Luthert, A. J. E. Foss and H. M. Byrne, Predictive Mathematical Models for the Spread and Treatment of Hyperoxia-induced Photoreceptor Degeneration in Retinitis Pigmentosa, *Invest. Ophthalmol. Visual Sci.* **59**(3) (2018) 1238–1249.

[123] R. E. Marc, B. W. Jones, C. B. Watt and E. Strettoi, Neural remodeling in retinal degeneration, *Progress Retinal Eye Res.* **22**(5) (2003) 607–655.

[124] J. Barrett, P. Degenaar and E. Sernagor, Blockade of pathological retinal ganglion cell hyperactivity improves optogenetically evoked light responses in rd1 mice, *Front. Cellular Neurosci.* **9** (2015) 330.

[125] J. Barrett, G. Hilgen and E. Sernagor, Dampening spontaneous activity improves the light sensitivity and spatial acuity of optogenetic retinal prosthetic responses, *Sci Rep.* **6** (2016) 33565.

https://doi.org/10.1142/9789811221903_0015

Chapter 15

Collective Sustained Oscillations in Complex Systems

Zhigang Zheng
Institute of Systems Science and College of Information Science and Engineering, Huaqiao University, Xiamen 361021, China
zgzheng@hqu.edu.cn

Collective dynamics of networks of coupled elements can be considered as the emergence of the order from microscopic self-organization at the macroscopic scale. In this chapter, we explore the mechanism of collective self-sustained oscillations in networks of coupled nonoscillatory units. The significance of network topology in leading to the collective oscillation is revealed by studying the excitable networks and gene regulated networks. In terms of the dominant phase-advanced driving method and the function-weight approach, the fundamental building blocks such as Winfree loops and motifs, which are responsible for generating sustained oscillations, are revealed, and the oscillation core and the propagating paths are identified. This topology reduction of complex systems gives a clear picture of different dynamical roles of topologies and nodes in governing the collective behaviors.

15.1. Biological Rhythms: An Introduction

Biological rhythm is an old question and can be found ubiquitously in various living systems [1]. Some fundamental issues such as the emergence and mechanism of various rhythms and their collective responses received potential concern during the past decades. It is well-known that circadian rhythms are driven by an internal biological clock that anticipates day–night cycles to optimize the physiology and the behavior of organisms.

There have been long-time and numerous records on the observations that organisms adapt their physiology and behavior to the time of the day in a circadian fashion. A typical example is the early observation of leaf and flower movements in plants. In 1729, the French scientist de Mairan observed that the leaves of a mimosa plant in the dark could still open and close rhythmically at the appropriate time of the day, implying an endogenous origin of the daily rhythm rather than external stimulus [2] (see Fig. 15.1).

The exploration of the emergence of circadian rhythms from the microscopic level (e.g., molecular or genetic levels) aroused a new tide of studies on biological oscillations [3]. The genetic mechanism responsible for the emergence of circadian rhythms was first explored by S. Benzer and R. Konopka from the 1960s. In 1971, their pioneering work identified mutants of the fruit fly Drosophila that

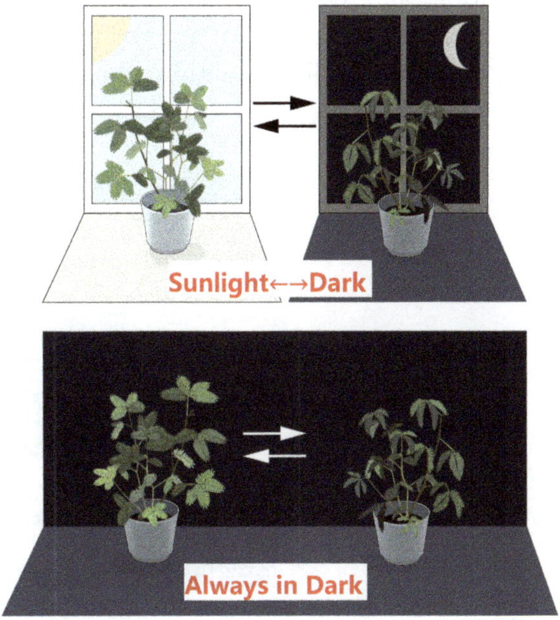

Fig. 15.1. The observation of de Mairan on the rhythmic daily opening–close in the sun and in the dark. (a) in the normal sunlight–dark environment and (b) always in dark. The environment-independent open–closing implies an endogenous origin of the daily rhythm. (Adapted from [11].)

displayed alterations in the normal 24-hour cycle of pupal eclosion and locomotor activity, which was named as period (PER) [4]. Later on, Hall and Rosbash at Brandeis University [5] and Young at Rockefeller University [6, 7] isolated and molecularly characterized the period gene.

People further realized that the emergence of the circadian clock does not depend not on a single gene but on the so-called transcription–translation feedback loop. It was revealed that the transcription of period and its partner gene timeless (TIM) are repressed by the PER and TIM proteins, which results in a self-sustained oscillation, as shown in Fig. 15.2 [8–10]. These explorations led to numerous potential studies that unraveled a series of coupled feedback loops together with a complex network of reactions. These biologically involve regulated protein phosphorylation and degradation of feedback-loop components, protein complex assembly, nuclear translocation and other post-translational modifications, generating oscillations with a period of approximately 24 hours. As a matter of fact, circadian oscillators within individual cells respond differently to entraining signals and control various physiological

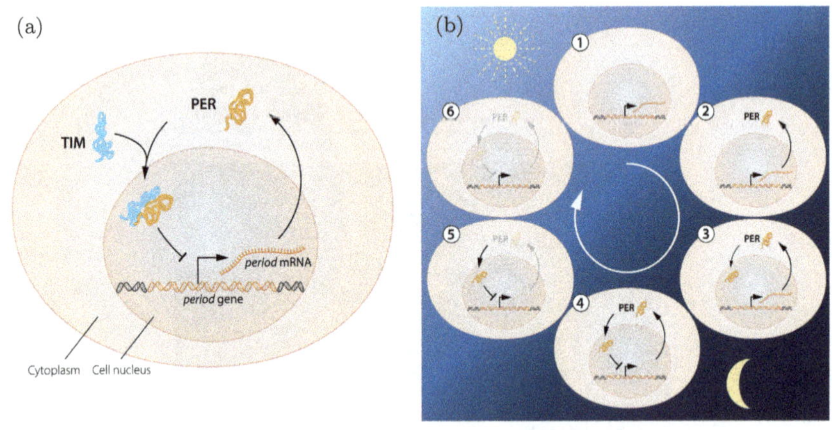

Fig. 15.2. (a) A simplified illustration of the biochemical process of the circadian clock. (b) The transcription–translation feedback loop (TTFL), i.e., the transcription of period and its partner gene timeless (TIM) are repressed by the PER and TIM proteins, generating a self-sustained oscillation. (Adapted from [11].)

outputs, such as sleep patterns, body temperature, hormone release, blood pressure, and metabolism. The seminal discoveries by Hall, Rosbash and Young have revealed a crucial physiological mechanism explaining circadian adaptation, with important implications for human health and disease [11]. The 2017 Nobel Prize in Physiology or Medicine was awarded to J. C. Hall, M. Rosbash and M.W. Young for their explorations of the genetic mechanism that controls circadian rhythms [11, 12].

Essentially, biological rhythms ubiquitously existing in various biological systems are physically oscillatory behaviors, which are indications of temporally periodic dynamics of nonlinear systems. In fact, rhythmic phenomena can be extensively observed in various situations, ranging from physics, and chemistry to biology [13, 14]. Therefore, these oscillatory behaviors with seemingly very different backgrounds can be universally studied in the framework of nonlinear dynamics.

In this chapter, we focus on the collective self-sustained oscillations in networks of coupled nonoscillatory units from the viewpoint of nonlinear dynamics. Though there are a large number of nodes and links in a network, only a small portion of them play a dominant role in the emergence of collective oscillations. We adopt excitable and gene-regulation dynamics as node dynamics, both of which are nonoscillatory. To reveal the dynamical role of each node in a network, we proposed the dominant phase-advanced driving method and the function-weight approach. We found that Winfree loops and motifs are responsible for generating sustained oscillations, which are subnetworks that are composed of only a few nodes and connections.

15.1.1. *Self-sustained oscillation in simple nonlinear systems*

Self-sustained oscillation is also called the *limit cycle*, and it is defined as a typical time-periodic nonlinear behavior. As a nontrivial nonlinear, nonstationary behavior, sustained oscillation has received much attention throughout the past century [13]. The essential mechanism of sustained oscillation in nonlinear systems is the existence of feedbacks, but the source of feedback depends strongly on different situations.

The early study of low-dimensional dynamical systems indicated that *the competition-balance mechanism between the positive feedback and the negative feedback can maintain a stable oscillation.* Taking the famous Van der Pol oscillator as an example [14], the governing dynamical equation of $x(t)$ can be written as

$$\ddot{x} + \varepsilon(x^2 - 1)\dot{x} + \omega_0^2 x = 0, \tag{15.1}$$

where ε and ω_0 are the damping coefficient and the natural frequency of the oscillator, respectively. The adjustable damping term $\varepsilon(x^2 - 1)$ provides the feedback mechanism for the energy compensation to sustain a stable limit cycle. It can be seen that the effective damping term becomes positive to dissipate energy when the amplitude $|x|$ is large ($|x| > 1$) [Fig. 15.3(a)] and becomes negative to input energy when the amplitude $|x|$ is small ($|x| < 1$) [Fig. 15.3(b)]. These two feedbacks eventually lead to the limit-cycle motion of the Van der Pol system.

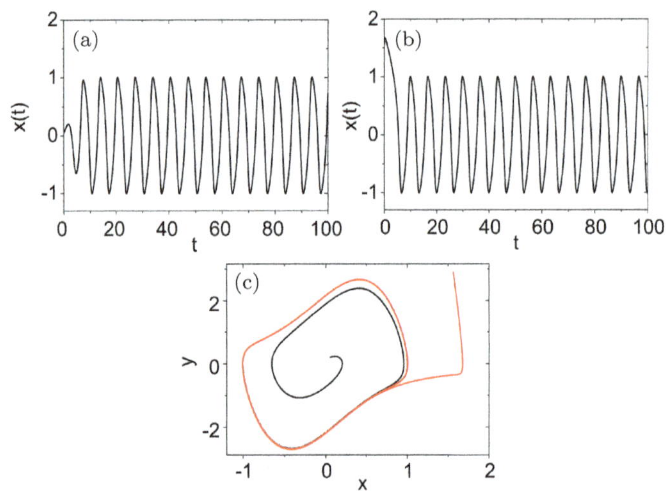

Fig. 15.3. Sustained oscillation of the Van der Pol system: (a) the oscillation starting from a small amplitude; (b) the oscillation starting from a large amplitude and (c) the limit-cycle motion presented in the space (x, y), where $y = \dot{x}$. Here $\varepsilon = 1$ and $\omega_0 = 1$.

In order to give an explicit analysis from the viewpoint of nonlinear dynamics, let us study a simple and analytically solvable nonlinear system with a limit cycle oscillation by adopting the following complex dynamical equation:

$$\dot{z} = (\lambda + \mathrm{i}\omega)z - bz|z|^2, \tag{15.2}$$

where the state variable $z(t) = x(t) + iy(t)$ is a complex order parameter. Physically the order parameter $z(t)$ can be obtained via the reduction procedure by eliminating the fast modes. Equation (15.2) presents two-dimensional dynamics in real space, and the only possible time-dependent solution of this equation is the limit cycle.

To obtain an analytical solution, it is convenient to introduce the polar coordinates

$$z(t) = r(t)e^{i\theta(t)},$$

and Eq. (15.2) can be decomposed into the following two-dimensional equations of the phase and the amplitude that are uncoupled to each other:

$$\dot{\theta}(t) = \omega, \tag{15.3a}$$

$$\dot{r}(t) = \lambda r - br^3. \tag{15.3b}$$

The phase Eq. (15.3a) indicates that the phase evolves uniformly with the phase velocity ω. The amplitude Eq. (15.3b) has two stationary solutions for $\lambda > 0$: the unstable solution $r = 0$ and the stable solution $r_0 = \sqrt{\lambda/b}$. By considering the dynamics of both the phase and the amplitude, the latter one represents the periodic solution of Eq. (15.2) as

$$z(t \to \infty) = r_0 e^{i(\omega t + \theta_0)}. \tag{15.4}$$

This is a stable and attractive limit cycle, and the relaxation process from an arbitrary initial state can lead to this solution. The stability of this sustained oscillator is the result of the competition between the positive feedback term λr and the negative feedback term $-br^3$.

15.1.2. *Self-sustained oscillations in complex systems*

For a system composed of many interacting elements, it is difficult to perform the above nonlinear dynamics analysis due to high-dimensional complexity. On the other hand, self-sustained oscillation in large-scale complex systems is a typical emergent behavior resulting from competitions and self-organizations of large number of degrees of freedom [15–17]. It is important to study how a number of nonoscillatory nodes organize themselves to emerge a collective oscillatory phenomenon [18–21].

The topic of self-sustained oscillations in complex systems was largely motivated by an extensive study and a strong background of system biology. People have extensively studied the possibility of collective sustained oscillations when these elements are coupled into a network and adjust themselves to emerge nontrivial oscillations. In fact, such common behaviors have been manifested themselves as oscillations in generegulatory networks [22–27], neural networks and brains [28–33].

From the viewpoint of feedback mechanism, collective oscillation should naturally come from the collaboration of units in a complex system. For simplicity, we assume that a single unit in a complex system is nonoscillatory, which means that its dynamical attractor is fixed point in the absence of external drivings, e.g., a gene or an excitable neuron. Therefore, the feedback mechanism of sustained oscillation of coupled nonoscillatory units should be related to their mutual interactions.

Recent progresses revealed that although the collective self-sustained oscillation emerges from the organization of large numbers of units, only some fundamental topologies or subnetworks with a small number or some of key units may dominate the emergence of sustained oscillations as typical building blocks [34–36]. We may call these organizing centers the *self-organization core* or the *oscillation source*. The center nodes and small skeletons may give rise to sustained oscillations and sustain target-wave-like patterns in large-scale networks [36–38], and this mechanism is related to the temporal information processing in neural systems [39]. The existence of multiple oscillation sources may lead further to interesting chaotic dynamics [40–42].

The system we are studying throughout this chapter is a network consisting of N nodes with nonoscillatory local dynamics described by well-defined coupled ordinary differential equations, and there are M ($M > N$) links among these different nodes. We are interested in the emergence of a global self-sustained oscillation on this network, where all nodes that are individually nonoscillatory become oscillatory. It is our motivation to find the mechanism supporting the oscillations in terms of the analysis of network topology and oscillation time series of each node and reveal the key topology of the organizing center, oscillation cores and further the propagation path. We will propose two useful methods, i.e., the dominant phase-advanced driving method and the functional weight approach, and then apply them to analyze and further pick up the key topologies from dynamics.

15.2. The Dominant Phase-Advanced Driving Method

The dominant phase-advanced driving (DPAD) method is a dynamical method that can find the strongest cross driving of the target node when a system is in an oscillatory state [34–37]. The basic idea of DPAD is a comparison and ordering of the significance of nodes in a network with sustained oscillations based on their phase dynamics.

The oscillatory behavior of an individually nonoscillatory node is apparently driven by signals from one or more interactions with advanced phases, if such a phase variable can be properly defined. We call such a signal the *phase-advanced driving*. Among all phase-advanced interactions, the interaction giving the most significant contribution to the given node can be defined as the *dominant phase-advanced driving* (DPAD). Based on this idea, the corresponding DPAD for each node can be identified. By applying this network reduction approach, the original oscillatory high-dimensional complex network of N nodes with M vertices/interactions can be reduced to a one-dimensional *unidirectional* network of size N with M' unidirectional dominant phase-advanced interactions.

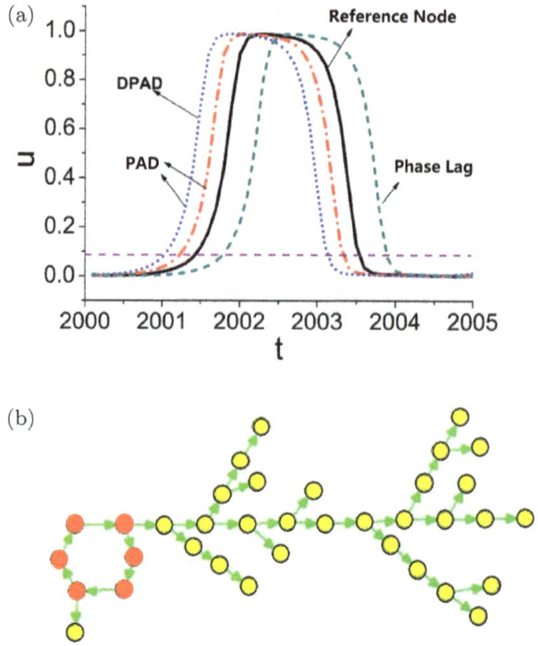

Fig. 15.4. (a) A schematic plot of the DPAD. As comparisons, the reference oscillatory time series, a usual phase-advanced driving and a phase-lagged node dynamics are also presented, respectively. (b) An example of simplified (unidirectional) network in terms of the DPAD scheme. (Adapted from [37].)

Taking the time series shown in Fig. 15.4(a) as an example. One has the black/solid curve denoting the oscillatory time series of a given node as the reference. Many nodes linking directly to this reference can be checked when a given network is proposed. Suppose there are three nodes whose dynamical time series labeled by green/dashed, blue/dotted and red/dash-dotted curves, respectively. The green/dashed curve exhibits a lagged oscillation to the reference curve, therefore one calls it the phase-lagged oscillation. The blue/dotted and red/dash-dotted curves provide the drivings for exciting the reference node, so these curves are identified as phase-advanced driving. The blue/dotted signal is the DPAD because it presents the earliest oscillation and makes the most significant contribution.

Figure 15.4(b) gives an illustration of a DPAD structure consisting of one loop and the nodes outside the loop radiated from the loop. For excitable node dynamics, as shown below, the red nodes form a unidirectional loop that acts as the oscillatory source, and yellow nodes beyond the loop form paths for the propagation of oscillations.

The DPAD structure reveals the dynamical relationship between different nodes. Based on this functional structure, we can identify the loops as the oscillation source, and illustrate the wave propagation along various branches. All the above ideas are generally applicable to diverse fields for self-sustained oscillations of complex networks consisting of individual nonoscillatory nodes.

15.3. The Functional-Weight Approach

The above DPAD approach provides a way in analyzing the phase relations embedded in dynamical data to unveil the information on unit connections. If one knows the detailed dynamics of coupled systems, it is convenient to study the topological information of dynamical organization. We start from the following dynamical system composed of N units labeled as $\boldsymbol{x} = (x_1, x_2, \ldots, x_N)$:

$$\frac{dx_i}{dt} = \lambda_i x_i + f_i(\boldsymbol{x}), \tag{15.5}$$

where x_i denotes the state variable of the ith node in a network. (For simplicity, one adopts the one-dimensional dynamics on the nodes.) For convenience, we separate the linear modes from the nonlinear coupling functions $f_i(\boldsymbol{x})$ at the right-hand side of Eq. (15.5). Because we explore the topological mechanism of collective oscillation in networks of coupled *nonoscillatory* units, the linear coefficients λ_i for nonstationary modes should be negative, and we set $\lambda_i = -1$ without losing generality.

The basic idea of the functional weight approach is to estimate the coupling significance of a node with other nodes. For all the nodes in a network that cannot oscillate independently, the oscillation of each node i is induced by the interactions from its neighbors represented by the nonlinear function $f_i(\boldsymbol{x})$ in Eq. (15.5). However, all the inputs from the neighbors to the target node are mixed together in $f_i(\boldsymbol{x})$

by nonlinear functions. To measure the degree of contributions of different neighbors to the oscillation of the target node, contributions from all these neighbors should be separated.

The cross-differential "force" between the ith and the jth nodes can be measured by

$$\delta \dot{f}_{ij}(\boldsymbol{x}) = \left(\frac{\partial f_i(\boldsymbol{x})}{\partial x_j}\right) \dot{x}_j, \tag{15.6}$$

where the total driving force is expressed as a simple summation of the contributions of all its neighbors

$$\delta \dot{f}_i(\boldsymbol{x}) = \sum_{j=1, j\neq i}^{N} \delta \dot{f}_{ij}(\boldsymbol{x}). \tag{15.7}$$

The above relation indicates that the differential form $\partial f_i(\boldsymbol{x})/\partial x_j$ rather than the nonlinear function $f_i(\boldsymbol{x})$ itself plays a crucial role in the emergence of collective oscillations, because the amount of variation of the target node i caused by the variation of a given neighbor j determines the functional driving relationship from j to i. At time t, the weight of the normalized contribution of the jth node can be easily computed from Eq. (15.7) as

$$w_{ij}(t) = \frac{|\delta \dot{f}_{ij}|}{\sum_{j=1, j\neq i}^{N} |\delta \dot{f}_{ij}(x)|} = \frac{|(\partial f_i(\boldsymbol{x})/\partial x_j)\dot{x}_j|}{\sum_{j=1, j\neq i}^{N} |(\partial f_i(\boldsymbol{x})/\partial x_j)\dot{x}_j|}, \tag{15.8}$$

which is nothing but the normalized Jacobian matrix at time t weighted by $\dot{x}_i$. The overall weight is integrated over T as

$$\overline{w}_{ij} = \lim_{T\to\infty} \frac{1}{T} \int_{t_0}^{t_0+T} w_{ij}(t)dt. \tag{15.9}$$

For periodic oscillatory dynamics,

$$\overline{w}_{ij} = \frac{1}{T} \int_{t_0}^{t_0+T} w_{ij}(t)dt, \tag{15.10}$$

where T is the period for periodic oscillations. The quantity w_{ij} introduced here represents the weight of neighbor node j in driving the target node i to oscillate and serves as the quantitative measure of the importance of the link from node j to node i. Here $0 \leq w_{ij} \leq 1$ and $\sum_{j=1, j\neq i}^{N} w_{ij} = 1$. A zero or small w_{ij} represents no or a weak functional interaction while a large or unity w_{ij} denotes a strong or dominant driving [40–42].

In the following, we will focus on self-sustained oscillations in excitable networks and gene-regulatory networks. One can find that both types of systems possess a common feature, that is, only a small number of units participate in the global oscillation, and some fundamental structures act as dominant roles in giving rising to oscillatory behaviors in the system, although the organizing cores may differ for these two types of networks.

15.4. Self-Sustained Oscillation in Excitable Networks

15.4.1. *The Bär–Eiswirth network*

Excitable dynamics is a typical feature in living systems such as neuron systems, biochemical reactions, cardiac dynamics, and so on [45–47]. Some well-known models include the integrate-and-fire model, the Hodgkin–Huxley model, the FitzHugh–Nagumo model, and the Bär model.

In this chapter, we use the following Bär–Eiswirth model [61] to describe the excitable dynamics of the nodes:

$$\dot{u} = -\frac{1}{\varepsilon}u(u-1)\left(u - \frac{v+b}{a}\right), \tag{15.11a}$$

$$\dot{v} = f(u) - v. \tag{15.11b}$$

Variables $u(t)$ and $v(t)$ describe the activator and the inhibitor dynamics of the node, respectively. The function $f(u)$ takes the following piecewise form:

$$f(u) = \begin{cases} 0 & \text{if } u < \frac{1}{3}, \\ 1 - 6.75u(u-1)^2 & \text{if } \frac{1}{3} \leq u < 1, \\ 1 & \text{if } u > 1. \end{cases} \tag{15.12}$$

The relaxation parameter $\varepsilon \ll 1$ represents the time ratio between the activator u and the inhibitor v. The dimensionless parameters

a and b denote the activator kinetics of the local dynamics and the ratio $u_T = b/a$ can effectively control the excitation threshold.

Moreover, network topology in many cases also plays an important role in governing collective behavior of the system. In the past two decades, the study of complex networks proposed a number of intriguing topics on random networks such as the small-world, scale-free, the clustering properties [48–56]. Here we consider the Erdős–Rényi (ER) random network for convenience. The network dynamics of N coupled Bär–Eiswirth excitable nodes is described by

$$\dot{u}_i = -\frac{1}{\varepsilon} u_i (u_i - 1) \left(u_i - d\frac{v_i + b}{a} \right) + D \sum_{j=1}^{N} A_{i,j} (u_j - u_i), \tag{15.13a}$$

$$\dot{v}_i = f(u_i) - v_i. \tag{15.13b}$$

Here D is the coupling strength between linking nodes. $\boldsymbol{A} = \{A_{i,j}\}$ is the adjacency matrix, which includes all the topological information of the network. For a symmetric and bidirectional network, the matrix is defined as $A_{i,j} = A_{j,i} = 1$ if there is a connection linking nodes i and j, and $A_{i,j} = A_{j,i} = 0$ otherwise. We study the ER random network shown in Fig. 15.5(a) as a typical example.

In the absence of couplings among nodes, each excitable node is nonoscillatory, i.e., they evolve asymptotically to the rest state $u = v = 0$ and will stay there perpetually unless some external force drives them away from this state. When a node is kicked from its rest state by a stimulus large enough, the unit can excite by its own internal excitable dynamics.

15.4.2. *Winfree loop and wave propagations*

For excitable networks, cooperation among units in the system leads to an ordered dynamical topology to maintain the oscillatory process [43, 44]. The loop structure is ubiquitous in real networks and plays an important role in network dynamics. Winfree constructed a one-dimensional (1D) excitable chain and gave an excitation to the first node. This excitable loop that supports a sustained oscillation

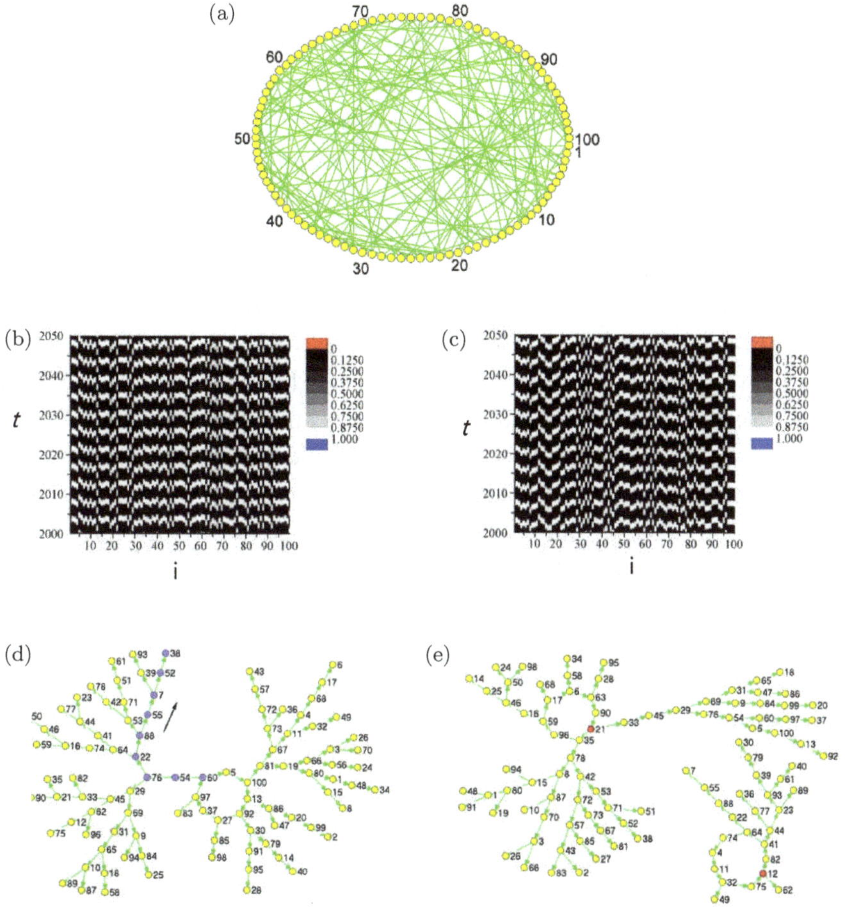

Fig. 15.5. (a) An example random network with $N = 100$ nodes, and each node connects to other nodes with the same degree $k = 3$. (b)–(c) The spatiotemporal evolution patterns of two different oscillatory states in the same network shown in (a) by starting from different initial conditions. Both patterns display the evolution of local variable u. The nodes are spatially arranged according to their indexes i. (d) The DPAD structure corresponding to the oscillation state in (b), where a loop and multiple chains are identified. (e) The DPAD structure corresponding to the oscillation state in (c), where two independent subgraphs are found with each subgraph containing a loop and numerous chains. (Adapted from [37].)

is called as the *Winfree loop*. A typical example can be found in Fig. 15.4(b), where the red nodes form a unidirectional Winfree loop that acts as an oscillatory source.

Recurrent excitation has been proposed to be the mechanism supporting self-sustained oscillations in neural networks. In fact, the loop topology provides a intrinsic feedback mechanism of recurrent excitation for realizing a global collective oscillation. First of all, an excitable node in a sustained oscillatory state must be driven by other nodes. Such drivings can be realized by a looped linking among some interacting nodes. A local excitation leads to a pulse and are propagated along the loop to drive the neighbor nodes in the loop, which forms a feedback mechanism of repeated driving.

Furthermore, the oscillation along the loop can be propagated by nodes outside the loop and spread throughout the system. The propagation of oscillation in the media gives rise to the wave patterns.

The DPAD method proposed in Sec. 2 can reveal the underlying Winfree loop in networks of excitable nodes and the oscillation source. In complex networks, numerous local regular connections coexist with some long-range links. The former plays an important role in target wave propagation and the latter are crucial for maintaining the self-sustained oscillations.

By using the above Bär–Eiswirth network model, one may study the dynamics by starting from different initial conditions. For some initial conditions, the system may exhibit global self-sustained oscillations. The spatio-temporal patterns given in Figs. 15.5(b) and 15.5(c) are two typical different examples of these oscillatory states.

By using the DPAD approach, in Figs. 15.5(d) and 15.5(e), the reduced directed networks corresponding to the oscillatory dynamics by using the DPAD method are obtained. For the dynamics shown in Fig. 15.5(b), the single Winfree loop $\{60 \to 5 \to 100 \to 13 \to 92 \to 27 \to 37 \to 97 \to 60\}$ plays the role of oscillation source, with cells in the loop exciting sequentially to maintain the self-sustained oscillation, as shown in Fig. 15.5(d). One may observe waves propagating downstream along several tree branches rooted at various cells in the Winfree loop. If we plot the spatio-temporal dynamics along these various paths by rearranging the node indices according to the sequence in the loop, regular and perfect wave propagation patterns can be found. This indicates that the DPAD structure well illustrates the wave propagation paths.

For the dynamics shown in Fig. 15.5(c), the corresponding DPAD structure given in Fig. 15.5(e) is a superposition of two sub-DPAD paths $\{6 \to 63 \to 90 \to 21 \to 35 \to 96 \to 59 \to 17 \to 6\}$ and $\{64 \to 74 \to 4 \to 11 \to 32 \to 75 \to 12 \to 82 \to 41 \to 64\}$, which means that there are two Winfree loops giving rise to two sustained oscillations and propagating along the trees.

The DPAD structures in Figs. 15.5(d) and 15.5(e) clearly show the distinctive significance of some units in the oscillation which cannot be observed in Figs. 15.5(b) and 15.5(c), where units evolve in the homogeneous and randomly coupled network, and no unit takes any priority over others in topology. Because the Winfree loop works as the oscillation source, units in the loop should be more important in the contributions of the oscillation.

15.4.3. *Minimum Winfree loop and self-sustained oscillations*

Studies on the emergence of self-sustained oscillations in excitable networks indicate that regular self-sustained oscillations can emerge. However, the intrinsic mechanism in determining the oscillations in networks is still unclear. For example, for random networks, whether the connection probability is related to sustained oscillations is an open topic.

In this section, we study the occurrence of sustained oscillation depending on the linking probability on excitable random networks, and find that the minimum Winfree loop is the intrinsic mechanism in determining the emergence of collective oscillations. Furthermore, the emergence of sustained oscillation is optimized at an optimal connection probability, which forms a one-to-one relationship with the minimum-Winfree-loop length. This relation is well understood that the connection probability interval and the optimal connection probability for supporting the oscillations in random networks are exposed to be determined by the minimum Winfree loop. These three important quantities can be approximately predicted by the network structure analysis, and have been verified in numerical simulations [57].

One adopts the Bär–Eiswirth model (13) on Erdös–Rényi random networks with N nodes. Each pair of nodes are connected with a given probability P, and the total number of connections is $PN(N-1)/2$. By manipulating P, one can produce a number of random networks with different detailed topologies for a given P.

We introduce the oscillation proportion

$$P_{\text{os}} = \frac{N_{\text{os}}}{N_{\text{ALL}}} \tag{15.14}$$

as the order parameter to quantitatively investigate the influence of system parameters on self-sustained oscillations in random networks, where N_{ALL} is the total number of tests starting from random initial conditions for each set of parameters, and N_{os} is the number of self-sustained oscillations counted in N_{ALL} dynamical processes.

In Figs. 15.6(a)–15.6(d), the dependence of the oscillation proportion P_{os} on the connection probability P for different parameters a, b, ε and D on random networks with $N = 100$ nodes are presented. It is shown from all these curves that the system can exhibit self-sustained oscillation in a certain regime of the connection probability, and no oscillations are presented at very small or very large P. Moreover, an optimal connection probability $P = P_{\text{OCP}}$ for supporting self-sustained oscillations can be expected on random networks. The number of self-sustained oscillations increases as the parameter a is increased (see Fig. 15.6(a)), while P_{os} decreases as b is increased as shown in Fig. 15.6(b). Moreover, P_{OCP} for supporting self-sustained oscillations is independent of the parameters a and b. Figure 15.6(c) reveals the dependence of P_{os} on the relaxation parameter ε. It is shown from Fig. 15.6(c) that as ε is increased, P_{os} decreases remarkably. Increasing the coupling strength D is shown to enhance the sustained oscillation (see Fig. 15.6(d)).

The nontrivial dependences of collective oscillations on various parameters such as the connection probability P are very interesting. As discussed above, the excitable wave propagating along an excitable loop can form a 1D Winfree loop, which serves as the

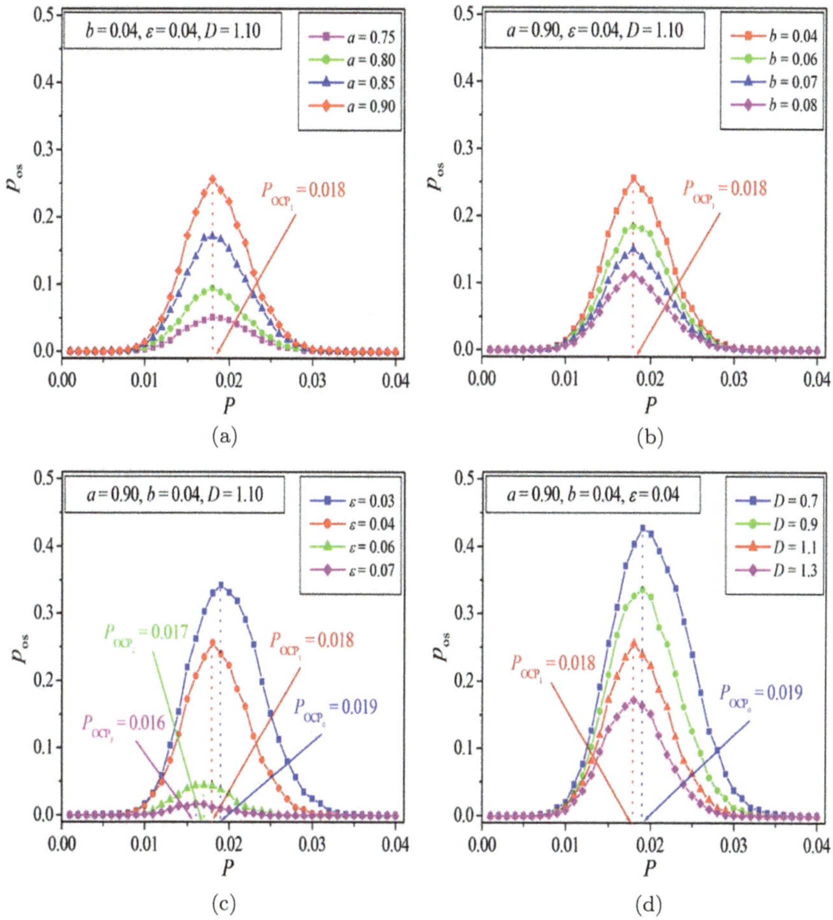

Fig. 15.6. The dependence of the oscillation proportion P_{os} on the connection probability P at different system parameters in excitable random networks. The optimal connection probabilities P_{OCP} for supporting self-sustained oscillations in random networks at different parameters are indicated. (Adapted from [57].)

oscillation source and maintain self-sustained oscillation in excitable complex networks. Figure 15.7(a) presents the dependence of the sustained-oscillation period T of the 1D Winfree loop on the loop length, where a shorter/longer period is expected for a shorter/longer loop length. However, due to the existence of the refractory period

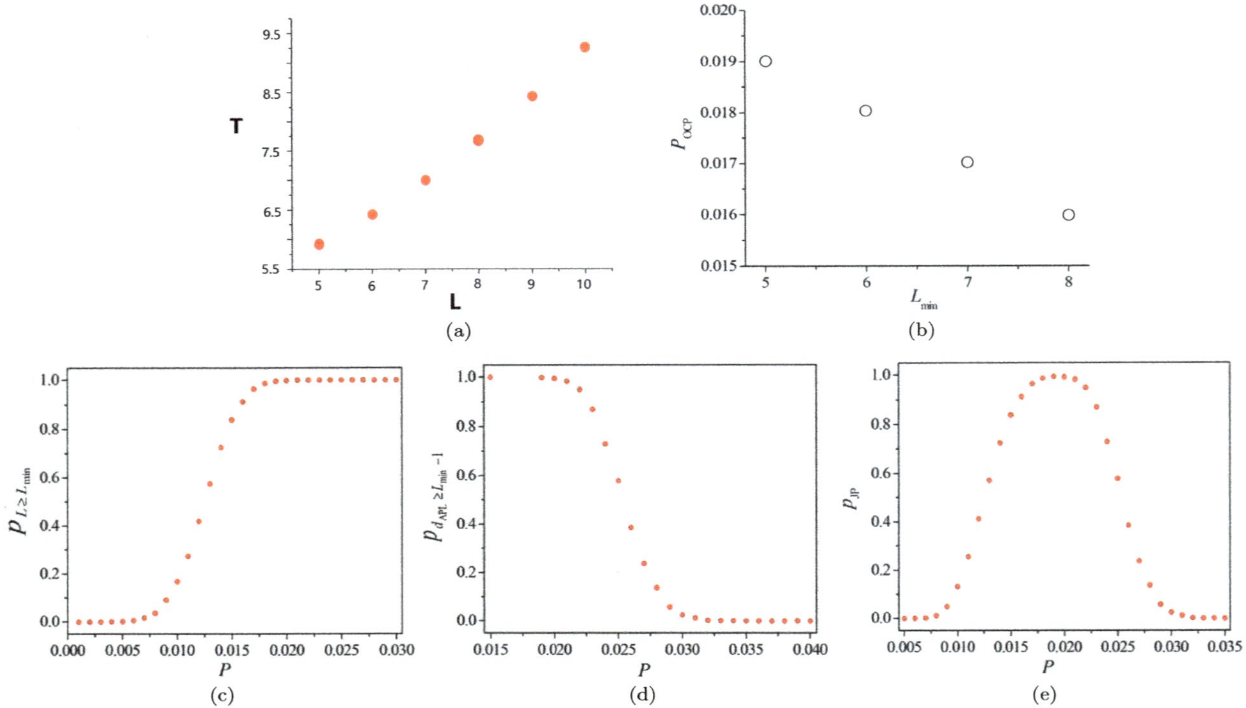

Fig. 15.7. (a) The dependence of the oscillation period of a Winfree loop against the loop length. When $L < L_{\min}$, the oscillation ceases. (b) The dependence of the optimal connection probability on $L_{\min}$. (c) The dependence of the proportion of network structures satisfying $L \geq L_{\min}$ on the connection probability P. (d) The dependence of the proportion of network structures with d_{APL} satisfying $d_{\text{APL}} \geq L_{\min} - 1$ on P. (e) The dependence of the joint probability on P. The minimum Winfree loop with the length $L_{\min} = 6$ is used as the example. (Adapted from [57].)

of excitable dynamics, a too short 1D Winfree loop cannot support sustained oscillations, implying a minimum Winfree loop length $L_{\min}$ for a given set of parameters. Moreover, sustained oscillation ceases for a shortest loop length, corresponding to $L_{\min}$.

In Fig. 15.7(b), P_{OCP} is found to build a one-to-one correspondence to $L_{\min}$, indicating that the emergence of collective oscillations is essentially determined by the minimum Winfree loop. This correspondence can be understood by analyzing the following two tendencies. First, as discussed above, a network must contain a topological loop with a length that is not shorter than the minimum Winfree loop, i.e.,

$$L \geq L_{\min}.$$

Second, the average path length d_{APL} of a given network should be large enough so that

$$d_{\mathrm{APL}} \geq L_{\min} - 1.$$

These two tendencies propose the necessary conditions for the formation of 1D Winfree loop supporting self-sustained oscillations. Moreover, the first condition leads to the lower critical connection probability, by violating which the network cannot support sustained oscillations. The second condition gives rise to the upper critical connection probability, and if d_{APL} is so short that the loops are too small to support oscillations. In Fig. 15.7(c), the probability of the loop length larger than $L_{\min}$ is computed against the connection probability P of the random network. An increasing dependence can be clearly seen, and a lower threshold exists. As shown in Fig. 15.7(d), the probability of the random network with $d_{\mathrm{APL}} \geq L_{\min} - 1$ is plotted against P. A decreasing relation and an upper threshold can be found. Necessary condition for sustained oscillation should be a joint probability satisfying both $L > L_{\min}$ and $d_{\mathrm{APL}} \geq L_{\min} - 1$. The dependence of this joint probability on the connection probability P is the product of Figs. 15.7(c) and 15.7(d), which naturally leads to a humped tendency shown in Fig. 15.7(e), where an optimal connection probability expected for the largest joint probability. This gives a perfect correspondence to the results proposed in Fig. 15.6.

15.4.4. *Recent progresses*

The above discussion indicates that self-sustained oscillation is related to the loop topology and dynamics, and are essentially determined by $L_{\min}$. The one-to-one correspondence between the optimal connection probability and $L_{\min}$ is revealed. The minimum Winfree loop is the key factor in determining collective oscillations on networks [58–60].

Topologically, there usually exist various loops with different lengths on a network, and we call them *topological loops*. For a given network, there are many topological loops that could be the potential candidates for the oscillation source. However, nodes in an excitable network must be excited in sequence and the excitable wave should propagate forward along the shortest path. Conforming to this rule, among the all possible topological loops, only some loops validate to serve as the oscillation source. We call this kind of loop as the *fundamental loop*, which is the unique candidate for the oscillation source in an excitable complex network.

Both the moderate fundamental loop and the minimum Winfree loop can effectively determine the oscillation period via different mechanisms. The moderate fundamental loop, which is created by the rewiring mechanism, can significantly influence the oscillation period in two ways. The first way is to turn itself into the oscillation source, while the other way is to behave as the competitor supervising the long oscillation source loop structure possessing the equivalent wavelength. The minimum Winfree loop, which is determined by the coupling strength, can profoundly affect the oscillation period by choosing the suitable fundamental loop to be the oscillation source at a given parameter setting. Corresponding to these two determinants, two control schemes that can effectively regulate the oscillation period have been proposed in our recent work [61].

15.5. Sustained Oscillation in Gene-Regulatory Networks

15.5.1. *Modeling gene-regulatory networks*

Gene-regulatory networks (GRNs), as a kind of biochemical regulatory networks in systems biology, can be well described by coupled differential equations and have been extensively explored in recent years. The equations describing biochemical regulation processes are strongly nonlinear and often have many degrees of freedom. We are concerned with the common features and network structures of GRNs.

Different from the above excitable networks, positive feedback loops and negative feedback loops have been identified in various biochemical regulatory networks and found to be important control modes in GRNs. Self-sustained oscillation, bi-rhythmicity, bursting oscillation and even chaotic oscillations are expected for these objects. Moreover, oscillations may occur in GRNs with a small number of units. The study of sustained oscillations in small GRNs is of great importance in understanding the mechanism of gene regulation processes in very large-scale GRNs. Network motifs as subgraphs can appear in some biological networks, and they are suggested to be elementary building blocks that carry out some key functions in the network. It is our motivation to unveil the relation between network structures and the existence of oscillatory behaviors in GRNs.

We consider the following GRN model:

$$\frac{dp_i}{dt} = f_i(\boldsymbol{p}) - p_i, \tag{15.15}$$

where $(\boldsymbol{p}) = (p_1, p_2, \ldots, p_N)$, and the function $f_i(\boldsymbol{p})$ satisfies the following form:

$$f_i(\boldsymbol{p}) = \begin{cases} A_i(\boldsymbol{p}) & \text{for activive regulation only,} \\ R_i(\boldsymbol{p}) & \text{for repressive regulation only,} \\ A_i(\boldsymbol{p})R_i(\boldsymbol{p}) & \text{for joint regulation.} \end{cases} \tag{15.16}$$

The active regulation function is

$$A_i(\boldsymbol{p}) = \text{act}_i^h/(\text{act}_i^h + K^h), \tag{15.17a}$$

and the repressive regulation function is written as

$$R_i(\boldsymbol{p}) = K^h/(\text{rep}_i^h + K^h), \tag{15.17b}$$

where

$$\mathrm{act}_i = \sum_{j=1}^{N} \alpha_{ij} p_j, \quad \mathrm{rep}_i = \sum_{j=1}^{N} \beta_{ij} p_j, \quad i, j = 1, 2, \ldots, N,$$

and p_i is the expression level of gene i, $0 < p_i < 1$. The adjacency matrices α, β determine the network structure, which are defined in such a way that $\alpha_{ij} = 1$ if gene j activates gene i, $\beta_{ij} = 1$ if gene j inhibits gene i and $\alpha_{ij} \cdot \beta_{ij} = 0$ for no dual-regulation of gene i by gene j. $\mathrm{act}_i(\mathrm{rep}_i)$ represents the sum of active (repressive) transcriptional factors to node i. The regulated expression of genes is represented by Hill functions with cooperative exponent h and the activation coefficient K, characteristic for many real genetic systems [40].

15.5.2. *Oscillating skeletons and cores*

Based on the known oscillation data of GRNs, we can explicitly compute all the functional weights w_{ij} between any two genes based on Eqs. (15.9)–(15.11) and draw the functional-weight maps for the oscillatory networks. Figure 15.8(a) shows a simple example of 6-node GRN with the dynamics given by Eq. (15.15), and computer simulations indicate that this GRN possesses an oscillatory attractor for a set of parameters given in [40]. From the oscillation data, we can draw the weight map in Fig. 15.9(b). An interesting as well as valuable feature of Fig. 15.8(b) is that the w_{ij} distribution is strongly heterogeneous, i.e., some weights $w_{ij} \approx 1$ indicate a significant control while many others $w_{ij} \approx 0$ represent weak functional links. The heterogeneity allows us to explore the self-organized functional structures supporting oscillatory dynamics by the strongly weighted links. In Fig. 15.8(c), we remove all the less important links with $w_{ij} \leq w_{th}$, where the threshold value $w_{th} = 0.30$. By removing any interaction and node in the network, we always mean to delete their oscillatory parts while keep the average influence of the deleted parts as control parameters in Eq. (15.10). The reduced subnetwork shown in Figs. 15.8(c) retaining only strongly weighted interactions is called the *skeleton* of the oscillatory GRN. Furthermore, we can reduce the

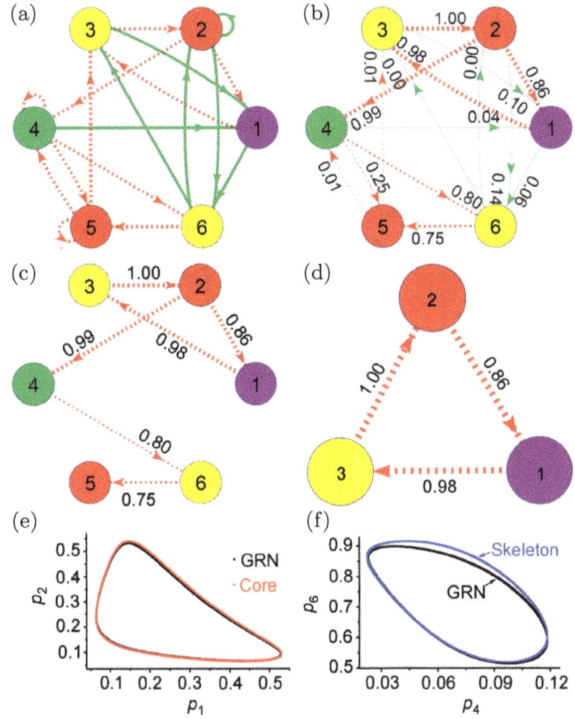

Fig. 15.8. An example of the functional-weight map for an oscillatory GRN. (a) An oscillatory GRN with $N = 6$ genes and $I = 15$ links. Different node colors represent different phases of the oscillatory nodes. Green full and red dotted arrowed lines represent active and repressive interactions, respectively. (b) The weight map of all interactions computed by Eqs. (15.10) and (15.11). (c) The reduced interaction skeleton obtained from (b) by deleting links with small weights. (d) An irreducible core structure obtained by only retaining interactions with both input and output in (c), which serves as the oscillation source propagating through the path of the skeleton (c). (e) A comparison of a dynamical orbit of the original GRN (a) with that of core (d) in a 2D phase plane. (f) A comparison of an orbit of the GRN (a) with that of skeleton (c) in another 2D phase plane. Good agreements between the orbits of the full GRN and that of the reduced subnetworks can be found. (Adapted from [42].)

skeleton network by removing the nodes without output one by one, and finally obtain an irreducible subnetwork where each node has both input and output which is defined as the *core* of the oscillation.

For the network Fig. 15.8(b) the core is given in Fig. 15.8(d). In Figs. 15.8(e) and 15.8(f), we plot the oscillation orbits of the

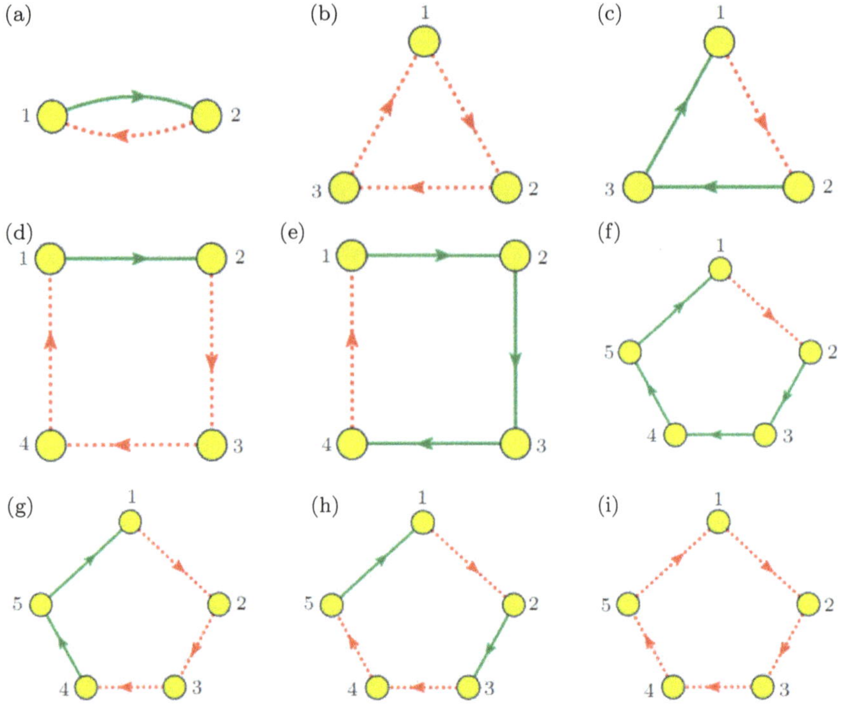

Fig. 15.9. All the oscillatory motifs with node number $N \leq 5$. All of them are the structures with single irreducible negative feedback loops. (Adapted from [42].)

original GRN and those of the core and skeleton in two-dimensional (p_1, p_2) and (p_4, p_6) phase planes, respectively. It is found that the dynamical orbits of the original network given in Fig. 15.8(a) can be well reproduced by the reduced structures in Figs. 15.8(c) and 15.8(d), conforming that the skeleton and core defined by the highly weighted links can dominate the essential dynamics of the original GRN. Considering Figs. 15.8(d), 15.8(e) and Figs. 15.8(c), 15.8(f) it is clear that the core (Fig 15.8(d)) serves as the oscillation source of the GRN while the skeleton (Fig. 15.8(c)) plays the role of main signal propagation paths from the core throughout the network.

The analysis presented in Fig. 15.8 can be well applied to more complicated GRNs with larger numbers of nodes and links to fulfill the reduction of network complexity for understanding and controlling network dynamics. Numerical works for a large network with many nodes and links indicate that the comparisons of the dynamics of the core and skeleton with those of the original network are in good agreement.

The above procedure can well predict all possible motifs of GRNs for a given number of genes. In Fig. 15.9, we give the fundamental motifs for $N = 2$, 3, and 4 nodes. It can be found that for $N = 2$ genes, there is only one motif that supports sustained oscillation, and two directional (one active and the other repressive) links are necessary. When the number of genes increases, more possible motifs exist and all of they can support sustained oscillations with different frequencies. In this case, the gene-regulatory relations become more complicated.

15.6. Concluding Remarks

Let us give a brief summary of the above discussions on the emergence of sustained oscillation dynamics in networks of nonoscillatory units. In this case, the feedback topology is the fundamental mechanism. For a very large network, not every node or link contributes to the collective oscillation, and only a small portion of nodes and their connections dominates, forming some fundamental building blocks such as motifs, loops, or cores. To dig out these basic topologies, an appropriate topology reduction is the key point. This meanwhile leads naturally to dynamics reduction, revealing the emergence of self-organized oscillation from collaborations of nonoscillatory node dynamics. It is a significant issue to make a topology reduction to reduce the dimensionality of dynamics of a complex system to obtain the essential structural ingredient of emergence behaviors. In recent years, we developed the related techniques and ways in revealing the embedded topologies [62, 63].

It should be stressed that there exist many fundamental topologies on a large network that may induce various possible sustained oscillations. Competitions and collaborations among these subnetworks and rhythms lead to complicated dynamics [64, 65].

References

[1] L. Glass and M. C. Mackay, *From Clocks to Chaos: The Rhythms of Life* (Princeton University Press, Princeton, NJ, 1988.

[2] De Mairan, Observation botanique, *Hist. Acad. Roy. Scien. Avec Mém. Mathé. Phys. Tirés Registres Decette Acad.* **35** (1729).

[3] Z. Mauro, C. Rodolfo, M. Giuseppe, F. Chiaki and T. Gianluca, Circadian clocks: What makes them tick? *Chronobiol. Int.* **17**(4) (2000) 433–451.

[4] R. J. Konopka and S. Benzer, Clock mutants of Drosophila melanogaster, *Proc. Natl. Acad. Sci. USA* **68** (1971) 2112–2116.

[5] P. Reddy, W. A. Zehring, D. A. Wheeler, V. Pirrotta, C. Hadfield, J. C. Hall and M. Rosbash, Molecular analysis of the period locus in Drosophila melanogaster and identification of a transcript involved in biological rhythms, *Cell* **38** (1984) 701–710.

[6] T. A. Bargiello and M. W. Young, Molecular genetics of a biological clock in Drosophila, *Proc. Natl. Acad. Sci. USA* **81** (1984) 2142–2146.

[7] T. A. Bargiello, F. R. Jackson and M. W. Young, Restoration of circadian behavioural rhythms by gene transfer in Drosophila, *Nature* **312** (1984) 752–754 .

[8] P. E. Hardin, J. C. Hall and M. Rosbash, Feedback of the Drosophila period gene product on circadian cycling of its messenger RNA levels, *Nature* **343** (1990) 536–540.

[9] M. H. Vitaterna, D. P. King, A. M. Chang, J. M. Kornhauser, P. L. Lowrey, J. D. McDonald, W. F. Dove, L. H. Pinto, F. W. Turek and J. S. Takahashi, Mutagenesis and mapping of a mouse gene, clock, essential for circadian behavior, *Science* **264**(5159) (1994) 719–725.

[10] J. L. Price, J. Blau, A. Rothenfluh, M. Abodeely, B. Kloss and M. W. Young, Double-time is a novel Drosophila clock gene that regulates PERIOD protein accumulation, *Cell* **94** (1998) 83–95.

[11] https://www.nobelprize.org/nobel_prizes/medicine/laureates/2017/press.html.

[12] Z. G. Zheng, Deciphering biological clocks, and reconstructing life rhythms, *Physics* **46**(12) (2017) 802–808.

[13] V. I. Nekorkin, *Introduction to Nonlinear Oscillations* (Wiley-VCH, 2015).

[14] J. Guckenheimer and P. Holmes, *Nonlinear Oscillations, Dynamical Systems, and Bifurcation of Vector Fields* (Springer, New York, 1983).

[15] L. Glass, Synchronization and rhythmic processes in physiology, *Nature* **410** (2001) 277–284.

[16] M. B. Elowitz and S. Leibler, A synthetic oscillatory network of transcriptional regulators, *Nature* **403** (2000) 335–338.
[17] G. Buzsaki and A. Draguhn, Neuronal oscillations in cortical networks, *Science* **304** (2004) 1926–1929.
[18] I. Prigogine and G. Nicolis, *Self-organization in Non-equilibrium Systems, from Dissipative Structures to Order Through Fluctuations* (Wiley, New Jersey, 1977).
[19] S. A. Kauffman, *The Origins of Order: Self-Organization and Selection in Evolution* (Oxford University Press, 1993).
[20] Z. G. Zheng, *Collective Behaviors and Spatiotemporal Dynamics in Coupled Nonlinear Systems* (Higher Education Press, 2004) (in Chinese).
[21] Z. G. Zheng, *Emergence Dynamics in Complex Systems: From Synchronization to Collective Transport,* Vols. I and II (Science Press, Beijing 2019).
[22] A. Goldbeter, Computational approaches to cellular rhythms, *Nature* **420** (2002) 238–245.
[23] M. B. Elowitz and S. Leibler, A synthetic oscillatory network of transcriptional regulators, *Nature* **403** (2000) 335–338.
[24] J. E. Ferrell, T. Y. Tsai and Q. Yang, Modeling the cell cycle: Why do certain circuits oscillate? *Cell* **144** (2011) 874–885.
[25] T. S. Gardner, C. R. Cantor and J. J. Collins, Construction of a genetic toggle switch in Escherichia coli, *Nature* **403**(6767) (2000) 339–342.
[26] E. Ozbudak *et al.*, Multistability in the lactose utilization network of Escherichia coli, *Nature* **427** (2004) 737–740.
[27] T. Y. Tsai, Y. S. Choi, W. Z. Ma, J. R. Pomerening, C. Tang and J. E. Ferrell Jr., Robust, tunable biological oscillations from interlinked positive and negative feedback loops, *Science* **321**(5885) (2008) 126–129.
[28] X. H. Huang, Z. G. Zheng, G. Hu, S. Wu and M. J. Rasch, Different propagation speeds of recalled sequences in plastic spiking neural networks, *New J. Phys.* **17** (2015) 035006.
[29] L. Chen and K. Aihara, Chaotic simulated annealing by a neural-network model with transient chaos, *Neur. Net.* **8** (1995) 915–930.
[30] A. V. M. Herz, T. Gollisch, C. K. Machens and D. Jaeger, Modeling single-neuron dynamics and computations: A balance of detail and abstraction, *Science* **314** (2006) 80–85.
[31] Y. Mandelblat-Cerf, I. Novick and E. Vaadia, Expressions of multiple neuronal dynamics during sensorimotor learning in the motor cortex of behaving monkeys, *PLoS One* **6**(7) (2011) e21626.

[32] N. Li, K. Daie, K. Svoboda and S. Druckmann, Robust neuronal dynamics in premotor cortex during motor planning, *Nature* **532** (2016) 459–464.
[33] J. F. Burke, *et al.*, Synchronous and asynchronous theta and gamma activity during episodic memory formation, *J. Neur.* **33** (2013) 292–304.
[34] Y. Qian, X. D. Huang, G. Hu and X. H. Liao, Structure and control of self-sustained target waves in excitable small-world networks, *Phys. Rev. E* **81** (2010) 036101.
[35] Y. Qian, X. H. Liao, X. D. Huang, Y. Y. Mi, L. S. Zhang and G. Hu, Diverse self-sustained oscillatory patterns and their mechanisms in excitable small-world networks, *Phys. Rev. E* **82** (2010) 026107.
[36] X. H. Liao, Q. Z. Xia, Y. Qian, L. S. Zhang, G. Hu and Y. Y. Mi, Pattern formation in oscillatory complex networks consisting of excitable nodes, *Phys. Rev. E* **83** (2010) 056204.
[37] X. H. Liao, Y. Qian, Y. Y. Mi, Q. Z. Xia, X. Q. Huang and G. Hu, Oscillation sources and wave propagation paths in complex networks consisting of excitable nodes, *Front. Phys.* **6** (2011) 124.
[38] Y. Y. Mi, L. S. Zhang, X. D. Huang, Y. Qian, G. Hu and X. H. Liao, Complex networks with large numbers of labelable attractors, *Europhys. Lett.* **95** (2011) 58001.
[39] Y. Y. Mi, X. H. Liao, X. H. Huang, L. S. Zhang, W. F. Gu, G. Hu and S. Wu, Long-period rhythmic synchronous firing in a scale-free network, *Proc. Natl. Acad. Sci. USA* **25** (2013) E4931–E4936.
[40] Z. Y. Zhang, W. M. Ye, Y. Qian, Z. G. Zheng, X. H. Huang and G. Hu, Chaotic motifs in gene regulatory networks, *PLoS ONE* **7** (2012) e39355.
[41] Z. Y. Zhang, Z. Y. Li, G. Hu and Z. G. Zheng, Exploring cores and skeletons in oscillatory gene regulatory networks by a functional weight approach, *Europhys. Lett.* **105** (2014) 18003.
[42] Z. Y. Zhang, X. H. Huang, Z. G. Zheng and G. Hu, Exploring cores and skeletons in oscillatory gene regulatory networks by a functional weight approach, *Sci. Sin.* **44** (2014) 1319 (in Chinese).
[43] W. Jahnke and A. T. Winfree, A survey of spiral wave behavior in the oregonator model;Int. *J. Bif. Chaos* **1** (1991) 445–466.
[44] M. Courtemanche, L. Glass and J. P. Keener, Instabilities of a propagating pulse in a ring of excitable media, *Phys. Rev. Lett.* **70** (1993) 2182–2185.
[45] M. Cross and P. Hohenberg, Pattern formation outside of equilibrium, *Rev. Mod. Phys.* **65** (1993) 851.
[46] M. Cross and H. Greenside, *Pattern Formation and Dynamics in Nonequilibrium Systems*, (Cambridge University Press, Cambridge, 2009).

[47] M. Gromov, V. Capasso, M. Gromov, A. Harel-Bellan, N. Morozova and L. L. Pritchard (eds.), *Pattern Formation in Morphogenesis: Problems and Mathematical Issues* (Springer, Berlin 2013).

[48] D. J. Watts and S. H. Strogatz, Collective dynamics of small-world networks, *Nature* **393**(6684) (1998) 440–442.

[49] A. L. Barabási and R. Albert, Emergence of scaling in random networks, *Science* **286** (1999) 509–512.

[50] R. Albert and A. L. Barabási, Statistical mechanics of complex networks, *Rev. Mod. Phys.* **74**(1) (2002) 47.

[51] E. Ben-Naim, H. Frauenfelder and Z. Torozckai (Eds.), *Complex Networks*, Lecture Notes in Physics, Vol. 650 (Springer, 2004).

[52] S. N. Dorogovtsev and J. F. F. Mendes, *Evolution of Networks: From Biological Nets to the Internet and WWW* (Oxford University Press, 2003).

[53] R. Pastor-Satorras and A. Vespignani, *Evolution and Stricture of the Internet: A Statistical Physics Approach* (Cambridge University Press, 2004).

[54] S. Boccaletti, V. Latora, Y. Moreno, M. Chavez and D. U. Hwang, Complex networks: Structure and dynamics, *Phys. Rep.* **424** (2006) 175–308.

[55] J. Q. Fang, X. F. Wang, Z. G. Zheng, Q. Bi, Z. R. Di and X. Li, New interdisciplinary science: Network science (I), *Progress Phys.* **27**(3) (2007) 239–343.

[56] J. Q. Fang, X. F. Wang, Z. G. Zheng, X. Li, Z. R. Di and Q. Bi, New interdisciplinary science: Network science (II), *Progress Phy.* **27**(4) (2007) 361–448.

[57] Y. Qian, X. H. Cui and Z. G. Zheng, Minimum Winfree loop determines self-sustained oscillations in excitable Erdos-Renyi random networks, Sci. Rep. **7** (2017) 5746.

[58] Z. G. Zheng and Y. Qian, Self-sustained oscillations in biological excitable media, *Chin. Phys. B* **27**(1) (2018) 018901.

[59] Y. Qian, G. Zhang, Y. F. Wang, C. G. Yao and Z. G. Zheng, Winfree loop sustained oscillation in two-dimensional excitable lattices: Prediction and realization, *Chaos* **29** (2019) 073106.

[60] Y. Qian, Y. Wang, G. Zhang, F. Liu and Z. G. Zheng, Collective sustained oscillations in excitable small-world networks: The moderate fundamental loop or the minimum Winfree loop? *Nonlinear Dynam.* **99**(2) (2020) 1415.

[61] Y. Qian, C. Zhang, Z. Wei, F. Liu, Y. Zhao and Z. G. Zheng, The optimal oscillation mode in excitable small-world networks, *Europhys. Lett.* **131** (2020) 38002.

[62] Z. Y. Zhang, Z. G. Zheng, H. J. Niu, Y. Y. Mi, S. Wu and G. Hu, Solving the inverse problem of noise-driven dynamic networks, *Phys. Rev. E* **91** (2015) 012814.

[63] Y. Chen, S. H. Wang, Z. G. Zheng, Z. Y. Zhang and G. Hu, Depicting network structures from variable data produced by unknown colored-noise driven dynamics, *Europhys. Lett.* **113** (2016) 18005.

[64] H. Haken, *Synergetics: An Introduction Nonequilibrium Phase Transitions and Self-Organization in Physics, Chemistry and Biology* (Springer, Berlin, 1978).

[65] H. Haken, *Advanced Synergetics: Instability Hierarchies of Self-Organizing Systems and Devices* (Springer, Berlin, 1983).

https://doi.org/10.1142/9789811221903_0016

Chapter 16

Stability of Anti-bunched Buses and Local Unidirectional Kuramoto Oscillators

Lock Yue Chew*,†,‡, Vee-Liem Saw*,† and Yi En Ian Pang*

*Division of Physics and Applied Physics,
School of Physical and Mathematical Sciences,
21 Nanyang Link, Nanyang Technological University,
Singapore 637371, Singapore

†Data Science and Artificial Intelligence Research Centre,
Block N4 #02a-32, Nanyang Avenue, Nanyang Technological University,
Singapore 639798, Singapore

‡Complexity Institute, 61 Nanyang Drive,
Nanyang Technological University,
Singapore 637335, Singapore

Inspired by our recent work that relates bus bunching as a phenomenon of synchronization of phase oscillators, we construct a model of Kuramoto oscillators that follows an analogous interaction mechanism of local unidirectional coupling. In the bus loop system, we can introduce a no-boarding policy as a form of kicking force to achieve a stable staggered (anti-bunched) state. For Kuramoto oscillators, it turns out that such stable anti-bunched states can exist (without any additional kicking force) if the number of oscillators is at least five. This correspondence between the bus loop system and the local unidirectional Kuramoto oscillators leads to the insight on how the bus loop system can remain staggered.

16.1. Introduction

In major cities worldwide, urban planning plays a crucial role in empowering development and sustainability of their growth. An important aspect of such urban planning is an effective public transportation system. This includes bus networks, overhead rail lines as well as subways. While rail and subway operators have their own dedicated lines for the carriages such that they can schedule the arrival of a train, its dwell time at designated stops and hence its departure time, the same is not true for buses which generally share common roads with other vehicles. Bus services interact with traffic from cars, motorbikes as well as heavy vehicles like lorries, and even traffic lights at road intersections. Hence unlike rail services, buses are susceptible to traffic perturbations. This has the effect of making a bus arriving slightly ahead of the intended schedule, such that it picks up less people and depart even earlier, which leaves more people to be picked up by the next bus and further slows it down — widening their headway. The repercussion of such complex interactions is the perennial phenomenon of bus bunching [1–6]. In fact, buses are also known to behave chaotically [7, 8].

Various approaches to overcome bus bunching have been studied, each with their own benefits depending on the particular route involved. A common way is to hold back buses which are deemed as too fast or ahead of schedule [9–28]. The opposite approaches which attempt to speed up a slow bus include skipping bus stops [17, 29–33], deadheading (which is to send an empty bus to certain designated bus stops) [30, 33–36], or limiting boarding (no-boarding) [28, 37–42]. Alternatively, some studies also considered buses with wide doors [43–45].

This chapter aims to explain why bus bunching is a natural consequence of the instability of buses to remain in a staggered configuration. By building on the framework developed in [1], which described buses serving a loop of bus stops as a system of coupled oscillators [46], we exploit the correspondence between this bus loop system and the local unidirectional Kuramoto oscillators [47] where the stability properties of the latter are known [48, 49].

We first provide some preliminaries in the following section, and then describe a Fourier expansion of the bus loop system in Sec. 16.3 which specifies the dynamics of buses serving bus stops in a loop. Section 16.4 relates the bus loop system to the corresponding local unidirectional interaction of Kuramoto oscillators where we derive a phase transition between partial and complete synchronization, analogous to the results derived in [1] for buses serving a loop of bus stops. Section 16.5 shows numerical results on partial synchronization of local unidirectional Kuramoto oscillators, which is again analogous to the interaction of buses serving a loop of bus stops for a certain range of the coupling strength between buses/oscillators [1]. A successful approach to overcome bus bunching is the no-boarding strategy, where a slow bus disallows boarding to speed itself up. However, this is known to only work effectively during busy periods when demands for service are high [38, 41]. Section 16.6 derives the critical coupling strength for the transition where the no-boarding strategy is barely successful and begins to backfire. Finally, Sec. 16.7 discusses the stability properties of local unidirectional Kuramoto oscillators, with Sec. 16.8 explaining how the no-boarding and holding strategies together create stable anti-bunched configurations which overcome bus bunching.

16.2. Preliminaries

In a bus loop service, N buses serve M (reasonably) staggered bus stops around the loop, going round and round picking up and alighting passengers at the bus stops. This loop can be isometrically mapped to the unit circle, and the positions of the buses on the loop are identifiable by their phases $\theta_i \in [0, 2\pi)$, $i = 1, \ldots, N$ on the unit circle. The buses have their own intrinsic natural frequencies $\omega_1 > \cdots > \omega_N$ (excluding time stopped at the bus stops). If the buses have two doors: one for passengers to alight and the other to board with alighting and boarding occurring simultaneously, then there is a critical transition defined by $k = k_c$ where the buses would exhibit *complete synchronization* if $k > k_c$. Here, $k := s/l$ is a parameter that determines the strength of coupling amongst the N buses due to the M bus stops, with s being the rate of passengers arriving at each of the M bus stops and l being the rate of loading/unloading of people onto/off the bus. Saw *et al.* [1] calculated this critical transition to be given by

$$k_c(N) = \frac{1}{M}\sum_{i=1}^{N-1}\left(1 - \frac{\omega_N}{\omega_i}\right). \tag{16.1}$$

Here, $\omega_i = 2\pi f_i = 2\pi/T_i$, for each $i = 1, \ldots, N$.

If $k > k_c$, then all N buses are synchronized and phase-locked with phase difference $\sim 0°$. By numerical simulations, Saw *et al.* [1] further found that if $\bar{k} < k < k_c(N)$ where $\bar{k} \approx k_c(2)$, with $k_c(2)$ being the corresponding system with $N = 2$ buses having natural frequencies ω_1 and ω_N, respectively, then these N buses are *partially synchronized* with some buses being phase-locked (phase difference $\sim$0°) while the others are able to pull away (opening up large phase differences). If $k < \bar{k}$, then all N buses are *unsynchronized* with no phase-locking. The phase $k < \bar{k}$ (note the dual usage of the word "phase") is referred to as *lull* since k is weak enough such that the demand for service is low and the buses do not persistently bunch (no synchronization). If $k > \bar{k}$ (i.e., both $\bar{k} < k < k_c$ and $k > k_c$), then this phase is referred to as *busy* since demand for service is high enough leading to the undesirable phenomenon of *persistent* bus bunching of at least one pair of buses due to phase-locking. Intriguingly, Saw *et al.* [1] empirically observed *real human-driven buses* serving $M = 12$ reasonably staggered bus stops in a loop service being: (a) in the lull phase with measured $k_{\text{measured}} < \bar{k}$, where those buses are observed to be not exhibiting persistent bunching; (b) in the busy phase with measured $k_{\text{measured}} > \bar{k}$, where some of those buses are indeed observed to be phase-locked and persistently bunched.

Incidentally, note that if boarding and alighting occur sequentially through one single door, i.e., people alight first and then followed by people boarding, there will be an overall factor of 1/2 in Eq. (16.1) as can be verified by direct calculations analogous to that carried out in [1] [or see the description below Eq. (16.27)], since only half of the time stopped at the bus stop is devoted to boarding the people waiting there.

16.3. Fourier Expansion for the Bus Loop System

Consider $N = 1$ bus serving $M = 1$ bus stop in a loop. This bus moves with its constant natural (angular) frequency ω when it is not at the bus stop, and stops for a duration of τ to embark/disembark passengers at the bus stop. This repeats over and over as it loops around the circle. Hence, we can write:

$$\frac{d\theta}{dt} = \omega - \omega f(t), \tag{16.2}$$

where

$$f(t) := \frac{\tau}{T} + \sum_{n=1}^{\infty} \frac{2}{n\pi} \sin\left(\frac{n\pi\tau}{T}\right) \cos\left(\frac{2n\pi}{T}t\right). \tag{16.3}$$

This function $f(t)$ is a pulse wave with period $T = 2\pi/\omega$ and pulse time τ.

In general with N buses, each bus's rate of change of θ_i takes the form of Eq. (16.2), with the width of the well τ_i depending on the headway from the bus immediately ahead of it. Therefore, τ_i is essentially proportional to $\theta_{i-1} - \theta_i$, i.e., the phase difference with respect to the bus immediately ahead of it. The larger the phase difference since the bus immediately ahead has left the bus stop, the more people would have accumulated at the bus stop to be picked up. These buses can overtake, and the identity of the bus immediately ahead must change accordingly. With M bus stops on the loop, there will be M such wells. Thus, we find this system to be governed by a complicated coupled set of nonlinear differential equations with feedback that is nonanalytic due to the change of identity corresponding to θ_{i-1} each time overtaking occurs.

16.4. Local Unidirectional Kuramoto Oscillators

Kuramoto studied a simple model of coupled oscillators where he specified only the first sine term in the Fourier expansion for the coupling function [47]. We would like to investigate the corresponding Kuramoto version for the bus loop system and find out which properties of the bus dynamics correspondingly exist in this simplified model with only the first sine term in the coupling function. Consider N Kuramoto oscillators with natural angular frequencies $\omega_1, \ldots, \omega_N$, subjected to local unidirectional coupling. This system differs from the original Kuramoto model (with the exception of $N = 2$ which is indeed equivalent), whereby here oscillator i, $i \in \{1, \ldots, N\}$ is only directly influenced by the oscillator that is immediately ahead of it. We are interested in this type of Kuramoto oscillators instead of the original model whereby the coupling is globally contributed by every oscillator, because this is the direct analogue to the bus system (a bus is only directly influenced by the bus immediately ahead).

In accordance to Refs. [46, 48], we say that a pair of adjacent oscillators are *phase-locked* if their phase difference remains constant in time. If the coupling is *strong enough*, i.e., $K \geq K_c$, then these N oscillators are *completely phase-locked*. The critical coupling K_c is the value of K such that the system is barely able to be completely phase-locked. Hence, the N equations governing the motion of the N oscillators in such a situation are:

$$\frac{d\theta_i}{dt} = \omega_i + K \sin(\theta_{i-1} - \theta_i), \tag{16.4}$$

where θ_i represents the phase of oscillator i on the unit circle, and $\theta_0 := \theta_N$. The constant K is the coupling strength between the oscillators. Note that we do not divide K by N (which is done in the original Kuramoto model), because there is only one sine term present (instead of $N-1$ sine terms in the original Kuramoto model). The condition for these oscillators being completely phase-locked is

$$\frac{d\theta_1}{dt} = \cdots = \frac{d\theta_N}{dt}, \tag{16.5}$$

giving $N - 1$ independent equations which are functions of $N - 1$ independent phase differences $\phi_1 = \theta_1 - \theta_2, \ldots, \phi_{N-1} = \theta_{N-1} - \theta_N$.

Let us work this out explicitly in the case of $N = 2$. The governing equations are as follows:

$$\frac{d\theta_1}{dt} = \omega_1 + K\sin(\theta_2 - \theta_1) = \omega_1 - K\sin(\theta_1 - \theta_2), \tag{16.6}$$

$$\frac{d\theta_2}{dt} = \omega_2 + K\sin(\theta_1 - \theta_2). \tag{16.7}$$

The phase-locked condition gives

$$\frac{d\theta_1}{dt} = \frac{d\theta_2}{dt}, \tag{16.8}$$

$$K = \frac{\omega_1 - \omega_2}{2\sin(\theta_1 - \theta_2)}. \tag{16.9}$$

This implies that the two oscillators can be phase-locked as long as there is a phase difference $\theta_1 - \theta_2$ which satisfies Eq. (16.9), given some value of K. If K is not strong enough, i.e., $K < K_c$, then no phase difference $\theta_1 - \theta_2$ can satisfy Eq. (16.9) and the two oscillators are not phase-locked. The sought after critical transition K_c between the unsynchronized and synchronized phases of the system occurs when Eq. (16.8) is satisfied for the smallest K, i.e., we need to minimize Eq. (16.9). This gives

$$K_c = \frac{1}{2}(\omega_1 - \omega_2), \tag{16.10}$$

which occurs when the phase difference between the two oscillators is $\theta_1 - \theta_2 = \pi/2$.

For $N = 3$, the governing equations are as follows:

$$\frac{d\theta_1}{dt} = \omega_1 + K\sin(\theta_3 - \theta_1) = \omega_1 - K\sin(\phi_1 + \phi_2), \tag{16.11}$$

$$\frac{d\theta_2}{dt} = \omega_2 + K\sin(\theta_1 - \theta_2) = \omega_2 + K\sin\phi_1, \tag{16.12}$$

$$\frac{d\theta_3}{dt} = \omega_3 + K\sin(\theta_2 - \theta_3) = \omega_3 + K\sin\phi_2, \tag{16.13}$$

where $\phi_1 = \theta_1 - \theta_2$, $\phi_2 = \theta_2 - \theta_3$, $\theta_3 - \theta_1 = -(\theta_1 - \theta_2 + \theta_2 - \theta_3) = -(\phi_1 + \phi_2)$. From the condition for complete phase-locking

$$\frac{d\theta_1}{dt} = \frac{d\theta_2}{dt} = \frac{d\theta_3}{dt}, \tag{16.14}$$

we have two independent equations

$$K = \frac{\omega_1 - \omega_2}{\sin\phi_1 + \sin(\phi_1 + \phi_2)}, \tag{16.15}$$

$$K = \frac{\omega_2 - \omega_3}{\sin\phi_2 - \sin\phi_1}. \tag{16.16}$$

This implies that the three oscillators can all be phase-locked if there are some phase differences ϕ_1 and ϕ_2 between these three oscillators which satisfy the above two equations, given some value of K. The sought after critical transition between completely phase-locked and partially phase-locked phases of the system occurs when K is minimized. Since K depends on the two variables ϕ_1, ϕ_2 which are subjected to the constraint

$$\frac{\omega_1 - \omega_2}{\sin\phi_1 + \sin(\phi_1 + \phi_2)} = \frac{\omega_2 - \omega_3}{\sin\phi_2 - \sin\phi_1}, \tag{16.17}$$

the problem of minimizing K can be solved by the method of Lagrange multipliers. Unfortunately, there does not appear to be a closed form solution for K_c. This applies to any N oscillators, where K is a function of $N-1$ independent phase differences which are subjected to $N-2$ constraint equations. The closed form solution is found for $N = 2$ in Eq. (16.10), but not for $N > 2$.

Interestingly for $N = 3$, we note that if $\omega_1 - \omega_2 = \omega_2 - \omega_3 = \Omega$, then $\phi_1 = 0$, $\phi_2 = \pi/2$ is the solution that minimizes K as can be verified by the method of Lagrange multipliers. In this case,

$$K_c = \Omega. \tag{16.18}$$

However, the corresponding situations with $\omega_1 - \omega_2 = \cdots = \omega_{N-1} - \omega_N$ for $N > 3$ do not seem to admit such a nice closed form solution. Nevertheless, these can be solved numerically as studied by Rogge and Aeyels [48] for the unidirectionally coupled case (like our formulation here where we are motivated by the bus loop system) and by Ochab and Góra [49] for a locally (nearest neighbor) coupled case.

16.5. Partial Synchronization of local Unidirectional Kuramoto oscillators

As mentioned in Sec. 16.2, the bus loop system has three phases: (a) complete phase-locking if $k > k_c(N)$; (b) partial phase-locking if $\bar{k} < k < k_c(N)$; (c) no phase-locking if $k < \bar{k}$; with $\bar{k} \sim k_c(2)$. The value of $\bar{k} \sim k_c(2)$ is a result found from simulations [1]. Here, we similarly carry out simulations for the local unidirectional Kuramoto model to investigate if partial synchronization occurs. We consider three situations:

(1) frequencies skewed towards lowest frequency;
(2) evenly spaced frequencies;
(3) frequencies skewed towards highest frequency.

These frequencies are selected with $\omega_1 = 1.39$, $\omega_N = 0.93$, and $\omega_2, \ldots, \omega_{N-1} \in (\omega_1, \omega_N)$. (*Note*: The frequency range of a university bus loop service studied in [1] was measured to be $[0.93, 1.39]$ mHz, i.e., a period range of $[12, 18]$ minutes, excluding time stopped at bus stops.) Simulations results as shown in Fig. 16.1 indicate that unlike the bus loop system [1], the local unidirectional Kuramoto oscillators only exhibit partial synchronization if the frequencies are skewed towards the highest frequency. Even so, the corresponding $\bar{K}$ where partial synchronization starts to occur does not seem to be near $K_c(2)$. The order parameter $r \in [0, 1]$ measures the degree of synchronicity amongst the N oscillators. If they are fully synchronized, then $r = 1$, while if they are fully unsynchronized then $r = 0$. Its definition is

$$r = \frac{1}{N}\sqrt{\left(\sum_{i=1}^{N} \cos\theta_i\right)^2 + \left(\sum_{i=1}^{N} \sin\theta_i\right)^2}. \tag{16.19}$$

The graphs measure the minimum r in the latter part of the simulations where the transients have weeded out. If there is some degree of phase-locking, then the minimum r should report a value larger than 0, while a minimum $r \sim 0$ implies that there is no synchronization since the oscillators overlap one another as well as open up large phase differences.

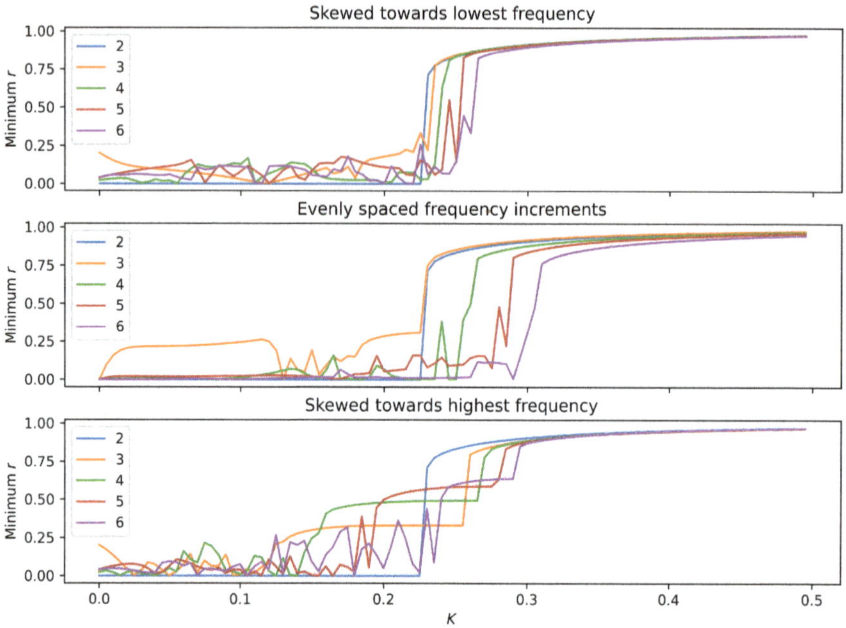

Fig. 16.1. Partial synchronization in the local unidirectional Kuramoto oscillators is observed when the natural frequencies are skewed towards the highest one. Shown here are simulation runs with $N = 2, 3, 4, 5, 6$ oscillators, respectively. Regions of partially phase-locked configurations with $0 \ll r \ll 1$ are seen in the bottom plot, whereas partial phase-locking is absent in the first two graphs since r essentially jumps from ~ 0 to a value near ~ 1.

16.6. Anti-bus Bunching

Several ways to overcome the notorious phenomenon of bus bunching have been proposed over the decades (see the introduction and references in [1]). A recent idea [41] explores a no-boarding strategy where a "slow" bus, i.e., one whose phase difference from the bus immediately behind it is way too small (or alternatively, if its phase difference from the bus immediately ahead is way too large), would disallow boarding and leave, leaving the remaining people to be picked up by the bus behind it who would be arriving soon. An analytical study in [41] showed that this strategy can significantly improve the efficiency of the bus system compared to leaving the buses on their own, in the case where all buses have an identical natural frequency. This is useful for the future when self-driving buses are the norm and can be programmed to maintain an identical natural frequency, unlike human-driven buses which tend to have different natural frequencies due to differing driving styles [1]. The no-boarding policy also works well for buses with frequency detuning when the demand level k is high, but surprisingly turns out to backfire when k is low. This section establishes the phase transition when this happens. Further Saw *et al.* [50] showed that self-learning buses without human instruction also end up discovering a no-boarding strategy as well as a holding strategy to maintain a staggered configuration around the loop and prevent bus bunching. On top of that, as a mandatory no-boarding policy is arguably stringent because it denies commuters who need service urgently, we have studied the situations where the commuters are allowed to cooperate or defect the implementation of no-boarding, and this turns out to be a reasonably viable approach [42].

Consider the simplest setup of $N = 2$ buses with natural angular frequencies $\omega_1 > \omega_2$ ($\omega_i = 2\pi f_i = 2\pi/T_i$) serving $M = 1$ bus stop in a loop. When a bus is at a bus stop, it would allow people to alight followed by allowing people to board, i.e., the two processes are sequential. According to the no-boarding policy, however, while boarding people it would disallow further boarding if its phase difference $\Delta\theta$ as measured from the bus behind it is less than a prescribed threshold θ_0. The intention of the no-boarding policy is to speed up this bus which is deemed as being too slow when $\Delta\theta < \theta_0$, leaving the people who are denied boarding to be picked up by the "faster" bus which would be approaching soon. Consequently, these two buses would reach a steady state whereby they both maintain a staggered configuration with an effective average phase difference of $\theta_{\text{eff}} > \theta_0$. The effective average phase difference θ_{eff} is greater than θ_0 where no-boarding is implemented since θ_0 serves as the lower bound to the evolving phase difference $\Delta\theta(t)$ which fluctuates around θ_{eff}. In an ideal implementation, the system would be fully synchronized and the N buses are phase-locked, with nearly staggered phase difference of $\theta_{\text{eff}} \sim 360°/N$ depending on the prescribed θ_0.

The no-boarding policy always works, regardless of k, if the two buses have the same natural frequency ω, provided that the prescribed θ_0 does not exceed an upper bound. This upper bound exists because if θ_0 is too large, then the buses would implement no-boarding too frequently and fail to board people at a rate faster than the rate of people arriving at the bus stop (see details in [41]). In contrast, this is not true if the buses have frequency detuning. The slow bus with ω_2 can only be sped up up to a certain extent, leaving the people denied boarding to be picked up by the fast bus with ω_1 which only slows it down up to a certain extent.

It turns out that there is a critical k_c [which we will determine below, and turns out to be the same k_c in Eq. (16.1) with a factor of 1/2] whereby if $k < k_c$, then the fast bus is not slowed down enough due to not that many people to pick up (i.e., the coupling strength is too weak). This situation can also be viewed as the frequency detuning being too large for the no-boarding policy to bridge. In this phase, the two buses would fail to maintain their staggered configuration with an effective average phase difference of θ_{eff}. Instead, the fast bus would periodically overlap the slow bus, with these two buses being unsynchronized and not phase-locked. On the other hand, if $k > k_c$, then the two buses are in a phase whereby they are phase-locked with an effective average phase difference of θ_{eff}. Thus, the critical k_c is defined by the condition that the slow bus has *zero* stoppage duration at the bus stop since this is the best possible way to speed it up by not having to stop at the bus stop at all. In this extreme no-boarding policy where the slow bus always implements it, all passengers are picked up by the fast bus.

16.6.1. *Direct calculation*

Let us directly calculate k_c. Figure 16.2 shows a cycle where the two buses are in steady state: (a) is the moment when the slow bus just leaves the bus stop after implementing the no-boarding policy as the phase difference from the bus behind it is $\Delta\theta = \theta_0$, (b) is the moment when the fast bus just arrives at the bus stop, (c) is the

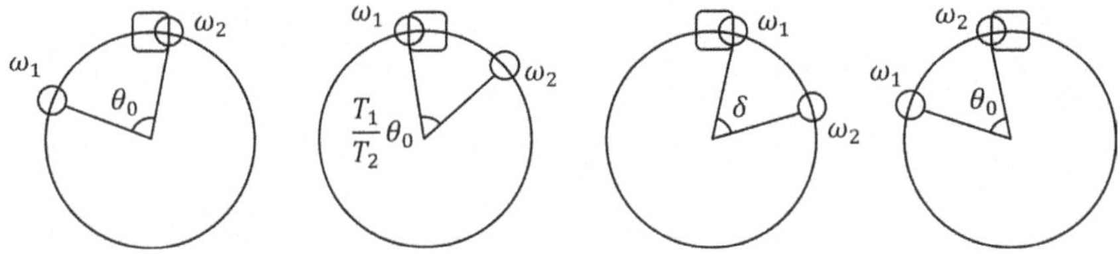

Fig. 16.2. $N = 2$ buses serving $M = 1$ bus stop in a loop, where k is strong enough such that they are phase-locked.

moment when the fast bus leaves the bus stop after allowing everyone to alight followed by everyone to board, and (d) is when the slow bus just arrives at the bus stop.

The time elapsed from (a) to (b) is θ_0/ω_1, so in (b) the phase difference of the slow bus as measured from the bus behind it is $T_1\theta_0/T_2$. The time elapsed from (b) to (c) is τ_1, the stoppage duration that the fast bus spends at the bus stop. This stoppage duration τ_1 at the critical k_c can be calculated as follows: the total number of people arriving at the bus stop per cycle is $s(T_1+\tau_1)$, since one cycle is the natural period of the fast bus, T_1, plus the time it stops at the bus stop, τ_1. Half of τ_1 is allocated for people to alight, with the other half for people to board. Hence $l\tau_1/2 = s(T_1+\tau_1)$, giving $\tau_1 = 2k_cT_1/(1-2k_c)$ with $k_c := s/l$. Consequently in (c), the phase difference of the slow bus as measured from the bus behind it is $\delta = T_1\theta_0/T_2 + \omega_2\tau_1$. From (c) to (d), the slow bus has to travel a phase difference of $2\pi - \delta$ which takes a time of $(2\pi-\delta)/\omega_2$, ending up with a phase difference of $\delta - (\omega_1-\omega_2)(2\pi-\delta)/\omega_2$ as measured from the bus behind it. At the critical k_c, the slow bus barely arrives at the bus stop and immediately implements no-boarding (note that in the previous round it picked up nobody and therefore has nobody to alight) since its phase difference from the bus behind it is θ_0. This gives the condition for obtaining k_c:

$$\delta - \frac{(\omega_1-\omega_2)(2\pi-\delta)}{\omega_2} = \theta_0. \tag{16.20}$$

Plugging in $\delta = T_1\theta_0/T_2 + \omega_2\tau_1$ and $\tau_1 = 2k_cT_1/(1-2k_c)$, we find that the terms involving θ_0 cancel out, giving:

$$k_c = \frac{1}{2}\left(1 - \frac{\omega_2}{\omega_1}\right). \tag{16.21}$$

16.6.2. *Alternative condition*

The direct calculation shows that regardless of the angle θ_0 where no-boarding is implemented, the resulting k_c is independent of θ_0. In other words, the two buses can remain phase-locked due to the no-boarding policy at any effective average phase difference $\theta_{\text{eff}} > \theta_0$ as long as $k > k_c$. There is a simpler alternative condition to Eq. (16.20). Since the critical transition is given by the condition whereby the slow bus has zero stoppage duration at the bus stop, thus in steady state, the natural period of the slow bus T_2 must equal the natural period of the fast bus T_1 plus the stoppage duration of the fast bus τ_1:

$$T_2 = T_1 + \tau_1, \tag{16.22}$$

where $\tau_1 = 2k_cT_1/(1-2k_c)$. The alternative condition (16.22) also leads to the same formula for k_c in Eq. (16.21), with no dependence on θ_0.

In fact, this approach is directly generalizable to any N buses with natural frequencies $\omega_1 > \cdots > \omega_N$ serving one bus stop in a loop. If $k > k_c$, then all buses can remain staggered and phase-locked with phase difference $\theta_{\text{eff}} > \theta_0$ (for any θ_0) due to the no-boarding policy. At the critical transition $k = k_c$, the slowest bus is sped up in the best possible way whereby its stoppage duration is zero. Therefore, we have $N-1$ (independent) equations when the N buses are in steady state:

$$T_N = T_i + \tau_i, \tag{16.23}$$

where $i = 1, 2, \ldots, N-1$, and τ_i are the stoppage durations for the other buses with natural frequencies $\omega_1 > \cdots > \omega_{N-1}$, respectively. These τ_i must satisfy:

$$\frac{1}{2}l\sum_{i=1}^{N-1}\tau_i = sT_N, \tag{16.24}$$

since half of all these stoppage durations would be used to board the people that arrive at the bus stop during one cycle of T_N (with the other half meant for alighting). Plugging in the $N-1$ equations for each τ_i from Eq. (16.23) into Eq. (16.24):

$$\frac{1}{2}\sum_{i=1}^{N-1}(T_N - T_i) = k_c T_N, \tag{16.25}$$

where $k_c := s/l$, gives the general formula for k_c:

$$k_c = \frac{1}{2}\sum_{i=1}^{N-1}\left(1 - \frac{\omega_N}{\omega_i}\right). \tag{16.26}$$

With M staggered bus stops in the loop each having a spawning rate of s, then each bus stop acts as a multiplier of the coupling strength. Hence, only one Mth of the coupling strength is required for the critical transition, i.e.,

$$k_c = \frac{1}{2M}\sum_{i=1}^{N-1}\left(1 - \frac{\omega_N}{\omega_i}\right), \tag{16.27}$$

where $\omega_N/\omega_i = f_N/f_i = T_i/T_N$.

Remarkably, this is the identical formula as Eq. (16.1) with the overall factor of 1/2 since alighting and boarding here occur sequentially via one door instead of simultaneously via two doors. This coincidence is due to the fact that at the critical transition to complete synchronization, the coupling strength is just enough to keep all N buses bunched together at the same bus stop. (See Ref. [1] for the original derivation which directly goes through the process for one cycle.) Therefore, the slowest bus would experience *zero stoppage duration* in steady state, i.e., we would arrive at the $N-1$ (independent) equations given by Eq. (16.23), as well as all the stoppage durations τ_i $(i = 1, \ldots, N-1)$ for the other $N-1$ buses satisfying Eq. (16.24). This would lead to the same formula for k_c in Eq. (16.27) for N buses serving M bus stops.

16.7. Stable Staggered Solutions of the Local Unidirectional Kuramoto Oscillators

Rogge and Aeyels [48] studied the stability of completely phase-locked unidirectional Kuramoto oscillators. Their results reveal the intriguing implications on the anti-bunched or staggered configuration of the oscillators:

A completely phase-locked solution with all phase differences between adjacent oscillators being less than $\pi/2$ is asymptotically stable.

To illustrate, consider a system of $N = 3$ Kuramoto oscillators with identical natural frequency ω. If these three oscillators are such that they are all 120° spaced apart, then they are completely phase-locked, i.e., they will always remain staggered. However, if they are slightly perturbed, then they will deviate away from this staggered configuration and end up in a fully synchronized configuration with $r = 1$ or $\theta_1 = \cdots = \theta_N$. Intriguingly if $N \geq 5$, then the staggered configuration is *stable* since all phase differences between adjacent oscillators are strictly less than $\pi/2$. If the system is slightly perturbed, then it will return to this staggered configuration. For more details and other stable/unstable conditions, see [48].

Figure 16.3 (videos are given as supplemental material) shows the situations with $N = 5$ oscillators and $K = 0.32$ where: (a) they have identical natural frequency of $\omega = 1$, and their initial positions are $0°, 100°, 144°, 216°, 288°$; and (b) four oscillators have identical natural frequency $\omega = 1$ with the other having $\omega_{\text{other}} = 1.1$, and their initial positions are perfectly staggered. In both cases, they remain anti-bunched with all phase differences less than $\pi/2$ — in accordance to the theorem by Rogge and Aeyels [48]. Simulations with these setups for the original globally coupled Kuramoto oscillators [47] reveal that they would all end up bunching with $r \sim 1$.

The intuition of this stability property can be grasped from the nature of the sine coupling function [see also Fig. 16.4(a)]: If the phase difference is less than $\pi/2$, the trailing oscillator is given a positive kick to catch up. The larger the phase difference, the stronger is the kick due to the coupling it receives to speed it up. However, $\pi/2$

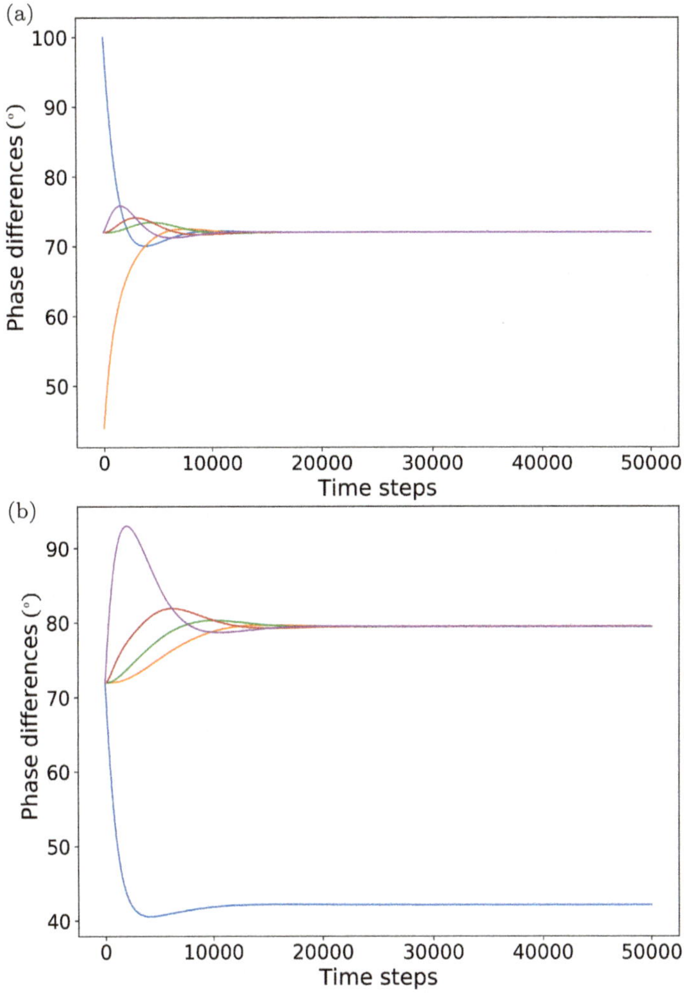

Fig. 16.3. Graphs of the phase differences for $N = 5$ local unidirectionally coupled Kuramoto oscillators. In both configurations, they remain anti-bunched with all phase differences less than $\pi/2$.

is a *turning point.* If the phase difference exceeds $\pi/2$, it lags even more but then receives *weaker kick* due to the coupling — which aggravates its lagginess. This is why the system with $N = 3$ staggered oscillators with identical natural frequency ω is unstable, because if one oscillator experiences a small lag, then the coupling it receives becomes weaker and it lags even more until they all crash into the fully synchronized configuration.

Rogge and Aeyels [48] further extended their results to a broader class of coupling functions which are continuous, 2π-periodic, odd, and has exactly one maximum at $\gamma \in (0, \pi)$. Then, a completely phase-locked solution with all phase differences between adjacent oscillators being less than γ is asymptotically stable. So instead of receiving the maximum boost when the phase difference is $\pi/2$, here it occurs at γ.

16.8. Creating Stability for a Staggered Configuration in the Bus Loop System

Equation (16.2) describes the instantaneous angular velocity experienced by the bus in terms of the pulse waveform where the bus moves with its constant natural frequency when it is not at a bus stop, and has zero angular velocity when it is at a bus stop over a dwell time of τ which depends on the headway (phase difference) from the bus immediately ahead of it.

A different view for the bus loop system may be seen from considering the average angular velocity over a time interval. For example, suppose there are N buses serving M staggered bus stops in a loop where each bus stop has people arriving at a rate of s people per second. When a bus just leaves a bus stop, suppose it takes a time of Δt before the next bus arrives at the bus stop and then leaves after disembarkation/embarkation. The number of people that this bus picks up is $s\Delta t$. If this bus spends a dwell time of τ, then $l\tau \sim 2s\Delta t$ where l is the rate of loading/unloading. The factor of 2 on the right-hand side is because one factor is expended to pick up $s\Delta t$ people with the other for those alighting. With $k := s/l$, we have $\tau \sim 2k\Delta t$. Since its natural frequency is ω_i, its natural period is $2\pi/\omega_i$. As there are M staggered bus stops in the loop, the time it takes to traverse one bus stop would be $2\pi/M\omega_i + \tau$, and it traverses an angular displacement of $2\pi/M$. Let $\eta_i = 2\pi/M\omega_i$, i.e., the time interval to traverse one bus stop excluding the dwell time τ. Thus, the average angular velocity is:

$$\left\langle \frac{d\theta_i}{dt} \right\rangle = \frac{\omega_i}{1 + 2k\Delta t/\eta_i} \tag{16.28}$$

$$= \omega_i \left(1 - \frac{1}{1 + \eta_i/2k\Delta t}\right). \tag{16.29}$$

We have in fact derived the result given by Eq. (2) of [1]. (Recall that in [1], people alight/board via two different doors simultaneously whereas here we have people first alighting and then boarding, sequentially, via one door. So here we have a factor of 2 in the expression for τ.) In this expression for the average angular velocity $\langle d\theta_i/dt \rangle$, the second term on the right-hand side $-\omega_i/(1+\eta_i/2k\Delta t)$ is the coupling amongst the buses due to the bus stop, which is directly analogous to the sine coupling of the Kuramoto oscillators.

The key point about Eq. (16.29) is that the average angular velocity to traverse a bus stop is a function which *monotonically decreases* with the time headway Δt for any $\Delta t > 0$, where the time headway is essentially the phase difference $\Delta\theta$ with respect to the bus immediately ahead of it [Fig. 16.4(b)]. In comparison with the Kuramoto oscillators, here the expression for $\langle d\theta_i/dt \rangle$ has no local maximum within $\Delta\theta \in (0, \pi)$ (the maximum at $\Delta\theta = 0$ does not count). Any perturbed lag on a trailing bus (i.e., a slight increase in Δt or $\Delta\theta$) would necessarily cause it to move slower, further lagging behind (i.e., opening up larger Δt or $\Delta\theta$), and move even slower ... until they bunch. This explains why the bus system cannot have a stable staggered configuration, leading to the perennial and notorious phenomenon of bus bunching which plagues bus systems worldwide.

The no-boarding strategy modifies the monotonically decreasing $\langle d\theta_i/dt \rangle$ with respect to Δt (or $\Delta\theta$) in Eq. (16.29) by speeding up a bus when it is too slow [Fig. 16.4(c)]. For instance, if its phase difference with respect to the bus immediately ahead of it gets too large, then it activates the no-boarding strategy to speed it up — in other words if $\Delta\theta$ grows then the coupling *increases* $\langle d\theta_i/dt \rangle$.

Note however, that the stability results from [48] only applies if there is such a completely phase-locked solution in the first place.

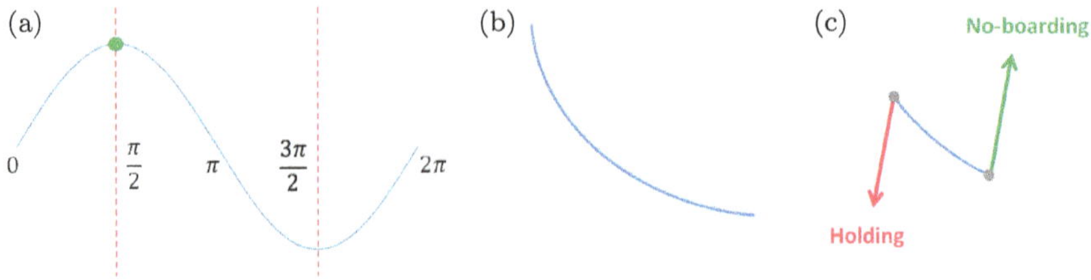

Fig. 16.4. (a) The local unidirectional Kuramoto oscillators experience a sine coupling with a local maximum at $\Delta\theta = \pi/2$. (b) The average angular velocity for the buses is monotonically decreasing with $\Delta\theta$. (c) The no-boarding strategy creates stability of the anti-bunched configuration by modifying this curve such that a slow bus gets a stronger kick to speed it up. The holding strategy requires a fast bus to slow down, preventing imminent bunching. Overall, the curve becomes sine-like.

Our calculations in Sec. 16.6 established that there is a critical k_c given by Eq. (16.27) such that the no-boarding strategy only works to create such an anti-bunched or staggered configuration if $k > k_c$. If k is too small or the frequency detuning is too great such that $k < k_c$, then there is no such anti-bunched solution, and according to Eq. (16.1), this is the lull period where the buses are completely unsynchronized and periodically lap one another. This lull period $k < k_c$ is thus a phase where it appears to defy the possibility for attaining the anti-bunched configuration.

Intriguingly, an approach where buses are allowed to learn on their own by reinforcement learning and without any human instruction turns out to discover novel solutions comprising the *no-boarding strategy* as well as a *holding strategy* which allows for stable anti-bunched configurations even in the lull period where $k < k_c$ [50]! The holding strategy is implemented by a bus when it is too fast. The way it works in the lull period where the buses have too much frequency detuning is that a fast bus with higher natural frequency slows itself down to reduce the effective frequency detuning of the buses — which, by Eq. (16.27) would reduce k_c, and therefore *allows for the existence of anti-bunched solutions*. The bus system is then able to maintain a staggered configuration. Also, holding makes a bus reduce its speed, when its phase difference from the bus ahead of it is too small. Thus overall, the holding and no-boarding strategies make the shape of the curve given by Eq. (16.29) to be sine-like instead of monotonically decreasing [Fig. 16.4(c)].

In conclusion, the no-boarding together with the holding strategies, when adaptively and appropriately applied in various situations, are able to allow for the buses to overcome the problem of bus bunching. The correspondence with the local unidirectional Kuramoto oscillators and their stability properties [48] has led us to understand why a typical bus system always has the bus bunching problem, and let us draw insightful physical understanding on how the bus system can achieve a stable anti-bunched solution. In particular, it provides an explanation to why the simulated buses arrive at the no-boarding and holding strategies by self-reinforcement learning [50], since these two strategies create stable anti-bunched configurations by modifying the shape of the effective average angular velocity to be sine-like — like that of the local unidirectionally coupled Kuramoto oscillators.

Supplementary Material

Videos on the $N = 5$ local unidirectionally coupled Kuramoto oscillators with stable anti-bunched configurations are found here: https://www.youtube.com/playlist?list=PLZIj25fISvwOALT0PADz_LiH0YbgKMkBz

Acknowledgments

This work was supported by the Joint WASP/NTU Programme (Project No. M4082189) and the DSAIR@NTU Grant (Project No. M4082418).

References

[1] V.-L. Saw, N. N. Chung, W. L. Quek, Y. E. I. Pang and L. Y. Chew, Bus bunching as a synchronization phenomenon, *Sci. Rep.* **9** (2019) 6887, doi: 10.1038/s41598-019-43310-7. URL https://www.nature.com/articles/s41598-019-43310-7.

[2] G. Newell and R. Potts, Maintaining a bus schedule, *2nd Australian Road Res. Board* **2** (1964) 388.

[3] R. A. Chapman and J. F. Michel, Modelling the tendency of buses to form pairs, *Transportation Science.* **12**(2) (1978) 165–175; doi: 10.1287/trsc.12.2.165.
[4] W. B. Powell and Y. Sheffi, A probabilistic model of bus route performance, *Transport. Sci.* **17**(4) (1983) 376–404; doi: 10.1287/trsc.17.4.376; URL https://doi.org/10.1287/trsc.17.4.376.
[5] C. Gershenson and L. A. Pineda, Why does public transport not arrive on time? the pervasiveness of equal headway instability, *PLoS One* **4**(10) (2009) 1–15; doi: 10.1371/journal.pone.0007292; URL https://doi.org/10.1371/journal.pone.0007292.
[6] G. Bellei and K. Gkoumas, Transit vehicles' headway distribution and service irregularity, *Public Transport* **2**(4), (2010) 269–289; doi: 10.1007/s12469-010-0024-7. URL https://doi.org/10.1007/s12469-010-0024-7.
[7] P. Ball, There's chaos on the bus, *Nature* (2003). URL https://doi.org/10.1038/news030929-2.
[8] T. Nagatani, Fluctuation of riding passengers induced by chaotic motions of shuttle buses, *Phys. Rev. E* **68** (2003) 036107; doi: 10.1103/PhysRevE.68.036107; URL https://link.aps.org/doi/10.1103/PhysRevE.68.036107.
[9] M. Abkowitz and I. Engelstein, Methods for maintaining transit service regularity, *Transport. Res. Record: J. Transport. Res. Board* **961** (1984) 1–8.
[10] M. D. Rossetti and T. Turitto, Comparing static and dynamic threshold based control strategies, *Transport. Res. Part A: Policy Practice* **32**(8) (1998) 607–620; doi: https://doi.org/10.1016/S0965-8564(98)00019-6.
[11] X. J. Eberlein, N. H. M. Wilson and D. Bernstein, The holding problem with real–time information available, *Transport. Sci.* **35**(1) (2001) 1–18; doi: 10.1287/trsc.35.1.1.10143.
[12] M. D. Hickman, An analytic stochastic model for the transit vehicle holding problem, *Transport. Sci.* **35**(3) (2001) 215–237; URL http://www.jstor.org/stable/25768957.
[13] L. Fu and X. Yang, Design and implementation of bus–holding control strategies with real-time information, *Transport. Res. Record* **1791**(1) (2002) 6–12; doi: 10.3141/1791-02.
[14] Y. Bin, Y. Zhongzhen and Y. Baozhen, Bus arrival time prediction using support vector machines, *J. Intelligent Transport. Syst.* **10**(4) (2006) 151–158; doi: 10.1080/15472450600981009. URL https://doi.org/10.1080/15472450600981009.
[15] N. Mukai, T. Watanabe and J. Feng, *Route Optimisation Using Q-learning for On-Demand Bus Systems* (Springer, Berlin, 2008).

[16] C. F. Daganzo, A headway-based approach to eliminate bus bunching: Systematic analysis and comparisons, *Transport. Res. Part B: Methodological* **43**(10) (2009) 913–921; URL http://www.sciencedirect.com/science/article/pii/S0191261509000484.

[17] C. E. Cortés, D. Sáez, F. Milla, A. Núñez and M. Riquelme, Hybrid predictive control for real-time optimization of public transport systems' operations based on evolutionary multi-objective optimization, *Transport. Res. Part C: Emerging Technol.* **18**(5) (2010) 757–769; doi: https://doi.org/10.1016/j.trc.2009.05.016. Applications of Advanced Technologies in Transportation: Selected papers from the 10th AATT Conference.

[18] O. Cats, A. N. Larijani, H. N. Koutsopoulos and W. Burghout, Impacts of holding control strategies on transit performance: Bus simulation model analysis, *Transport. Res. Record* **2216**(1) (2011) 51–58; doi: 10.3141/2216-06. URL https://doi.org/10.3141/2216-06.

[19] C. Gershenson, Self-organization leads to supraoptimal performance in public transportation systems, *PLoS One* (2011) **6**(6) 1–6; doi: 10.1371/journal.pone.0021469. URL https://doi.org/10.1371/journal.pone.0021469.

[20] J. J. Bartholdi and D. D. Eisenstein, A self-coordinating bus route to resist bus bunching, *Transport. Res. Part B: Methodological* **46**(4) (2012) 481–491; doi: 10.1016/j.trb.2011.11.001. URL http://www.sciencedirect.com/science/article/pii/S0191261511001676.

[21] C. Chen, W. Chen and Z. Chen, A multi-agent reinforcement learning approach for bus holding control strategies, *Adv. Transport. Studies Special Issue* **2** (2015) 41.

[22] O. Ibarra-Rojas, F. Delgado, R. Giesen and J. Munoz, Planning, operation, and control of bus transport systems: A literature review, *Transport. Res. Part B: Methodological* **77**(38) (2015).

[23] W. Chen, K. Zhou and C. Chen, Real-time bus holding control on a transit corridor based on multi-agent reinforcement learning, in *IEEE 19th Int. Conf. Intelligent Transportation Systems (ITSC)* (2016), p. 100.

[24] L. Moreira-Matias, O. Cats, J. Gama, J. Mendes-Moreira and J. F. de Sousa, An online learning approach to eliminate bus bunching in real-time, *Appl. Soft Comput.* **47** (2016) 460–482; URL http://www.sciencedirect.com/science/article/pii/S1568494616303118.

[25] P. Wang, X. Chen, W. Chen, L. Cheng and D. Lei, Provision of bus real-time information: Turning passengers from being contributors of headway irregularity to controllers, *Transport. Res. Record* (2018); doi: 10.1177/0361198118798722.

[26] F. Alesiani and K. Gkiotsalitis, Reinforcement learning-based bus holding for high-frequency services, in *21st Int. Conf. on Intelligent Transportation Systems (ITSC)* (2018).
[27] K. Menda, Y. Chen, J. Grana, J. W. Bono, B. D. Tracey, M. J. Kochenderfer and D. Wolpert, Deep reinforcement learning for event-driven multi-agent decision processes, *IEEE Trans. Intelligent Transport. Syst.* **20**(4) (2019) 1259–1268.
[28] W. L. Quek, N. N. Chung, V.-L. Saw and L. Y. Chew, Analysis and simulation of intervention strategies against bus bunching by means of an empirical agent-based model, preprint (2020); https://arxiv.org/abs/2004.13022.
[29] Y. Li, J.-M. Rousseau and M. Gendreau, Real-time scheduling on a transit bus route: A 0–1 stochastic programming model, *Proc. 33rd Annual Meeting, Transportation Research Forum* (1991), pp. 157–166.
[30] X. J. Eberlein, Real-time control strategies in transit operations: models and analysis, Ph.D. dissertation, Department of Civil and Environmental Engineering, Massachusetts Institute of Technology (1995).
[31] L. Fu, Q. Liu and P. Calamai, Real-time optimization model for dynamic scheduling of transit operations, *Transport. Res. Record* **1857**(1) (2003) 48–55; doi: 10.3141/1857-06.
[32] A. Sun and M. Hickman, The real-time stop-skipping problem, *J. Intelligent Transport. Syst.* **9**(2) (2005) 91–109; doi: 10.1080/15472450590934642.
[33] Z. Liu, Y. Yan, X. Qu and Y. Zhang, Bus stop-skipping scheme with random travel time, *Transport. Res. Part C: Emerging Technologies* **35** (2013) 46–56; doi: https://doi.org/10.1016/j.trc.2013.06.004. URL http://www.sciencedirect.com/science/article/pii/S0968090X13001265.
[34] P. G. Furth, Alternating deadheading in bus route operations, *Transport. Sci.* **19**(1) (1985) 13–28; doi: 10.1287/trsc.19.1.13.
[35] P. Furth and F. Day, Transit routing and scheduling strategies for heavy-demand corridors, *Transport. Res. Record* **1011** (1985) 23–26.
[36] X. J. Eberlein, N. H. Wilson, C. Barnhart and D. Bernstein, The real-time deadheading problem in transit operations control, *Transport. Res. Part B: Methodological* **32**(2) (1998) 77–100; URL http://www.sciencedirect.com/science/article/pii/S0191261597000131.
[37] F. Delgado, J. C. Muñoz, R. Giesen and A. Cipriano, Real-time control of buses in a transit corridor based on vehicle holding and boarding limits, *Transport. Res. Record* **2090**(1) (2009) 59–67; doi: 10.3141/2090-07. URL https://doi.org/10.3141/2090-07.

[38] F. Delgado, J. C. Munoz and R. Giesen, How much can holding and/or limiting boarding improve transit performance? *Transport. Res. Part B: Methodological* **46**(9) (2012) 1202–1217; URL http://www.sciencedirect.com/science/article/pii/S0191261512000653.

[39] S. Z. Zhao, C. X. Lu, S. D. Liang and H. S. Liu, A self-adjusting method to resist bus bunching based on boarding limits, *Math. Problems Eng.* **2016** (2016) 8950209. URL https://doi.org/10.1155/2016/8950209.

[40] W. Sun and J.-D. Schmöcker, Considering passenger choices and overtaking in the bus bunching problem, *Transportmetrica B: Transport Dynam.* **6**(2) (2018) 151–168; doi: 10.1080/21680566.2017.1387876. URL https://doi.org/10.1080/21680566.2017.1387876.

[41] V.-L. Saw and L. Y. Chew, No-boarding buses: Synchronisation for efficiency, *PLoS One* **15**(3) (2020) e0230377. URL https://doi.org/10.1371/journal.pone.0230377.

[42] V.-L. Saw and L. Y. Chew, No-boarding buses: Agents allowed to cooperate or defect, *J. Phys. Complexity* **1**(1) (2020) 015005; doi: 10.1088/2632-072x/ab4af5. URL https://doi.org/10.1088%2F2632-072x%2Fab4af5.

[43] S. Jara-Díaz and A. Tirachini, Urban bus transport, *J. Transport Econom. and Policy* **47** (2013) 91.

[44] C. Stewart and A. El-Geneidy, All aboard at all doors: Route selection and running-time savings estimation for multiscenario all-door bus boarding, *Transport. Res. Record* **2418**(1) (2014) 39–48; doi: 10.3141/2418-05. URL https://doi.org/10.3141/2418-05.

[45] A. El-Geneidy, D. van Lierop, E. Grisé, G. Boisjoly, D. Swallow, L. Fordham, and T. Herrmann, Get on board: Assessing an all-door boarding pilot project in Montreal, Canada, *Transport. Res. Part A: Policy Practice* **99** (2017) 114–124; URL http://www.sciencedirect.com/science/article/pii/S0965856416306139.

[46] A. Pikovsky, M. Rosenblum and J. Kurths, *Synchronization: A Universal Concept in Nonlinear Sciences* (Cambridge Nonlinear Science Series Cambridge University Press, Cambridge, 2003). URL http://www.cambridge.org/sg/academic/subjects/physics/nonlinear-science-and-fluid-dynamics/synchronization-universal-concept-nonlinear-sciences?format=PB&isbn=9780521533522.

[47] Y. Kuramoto, *Chemical Oscillations, Waves, and Turbulence* (Springer, New York, 1984); doi: 10.1007/978-3-642-69689-3. URL https://www.springer.com/gp/book/9783642696916.

[48] J. A. Rogge and D. Aeyels, Stability of phase locking in a ring of unidirectionally coupled oscillators, *J. Phys. A: Math. Gen.* **37**(46) (2004) 11135–11148; doi: 10.1088/0305-4470/37/46/004. URL https://doi.org/10.1088%2F0305-4470%2F37%2F46%2F004.

[49] J. Ochab and P. F. Góra, Synchronization of coupled oscillators in a local one-dimensional kuramoto model, *Acta Physica Polon. B Proc. Suppl.* **3**(2) (2010) 453.

[50] V.-L. Saw, L. Vismara and L. Y. Chew, Intelligent buses in a loop service: Emergence of no-boarding and holding strategies, *Complexity* Special Issue on: New Models, New Technologies, New Data and Applications of Urban Complexity from Spatio-temporal Perspectives. **2020** (2020), Article ID 7274254. URL https://doi.org/10.1155/2020/7274254.

https://doi.org/10.1142/9789811221903_0017

Chapter 17

Feedback Delay as a Control Tool: The Complex Ginzburg–Landau Equation with Local and Nonlocal Delayed Perturbations

Alfonso Carlos Casal[*,‡] and Jesús Ildefonso Díaz[†,§]

[*]*Departamento de Matemática Aplicada, E.T.S. de Arquitectura, Universidad Politécnica de Madrid, 28040 Madrid, Spain*

[†]*Instituto de Matemática Interdisciplinar, Universidad Complutense de Madrid, 28040 Madrid, Spain*

[‡]*alfonso.casal@upm.es*

[§]*jidiaz@ucm.es*

This survey collects several works by the authors dealing with the role of suitable feedback delayed terms as a suitable control, to different purposes, concerning the behavior of solutions to the complex Ginzburg–Landau

$$\frac{\partial u}{\partial t} - (1 + i\epsilon)\,\Delta u + (1 + i\beta)\,|u|^2\,u - (1 - i\omega)\,u = F\,(u(x, t - \tau))$$

for $t > 0$, with

$$F\,(u(x, t - \tau)) = e^{i\chi_0}\left\{\frac{\mu}{|\Omega|}\int_\Omega u\,(x, t - \tau)\,dx + \nu u\,(x, t - \tau)\right\},$$

where μ, $\nu \geq 0$, $\tau > 0$ but the rest of real parameters ϵ, β, ω and χ_0 do not have a prescribed sign. Besides the existence and uniqueness of weak solutions of the problem, we analyze the role of delayed feedback perturbations to control the chemical turbulence. The Hopf bifurcation is also considered in presence of the delayed terms.

17.1. Introduction

This survey collects several works by the authors dealing with the role of suitable feedback delayed terms as a suitable control, to different purposes, concerning the behavior of solutions to the complex Ginzburg–Landau equation (CGLE)

$$\frac{\partial \mathbf{u}}{\partial t} - (1 + i\epsilon)\,\Delta\mathbf{u} + (1 + i\beta)\,|\mathbf{u}|^2\,\mathbf{u} - (1 - i\omega)\,\mathbf{u} = \mathbf{F}\,(\mathbf{u}(x, t - \tau))\,.$$

Some introductory words on the complex Ginzburg–Landau equations: Before dealing with the role of delayed terms, it is important to point out that the complex Ginzburg–Landau is a very relevant generalization of the nonlinear Schrödinger equations used for many modeling purposes after the pioneering work by Ginzburg and Landau [40] in 1950 in superconductivity. Of course, the *Ginzburg–Landau equation* has been systematically used to study different types of phenomena in superconductor theory. Moreover, a rich variety of mathematical models of PDEs have also been inspired by the original model of Ginzburg and Landau to study a large number of physical phenomena (see, for instance, [48, 50, 65] and references therein). Notice that the presence of complex coefficients introduces important differences with the classical Ginzburg–Landau equations arising in superconductivity [24]. Let us see with some detail how this equation arises in the study of some apparently different problems as some reaction–diffusion systems. The evolution of a chemical system consisting of n species which are reacting with each other and allowed to diffuse in a spatially extended medium, is generally described by an n-component reaction–diffusion equation for the n-concentrations $\mathbf{c}(x, t)$

$$\frac{\partial \mathbf{c}}{\partial t} = \mathbf{F}(\mathbf{c}; p) + \mathbf{D}\Delta\mathbf{c}, \tag{17.1}$$

where $\mathbf{F}$ denotes the typically nonlinear reaction term representing chemical kinetics, $\mathbf{D}\Delta\mathbf{c}$ the diffusion term (being $\mathbf{D}$ the diffusion matrix) and p a scalar control parameter. Let us assume that this system has a homogeneous, stationary solution $\mathbf{c_s}$ which undergoes a *Hopf bifurcation* at $p = p_0$, i.e., for $p \in (p_0, p_0 + \varepsilon)$ the stationary solution $\mathbf{c_s}$ becomes a time periodic solution, at least for $\varepsilon > 0$ small enough. It has been shown by Kuramoto and others that the dynamics of any reaction–diffusion system (17.1) in the vicinity of a Hopf bifurcation is described, by means of suitable parameterizations, by a nonlinear parabolic equation with complex coefficients, the so-called *complex Ginzburg–Landau equation* (CGLE), see, e.g., [36, 48]. The relation between reaction–diffusion systems and the CGLE has been treated in many texts, here we will follow the presentation of [46]. After a convenient choice of variables $\mathbf{X} = \mathbf{c} - \mathbf{c_s}$ (the *concentration deviations*) and $\epsilon = p - p_0$, the system can be reformulated as

$$\frac{\partial \mathbf{X}}{\partial t} = \mathbf{JX} + \mathbf{f}(\mathbf{x}, \epsilon) + \mathbf{D}\Delta\mathbf{X},$$

where $\mathbf{J}$ is the Jacobian matrix for the homogeneous system evaluated at $\mathbf{X_s} = \mathbf{0}$, i.e. $\mathbf{F}(\mathbf{c}; p) - \mathbf{F}(\mathbf{c}_s; p_0) = \mathbf{JX} + \mathbf{f}(\mathbf{x}, \epsilon)$. At the *bifurcation point*, $\mathbf{J}$ has two imaginary eigenvalues $\pm i\omega_0$, being ω_0 the so-called *Hopf frequency*. The corresponding *right eigenvectors* $\mathbf{e}_1$ and $\mathbf{e}_2 = \bar{\mathbf{e}}_1$ (normalized with left eigenvectors $\mathbf{e}_i^+$ according to $\mathbf{e}_i^+ \mathbf{e}_j = \delta_{ij}$) span the *center subspace* E^c of the homogeneous solution. The *center manifold* W^c is tangent to E^c at $\mathbf{X} = \mathbf{0}$, $\epsilon = 0$. The other $n - 2$ eigenvalues are all assumed to be large and negative. This assures that a homogeneous solution converges fast towards W^c provided that $\mathbf{X}$ and ϵ are sufficiently small (for details and further references, see [46]). This allows us to express the concentration deviations $\mathbf{X}$ in terms of *amplitude coordinates* $\mathbf{Y} \in E^c$ by

$$\mathbf{X} = \mathbf{Y} + \mathbf{h}(\mathbf{Y}, \epsilon).$$

This equation describes a mapping from coordinates in the center subspace E^c onto the center manifold W^c. The function $\mathbf{h}(\mathbf{Y}, \epsilon)$ is selected in such a way to successively eliminate as many nonlinear terms as possible from the kinetic equations starting from the lowest order [46]. Each kind of bifurcation is characterized by the specific terms which cannot be eliminated (the so-called *resonant terms*). In this way, we obtain a general equation valid for all reaction–diffusion equations undergoing a given bifurcation. In the case of the Hopf bifurcation, neglecting the diffusion term, to third order we obtain the so-called *Stuart–Landau equation*

$$\frac{d\mathbf{Y}}{dt} = (i\omega_0 + \sigma_1\epsilon)\mathbf{Y} - g|\mathbf{Y}|^2\mathbf{Y},$$

where $\mathbf{Y}$ is a complex amplitude given by $\mathbf{Y} = Y\mathbf{e}_1 + \overline{Y}\mathbf{e}_2$. The parameters σ_1 and g are complex and given by solutions of lengthy equations given in [46]. The Stuart–Landau equation represents the *normal form* of a homogeneous system close to a Hopf bifurcation. Performing a similar derivation, but including diffusion, we arrive to

$$\frac{\partial\mathbf{Y}}{\partial t} = (i\omega_0 + \sigma_1\epsilon)\mathbf{Y} - g|\mathbf{Y}|^2\mathbf{Y} + d\Delta\mathbf{Y},$$

with $d = \mathbf{e}_1^+ \cdot \mathbf{D}\mathbf{e}_1$. After rescaling of space, time, and introducing $\mathbf{A}$ for $\mathbf{Y}$, we finally arrive to the rescaled complex Ginzburg–Landau equation

$$\frac{\partial\mathbf{A}}{\partial t} = (1 - i\omega)\mathbf{A} - (1 + \mathrm{i}\alpha)|\mathbf{A}|^2\mathbf{A} + (1 + i\beta)\Delta\mathbf{A}, \qquad (17.2)$$

where $\mathbf{A}$ is the *complex oscillation amplitude*, ω the *linear frequency parameter*, α the *nonlinear frequency parameter*, and β the *linear dispersion coefficient*. All reaction–diffusion systems sufficiently close to a Hopf bifurcation are described by the complex Ginzburg–Landau equation. The specific details of the original system are incorporated in the parameter values. If one wishes to express the solution of the CGLE in the original variables, to first order the concentrations of the chemical species are expressed by

$$\mathbf{c} = \mathbf{c_s} + \sqrt{\epsilon}(Y(x,t)\mathbf{e}_1 + \overline{Y}(x,t)\mathbf{e}_2).$$

Different scalings of the CGLE are considered in the literature [6]. Here, we assume that the Hopf frequency is not scaled out, and hence contributes to ω in Eq. (17.2). We also refer the reader to [48, Appendix B] for the detailed derivation of the CGLE associated to the Brusselator model. This bifurcation is omnipresent in reaction–diffusion systems like the Belousov–Zhabotinsky reaction or the oxidation of CO on platinum. The basic solution of CGL, uniform oscillations, are unstable and lead to *space–time chaos* if $1 + \alpha\beta < 0$. Different space–time chaos control schemes have been explored over the last years, both for the CGLE [11, 12, 22], and for the CO reaction model [23], also validated in the corresponding experiment [47].

Some introductory words on the delay as control. It is well known that feedback delayed term can be introduced to control very complex phenomena (see, e.g., the expositions made in [9, 39, 67]). In recent years, many studies have investigated the mechanisms capable of controlling the behavior of dynamic systems, in particular, of reaction–diffusion systems in a regime of space–time chaos. A method widely studied for its efficiency is that of *time-delay autosynchronization* (TDAS), based on a work by Pyragas [58]. The idea of this method is to apply a signal to the system (*feedback* $\mathbf{F}$) that is proportional to the difference between the current state of the system (measured by an appropriate variable $\mathbf{A}$) at a certain time and the state of the system at a previous time $t-\tau$, where τ represents an adjustable delay: $\mathbf{F}(t,\tau) \propto \mathbf{A}(t-\tau)-\mathbf{A}(t)$. In the case of some finite-dimensional dynamical systems, it was proved that using this method it is possible to manipulate the dynamics of a system and create and stabilize a multitude of spatio-temporal patterns, not present or unstable without TDAS. We will collect here several works dealing with the application of this general philosophy to the case of the complex Ginzburg–Landau equation. Sometimes the feedback consists of a global variable or a spatial average of a local variable. This type of feedback is easy to implement in a real system and therefore has been applied frequently. In the following section, we extend the concept of feedback by admitting local components.

It is remarkable that even if the Hopf bifurcation is supercritical, and hence the limit cycle is a stable solution of the Stuart–Landau equation, the oscillations in the spatially extended system may be unstable. The resulting states of spatio-temporal chaos appear if the Benjamin–Feir–Newell criterion $1+\alpha\beta < 0$ is fulfilled, a phenomenon that is induced by the diffusive coupling and that is therefore genuine to a system with spatial degrees of freedom. Considerable efforts have been made to understand this type of chaotic behavior and to apply methods to suppress this kind of turbulence and replace it by regular dynamics. In the context of the reaction–diffusion systems, the introduction of forcing terms or global feedback terms have been shown to be efficient ways to control turbulence [47, 54]. Still, control of chaotic states in nonlinear systems is a wide field of research that we cannot review here [60]. Global feedback methods, where a spatially independent quantity (or, e.g., a spatial average of a space-dependent quantity) is coupled back to the system dynamics, have attracted much attention since in many cases the models are simpler and easier to be carried out experimentally. Nevertheless, local methods have gained interest in recent years since they allow to access to other solutions of the systems and may also be implemented, such as in the light-sensitive BZ reaction or in neurophysiological experiments [54]. Feedback methods with an explicit time delay amplify the range of possibilities of control that can be applied to the system and provide the researcher with an additional adjustable parameter. On the level of the mathematical description, the model equations become delay differential equations [9, 39]. Obviously, time-delay feedback can be applied to any solution of the dynamics, not necessarily to a chaotic one.

In 1996, Battogtokh and Mikhailov [12] introduced a nonlocal delayed term in the generalized equation in order to control the system and suppress turbulence (see also [11]). The equation appears in the study of some chemical reactions and models the concentration of various reacting species. Battogtokh and Mikhailov analyze numerically this model and control the turbulence thanks to the delayed term. The idea is to adjust two real parameters: the feedback intensity μ and the delay time τ. The results were made rigorous later in a series of articles, which we indicate below. This work is a natural companion of those rigorous studies. For instance, a first rigorous approach was presented in [30], where the control of turbulence in oscillatory reaction–diffusion systems is made through a combination of global and local feedback by means of a pseudo-linearization technique (see also [28–33]). In [38], we consider weaker assumptions on the initial data and parameters than in the above mentioned papers and others results in the literature (see, e.g., [4]).

On some collected researches: Different parts of this exposition. In Sec. 2, we will analyze the basic questions of the existence and uniqueness of solutions to the associate initial boundary value problem such as developed in the paper [38]. Let us mention that different types of boundary condition can be imposed to the solutions of the complex nonlinear PDE. The mathematical treatment is different in each case but there are many common points in the diverse approaches. In Sec. 2 of this survey we will consider the case of a global delayed problem in which two real parameters play a fundamental role: the feedback intensity, μ, and the delay time, τ. The problem is reduced to find a complex-valued field $\mathbf{u}$ in $Q := \Omega \times (0, T)$, where $\Omega \subset \mathbb{R}^N$ is a bounded domain for $N \leq 3$ with regular boundary $\partial\Omega$ and $t > 0$.

$$\begin{cases} \dfrac{\partial \mathbf{u}}{\partial t} - (1 + i\epsilon)\,\Delta \mathbf{c} + (1 + i\beta)\,|\mathbf{u}|^2\,\mathbf{u} - (1 - i\omega) & \\ \quad \mathbf{u} = \mathbf{F}\,(\mathbf{u}(x, t - \tau)) & \text{in } Q, \\ \dfrac{\partial \mathbf{u}}{\partial \mathbf{n}} = \mathbf{0} & \text{on } \partial\Omega \times (0, T), \\ \mathbf{u}\,(x, 0) = \mathbf{u}_0\,(x) & \text{on } \Omega, \\ \mathbf{F}\,(u(s)) = \mathbf{F}_0(s) & s \in (-\tau, 0), \end{cases} \tag{17.3}$$

where the global delayed feedback term is given by

$$\begin{aligned} \mathbf{F}\,(\mathbf{u}(x, t - \tau)) &= F_1\,(\mathbf{u}(x, t - \tau)) + iF_2\,(\mathbf{u}(x, t - \tau)) \\ &:= \mu e^{i\chi_0} \left\{ \frac{1}{|\Omega|} \int_\Omega \mathbf{u}\,(x, t - \tau)\,dx \right\}, \end{aligned} \tag{17.4}$$

here ω, β, ϵ, τ, μ and χ_0 are given real numbers without prescribed sign, $\mathbf{u}_0(x)$ and $\mathbf{F}_0(s)$ are given complex functions and $\mathbf{n}$ is the outward normal vector to $\partial\Omega$. We point out that, in contrast with most of the delayed problems, here the *initial past history* is composed of a pointwise information at $t = 0$ (the usual initial condition $\mathbf{u}\,(x, 0) = \mathbf{u}_0\,(x)$ on Ω) and only a partial information on the function $\mathbf{u}(s)$ when $s \in (-\tau, 0)$: only the integral of the unknown is prescribed for $s \in (-\tau, 0)$. Under suitable conditions on $\mathbf{u}_0\,(x)$ and $\mathbf{F}_0(s)$, we prove (in Theorems 1 and 2) that there exists a unique solution of (17.3). A second model concerns the case, already used in [11, 12, 30], in which the delayed feedback term involves the own unknown

$$\mathbf{F}\left(\mathbf{u}(x,t-\tau)\right)=e^{i\chi_0}\left\{\frac{\mu}{|\Omega|}\int_{\Omega}\mathbf{u}\left(x,t-\tau\right)dx+\nu\mathbf{u}\left(x,t-\tau\right)\right\}. \tag{17.5}$$

In this case, it is clear that the required *initial past history* must be more complete and so the new formulation is the usual one for delayed problems. As a matter of fact, as we will mention later, the nonlinear perturbation can be easily treated under a more general growth condition of the type $(1+i\beta)\left|\mathbf{u}\right|^{m-1}\mathbf{u}$, for any $m > 0$. In particular, when $m \in (0,1)$ we comment how to apply the techniques introduced in a series of works concerning the pure Schrödinger equation with a non-Lipschitz perturbation to our case (see [13, 14, 16, 17]). See also the study made in [4], for complex Ginzburg–Landau equations without any delayed term. Thus our second problem can be formulated as

$$\begin{cases}\dfrac{\partial\mathbf{u}}{\partial t}-(1+i\epsilon)\,\Delta\mathbf{u}+(1+i\beta)\left|\mathbf{u}\right|^{m-1} & \\ \quad\mathbf{u}-(1-i\omega)\,\mathbf{u}=\mathbf{F}\left(\mathbf{u}(x,t-\tau)\right) & \text{in } Q,\\ \dfrac{\partial\mathbf{u}}{\partial\mathbf{n}}=\mathbf{0} & \text{on } \partial\Omega\times(0,T),\\ \mathbf{u}\left(x,s\right)=\mathbf{u}_0\left(x,s\right), & s\in[-\tau,0],\ x\in\Omega.\end{cases} \tag{17.6}$$

In the special case of $m \in (0,1)$ and $\mathbf{F}$ given by (17.5) with $\mu = 0$ (i.e., with only local delayed feedback terms), we prove that several qualitative properties as the finite speed of propagation or the finite extinction time property obtained previously in the literature for complex formulations problems without delayed term (see [4, 13, 14, 16, 17]) can be easily extended to the mentioned delayed formulation.

The sections are organized as follows: the existence and uniqueness of solutions for problems (17.3) and (17.6) is obtained in Sec. 2. The proof of the existence of solutions use an iterative argument as well as a Galerkin method when $t \in [0,\tau)$ jointly with suitable a priori estimates which allow to justify the passing to the limit. The uniqueness of solutions is given for $N \leq 3$. Finally, the study of some qualitative properties, for $m \in (0,1)$ and $\mathbf{F}$ given by (17.5) with $\mu = 0$, will be collected, where some energy methods will be applied.

In Sec. 3, we will study the stabilization of the uniform oscillations for the *complex Ginzburg–Landau equation* by means of some global delayed feedback according to the papers [28, 30, 34]. Our main interest in this section concerns the control of chemical turbulence in oscillatory reaction–diffusion systems made through a combination of global and local delayed feedback. Now, we will assume the domain given by $\Omega = (0, L_1) \times (0, L_2)$. We define the faces of the boundary.

$$\Gamma_j = \partial\Omega \cap \{x_j = 0\}, \Gamma_{j+2} = \partial\Omega \cap \{x_j = L_j\}, \quad j = 1, 2,$$

on which we assume periodic boundary conditions and, so, the problem under study can be formulated as

$$(P_1) \begin{cases} \dfrac{\partial \mathbf{u}}{\partial t} - (1 + i\epsilon)\Delta \mathbf{u} = (1 - i\omega) \\ \quad \mathbf{u} - (1 + i\beta)\left|\mathbf{u}\right|^2 \mathbf{u} + \mu e^{i\chi_0} \mathbf{F}(\mathbf{u}, t, \tau) & \Omega \times (0, +\infty), \\ \mathbf{u}|_{\Gamma_j} = \mathbf{u}|_{\Gamma_{j+2}}, \left(- \left.\dfrac{\partial \mathbf{u}}{\partial \mathbf{n}}\right|_{\Gamma_j} = \right) \left.\dfrac{\partial \mathbf{u}}{\partial x_j}\right|_{\Gamma_j} \\ \quad = \left.\dfrac{\partial \mathbf{u}}{\partial x_j}\right|_{\Gamma_{j+2}} \left(= \left.\frac{\partial \mathbf{u}}{\partial \mathbf{n}}\right|_{\Gamma_{j+2}} \right), & \partial\Omega \times (0, +\infty), \\ \mathbf{u}(x,s) = \mathbf{u}_0(x, s) & \Omega \times [-\tau, 0], \end{cases} \tag{17.7}$$

where $\mathbf{n}$ is the outpointing normal unit vector, and

$$\mathbf{F}(\mathbf{u}, t, \tau) = [m_1 \mathbf{u}(t) + m_2 \overline{\mathbf{u}}(t) + m_3 \mathbf{u}(t - \tau, x) + m_4 \overline{\mathbf{u}}(t - \tau)]$$

with

$$\overline{\mathbf{u}}(s) = \frac{1}{|\Omega|} \int_\Omega \mathbf{u}(s, x) dx.$$

Here the parameters $\epsilon, \beta, \omega, \mu, \chi_0, m_i$ and τ are real numbers, in contrast with the solution $\mathbf{u}(x,t) = u_1(x,t) + iu_2(x,t)$. We point out that most of our results remain true for N-dimensional domains (with $N > 2$) as well as for Neumann boundary conditions (a previous study dealing with the one-dimensional case was carried out in [28]).

With the basis of a sound experimental work, many recent studies of a more descriptive nature, but of a great originality and interest have been written. In those studies the delay term $\mathbf{F}(\mathbf{u},t,\tau)$ has been taken corresponding to $m_4=1, m_i=0$ for $i=1,2,3$ and introduced as a control mechanism (see, e.g., [11, 12, 53, 61]). Our main goal is to carry out a rigorous analysis of those studies We also want to investigate the possibility of controlling the turbulence by using other terms (see Remark 4). In particular our treatment does not use the Fourier transform, apparently hard to be rigorously justified in this setting.

We focus our attention on the so-called *slowly varying complex amplitudes* defined by $\mathbf{u}(x,t)=\mathbf{v}(x,t)e^{-i\omega t}$. Thus, $\mathbf{v}$ satisfy (P_2) :

$$
(P_2)\begin{cases}
\dfrac{\partial \mathbf{v}}{\partial t}-(1+i\epsilon)\Delta\mathbf{v}=\mathbf{v}-(1+i\beta)|\mathbf{v}|^{\mathbf{2}}\,\mathbf{v} & \\
\quad +\mu e^{i\chi_0}[m_1\mathbf{v}+m_2\overline{\mathbf{v}}+e^{i\omega\tau}(m_3\mathbf{v}(t-\tau,x) & \\
\quad +m_4\overline{\mathbf{v}}(t-\tau))] & \text{in } \Omega\times(0,+\infty),\\
\mathbf{v}|_{\Gamma_j}=\mathbf{v}|_{\Gamma_{j+2}},\left(-\frac{\partial\mathbf{v}}{\partial\mathbf{n}}\Big|_{\Gamma_j}=\right)\dfrac{\partial\mathbf{v}}{\partial x_j}\Big|_{\Gamma_j} & \\
\quad =\dfrac{\partial\mathbf{v}}{\partial x_j}\Big|_{\Gamma_{j+2}}\left(=\dfrac{\partial\mathbf{v}}{\partial\mathbf{n}}\Big|_{\Gamma_{j+2}}\right) & \text{on } \partial\Omega\times(0,+\infty),\\
\mathbf{v}(x,s)=\mathbf{u}_0(x,s)e^{i\omega s} & \text{on } \Omega\times[-\tau,0].
\end{cases}
\tag{17.8}
$$

We study in this section the stability of *uniform oscillations*, i.e., special solutions of (P_2) of the form $\mathbf{v}_{uosc}(x,t)=\rho_0 e^{-i\theta t}$ which determines completely ρ_0 and θ. As we shall see, the only effect of the delay τ is that it controls the effective phase shift $\chi(\tau)$.

In absence of delay ($\tau=0$), and for $|\Omega|=+\infty$ and $\mu=0$, it is known (see [48, 53]) that the *Benjamin–Feir condition* $\beta<-\frac{1}{\epsilon}$ implies the instability of such uniform oscillations. Here we shall assume merely that

$$\beta\leq 0 \quad \text{and} \quad \varepsilon\geq 0 \tag{17.9}$$

and we shall prove that this instability holds, in absence of delay, for $L<+\infty$ once $\chi_0\in(\frac{\pi}{2},\frac{3\pi}{2})$ and $\mu>\frac{1}{|\cos\chi_0|}$. Moreover, we shall also prove that when $\tau>0$ is suitably chosen then the uniform oscillation becomes linearly stable. We point out that the above stabilization phenomenon requires a nonzero complex component perturbation (notice that χ_0 can not be zero) and that it applies to the case of $\mu > 0$ and $\epsilon = \beta = \omega = 0$. In this approach, we will use the *pseudo-linearization principle* introduced by the authors in [29].

Finally, in Sec. 4 we will analyze several bifurcation effects produced by the delay time in the behavior of solutions of the complex Ginzburg–Landau equation with this type of feedback as shown in [33]. We prove a Hopf bifurcation result for the equation without diffusion (the Stuart–Landau equation) when the amplitude of the delayed term is suitably chosen. This simplified formulation has the advantage that closed analytical solutions are possible and the necessary eigenvalue computations can be carried out in full. The diffusion case is considered firstly in the case of the whole space and later on a bounded domain with periodicity conditions. In the case in which the space is the whole $\mathbb{R}$ (we consider here the one-dimensional case) we performed a linear stability analysis of uniform oscillations with respect to spatio-temporal perturbations following the treatment made in [63]: we express the complex oscillation amplitude $\mathbf{A}$ as the superposition of a homogeneous mode $\mathbf{H}$ (corresponding to uniform oscillations) with spatially inhomogeneous perturbations,

$$\mathbf{A}(x,t) = \mathbf{H}(t) + \mathbf{A}_+(t)e^{i\kappa x} + \mathbf{A}_-(t)e^{-i\kappa x}.$$

With the help of computational arguments, we get several bifurcation diagrams where besides the delay time it is possible to use the feedback magnitude term. Among many other detailed informations, we obtain numerical evidence of the fulfillment of the delicate transversality condition. The section ends by analyzing the case in which the bifurcation takes place starting from a uniform oscillation and originating a path over a torus. This time the study is carried out in two spatial dimensions over a rectangle in which we impose periodic boundary conditions. We show the applicability of an abstract result (see [67]) to our formulation thanks to a suitable choice of the involved functional spaces. In this way, the spatial perturbations can be considered in their greatest generality.

17.2. Existence and Uniqueness of Solutions: Problems (17.3) and (17.6)

In this section, we will use the following notations: $W^{s,p}(D)$ and $H^s(D)$ denote the standard Sobolev spaces which consist of real scalar-(or vector-) valued functions defined on D (an open subset of $\mathbb{R}^N$ or $\mathbb{R}^{N+1}$). Sobolev spaces of complex-valued functions are denoted by $\mathcal{W}^{s,p}(D)$ and $\mathcal{H}^s(D)$ with calligraphic letters, as well, as continuous functions $\mathcal{C}(D)$ defined over a domain D. We use $\|\cdot\|$ and $(\cdot,\cdot)$ for the usual norm and the inner product of $L^2(D)$ or $\mathcal{L}^2(D)$, respectively. Given a general Banach space B, $\|\cdot\|_B$ denotes the norm of the Banach space B. Its topological dual space will be denoted by B'. By $\langle\cdot,\cdot\rangle_{B',B}$, we denote the duality product between B' and B.

We first introduce the notion of weak solution of problem (17.3).

Definition 1. Let $T \leq \infty$, and assume $\mathbf{u}_0 \in \mathcal{L}^4(\Omega) \cap \mathcal{H}^1(\Omega)$ and $\mathbf{F}_0 \in \mathcal{L}^2(-\tau, 0)$. A function $\mathbf{u} : \Omega \times (-\tau, T) \to \mathbb{C}$ is called a *weak solution* of problem (17.3) if

$$\begin{gathered}\mathbf{u} \in C(0,T : \mathcal{L}^2(\Omega)) \cap \mathcal{L}^2(0,T : \mathcal{H}^1(\Omega)) \\ \cap \mathcal{L}^4(0,T : \mathcal{L}^4(\Omega)) \cap \mathcal{L}^2\left(-\tau, 0 : \mathcal{L}^1(\Omega)\right), \\ \mathbf{u}_t \in \mathcal{L}^2(0,T : (\mathcal{H}^1(\Omega))'),\end{gathered}$$

for every $t \in (0,T)$

$$\begin{aligned}\left\langle \frac{\partial}{\partial t}\mathbf{u}, \varphi \right\rangle_{(\mathcal{H}^1(\Omega))' \times \mathcal{H}^1(\Omega)} &= (1 - i\omega)\int_\Omega \mathbf{u}\bar{\varphi}dx - (1 + i\beta)\int_\Omega |\mathbf{u}|^2 \mathbf{u}\bar{\varphi}dx \\ &\quad - (1 + i\epsilon)\int_\Omega \nabla\mathbf{u} \cdot \nabla\bar{\varphi}dx + \mathbf{F}\left(\mathbf{u}(t - \tau)\right) \\ &\quad \times \int_\Omega \bar{\varphi}dx, \ \forall\varphi \in \mathcal{H}^1(\Omega), \\ \mathbf{u}(x, 0) &= \mathbf{u}_0(x) \ \text{ in } \mathcal{L}^2(\Omega) \qquad (17.10)\end{aligned}$$

and

$$\mathbf{F}(\mathbf{u}(\cdot)) = \mathbf{F}_0(\cdot) \text{ in } \mathcal{L}^2(-\tau, 0),$$

where $\mathbf{F}(\mathbf{u}(t-\tau))$ is given by (17.4).

In the case of problem (17.6), a stronger notion of weak solution must be introduced.

Definition 2. Let $T \leq \infty$, and assume $\mathbf{u}_0 \in \mathcal{C}([-\tau, 0] : \mathcal{L}^2(\Omega))$, $\mathbf{u}_0(\cdot, 0) \in \mathcal{L}^{m+1}(\Omega) \cap \mathcal{H}^1(\Omega)$ $(m > 0)$. A function $\mathbf{u} : \Omega \times (-\tau, T) \to \mathbb{C}$ is called a *weak solution* of problem (17.6) if

$$\mathbf{u} \in \mathcal{L}^2(0, T : \mathcal{H}^1(\Omega)) \cap \mathcal{L}^{m+1}(0, T : \mathcal{L}^{m+1}(\Omega)) \cap \mathcal{L}^2(-\tau, 0 : \mathcal{L}^2(\Omega)),$$
$$\mathbf{u}_t \in \mathcal{L}^2(0, T : (\mathcal{H}^1(\Omega))'),$$

for every $t \in (0, T)$

$$\begin{aligned} \left\langle \frac{\partial}{\partial t}\mathbf{u}, \varphi \right\rangle_{\mathcal{H}^{-1}(\Omega) \times \mathcal{H}^1(\Omega)} &= (1 - i\omega) \int_\Omega \mathbf{u}\bar{\varphi} dx - (1 + i\beta) \\ &\quad \times \int_\Omega |\mathbf{u}|^{m-1} \mathbf{u}\bar{\varphi} dx \\ &\quad - (1 + i\epsilon) \int_\Omega \nabla \mathbf{u} \cdot \nabla \bar{\varphi} dx \\ &\quad + \int_\Omega \mathbf{F}(\mathbf{u}(x, t-\tau)) \bar{\varphi} dx, \ \forall \varphi \in \mathcal{H}^1(\Omega) \end{aligned} \tag{17.11}$$

and

$$\mathbf{u} = \mathbf{u}_0 \text{ in } \mathcal{C}([-\tau, 0] : \mathcal{L}^2(\Omega)),$$

where $\mathbf{F}(\mathbf{u}(x, t-\tau))$ is given by (17.5).

It is useful to rewrite the complex Gingzburg–Landau problem (17.3) in terms of the real components (u_1, u_2) of the solution $\mathbf{u}$, i.e.,

$\mathbf{u} = u_1 + iu_2$. The associated real system in Q is as follows:

$$\begin{cases} \dfrac{\partial u_1}{\partial t} = \Delta u_1 - \epsilon \Delta u_2 + \left(u_1^2 + u_2^2\right)\left(-u_1 + \beta u_2\right) & \\ \quad + u_1 + \omega u_2 + F_1(u(x, t-\tau)) & \text{in } Q, \\ \dfrac{\partial}{\partial t} u_2 = \epsilon \Delta u_1 + \Delta u_2 - \left(u_1^2 + u_2^2\right)\left(\beta u_1 + u_2\right) & \\ \quad + u_2 - \omega u_1 + F_2(u(x, t-\tau)) & \text{in } Q, \\ u_1(x,t) = \mathcal{R}\mathrm{eal}(U_0(x,t)) & \\ \quad \text{and } u_1(x,t) = \mathcal{I}\mathrm{m}(U_0(x,t)) & \text{in } (-\tau, 0) \times \Omega, \\ \dfrac{\partial u_1}{\partial \mathbf{n}} = \dfrac{\partial u_2}{\partial \mathbf{n}} = 0, & \text{on } \partial\Omega \times (0, T) \end{cases}$$

for F_1 and F_2 defined in (17.4) as the real and imaginary part of F, respectively.

The main results of this section is enclosed in the following theorem.

Theorem 1. (i) *Assume* $\mathbf{F}_0 \in \mathcal{L}^2(-\tau, 0)$ *and let* $\mathbf{u}_0$ *be such that*

$$\mathbf{u}_0 \in \mathcal{L}^4(\Omega) \cap \mathcal{H}^1(\Omega).$$

Then, there exists at least a weak solution to (17.3) *in* $(0, \infty)$.

(ii) *Assume* $\mathbf{u}_0 \in \mathcal{C}\left([-\tau, 0] : \mathcal{L}^2(\Omega)\right)$, $\mathbf{u}_0(x, 0) \in \mathcal{L}^{m+1}(\Omega) \cap \mathcal{H}^1(\Omega)$. *Then, there exists at least a weak solution to* (17.6) *in* $(0, \infty)$.

Remark 1. Although there are some works in the literature dealing with a partial information on the initial history (see, e.g., [3] and its references), we point out that the initial information required in problem (17.3) is weaker than in those series of works.

To prove the existence of a weak solution of (17.3), we first obtain some a priori estimates in the following lemma.

Lemma 1. *Let $T < \infty$ and assume $\mathbf{F}_0 \in \mathcal{L}^2(-\tau, 0)$ and let $\mathbf{u}_0$ be such that*

$$\mathbf{u}_0 \in \mathcal{L}^4(\Omega) \cap \mathcal{H}^1(\Omega).$$

Let $\mathbf{u} \in \mathcal{L}^2\left(0, T : \mathcal{L}^4(\Omega)\right)$ be a weak solution of (17.3). Then

$$\mathbf{u} \in \mathcal{L}^\infty\left(0, T : \mathcal{L}^2(\Omega)\right). \tag{17.12}$$

Moreover, the norm of $\mathbf{u}$ in this space, as well as in the spaces $\mathcal{L}^2\left(0, T : \mathcal{H}^1(\Omega)\right)$ and $\mathcal{L}^2\left(0, T : \mathcal{L}^4(\Omega)\right)$, has a bound only depending on $\mathbf{F}_0, \mathbf{u}_0$, μ, τ, β and T.

Lemma 2. *Let $T < \infty$ and assume*

$$\mathbf{u}_0 \in \mathcal{L}^4(\Omega) \cap \mathcal{H}^1(\Omega)$$

and

$$\mathbf{F}_0 \in \mathcal{L}^2(-\tau, 0).$$

Let $\mathbf{u}$ be a "strong" solution of (17.3). Then

$$\mathbf{u} \in \mathcal{H}^1\left(0, T : \mathcal{L}^2(\Omega)\right) \cap \mathcal{L}^\infty\left(0, T : \mathcal{H}^1(\Omega)\right) \cap \mathcal{L}^\infty(0, T : \mathcal{L}^4(\Omega)).$$

Moreover, there exists $K > 0$ such that

$$\begin{aligned}\int_0^t \int_\Omega |\mathbf{u}_t|^2 dx + \frac{1}{4}\int_\Omega |\mathbf{u}(t)|^4 dx + \frac{1}{4}\int_\Omega |\nabla \mathbf{u}(t)|^2 dx \\ \leq K \int_{-\tau}^0 |\mathbf{F}_0(s)|^2 ds + \frac{1}{4}\int_\Omega |\mathbf{u}_0(x)|^4 dx \\ + \frac{1}{4}\int_\Omega |\nabla \mathbf{u}_0(x)|^2 dx, \ a.e.\ t \in (0, T).\end{aligned} \tag{17.13}$$

The proof of Theorem 1 uses a Galerkin approximation expansion. The above lemmas supply some uniform estimates and we pass them to the limits by using the compactness Aubin–Lions lemma (see [38] for details).

The proof of the uniqueness of solutions follows a contradiction argument using the previous estimates. We recall that in this section we are assuming that $N \leq 3$.

Theorem 2. *Assume the conditions on* $\mathbf{F}_0$, $\mathbf{u}_0$ *and* $\mathbf{u}_0$ *given in parts (i) and (ii) of Theorem* 1. *Then, problems* (17.3) *and* (17.6) *have at most one weak solution for the following cases:*

- $m \in [1, \infty)$ *if* $N = 1, 2$
- $m \in [1, 5)$ *if* $N = 3$,
- $m \in (0, 1)$ *if* $N = 1, 2, 3$ *provided* β *satisfies*

$$|\beta| \leq \frac{1-m}{2m^{\frac{1}{2}}}. \tag{17.14}$$

Remark 2. The above two theorems also hold, with minor changes, for other type of boundary conditions such as, Dirichlet boundary conditions or periodic boundary conditions (as considered in [11, 12, 30]). We also point out that the assumption $(1 + i\epsilon)$ on the coefficient of the complex diffusion operator is absolutely crucial since when the real part of such a coefficient vanishes the equation becomes a nonlinear Schrödinger delayed equation and some additional conditions on the coefficient of the nonlinear part (in this paper assumed of the form $(1 + i\beta)$) are required (see, e.g., the existence and uniqueness results for the case $m \in (0, 1)$ given in [18]).

It is possible to get some qualitative properties of solutions of the delayed problem when $m \in (0, 1)$ by adapting to the case of delayed problems the energy methods presented in [8] and, more concretely, their adaptation to complex Ginzburg–Landau equations with absorption made in [4]. Due to the presence of the "bad term" $-(1 - i\omega)\,\mathbf{u}$ in the equation in all this section we shall need an extra information on the solutions: we will always assume that the solution is bounded. This condition could be avoided in absence of such a term in the equation.

A first qualitative result concerns the so-called *finite extinction time*. This property is of interest in many different contexts. For instance, in Control Theory it usually associated to the "zero exact controllability property".

Theorem 3. *Let* $m \in (0, 1)$ *and* $\alpha \in (0, 1)$ *such that*

$$\|\mathbf{u}\|_{L^\infty(Q)} \leq |1 - \alpha|^{\frac{1}{1-m}}. \tag{17.15}$$

(i) *Assume*

$$\|\mathbf{u}_0\|^2_{\mathcal{L}^2(\Omega)} \text{ small enough.} \tag{17.16}$$

Assume also that there exists $t^* \in (0, \tau)$ *such that*

$$|\mathbf{F}_0(s)|^{\frac{m+1}{m}} \leq c\,[(t^* - \tau) - s]_+^{\frac{\delta}{1-\delta}} \text{ for a.e. } s \in (-\tau, 0), \tag{17.17}$$

for some $c > 0$ *and some* $\delta \in (0,1)$. *Then any bounded solution of the nonlocal problem* (17.6) (*i.e. with* $\nu = 0$) *satisfies*

$$\|\mathbf{u}(\cdot,t)\|^2_{\mathcal{L}^2(\Omega)} \leq c\kappa\,[t^* - t]_+^{\frac{1}{1-\delta}} \quad \text{for any } t \in [0, \tau)$$

for some $c > 0$. *In particular* $\mathbf{u}(\cdot,t) \equiv \mathbf{0}$ *in* Ω *for any* $t \in [t^*, \tau)$. *In addition,* $\mathbf{u}(\cdot,t) \equiv 0$ *in* Ω *for any* $t \in [nt^*, n\tau)$ *for any* $n \in \mathbb{N}$.

(ii) *Assume*

$$\|\mathbf{u}_0(\cdot,s)\|^{\frac{2(m+1)}{m}}_{\mathcal{L}^2(\Omega)} \leq \kappa\,[(t^* - \tau) - s]_+^{\frac{\delta}{1-\delta}} \quad \text{for a.e. } s \in (-\tau, 0), \tag{17.18}$$

for some $\kappa > 0$ *and some* $\delta \in (0,1)$. *Then any bounded solution of the* (17.6), *with* $\nu > 0$, *satisfies*

$$\|\mathbf{u}(\cdot,t)\|^2_{\mathcal{L}^2(\Omega)} \leq c\kappa\,[t^* - t]_+^{\frac{1}{1-\delta}} \quad \text{for any } t \in [0, \tau)$$

for some $c > 0$. *In particular,* $\mathbf{u}(\cdot,t) \equiv \mathbf{0}$ *in* Ω *for any* $t \in [t^*, \tau)$. *In addition,* $\mathbf{u}(\cdot,t) \equiv \mathbf{0}$ *in* Ω *for any* $t \in [nt^*, n\tau)$, *for any* $n \in \mathbb{N}$.

Roughly speaking, for the proof of these results we follow the energy method presented in [8, Sec. 6.2] (see the applications to complex equations made in [4, 18]). In fact, we will use the following improvement of a suitable energy inequality.

Lemma 3 ([18]). *Let* $y \in W^{1,1}_{\text{loc}}([0,\infty);\mathbb{R})$ *with* $y \geq 0$ *over* $[0,\infty)$, $\delta \in (0,1)$, $\alpha, T_0 > 0$ *and,*

$$y_\star = \left(\kappa\,\delta^\delta(1-\delta)\right)^{\frac{1}{1-\delta}}, \tag{17.19}$$

$$x_\star = (\kappa\,\delta\,(1-\delta)\,T_0)^{\frac{1}{1-\delta}}. \tag{17.20}$$

If,

$$y(0) \leq x_\star,$$

and if for almost every $t > 0$,

$$y'(t) + \kappa\, y(t)^\delta \leq y_\star\,(T_0 - t)_+^{\frac{\delta}{1-\delta}},$$

then there exists $k^* > 0$ *such that*

$$y(t) \leq k^*(T_0 - t)_+^{\frac{1}{1-\delta}} \quad \textit{for any } t > 0. \tag{17.21}$$

Remark 3. If the initial history $\mathbf{F}_0(s)$ and $\mathbf{u}_0$ (respectively $\mathbf{u}_0(\cdot, s)$) does not vanish also for $s \in [-\tau, \underline{t}) \cup \{0\}$, for some $\underline{t} < 0$, then Theorem 3 proves that the solution of the nonlocal (17.6), i.e., with $\nu = 0$ (respectively, the local case, with $\nu > 0$) is discontinuous at the times $t = n\tau$ for any $n \in \mathbb{N}$. The technique of proof of Theorem 3 could also be used to prove that the solutions may vanish too on intervals of the form $[n\tau, n\tau + \varepsilon]$ avoiding the above-mentioned discontinuity.

Remark 4. The detailed analysis made in [4] shows that, in fact,

$$\delta = \frac{m+1}{\theta(m+1) + 2(1-\theta)} \quad \text{with} \quad \theta = \frac{N(1-m)}{N(1-m) + 2(1+m)}.$$

Remark 5. The above assumptions are, in some sense, necessary. Indeed, if for instance $\|\mathbf{u}_0\|^2_{\mathcal{L}^2(\Omega)}$ is big enough then it is possible to take the parameters such that any function $y(t)$ satisfying the ordinary differential inequality with zero in the right-hand side satisfies $y(\tau) > 0$ and thus the contribution of the global initial memory cannot be of any help in the rest of values of time $t \in [\tau, +\infty)$. Analogously, the decay condition indicated in the assumption (17.17) is the optimal decay which is compatible with the decay of any function $y(t)$ satisfying the ordinary differential inequality with zero in the right-hand side and with an exponent $\delta \in (0, 1)$. On the other hand, it is well known that if $\delta \geq 1$ then $y(t) > 0$ for any $t \in [0, +\infty)$.

Remark 6. Assumption (17.15) is used to obtain the finite time extinction of the solution. If assumption (17.15) is not satisfied, for some initial data, the solution reaches a nonzero constant value in finite time. Such a constant is, in fact, the average (in space) of the solution and its behavior is determined by an ordinary differential equation. See, for instance, [8, Chap. 2, Sec. 7.2] and references therein, where finite time convergence to the average of the solution is studied for a porous-media type equation.

Theorem 4. *Let $m \in (0, 1)$ and $\alpha \in (0, 1)$ such that*

$$\|\mathbf{u}\|_{L^\infty(Q)} \leq |1 - \alpha|^{\frac{1}{1-m}}. \tag{17.22}$$

(i) *Assume that there exists $\rho_0 > 0$ such that*

$$\mathbf{u}_0 = 0 \quad \textit{in } B_{\rho_0}. \tag{17.23}$$

Assume also that there exists $s_{\mathbf{F}_0} \in (0, \tau)$ such that

$$\mathbf{F}_0(s) = \mathbf{0} \ \textit{a.e. } s \in (-\tau, -s_{\mathbf{F}_0}). \tag{17.24}$$

Let $\mathbf{u}$ be a bounded solution of the nonlocal (17.6) *(i.e., with $\nu = 0$). Then, there exist $\rho_1 \in (0, \rho_0)$ and $t_1 \in (0, \tau)$ (both depending of the energies associated to $\mathbf{u}$) such that*

$$\mathbf{u} = \mathbf{0} \quad \textit{in } Q_{\rho_1, t_1}.$$

(ii) *Assume that there exist $s_{\mathbf{F}_0} \in (0, \tau)$ and $\rho_0 > 0$ such that*

$$\mathbf{u}_0(\cdot, s) = \mathbf{0} \quad \textit{on } B_{\rho_0} \textit{ for } s = 0 \textit{ and for a.e. } s \in (-\tau, -s_{\mathbf{F}_0}). \tag{17.25}$$

Let $\mathbf{u}$ be a bounded solution of the local (17.6) *(i.e. with $\nu > 0$). Then, there exist $\rho_1 \in (0, \rho_0)$ and $t_1 \in (0, \tau)$ (both depending of the energies associated to $\mathbf{u}$) such that*

$$\mathbf{u} = \mathbf{0} \quad \textit{in } Q_{\rho_1, t_1}.$$

Remark 7. As in [4, Theorem 5.3], it is possible to show a "waiting time property"(showing that in fact $\rho_1 = \rho_0$ for any $t \in (0, t_0)$, for some $t_0 \in (0, \tau)$) for the solution $\mathbf{u}$ of the nonlocal (17.6) (i.e. with $\nu = 0$), if we make the decay stronger assumption

$$\int_{B_\rho} |\mathbf{u}_0(x)|^2 \, dx \leq \delta(\rho - \rho_0)_+^{\frac{1}{1-\xi}} \text{ for } a.e.\ \rho \in (0, \rho_0 + \varepsilon) \tag{17.26}$$

for some $\delta, \varepsilon > 0$ and with $\xi \in (0, 1)$ the exponent arises in (17.26). In the case of the local (17.6) (i.e. with $\nu > 0$) it must be required a similar decay stronger assumption now on $\mathbf{u}_0(\cdot, s)$. To be more precise, we must assume that there exist $s_{\mathbf{F}_0} \in (0, \tau]$, $\rho_0 > 0$ and $\delta, \varepsilon > 0$ such that

$$\int_{B_\rho} |\mathbf{u}_0(x, 0)|^2 \, dx + \int_{-\tau}^{-s_{\mathbf{F}_0}} \int_{B_\rho} |\mathbf{u}_0(x, s)|^{\frac{m+1}{m}} \, dx ds$$
$$\leq \delta(\rho - \rho_0)_+^{\frac{1}{1-\xi}} \text{ for } a.e.\ \rho \in (0, \rho_0 + \varepsilon). \tag{17.27}$$

Remark 8. As in the case of the equation without delay terms, it remains an open question to know if the above finite speed of propagation also holds for the pure Schrödinger equation with the same absorption perturbation term. A partial answer was given in [17].

17.3. Turbulence Control by Feedback Delay Perturbations

We start by pointing out that the existence and uniqueness of a solution of (P_1) can be proven, as in Sec. 2, once we assume that $\mathbf{u}_0 \in \mathbf{C}([-\tau, 0] : \mathbf{L}^2(\Omega))$. As mentioned in Sec. 1, we are now interested in the stability analysis of the time-periodical function $\mathbf{v}_{\text{uosc}}(x, t) = \rho_0 e^{-i\theta t}$. In order to avoid the application of techniques for the study of the stability of periodic solutions, we can reduce the study to the stability of stationary solutions of some auxiliary problem by introducing the change of unknown $\mathbf{z}(x, t) = \mathbf{v}(x, t)e^{i\theta t}$ where $\mathbf{v}(x, t)$

is a solution of (P_2). Thus, $\mathbf{z}(x,t)$ satisfies

$$(P_3)\begin{cases} \dfrac{\partial \mathbf{z}}{\partial t} - (1+i\epsilon)\Delta \mathbf{z} = (1+i\theta)\mathbf{z} - (1+i\beta) \\ \quad |\mathbf{z}|^2\,\mathbf{z} + \mu e^{i\chi_0}[m_1\mathbf{z} + m_2\overline{\mathbf{z}} + e^{i(\omega+\theta)\tau} \\ \quad (m_3\mathbf{z}(t-\tau,x) + m_4\overline{\mathbf{z}}(t-\tau))] & \text{in } \Omega\times(0,+\infty), \\ \mathbf{z}|_{\Gamma_j} = \mathbf{z}|_{\Gamma_{j+2}}, \left(-\left.\dfrac{\partial \mathbf{z}}{\partial \mathbf{n}}\right|_{\Gamma_j} = \right) \left.\dfrac{\partial \mathbf{z}}{\partial x_j}\right|_{\Gamma_j} \\ = \left.\dfrac{\partial \mathbf{z}}{\partial x_j}\right|_{\Gamma_{j+2}} \\ \left(= \left.\dfrac{\partial \mathbf{z}}{\partial \mathbf{n}}\right|_{\Gamma_{j+2}}\right) & \text{on } \partial\Omega\times(0,+\infty), \\ \mathbf{z}(x,s) = \mathbf{u}_0(x,s)e^{i(\omega-\theta)s} & \text{on } \Omega\times[-\tau,0]. \end{cases} \tag{17.28}$$

Now, $\mathbf{v}_{\text{uosc}}(x,t) = \rho_0 e^{-i\theta t}$ is a uniform oscillation if and only if $\mathbf{z}(x,t) = \mathbf{v}_{\text{uosc}}(x,t)e^{i\theta t} = \mathbf{z}_\infty = \rho_0$ is a stationary solution of (P_3), i.e.,

$$\begin{aligned} \mathbf{0} = (1+i\theta)\mathbf{z}_\infty - (1+i\beta)\,|\mathbf{z}_\infty|^2\,\mathbf{z}_\infty + \mu e^{i\chi_0}[m_1 \\ + m_2 + e^{i(\omega+\theta)\tau}(m_3+m_4)]\mathbf{z}_\infty. \end{aligned} \tag{17.29}$$

In order to keep some resemblance with [11, 12], we shall assume that

$$m_1 + m_2 = 0 \quad \text{and} \quad m_3 + m_4 = 1. \tag{17.30}$$

Then, we get the expressions $\rho_0(\tau) = (1+\mu\cos\chi(\tau))^{1/2}$, where $\chi(\tau) = \chi_0 + (\omega+\theta(\tau))\tau$ and with $\theta(\tau)$ given as the solution of the implicit equation

$$\theta = \beta - \mu(\sin(\chi_0 + (\omega+\theta)\,\tau) - \beta\cos(\chi_0 + (\omega+\theta)\,\tau)). \tag{17.31}$$

Notice that if $\mu = 0$ we deduce that $\rho_0(\tau) = 1$ and that $\theta(\tau) = \beta$ for any τ and that $\rho_0(0) = (1+\mu\cos\chi_0)^{1/2}, \theta(0) = \beta - \mu(\sin\chi_0 - \beta\cos\chi_0)$. It is not difficult to prove (see Sec. 3.3 below) the existence and uniqueness of such a function $\theta(\tau)$ and that $\theta \in C^1$.

Our main stabilization result is as follows:

Theorem 5. *Assume* (17.9), (17.30), $\chi_0 \in (\pi, \frac{3\pi}{2})$,

$$3 - m_1 - 2m_3 \geq 0, \quad m_1 + m_3 \geq 0, \quad 3 + 2m_3 > 0,$$

$$\mu > \max \left\{ \frac{1}{|\cos \chi_0|}, \frac{3\beta - \omega + 3(\omega + \beta) \sin \chi_0 + \cos \chi_0}{5(-\beta) \sin \chi_0 \cos \chi_0 + 1}, \right.$$

$$\left. \frac{m_3 \left(3\beta - \omega - \varepsilon \frac{\pi^2}{L^2}\right) + 3(\omega + \beta) \sin \chi_0 + (m_1 + m_3) \cos \chi_0}{(3 - m_1 - 2m_3) \sin^2 \chi_0 + (m_1 + m_3) \cos^2 \chi_0 + (-\beta)(3 + 2m_3) \sin \chi_0 \cos \chi_0} \right\}. \tag{17.32}$$

Then, there exists some $\tau_0 \in (0, 1)$ *such that if we assume* $\tau \in (\tau_0, 1)$ *we get that*

$$|\mathbf{v}(x,t) - \rho_0| \leq M e^{-\alpha t} \left\| \mathbf{u}_0(\cdot, \cdot) e^{i\omega \cdot} - \rho_0 \right\|.$$

For the proof, we shall first introduce a new and quite general *pseudo-linearization principle.* Then, we shall show the applicability of it to the delayed problem and, at the end, we shall study the eigenvalues of the linear part to find the range of parameters for the stability of the linear part.

17.3.1. *The pseudo-linearization principle*

It is useful to analyze the stabilization, as $t \to \infty$, of the solutions in a more general framework: the nonlinear abstract functional differential equation

$$\begin{cases} \dfrac{du}{dt}(t) + Au(t) + Bu(t) \ni F(u_t(.)) & \text{in } X, \\ u(s) = u_0(s), & s \in [-\tau, 0], \end{cases} \tag{17.33}$$

on a Banach space X, where

$$u_t(\theta) = u(t + \theta), \theta \in [-\tau, 0],$$

to the associated equilibria: $w \in D(A) \subset D(B) \subset X$ such that

$$Aw + Bw \ni F(\widehat{w}(\cdot)),$$

where $\widehat{w} \in C := C([-\tau, 0] : X)$ is the function which takes constant values equal to w. Our main goal in this subsection is to extend, to a broad class of nonlinear operators A, the usual linearized stability principle saying, roughly speaking, that for the special case of A linear (single valued) and B and F differentiable, the asymptotic stability of the zero solution of the linearized equation,

$$\begin{cases} \dfrac{dv}{dt}(t) + Av(t) + \mathrm{D}B(w)v(t) = \mathrm{D}F(\widehat{w})v_t(.) & \text{in } X, \\ v(s) = u_0(s), & s \in [-\tau, 0]. \end{cases} \tag{17.34}$$

implies that $u(t : u_0) \to w$ as $t \to \infty$, at least if $u_0(.)$ is close enough to $\widehat{w}$. We point out that some relevant examples of nonlinear functional equations arise in the most different contexts (see, for instance, [37] for one example in [35] for a family of examples dealing with the wealth of nations and the general exposition made in [44]).

The motivation to keep A nonlinear after the process of linearization (reason why we used the term of *pseudo-linearization principle*) comes from the fact that if we use the representation for the unknown of the delayed nonlinear equation (P_3) as $\mathbf{z}(x, t) = \rho(x, t)e^{i\phi(x,t)}$ then we arrive to a coupled nonlinear system of delayed equations for ρ and ϕ which can be described in terms of the representation operator given by $\mathbf{P} : \mathbb{R}^2 \to \mathbb{C}$, $\mathbf{P}(\rho,\phi) = \rho e^{i\phi}$. Indeed, notice that $\mathbf{P}$ is nonlinear and that if $\mathbf{q} = (\rho,\phi)$ then $\mathbf{z}(x, t) = \mathbf{P}(\mathbf{q}(x, t))$ and the (P_3) can be formulated as $\frac{d\mathbf{P}(\mathbf{q}(\cdot,t))}{dt} + A\mathbf{P}(\mathbf{q}(\cdot, t)) + B\mathbf{P}(\mathbf{q}(\cdot, t)) = F(\mathbf{P}(\mathbf{q}(\cdot))_t)$. By using that the matrix $\mathbf{C}(\mathbf{q}(\cdot, t)) = \operatorname{grad} \mathbf{P}(\mathbf{q}(\cdot, t))$ is not singular, we can arrive to the simpler formulation

$$\begin{aligned} &\frac{d\mathbf{q}}{dt}(\cdot, t) + \mathbf{C}(\mathbf{q}(\cdot,t))^{-1}[A\mathbf{P}(\mathbf{q}(\cdot, t)) + B\mathbf{P}(q(\cdot, t))] \\ &\quad = \mathbf{C}(\mathbf{q}(\cdot,t))^{-1}F(\mathbf{P}(\mathbf{q}(\cdot))_t). \end{aligned} \tag{17.35}$$

Notice that, although this delayed system can also be (formally) linearized (this is the procedure followed in [12, 53], the above diffusion operator $\mathbf{C}(\mathbf{q}(\cdot,t))^{-1}A\mathbf{P}(\mathbf{q}(\cdot,t))$ becomes now quasilinear on $\mathbf{q}$ and thus the mathematical justification is much more delicate.

There are some others linearization principles in the literature. Their motivation is usually a particular problem, but its applicability is wider. Close to ours we can mention that of Ruess [59], although the formulation, scope and proof are different. Besides its applicability to the problem in this work, ours can also be applied to the case in which A is nondifferentiable and nonlinear, among many others (see [34]).

Coming back to the abstract formulation, the structural assumptions we shall assume in this paper are the following:

(H1): $A \in \mathcal{A}(\omega : X)$, for some $\omega \in \mathbb{C}$, with

$$\mathcal{A}(\omega : X) = \{A : D_X(A) \subset X \to \mathcal{P}(X) \text{ such that } A + \omega I \text{ is a m-accretive operator}\}$$

(see [26] for the case of $X = H$ a Hilbert space and the works by Benilan, Crandall, Pazy and others for the case of a general Banach space: see [20, 66]);

(H2): the operators' semigroup $T(t) : \overline{D_x(A)}^X \to X$, $t \geq 0$, generated by A, is compact (see [66]);

(H3): $B \in \mathcal{A}(0 : X)$, B is single-valued, Fréchet differentiable, and B is dominated by A; i.e.,

$$\begin{aligned} &D_X(A) \subset D_X(B) \text{ and } |Bu| \leq k \left|A^0 u\right| + \sigma(|u|) \\ &\quad \text{for any } u \in D_X(A) \text{ and for some } k < 1 \\ &\quad \text{and some continuous function } \sigma : \mathbb{R} \to \mathbb{R}, \end{aligned} \tag{17.36}$$

where, here and in what follows, $|\cdot|$ denotes the norm in the space X (in contrast with the norm in space C which will be denoted by $\|.\|$ if there is no ambiguity, when handling two spaces X and Y the corresponding norms will be indicated), $\left|A^0 u\right| := \inf\{|\xi| : \xi \in Au\}$ for $u \in D_X(A)$;

(H4): $F : C \to X$ satisfies a local Lipschitz condition, i.e.,

$$\begin{cases} \text{for any } R > 0 \text{ there exists } L(R) > 0 \text{ such that} \\ |F(\phi) - F(\psi)| \leq L(R) \|\phi - \psi\| \text{ for any } \phi, \psi \in C \\ \text{and } \|\phi\|, \|\psi\| \leq R; \end{cases} \tag{17.37}$$

(H5): there exists $\delta^F > 0$ such that $F : B^X_{\delta^F}(\widehat{w}) \to X$ is Fréchet differentiable with the Fréchet derivative $DF(\widehat{w})$ given by

$\mathrm{D}(F(\widehat{w}))\phi = \int_{-\tau}^{0} d\eta(\theta)\phi(\theta), \phi \in C$, for $\eta : [-\tau, 0] \to B(X, X)$ of bounded variation and the Fréchet derivative is locally Lipschitz continuous, where $B_{\delta^F}^X(\widehat{w}) = \left\{\phi \in C; \|\phi - \widehat{x}\| < \delta^F\right\}$.

We further assume the main condition of our arguments:

(H6): the operator $y \to Ay + By - \mathrm{D}F(\widehat{w})\,(e^{\omega \cdot}y)$ belongs to $\mathcal{A}(\omega : X)$, for some $\omega \in \mathbb{C}$ with $\mathrm{Re}\,\omega = \gamma < 0$ where $e^{\omega \cdot}v \in C$ is defined by

$$(e^{\omega \cdot}v)(s) = e^{\omega s}\widehat{v}(s), \text{ with } \widehat{v}(s) = v, \qquad \text{for any } s \in [-\tau, 0], \text{ for } v \in X. \tag{17.38}$$

In order to treat the case in which B is differentiable we introduce the following conditions:

(H7): there exist a Banach space Y and $\delta^B > 0$ such that B is Fréchet differentiable as function from $B_{\delta^B}(w) = \left\{z \in D(B);\ |w - z| < \delta^B\right\}$ into Y, with the Fréchet derivative $\mathrm{D}B(w)$ locally Lipschitz continuous;

(H8): the operator $y \to Ay + \mathrm{D}B(w)y - \mathrm{D}F(\widehat{w})\left(e^{\omega^* \cdot}y\right)$ belongs to $\mathcal{A}(\omega^* : Y)$, for some $\omega^* \in \mathbb{C}$ with $\mathrm{Re}\,\omega^* = \gamma^* < 0$.

A concrete statement of the *pseudo-linearization principle is as follows:*

Theorem 6. *Assume* (H1)–(H6)*. Then there exist* $\alpha > 0$, $\epsilon > 0$ *and* $M \geq 1$ *such that if* $u_0 \in B_\epsilon^X(\widehat{w})$, $u_0(s) \in D_X(B)$ *for any* $s \in [-\tau, 0]$ *then the solution* $u(\cdot : u_0)$ *of* (17.59) *exists on* $[-\tau, +\infty)$ *and*

$$|u(t : u_0) - w| \leq Me^{-\alpha t}\|u_0 - \widehat{w}\|, \quad \textit{for any } t > 0. \tag{17.39}$$

Moreover, if we also assume that (H7), (H1)–(H5) *hold on the space* Y *and* (H8), *then there exist* $\alpha^* > 0$, $\epsilon^* \in (0, \epsilon]$ *and* $M^* \geq 1$ *such that if* $u_0 \in B_{\epsilon^*}^{X \cap Y}(\widehat{w})$, $u_0(s) \in D_X(B) \cap D_Y(B)$ *for any* $s \in [-\tau, 0]$, *then*

$$\begin{aligned}&|u(t:u_0)-w|_X+|u(t:u_0)-w|_Y\\&\quad\le M^*e^{-\alpha^*t}(\|u_0-\widehat{w}\|_X+\|u_0-\widehat{w}\|_Y),\\&\quad \textit{for any } t>0.\end{aligned}\tag{17.40}$$

Remark 9. It is not difficult to show that the assumption (H8) is implied (when A is linear) by the condition: "if $\lambda\in\mathbb{C}$ is given so that there exists $y\in D(B)\setminus\{0\}$ such that $Ay+\mathrm{D}B(w)y-\lambda y\ni \mathrm{D}F(\widehat{w})\left(e^{\lambda\cdot}y\right)$ then $\mathrm{Re}\,\lambda>0$". This allows to see Theorem 4.1 of [67] (see also [57] and its references) as a special case of our abstract result with $B=0$. In that case, the "variation of the constants formula" can be used to get a different proof of the theorem since A is linear. Notice that if $B\neq 0$ and $D(B)\subsetneq X$ then the arguments of the proof of Wu [67] do not work (in spite of what is claimed in the Example 4.8 given there).

Remark 10. When A is linear, as in the case without delay, assumption (H7) implies that the zero solution of the linearized problem $\frac{dU}{dt}(t)+AU(t)+\mathrm{D}B(w)U(t)$-$\mathrm{D}F(\widehat{w})U_t(\cdot)=0$ in X is locally asymptotically stable (see [67]).

Remark 11. It is possible to prove the existence of global solutions for a general class of initial data (not necessarily near $\widehat{w}$) by using that $A+B\in\mathcal{A}(\omega:X)$, for some $\omega\in\mathbb{C}$, some truncation of the nonlocal term $F(u_t)$ and passing to the limit by the compactness of the semigroup generated by A (see [66] for some related results).

An easy adaptation of the above proof leads to the following linearization result (now on a possibly smaller neighborhood of w) when A is differentiable.

Theorem 7. *The conclusion of the above result remains true if we assume, additionally, that condition* (H7) *also holds for A and we replace condition* (H8) *by*

(H9) *the operator* $y\to \mathrm{D}A(w)y+\mathrm{D}B(w)y-\mathrm{D}F(\widehat{w})\left(e^{\omega\cdot}y\right)$ *belongs to* $\mathcal{A}(\omega)$, *for some* $\omega\in\mathbb{C}$ *with* $\mathrm{Re}\,\omega=\gamma<0$.

Remark 12. We claim that our arguments keeping A nonlinear after linearizing the rest of the terms (and in particular the way in which we apply Gronwall inequality) allow to extend, to the case of quasilinear equations, the so-called *method of quasilinearization* which, introduced by Bellman and Kalaba [19], we used to find solutions of a parabolic semilinear problem through the iteration of solutions of the linearized equation when starting in a super and a subsolution of the original semilinear problem (see, e.g., [27, 49] and their references). This will be the subject of a future work by the authors.

17.3.2. *Applications of the abstract result to the complex Ginzburg–Landau equation*

Motivated by the special form of the nonlinear term of the equation in (P_3) we shall take $X = \mathbf{L}^4(\Omega)$ and $Y = \mathbf{L}^{4/3}(\Omega)$ (notice that, in contrast with the case of scalar equations (see [57]) the space $\mathbf{L}^\infty(\Omega)$ is not a suitable space to check assumption (H1): see [10]). A detailed analysis of the associated diffusion operator is consequence of some previous results in the literature: see, for instance, [5]. Notice that the operator $A\mathbf{u}$ can be formulated matricially as

$$\begin{pmatrix} u_1 \\ u_2 \end{pmatrix} \to \begin{pmatrix} \Delta & -\epsilon\Delta \\ \epsilon\Delta & \Delta \end{pmatrix} \begin{pmatrix} u_1 \\ u_2 \end{pmatrix}.$$

So, if $\epsilon \neq 0$ the diffusion matrix has a nonzero antisymmetric part. In particular, A is the generator of a semigroup of contractions $\{T(t)\}_{t\geq 0}$ on X and the compactness of the semigroup is a consequence of the compactness of the inclusion $D(A) \subset X$ (notice that, since $N = 2$, $\mathbf{W}^{1,4}(\Omega) \subset \mathbf{W}^{1,4/3}(\Omega) \subset \mathbf{C}(\overline{\Omega})$ with compact imbedding) and some regularity results for nonsymmetric systems.

Concerning the rest of the terms of the equation in (P_3), we define $B\mathbf{u} = (1 + i\beta)\,|\mathbf{u}|^2\,\mathbf{u}$ with $D(B) = \mathbf{L}^{12}(\Omega)$. By using the characterization of the semi inner-bracket $[,]$ for the spaces $L^p(\Omega)$ (see, for instance, [20]), it is easy to see that $\mathbf{B}$ verifies (H3). Moreover, by the results on the Frechet differentiability of Nemitsky operators (see [7, Theorem 2.6]) (with $p = 4$) we get that (H7) holds, with $\mathrm{D}B(\mathbf{y})\mathbf{v} = 3(1+i\beta)\left|\mathbf{y}\right|^2\mathbf{v}$, if we take $Y = \mathbf{L}^{4/3}(\Omega)$. It can be found in the above-mentioned reference that assumption (H7) does not hold if we take $X = Y = \mathbf{L}^2(\Omega)$.

The nonlocal term is defined by

$$\begin{aligned}\mathbf{F}(\mathbf{u}_t) = (1+i\theta)\mathbf{u}(t) + \mu e^{i\chi_0}[&m_1\mathbf{u}(t)+m_2\overline{\mathbf{u}}(t)\\ &+e^{i(\omega+\theta)\tau}(m_3\mathbf{u}(t-\tau)+m_4\overline{\mathbf{u}}(t-\tau))],\end{aligned}$$

is locally Lipschitz continuous and its Frechet derivative is given by

$$\begin{aligned}\mathrm{D}\mathbf{F}(\widehat{\mathbf{y}})\mathbf{v}(t) = -(1+i\theta)\mathbf{v}(t) - \mu e^{i\chi_0}[&m_1\mathbf{v}(t)+m_2\overline{\mathbf{v}}(t)\\ &-e^{i(\omega+\theta)\tau}(m_3\mathbf{v}(t-\tau)-m_4\overline{\mathbf{v}}(t-\tau))] \quad (17.41)\end{aligned}$$

since for any $\phi \in C$, the nonlocal operator $\phi \to \frac{1}{|\Omega|}\int_\Omega \phi(s)dx$ is linear and we can write $\mathrm{D}\mathbf{F}(\widehat{\mathbf{y}})\phi = \int_{-\tau}^{0} d\eta(s)\phi(s)$, with

$$\begin{aligned}d\eta(s)\mathbf{v}(s) = \delta_0(s)(1+i\theta)\mathbf{v}(s)+\mu e^{i\chi_0}[&\delta_0(s)(m_1\mathbf{v}(s)+m_2\overline{\mathbf{v}}(s))\\ &+e^{i(\omega+\theta)\tau}\delta_{-\tau}(s)(m_3\mathbf{v}(s)+m_4\overline{\mathbf{v}}(s))] \quad (17.42)\end{aligned}$$

for any $\mathbf{v} \in C([-\tau,\infty)\colon \mathbf{L}^4(\Omega))$ and any $s \in [-\tau,\infty)$, where $\delta_0(s), \delta_{-\tau}(s)$ denote the Dirac delta at the points $s=0$ and $s=-\tau$, respectively. By well-known results, we have that $\eta : [-\tau, 0] \to B(X, X)$ has a bounded variation and so, conditions (H4) and (H5) hold (and analogously replacing X by Y).

Finally, assumption (H6) can be read as a condition on the stationary state $\mathbf{y}$ (a study of the eigenvalue of operator A can be found, for instance, in [65]).

Remark 13. By introducing the representation operator $\mathbf{P} : \mathbb{R}^2 \to \mathbb{C}$, $\mathbf{P}(\rho,\phi) = \rho e^{i\phi}$, it is clear that the quasilinear operator $A\mathbf{P}(\mathbf{q})$ obtained from the operator $A\mathbf{u}$=-$(1+i\epsilon)\Delta\mathbf{u}$ satisfies also condition $A \in \mathcal{A}(\omega)$ (since $\mathbf{P}$ is merely a change of variables). We point out that

$$A\mathbf{P}(\mathbf{q}) = -(1+i\epsilon)[\Delta\rho - \rho\left|\nabla\phi\right|^2 + i(2\nabla\rho\cdot\nabla\phi + \rho\Delta\phi)]e^{i\phi}.$$

Then, the "formal linearization" of the operator $\mathbf{E}(\mathbf{q}) := A\mathbf{P}(\mathbf{q})$ at $\mathbf{q}^*(x, y) := \mathbf{y} \equiv \rho_0$ becomes

$$DE(\mathbf{q}^*)(\rho e^{i\phi}) = -(1 + i\epsilon)[\Delta\rho + i\rho_0\Delta\phi]e^{i\phi}.$$

Notice that the linearization of $\mathbf{C}(\mathbf{q})^{-1}A\mathbf{P}(\mathbf{q})$ needs a slight modification of the above linear expression.

17.3.3. *Study of the eigenvalues of the linearized problem*

In this subsection, we shall study the eigenvalues $\lambda \in \mathcal{C}$, $\lambda = a + ib$ of the linearized problem and, which is crucial, we look for

$$\begin{cases} \text{any } \lambda \in \mathcal{C} \text{ such that } \exists\ v \in D(A),\ v \neq 0, \text{ such that} \\ 0 = \lambda v + Av + \mathrm{D}B(w)v - \mathrm{D}F(\widehat{w})(e^{\lambda \cdot}v), \text{ and } Re\lambda < 0, \end{cases} \tag{17.43}$$

where $e^{\lambda \cdot}v \in C$ is defined by

$$(e^{\lambda \cdot}v)(s) = e^{\lambda s}\widehat{v}(s), \quad \text{with } \widehat{v}(s) = v, \quad \text{for any } s \in [-\tau, 0]. \tag{17.44}$$

As in the case without delay, Eq. (17.43) implies that the zero solution of the linearized problem $\frac{dU}{dt}(t) + AU(t) + \mathrm{D}B(w)U(t)$-$\mathrm{D}F(\widehat{w})U_t(.) = 0$ in X is locally asymptotically stable (see [67]).

We go back now to the problems (17.8) and (17.27), and recall the expressions (17.29), (17.30) and (17.31)

$$\theta = \beta - \mu(\sin(\chi_0 + (\omega + \theta)\,\tau) - \beta\cos(\chi_0 + (\omega + \theta)\,\tau)). \tag{17.45}$$

Notice that if $\mu = 0$ we deduce that $\rho_0(\tau) = 1$ and that $\theta(\tau) = \beta$ for any τ and that $\rho_0(0) = (1 + \mu\cos\chi_0)^{1/2}, \theta(0) = \beta - \mu(\sin\chi_0 - \beta\cos\chi_0)$. It is not difficult to prove (see the following proposition) the existence and uniqueness of such a function $\theta(\tau)$ and that $\theta \in C^1$.

Proposition 1. *There exists a unique function $\theta(\tau)$ such that*

$$\theta(\tau) - \beta + \mu(\sin(\chi_0 + (\omega + \theta(\tau))\,\tau) - \beta\cos(\chi_0 + (\omega + \theta(\tau))\,\tau)) = 0$$

for any $\tau \in [0, 1]$. Moreover $\theta \in C^1$.

Proof. It is enough to see, by the implicit function theorem, that $\theta(\tau)$ is characterized as the (unique) solution of the Cauchy problem associated to the ODE

$$\frac{d\theta}{d\tau}(\tau) = \frac{\begin{array}{c} -[\mu(\cos(\chi_0 + (\omega + \theta(\tau))\tau)(\omega + \theta) \\ + \beta \sin(\chi_0 + (\omega + \theta(\tau))\tau))](\omega + \theta(\tau)) \end{array}}{\begin{array}{c} 1 + \mu(\cos(\chi_0 + (\omega + \theta(\tau))\tau)\tau \\ + \beta \sin(\chi_0 + (\omega + \theta(\tau))\tau))\tau \end{array}}.$$

We recall that in our case, $\mathbf{z}_\infty = \rho_0$ and so we can arrive to the linear problem

$$(P_4) \begin{cases} -(1 + i\epsilon)\Delta \mathbf{z} = -(a + ib)\mathbf{z} \\ \quad +[(1 + i\theta) - 3(1 + i\beta)\rho_0^2)]\mathbf{z} \\ \quad +\mu e^{i\chi_0} \left[m_1\mathbf{z} + m_2\overline{\mathbf{z}} + e^{-a\tau + i(\omega+\theta-b)\tau}(m_3\mathbf{z} + m_4\overline{\mathbf{z}})\right] & \text{in } \Omega, \\ \dfrac{\partial \mathbf{z}}{\partial \overrightarrow{n}} = \mathbf{0} & \text{on } \partial\Omega. \end{cases}$$

As usual, the linear structure of the equation leads to the search of nontrivial solutions $\mathbf{z}(x)$ of the form $\mathbf{A_k} w_{\mathbf{k}}^j(x)$, with $j = 1, 2$, where $w_{\mathbf{k}}^j(x)$ are the eigenfunctions for the usual Laplacian operator Δ with periodic boundary conditions on $\Omega = (0, L_1) \times (0, L_2)$. We recall that the eigenvalues of this problem are given by

$$\lambda_0^0 = 0, \lambda_{\mathbf{k}}^0 = 4\pi \left(\frac{k_1^2}{L_1^2} + \frac{k_2^2}{L_2^2}\right), \quad k_1, k_2 \in \mathbb{N}$$

with the associate eigenfunctions

$$w_0 = \frac{1}{\sqrt{|\Omega|}}, w_{\mathbf{k}}^1 = \sqrt{\frac{2}{|\Omega|}} \cos 2\pi \mathbf{kx}, w_{\mathbf{k}}^2$$
$$= \sqrt{\frac{2}{|\Omega|}} \sin 2\pi \mathbf{kx}, \text{ with } |\Omega| = L_1 L_2,$$

where we have written $\mathbf{kx} := \left(\frac{k_1}{L_1}x_1 + \frac{k_2}{L_2}x_2\right)$ (see, e.g., [65]).

The following general lemma will be used in the study of $\mathbf{z}(x)$. □

Lemma 4. *Let A be a self-adjoint operator on $L^2(\Omega)$ and let $\{\varphi_n\}$ be a family of eigenfunctions associated to the different eigenvalues $\{\lambda_n^0\}$. Assume that $\lambda_0^0 = 0$ is an eigenvalue and that $\varphi_0 = 1$ is an eigenfunction associated to λ_0. Then*

$$\int_\Omega \varphi_n = 0 \textit{ for any } n \neq 0.$$

Proof. It is enough to recall that $\int_\Omega \varphi_n \varphi_m = 0$ *for any* $n \neq m$ since $\lambda_n^0 \neq \lambda_m^0$

$$\lambda_n^0 \int_\Omega \varphi_n \varphi_m = \int_\Omega A\varphi_n \varphi_m = \int_\Omega \varphi_n A\varphi_m = \lambda_m^0 \int_\Omega \varphi_n \varphi_m.$$

Then taking $m = 0$ we get the conclusion. □

In order to keep a coherent notation with the one used in [11, 12] we introduce the notation $\lambda_{\mathbf{k}} = a_{\mathbf{k}} + ib_{\mathbf{k}}$ for the real and imaginary parts of the eigenvalues of the problem stated in (H8). Notice that, by the previous lemma, $\int_\Omega w_{\mathbf{k}}^j = 0$ for any $\mathbf{k} \neq 0$ and $j = 1, 2$. Then we get that

$$\begin{aligned}
&(a_{\mathbf{k}} + ib_{\mathbf{k}}) - (1 + i\epsilon)(-\lambda_{\mathbf{k}}) \\
&\quad = (1 + i\theta) - 3(1 + i\beta)\rho_0^2 \\
&\qquad + \mu e^{i\chi_0}[m_1 + m_2\delta_{0\mathbf{k}} + e^{-a\tau + i(\omega+\theta-b)\tau}(m_3 + m_4\delta_{0\mathbf{k}})],
\end{aligned}$$

where $\delta_{0\mathbf{k}}$ denotes the Kronecker delta function. We arrive to

$$\begin{cases}
a_{\mathbf{k}} = -\lambda_{\mathbf{k}}^0 - 2 - 3\mu\cos\chi(\tau) + \mu(m_1 + m_2\delta_{0k})\cos\chi_0 \\
\quad + \mu e^{-a_{\mathbf{k}}\tau}(m_3 + m_4\delta_{0k})\cos(\chi_0 + (\omega + \theta - b_{\mathbf{k}})\tau), \\
b_{\mathbf{k}} = \theta - \epsilon\lambda_{\mathbf{k}}^0 - 3\beta(1 + \mu\cos\chi) + \mu(m_1 + m_2\delta_{0k})\sin\chi_0 \\
\quad + \mu e^{-a_{\mathbf{k}}\tau}(m_3 + m_4\delta_{0k})\sin(\chi_0 + (\omega + \theta - b_{\mathbf{k}})\tau).
\end{cases} \tag{17.46}$$

The previous equations are transcendent and we cannot get an explicit expression for the real and imaginary part of the eigenvalues (for some similar transcendent equations arising in delayed ODEs, see [44]).

Now, we focus our attention on the dependence of $a_{\mathbf{k}}$ and $b_{\mathbf{k}}$ with respect to τ. So, by the regularity of the involved functions we can assume

$$a_{\mathbf{k}} = a_{\mathbf{k}0} + a_{\mathbf{k}1}\tau + o(\tau),\ b_{\mathbf{k}} = b_{\mathbf{k}0} + b_{\mathbf{k}1}\tau + o(\tau),$$

as we get, for instance, by a "formal" series development in powers of the τ argument. Here we used the Landau notation ($f(\tau) = o(\tau)$ means that $\frac{f(\tau)}{\tau} \to 0$ when $\tau \to 0$).

The terms of order zero in τ are obtained by making $\tau = 0$ in (17.46)

$$\begin{cases} a_{\mathbf{k}0} = -\left(2+\lambda^0_{\mathbf{k}}\right) + \mu\cos\chi_0(m_1+m_2\delta_{0\mathbf{k}}+m_3+m_4\delta_{0\mathbf{k}}), \\ b_{\mathbf{k}0} = 4\beta - \epsilon\lambda^0_{\mathbf{k}} + 3\mu\beta\cos\chi_0 + \mu\sin\chi_0(m_1+m_2\delta_{0\mathbf{k}} + m_3+m_4\delta_{0\mathbf{k}}). \end{cases} \tag{17.47}$$

So, we can state a first result concerning the case without any delay.

Proposition 2. *Assume $\tau = 0$, $\chi_0 \in (\frac{\pi}{2}, \frac{3\pi}{2})$, and $\mu > \frac{1}{|\cos\chi_0|}$. Then, the uniform oscillation $v_{\text{uosc}}(x,t) = \rho_0 e^{-i\theta t}$ is linearly unstable.*

Proof. From (17.47), we see that $a_{00} > 0$ and since $\tau = 0$ we get the existence of at least one eigenvalue λ of the linearized problem with $Re(\lambda) > 0$ which implies the result.

The first-order terms in τ are calculated below. □

Lemma 5. *We have*

$$\begin{aligned} a_{\mathbf{k}1} = \left[\frac{da_{\mathbf{k}}}{d\tau}\right]_{\tau=0} &= \left(2+\lambda^0_{\mathbf{k}}\right) + \mu\left[3\left(\omega+\beta\right)\sin\chi_0\right. \\ &\quad \left. + \left(m_3+m_4\delta_{0\mathbf{k}}\right)\left(3\beta - \epsilon\lambda^0_{\mathbf{k}} - \omega\right)\right] \\ &\quad + \mu^2\{-3\sin^2\chi_0 + 3\beta\sin\chi_0\cos\chi_0 \\ &\quad + \left(m_3+m_4\delta_{0\mathbf{k}}\right)\left[\sin^2\chi_0 + 2\beta\sin\chi_0\cos\chi_0 + \right. \\ &\quad \left. + \left(m_1+m_2\delta_{0\mathbf{k}}+m_3+m_4\delta_{0\mathbf{k}}\right)\right]\left(\sin^2\chi_0 - \cos^2\chi_0\right)\}. \end{aligned} \tag{17.48}$$

Differentiating in (17.46) we get that

$$
\begin{aligned}
a_{\mathbf{k}1} &= \left[\frac{da_{\mathbf{k}}}{d\tau}\right]_{\tau=0} = \left[3\mu\sin\chi(\tau)\frac{d\chi}{d\tau}\right]_{\tau=0} \\
&\quad + \left[(-a_{\mathbf{k}})\,\mu e^{-a_{\mathbf{k}}\tau}(m_3{+}m_4\delta_{0\mathbf{k}})\cos(\chi_0 + (\omega+\theta-b_{\mathbf{k}})\tau)\right]_{\tau=0} \\
&\quad - \left[\mu e^{-a_{\mathbf{k}}\tau}(m_3{+}m_4\delta_{0\mathbf{k}})\sin(\chi_0 + (\omega+\theta-b_{\mathbf{k}})\tau)\right]_{\tau=0} \\
&\quad \times \left[\frac{d(\omega+\theta-b_{\mathbf{k}})\tau}{d\tau}\right]_{\tau=0} \\
&= (3\mu\sin\chi_0)\,(\omega+\beta-\mu(\sin\chi_0-\beta\cos\chi_0)) \\
&\quad - \left(-\left(2+\lambda_{\mathbf{k}}^0\right) + \mu\cos\chi_0(m_1{+}m_2\delta_{0\mathbf{k}}{+}m_3{+}m_4\delta_{0\mathbf{k}})\right) \\
&\quad \times \mu(m_3{+}m_4\delta_{0\mathbf{k}})\cos\chi_0 \\
&\quad - \mu(m_3{+}m_4\delta_{0\mathbf{k}})(\omega+\beta-\mu(\sin\chi_0-\beta\cos\chi_0)-b_{\mathbf{k}})\sin\chi_0.
\end{aligned}
$$

Thus, by using the expression for $b_{\mathbf{k}}$ *(see* (17.46)*) we obtain that*

$$
\begin{aligned}
a_{\mathbf{k}1} &= (3\mu\sin\chi_0)\,(\omega+\beta-\mu(\sin\chi_0-\beta\cos\chi_0)) \\
&\quad - \left(-\left(2+\lambda_{\mathbf{k}}^0\right) + \mu\cos\chi_0(m_1{+}m_2\delta_{0\mathbf{k}}{+}m_3{+}m_4\delta_{0\mathbf{k}})\right) \\
&\quad \mu(m_3{+}m_4\delta_{0\mathbf{k}})\cos\chi_0 \\
&\quad -\mu(m_3{+}m_4\delta_{0\mathbf{k}})(\omega+\beta-\mu(\sin\chi_0-\beta\cos\chi_0))\sin\chi_0 \\
&= (3\mu\sin\chi_0)\,(\omega+\beta-\mu(\sin\chi_0-\beta\cos\chi_0)) \\
&\quad - \left(-\left(2+k^2\right) + \mu\cos\chi_0(m_1{+}m_2\delta_{0\mathbf{k}}{+}m_3{+}m_4\delta_{0\mathbf{k}})\right) \\
&\quad \times \mu(m_3{+}m_4\delta_{0\mathbf{k}})\cos\chi_0 \\
&\quad - \mu(m_3{+}m_4\delta_{0\mathbf{k}})(\omega+\beta-\mu(\sin\chi_0-\beta\cos\chi_0))\sin\chi_0 \\
&\quad + \mu(m_3{+}m_4\delta_{0\mathbf{k}})(4\beta-\epsilon\lambda_{\mathbf{k}}^0+3\mu\beta\cos\chi_0+\mu\sin\chi_0 \\
&\quad \times (m_1{+}m_2\delta_{0\mathbf{k}}+m_3{+}m_4\delta_{0\mathbf{k}}))\sin\chi_0.
\end{aligned}
$$

In consequence

$$
\begin{aligned}
a_{\mathbf{k}1} &= \left(2+\lambda_{\mathbf{k}}^0\right) + \mu\left(3\,(\omega+\beta)\sin\chi_0 - (m_3{+}m_4\delta_{0\mathbf{k}})\,(\omega+\beta)\right. \\
&\quad \left. + \left(4\beta-\epsilon\lambda_{\mathbf{k}}^0\right)(m_3{+}m_4\delta_{0\mathbf{k}})\right)
\end{aligned}
$$

$$
\begin{aligned}
&- \mu^2 \left(3 \sin \chi_0 (\sin \chi_0 - \beta \cos \chi_0) + \cos^2 \chi_0\right. \\
&\times \left.(m_1{+}m_2\delta_{0\mathbf{k}}{+}m_3{+}m_4\delta_{0\mathbf{k}})(m_3{+}m_4\delta_{0\mathbf{k}})\right) \\
&+ \mu^2 (m_3{+}m_4\delta_{0\mathbf{k}}) \left[(\sin \chi_0 - \beta \cos \chi_0) \sin \chi_0 + (3\beta \cos \chi_0\right. \\
&+ \left.\sin \chi_0 (m_1{+}m_2\delta_{0\mathbf{k}} + m_3{+}m_4\delta_{0\mathbf{k}})) \sin \chi_0\right],
\end{aligned}
$$

which proves the result.

Proposition 3. *Assume* (17.9), $\chi_0 \in (\pi, \frac{3\pi}{2})$, (17.30) *and*

$$
\mu > \max \left\{ 0, \frac{3\beta - \omega + 3(\omega + \beta) \sin \chi_0 + \cos \chi_0}{5(-\beta) \sin \chi_0 \cos \chi_0 + 1} \right\}.
$$

Then $a_{00} + a_{01} < 0$.

Proof. By using (17.47), (17.48), and (17.30), we get

$$
\begin{aligned}
a_{00} + a_{01} = \mu[&(3\beta - \omega + 3(\omega + \beta) \sin \chi_0 + \cos \chi_0) \\
&- \mu(5(-\beta) \sin \chi_0 \cos \chi_0 + 1)].
\end{aligned}
$$

Then, the assumptions imply the positivity of the coefficient of μ^2 and the result holds. □

Proposition 4. *Assume* (17.9), $\chi_0 \in (\pi, \frac{3\pi}{2})$, (17.30) *and*

$$
\mu > \max \left\{ 0, \frac{\begin{array}{c} m_3 \left(3\beta - \omega - \varepsilon 4\pi \left(\frac{1}{L_1^2} + \frac{1}{L_2^2}\right)\right) \\ + 3(\omega + \beta) \sin \chi_0 + (m_1 + m_3) \cos \chi_0 \end{array}}{\begin{array}{c} (3 - m_1 - 2m_3) \sin^2 \chi_0 + (m_1 + m_3) \cos^2 \chi_0 \\ + (-\beta)(3 + 2m_3) \sin \chi_0 \cos \chi_0 \end{array}} \right\}.
$$

Then, for any $k, a_{\mathbf{k}0} + a_{\mathbf{k}1} < 0$. *Moreover, for any* $k \neq 0$ *and any* $\tau \in (0, 1]$,

$$
a_{k(n)0} + a_{k(n)1}\tau < a_{k(1)0} + a_{k(1)1}\tau.
$$

By using (17.47), (17.48) *and that* $0 < \lambda^0_{(1,1)} < \lambda^0_{\mathbf{k}}$ *for any* $k \in N^2$, $k \neq (1,1)$, *we obtain that*

$$\begin{aligned} a_{\mathbf{k}0} + a_{\mathbf{k}1} = \mu \Bigg[\Bigg(& m_3 \left(3\beta - \omega - \varepsilon 4\pi \left(\frac{1}{L_1^2} + \frac{1}{L_2^2}\right)\right) \\ & + 3(\omega+\beta)\sin\chi_0 + (m_1+m_3)\cos\chi_0\Bigg) \\ & - \mu((3 - m_1 - 2m_3)\sin^2\chi_0 + (m_1+m_3)\cos^2\chi_0 \\ & + (-\beta)(3+2m_3)\sin\chi_0\cos\chi_0)\Bigg]. \end{aligned}$$

Again, the assumptions made on the parameters imply the positivity of the coefficient of μ^2 and the result holds. Moreover,

$$\begin{aligned} & a_{k(n)0} - a_{k(1)0} + (a_{k(n)1} - a_{k(1)1})\tau \\ & \quad = -k(n)^2 + k(1)^2 - (m_3\epsilon k(n)^2 - m_3\epsilon k(1)^2)\tau < 0. \end{aligned}$$

The proof of Theorem 5 is now complete since from Propositions 3 and 4 we deduce the existence of some $\tau_0 \in (0,1)$ (independent of $\mathbf{k} \in \mathbb{N}^2$) such that for any $|\mathbf{k}| \geq 0$ we have $a_{\mathbf{k}0} + a_{\mathbf{k}1}\tau < 0$ for any $\tau \in (\tau_0, 1)$. This implies the hypothesis of the abstract result and the conclusion follows.

Remark 14. Notice that Theorem 5 applies to the case $m_1 = m_2 = m_3 = 0$ which corresponds to a formulation similar to the one of [11]. Moreover, it also applies to the choice $m_1 = \kappa, m_2 = -1-\kappa$, $m_3 = 0$ and $m_4 = 1$, for any $\kappa \in (0,1)$ which corresponds to a formulation quite close to the pioneering paper [58] (concerning chaotic ODEs).

Remark 15. Since the eigenvalue $\lambda^0_0 = 0$, using Lemma 4 it is possible to obtain the same result for Neumann conditions.

Remark 16. Numerical simulations show that while a purely local control is unsuitable to produce uniform oscillations, a mixed local and global control can be efficient and also able to create other patterns such as standing waves, amplitude death, or traveling waves (see [62] and its references).

17.4. Hopf Bifurcation and Delay Terms

Before considering the case of the complex Ginzburg–Landau equation (CGLE), for the purposes of clarity and ease of understanding, we start by considering a very simplified version of the general model to be given later which has the advantage that closed analytical solutions are possible and the necessary eigenvalue computations can be carried out in full. Unfortunately, such precise calculations are not available for the general model and a fairly complete graphical-numerical study will be given in exchange.

17.4.1. *Hopf bifurcation for the Stuart–Landau equation with a time delay feedback*

In the Stuart–Landau equation, the diffusion term is absent, which amounts to restricting our study to the *spatially homogeneous* solutions (which always satisfy periodic boundary conditions). On the other hand, we assume that a *delayed linear feedback term* is added, so the equation under study in this section will be

$$\frac{\partial \mathbf{A}}{\partial t} = (1 - i\omega)\mathbf{A} - (1 + i\alpha)\left|\mathbf{A}\right|^2 \mathbf{A} + m_1\mathbf{A} + m_3\mathbf{A}(t - \tau). \quad (17.49)$$

More general control terms will be considered later. The change of variables $\mathbf{w}(t) = e^{-i\phi t}\mathbf{A}(t)$ gives

$$\frac{\partial \mathbf{w}}{\partial t} = (1 - i\omega - i\phi)\mathbf{w} - (1 + i\alpha)\left|\mathbf{w}\right|^2 \mathbf{w} + m_1\mathbf{w} + m_{\bar{3}}\, e^{i\phi\tau}\mathbf{w}(t - \tau). \quad (17.50)$$

We now choose $\phi = -\alpha - \omega$ and $m_3 = -ei^{i\phi\tau}m_1$ and denote the stationary solution of

$$\frac{\partial \mathbf{w}}{\partial t} = (1 + i\alpha)(\mathbf{w} - \left|\mathbf{w}\right|^2 \mathbf{w}) + m_1\left[\mathbf{w} - \mathbf{w}(t - \tau)\right] \quad (17.51)$$

by $\mathbf{w}_0$. In order to check if at some critical value of the delay $\tau = \tau^*$ a Hopf bifurcation takes place, we linearize the equation around $\mathbf{w}_0 = 1$ and check whether a pair of complex eigenvalues $\lambda(\tau) = a(\tau) \pm ib(\tau)$

of the linearization cross transversally the imaginary axis away from the origin, i.e., they satisfy $a(\tau^*) = 0,\ b(\tau^*) \neq 0$ and $a'(\tau^*) \neq 0$ (see, e.g., [67]).

Observe now that the complex term $|v|^2 v$, although perfectly differentiable from the real point of view (in fact, the complex map $z \longmapsto |z|^2 z = z^2\bar{z}$ is *real-analytic*), is *not an analytic* (or holomorphic) function from the complex viewpoint. Therefore, it becomes convenient at this point to abandon the complex notation and write the system in real form ($\mathbf{w} = u + iv$) as follows:

$$\partial_t \begin{pmatrix} u \\ v \end{pmatrix} = \begin{pmatrix} 1 & -\alpha \\ \alpha & 1 \end{pmatrix} (1 - (u^2 + v^2)) \begin{pmatrix} u \\ v \end{pmatrix} + m_1 \begin{pmatrix} u - u(t-\tau) \\ v - v(t-\tau) \end{pmatrix}.$$

Let us fix our attention to the stationary solution $\mathbf{w}_0 = (u_0, v_0) = (1, 0)$. The linearization around $\mathbf{w}_0$ is given by

$$\partial_t \begin{pmatrix} U \\ V \end{pmatrix} = \begin{pmatrix} 1 & -\alpha \\ \alpha & 1 \end{pmatrix} \begin{pmatrix} -2 & 0 \\ 0 & 0 \end{pmatrix} \begin{pmatrix} U \\ V \end{pmatrix} + m_1 \begin{pmatrix} U - U(t-\tau) \\ V - V(t-\tau) \end{pmatrix} \tag{17.52}$$

and the eigenvalue–eigenvector pairs associated to this vector equation are the solutions of (17.52) of the special form $U(t) = e^{\lambda t} U_0,\ V(t) = e^{\lambda t} V_0$ where $\lambda \in \mathbb{C}$ and U_0, V_0 are (possibly complex) constant (nonzero) 2-vectors. One thus easily finds

$$\lambda \begin{pmatrix} U_0 \\ V_0 \end{pmatrix} = \begin{pmatrix} -2 + m_1 & 0 \\ -2\alpha & m_1 \end{pmatrix} \begin{pmatrix} U_0 \\ V_0 \end{pmatrix} - m_1 e^{-\lambda\tau} \begin{pmatrix} U_0 \\ V_0 \end{pmatrix},$$

thus arriving to the *characteristic equation*

$$\begin{vmatrix} \lambda + 2 - m_1 + m_1 e^{-\lambda\tau} & 0 \\ 2\alpha & \lambda - m_1 + m_1 e^{-\lambda\tau} \end{vmatrix} = 0.$$

This means that we have a double collection of eigenvalues: those satisfying $\lambda - m_1 + m_1 e^{-\lambda\tau} = 0$ and those satisfying $\lambda + 2 - m_1 + m_1 e^{-\lambda\tau}$. Denoting $\lambda = a + ib$, we identify two classes of eigenvalues:

$$\lambda - m_1 + m_1 e^{-\lambda\tau} = 0$$
$$\iff \begin{cases} a - m_1 + m_1 e^{-a\tau}\cos b\tau = 0 \\ b - m_1 e^{-a\tau}\sin b\tau \end{cases} \quad \text{(Class 1)},$$
$$\lambda + 2 - m_1 + m_1 e^{-\lambda\tau} = 0$$
$$\iff \begin{cases} a + 2 - m_1 + m_1 e^{-a\tau}\cos b\tau = 0 \\ b - m_1 e^{-a\tau}\sin b\tau \end{cases} \quad \text{(Class 2)}.$$

We now look for values $\tau = \tau^*$ for which $a = 0$ and $b \neq 0$. We find no eigenvalues of this kind for Class 1, since $-1 + \cos b\tau = 0$ implies $\sin b\tau = 0$, and hence $b = 0$ from the second equation.

However, Class 2 does give us some useful values:

$$2 - m_1 + m_1 \cos b\tau = 0 \Longrightarrow \cos b\tau = \frac{m_1 - 2}{m_1},$$
$$b - m_1 \sin b\tau = 0 \Longrightarrow \sin b\tau = \frac{b}{m_1}.$$

Thus,

$$1 = \cos^2 b\tau + \sin^2 b\tau = \left(\frac{m_1 - 2}{m_1}\right)^2 + \frac{b^2}{m_1^2}$$
$$\Longrightarrow b^2 = m_1^2 - (m_1 - 2)^2 = 4(m_1 - 1).$$

Hence, if $m_1 > 1$, we have

$$\cos b\tau = \frac{m_1 - 2}{m_1} \Longrightarrow b\tau = \arccos\left(\frac{m_1 - 2}{m_1}\right)$$

which is well defined for every $m_1 > 1$.

Summarizing, the set of values

$$b^* = 2\sqrt{m_1 - 1}, \quad \tau^* = \frac{1}{b^*}\left[\arccos\left(\frac{m_1 - 2}{m_1}\right) + 2k\pi\right]$$

corresponds to a (possible) bifurcation point of Hopf type. For instance, for $m_1 = 2$ we have $b^* = 2$ and $\tau^* = k\pi + \pi/4$.

We now need to compute the derivative $a'(\tau^*)$. It is easier now to go back to the complex formulation of Class 2 eigenvalues

$$\lambda + 2 - m_1 + m_1 e^{-\lambda\tau} = 0,$$

and find $d\lambda/d\tau$ by implicit differentiation:

$$\frac{d\lambda}{d\tau} + m_1 e^{-\lambda\tau}\left(-\frac{d\lambda}{d\tau}\tau - \lambda\right) = 0$$

$$\Longrightarrow \frac{d\lambda}{d\tau} = \frac{\lambda e^{-\lambda\tau}}{1 - m_1 e^{-\lambda\tau}\tau} = \frac{\lambda}{1 - m_1 e^{\lambda\tau}\tau}.$$

Concentrating on the specific values $b^* = 2$ and $\tau^* = \pi/4$ we find, at the bifurcation values τ^*, $\lambda^* = ib^*$, that

$$\left.\frac{d\lambda}{d\tau}\right|_{(\tau^*,\lambda^*)} = \frac{ib^*}{1 - m_1 e^{ib^*\tau^*}\tau^*} = -\frac{4\pi}{\pi^2+4} + \frac{8}{\pi^2+4}i.$$

Hence

$$\frac{da}{d\tau}(\tau^*) = -\frac{4\pi}{\pi^2+4} \neq 0$$

and the transversality condition is satisfied. Therefore, a Hopf bifurcation occurs, and a periodic orbit of approximate period

$$T \simeq \frac{2\pi}{b(\tau^*)} = \pi$$

exists for delay values τ near τ^*.

Remark 17. To decide the sub- or supercritical character of the bifurcation a much longer analysis is necessary. On the other hand, for $\tau > 1/2$ there are always positive *real* eigenvalues coming from the first class, which means that the stationary point has become already unstable *before* the delay reaches $\tau^* = \pi/4$ value. Hence, the periodic orbit cannot capture the stability lost by the stationary point, since that stability was already lost.

17.4.2. *Hopf bifurcation for the complex Ginzburg–Landau equation on the whole space and with delayed time feedback*

We come back to the consideration of the complex Ginzburg–Landau equation subjected to a time-delay feedback with local and global terms but now for the case of a spatial domain given by the whole space:

$$\begin{aligned}\partial_t \mathbf{A} &= (1 - \mathrm{i}\omega)\mathbf{A} - (1 + \mathrm{i}\alpha)|\mathbf{A}|^2\mathbf{A} + (1 + \mathrm{i}\beta)\partial_{xx}\mathbf{A} + \mathbf{F},\\ \mathbf{F} &= \mu e^{\mathrm{i}\xi}\left[m_1\mathbf{A} + m_2\langle\mathbf{A}\rangle + m_3\mathbf{A}(t-\tau) + m_4\langle\mathbf{A}(t-\tau)\rangle\right],\end{aligned} \tag{17.53}$$

where

$$\langle\mathbf{A}\rangle = \frac{1}{L}\int_0^L \mathbf{A}(x,t)dx$$

denotes the spatial average of $\mathbf{A}$ over a one-dimensional medium of length L. There are many previous works in the literature dealing with such type of formulations: [30, 31, 62, 63].

Extensive simulations [31] and an analytical stability analysis [63] for a special case representing a Pyragas type feedback [58] ($m_3 = -m_1 = m_l$, $m_4 = -m_2 = m_g$) showed the range of patterns that can be stabilized as function of the local and global feedback terms. If the feedback is global, uniform oscillations can be stabilized for a large range of feedback parameters, while as the contribution of the local feedback term becomes larger, the parameter regions increase where the homogeneous fixed point solution, standing waves and traveling waves are found.

Uniform oscillations $\mathbf{A}(t) = \rho_0 \exp(-\mathrm{i}\theta t)$ are a solution of Eq. (17.53) with amplitude and frequency given by

$$\begin{aligned}\rho_0 &= \sqrt{1 + \mu(m_g + m_l)(\cos(\xi + \theta\tau) - \cos\xi)},\\ \theta &= \omega + \alpha + \mu(m_g + m_l)[\alpha(\cos(\xi + \theta\tau)\\ &\quad - \cos\xi) - (\sin(\xi + \theta\tau) - \sin\xi)].\end{aligned} \tag{17.54}$$

In [63], it is performed a linear stability analysis of uniform oscillations with respect to spatio-temporal perturbations. There, we expressed the complex oscillation amplitude $\mathbf{A}$ as the superposition of a homogeneous mode $\mathbf{H}$ (corresponding to uniform oscillations) with spatially inhomogeneous perturbations,

$$\mathbf{A}(x,t) = \mathbf{H}(t) + \mathbf{A}_+(t)e^{icx} + \mathbf{A}_-(t)e^{-icx}. \tag{17.55}$$

Notice that here we are using the fact that the equation takes place on the whole space, which allows the justification of the spatially inhomogeneous perturbations of the form $\mathbf{A}_+(t)e^{icx} + \mathbf{A}_-(t)e^{-icx}$. Inserting (17.55) into (17.53), and assuming that the amplitudes $\mathbf{A}_\pm$ are small, we obtain a set of equations for $\mathbf{H}$, $\mathbf{A}_+$, and $\mathbf{A}_-^*$ (see [63] for details of this derivation). To investigate linear stability of uniform oscillations with respect to spatio-temporal perturbations, we make the ansatz

$$\begin{aligned}\mathbf{A}_+ &= A_+^0 \exp(-i\theta t)\exp(\lambda t),\\ \mathbf{A}_-^* &= A_-^{*0} \exp(i)\exp(\lambda t),\end{aligned} \tag{17.56}$$

where $\lambda = \lambda_1 + \mathrm{i}\lambda_2$ is a complex eigenvalue. Using ansatz (17.56), we arrive at the following eigenvalue equation:

$$\mathbf{F} = (A + iB - i\lambda_2 + D_1 + iD_2)(A - iB - i\lambda_2 + C_1 + iC_2), \tag{17.57}$$

where we have defined

$$\begin{aligned}
F &= (1+\alpha^2)\rho_0^4,\\
A &= 1 - \lambda_1 - 2\rho_0^2 - \kappa^2,\\
B &= \theta - \omega - 2\alpha\rho_0^2 - \beta\kappa^2,\\
C_1 &= \mu m_l e^{-\lambda_1\tau}\cos(\xi + \theta\tau + \lambda_2\tau) - \mu m_l \cos\xi,\\
C_2 &= -\mu m_l e^{-\lambda_1\tau}\sin(\xi + \theta\tau + \lambda_2\tau) + \mu m_l \sin\xi,\\
D_1 &= \mu m_l e^{-\lambda_1\tau}\cos(\xi + \theta\tau - \lambda_2\tau) - \mu m_l \cos\xi,\\
D_2 &= \mu m_l e^{-\lambda_1\tau}\sin(\xi + \theta\tau - \lambda_2\tau) - \mu m_l \sin\xi.
\end{aligned}$$

We point out that the above eigenvalue equation can be obtained also by a formal linearization argument involving the Fréchet derivatives as in the following section. There is no general analytic solution to Eq. (17.57) for $\lambda_{1,2}$. Thus, Eq. (17.57) must be solved numerically for a given set of parameters. We keep the CGLE parameters α, β, ω and the feedback parameters m_l, m_g, and ξ constant and solve Eq. (17.57) with the FindRoot routine of the Mathematica package [52]. We then find, for each point in the (τ, μ)-space, the functional dependence of λ_1 and λ_2 on κ. Notice that if we assume $\kappa = 0$ the study can be applied to the case of the Stuart–Landau equation, as before.

In general, Eq. (17.57) has multiple solutions, reflected by multiple branches in the dispersion relation. Stability is determined by the sign of λ_1. The curves $\lambda_1(\kappa)$ either lie below $\lambda_1 = 0$, so that uniform oscillations are stable, or they display an interval of κ-values, where $\lambda_1 > 0$, so that uniform oscillations are unstable. At criticality, we have $\lambda_1 = 0$, $\partial_\epsilon \lambda_1 \neq 0$, where ϵ stands for either μ or τ. For the critical wavenumber κ_c, there are two possibilities: $\kappa_c = 0$ or $\kappa_c \neq 0$ ($\pm\kappa_c$ are solutions, although below, we consider only $\kappa_c > 0$ without loss of generality).

Two instabilities are particularly important in our system: the first one is associated with $\kappa_c > 0$ and $\lambda_2(\kappa_c) = 0$, and the second one with $\kappa_c = 0$ and $\lambda_2(\kappa_c) \neq 0$. In Fig. 17.1, we show as an example the control diagram in (μ, τ)-space for $m_l = 0.4$, $m_g = 0.6$. Stable uniform oscillations are observed above the solid curve and to

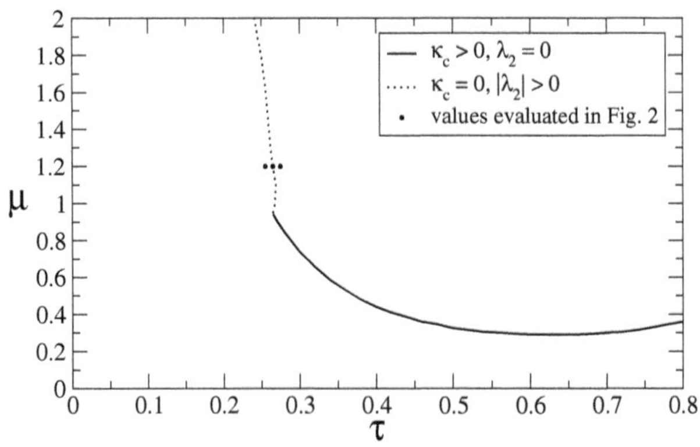

Fig. 17.1. Control diagram in (μ, τ)-space for $m_l = 0.4$, $m_g = 0.6$. The other parameters are $\alpha = -1.4$, $\beta = 2$, $\omega = 2\pi - \alpha$, $\xi = \pi/2$. At the solid curve, uniform oscillations become unstable with respect to perturbations with $\kappa_c > 0$ and $\lambda_2(\kappa_c) = 0$, at the dotted curve, with $\kappa_c = 0$ and $\lambda_2(\kappa_c) \neq 0$. The dots indicate parameter values are further studied in Fig. 17.2.

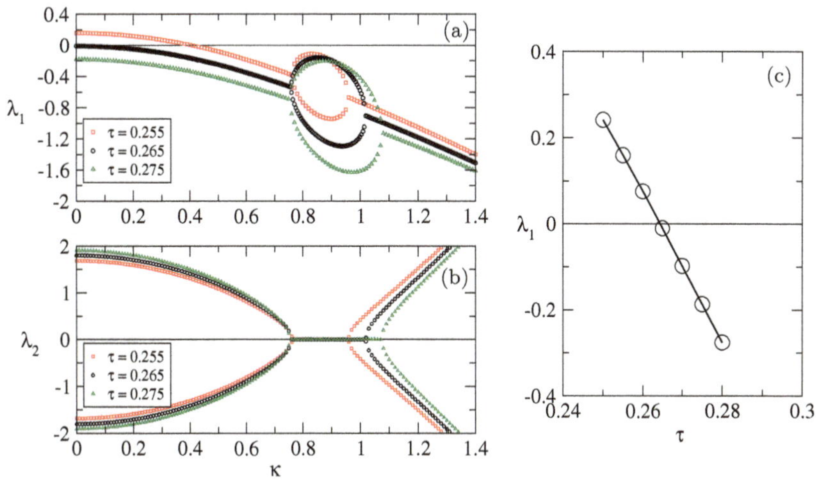

Fig. 17.2. Dispersion relations for three parameter sets close to criticality: $\tau = 0.255$ (red squares), $\tau = 0.265$ (black circles), $\tau = 0.275$ (green triangles). (a) Real part of the eigenvalue as function of the wavenumber κ. (b) Imaginary part of the eigenvalue. The instability is characterized by $\kappa_c = 0$ and $\lambda_2(\kappa_c) \neq 0$ and occurs for $\mu = 1.2$ at $\tau = 0.264399$. (c) Real part of the eigenvalue as function of τ, demonstrating transversality.

the right of the dotted curve. At the solid curve, uniform oscillations become unstable with respect to perturbations with $\kappa_c > 0$ and $\lambda_2(\kappa_c) = 0$, at the dotted curve, with $\kappa_c = 0$ and $\lambda_2(\kappa_c) \neq 0$. In Figs. 17.2(a) and 17.2(b), the dispersion relations $\lambda_{1,2} = \lambda_{1,2}(\kappa)$ are shown for three τ values close to criticality, demonstrating clearly the nature of the underlying instability. In Fig. 2(c), we show that λ_1 crosses $\lambda_1 = 0$ as τ is varied, hence demonstrating transversality. As the uniform oscillations become unstable with respect to a mode with complex conjugated eigenvalues and since ρ_0 remains finite, we infer the presence of a secondary Hopf bifurcation.

17.4.3. *Hopf bifurcation for the delayed CGLE in a bounded domain*

We consider now the case of two spatial dimensions varying on the domain $\Omega = (0, L_1) \times (0, L_2)$. Our goal is to show a bifurcation phenomenon near uniform oscillations for the CGLE in terms of the delay term as parameter. We define the faces of the boundary

$$\Gamma_j = \partial\Omega \cap \{x_j = 0\}, \Gamma_{j+2} = \partial\Omega \cap \{x_j = L_j\}, \quad j = 1, 2,$$

on which we assume periodic boundary conditions and, hence, the problem under study can be formulated as

$$(P_1)\begin{cases} \partial_t \mathbf{u} - (1+i\beta)\Delta\mathbf{u} = (1-i\omega)\mathbf{u} - (1+i\alpha)|\mathbf{u}|^2\mathbf{u} & \\ \qquad\qquad + \mu e^{i\xi}\mathbf{F}(\mathbf{u},t,\tau), & \Omega\times(0,\infty), \\ \left.\begin{aligned} &\mathbf{u}|_{\Gamma_j} = \mathbf{u}|_{\Gamma_{j+2}}, \\ &\left(-\left.\frac{\partial\mathbf{u}}{\partial\mathbf{n}}\right|_{\Gamma_j} =\right)\left.\frac{\partial\mathbf{u}}{\partial x_j}\right|_{\Gamma_j} \\ &\quad = \left.\frac{\partial\mathbf{u}}{\partial x_j}\right|_{\Gamma_{j+2}}\left(=\left.\frac{\partial\mathbf{u}}{\partial\mathbf{n}}\right|_{\Gamma_{j+2}}\right)\end{aligned}\right\}, & \partial\Omega\times(0,\infty), \\ \mathbf{u}(x,s) = \mathbf{u}_0(x,s), & \Omega\times[-\tau,0], \end{cases}$$

where $\mathbf{n}$ is the outpointing normal unit vector, and

$$\mathbf{F}(\mathbf{u},t,\tau) = [m_1\mathbf{u}(x,t) + m_2\langle\mathbf{u}(t)\rangle + m_3\mathbf{u}(x,t-\tau) + m_4\langle\mathbf{u}(t-\tau)\rangle]$$

with

$$\langle\mathbf{u}(s)\rangle = \frac{1}{|\Omega|}\int_\Omega \mathbf{u}(x,s)dx.$$

Again, the parameters $\alpha, \beta, \omega, \mu, \xi, m_i$ and τ are real, while $\mathbf{u}(x,t) = \mathbf{u}_1(x,t) + i\mathbf{u}_2(x,t)$ is complex.

We study the stability of *uniform oscillations*, i.e., solutions of (P_1) of the form $\mathbf{v}_{\text{uo}}(t) = \rho_0 e^{-i\theta t}$ which determines completely ρ_0 and θ. We are interested in the Hopf bifurcation close to $\mathbf{v}_{\text{uo}}(t)$ which gives rise to some paths on a suitable torus (for a different study dealing with invariant tori, see [64]).

In order to avoid the application of very sophisticated techniques (dealing with periodic solutions), we can reduce the study to the Hopf bifurcation near a stationary solution of some auxiliary problem by introducing the change of unknown $\mathbf{z}(x,t) = \mathbf{v}(x,t)e^{i\theta t}$ where $\mathbf{v}(x,t)$

is a solution of (P_1). Thus, $\mathbf{z}(x,t)$ satisfies

$$(P_2)\begin{cases} \partial_t \mathbf{z} - (1+i\beta)\Delta\mathbf{z} = (1+i\theta)\mathbf{z} \\ \quad -(1+i\alpha)|\mathbf{z}|^2\mathbf{z} + \mu e^{i\xi} \\ \quad \times[m_1\mathbf{z} + m_2\langle\mathbf{z}\rangle + e^{i(\omega+\theta)\tau} \\ \quad \times(m_3\mathbf{z}(t-\tau) + m_4\langle\mathbf{z}(t-\tau)\rangle)], & \Omega\times(0,\infty), \\ \left.\begin{array}{l} \mathbf{z}|_{\Gamma_j} = \mathbf{z}|_{\Gamma_{j+2}}, \\ \left(-\left.\dfrac{\partial\mathbf{z}}{\partial\mathbf{n}}\right|_{\Gamma_j} =\right)\left.\dfrac{\partial\mathbf{z}}{\partial x_j}\right|_{\Gamma_j} = \left.\dfrac{\partial\mathbf{z}}{\partial x_j}\right|_{\Gamma_{j+2}} \\ \quad\times\left(=\left.\dfrac{\partial\mathbf{z}}{\partial\mathbf{n}}\right|_{\Gamma_{j+2}}\right) \end{array}\right\}, & \partial\Omega\times(0,\infty), \\ \mathbf{z}(x,s) = \mathbf{u}_0(x,s)e^{i(\omega-\theta)s}, & \Omega\times[-\tau,0]. \end{cases}$$

Now, $\mathbf{v}_{\text{uo}}(t) = \rho_0 e^{-i\theta t}$ is a uniform oscillation if and only if $\mathbf{z}(x,t) = \mathbf{v}_\infty(t)e^{i\theta t} = \mathbf{z}_\infty = \rho_0$ is a stationary solution of (P_2), i.e.,

$$\begin{aligned} \mathbf{0} = (1+i\theta)\mathbf{z}_\infty - (1+i\alpha)\left|\mathbf{z}_\infty\right|^2\mathbf{z}_\infty + \mu e^{i\xi} \\ \times[m_1 + m_2 + e^{i(\omega+\theta)\tau}(m_3+m_4)]\mathbf{z}_\infty. \end{aligned}$$

17.4.3.1. *An abstract Hopf bifurcation theorem for semilinear functional equations*

We shall apply to our setting an abstract result due to Wu (see [67, Theorem 2.1]) stated for problems of the type

$$\begin{cases} \dfrac{du}{dt}(t) + Au(t) = L(\mu, u_t(\cdot)) + g(u_t(\cdot)) \text{ in } X, \\ u(s) = u_0(s), & s\in[-\tau,0] \end{cases}$$

on a Banach space X, where $u_t : [-\tau,0] \to X$, under the following list of conditions:

- (H_1) A generates an analytic compact semigroup $\{T(t)\}_{t\geq 0}$.
- (H_2) The point spectrum of A consists of a sequence of real number $\{\mu_k\}_{k\geq 1}$ with the corresponding eigenspace M_k and the projection $P_k : X \to M_k$. Moreover, if $\sum_{k=1}^{\infty} x_k = 0$ for $x_k \in M_k$ then each x_k must be zero;

(H_3) Every $x \in D(A)$ has a unique expression $x = \sum_{k=1}^{\infty} P_k x$ and $Ax = \sum_{k=1}^{\infty} \mu_k P_k x$.

(H_4) The mapping $L : \mathbb{R} \times C \to X$ (with $C := C([-\tau, 0] : X)$) is C^k-smooth ($k \geq 4$) and is given by

$$L(\mu, \phi) = \int_{-\tau}^{0} \phi(\theta) d\eta(\mu, \theta)$$

for any $(\mu, \phi) \in \mathbb{R} \times C$, for a function $\eta(\mu, .)$: $[-\tau, 0] \to B(X, X)$ of bounded variation. Moreover, $L(\mu, P_k\phi) \in M_k$, $k \geq 1$, $\phi \in C$ and $L(\mu, \sum_{k=1}^{\infty} P_k\phi) = \sum_{k=1}^{\infty} L(\mu, P_k\phi)$ for any $\phi \in C$ such that $\sum_{k=1}^{\infty} P_k\phi \in C$, where $P_k\phi$ is defined by $(P_k\phi)(\theta) = P_k\phi(\theta)$ for $\theta \in [-\tau, 0]$;

(H_5) $g : \mathbb{R} \times C \to X$ has kth-continuous Fréchet derivatives with $g(\mu, 0) = 0$ and $Dg(\mu, 0) = 0$ for $\mu \in \mathbb{R}$.

(H_6) There exist $\mu_0 \in \mathbb{R}$ and $\omega_0 > 0$ such that $\pm i\omega_0$ are simple characteristic values of the linear equation

$$\dot{u}(t) + Au(t) = L(\mu_0, u_t(\cdot)) \tag{17.58}$$

and all other characteristic values have negative real parts;

(H_7) *Transversality condition*. If μ is near μ_0 the eigenvalues of the corresponding problem (17.58) are given by $\lambda(\mu) = \alpha(\mu) + i\omega(\mu)$, $\lambda(\mu_0) = i\omega_0$, $\lambda(\mu)$ is C^k-smooth in μ and

$$\alpha'(\mu_0) \neq 0.$$

Remark 18. A careful reading of the proof of [67, Theorem 2.1] allows to see that the use of the same notation u_t in the terms $L(\mu, u_t(\cdot))$ and $g(u_t(\cdot))$ does not need that the kernels involved in each of the possible nonlocal terms be exactly the same. So, in particular, the conclusion remains valid in the special case in which $g(u_t(\cdot)) = g(u(\cdot))$, i.e., without delay or neutral term.

17.4.3.2. *Application to the delayed CGLE on a bounded domain*

Motivated by the special form of the nonlinear term of the equation in (P_2) we shall take $X = \mathbf{L}^4(\Omega)$ and $Y = \mathbf{L}^{4/3}(\Omega)$. A detailed analysis of the associated diffusion operator is consequence of some previous results in the literature: see, e.g., [5]. Notice that the operator $A\mathbf{u}$ can be formulated matricially as

$$\begin{pmatrix} u_1 \\ u_2 \end{pmatrix} \rightarrow \begin{pmatrix} \Delta & -\beta\Delta \\ \beta\Delta & \Delta \end{pmatrix} \begin{pmatrix} u_1 \\ u_2 \end{pmatrix}.$$

So, if $\beta \neq 0$ the diffusion matrix has a nonzero antisymmetric part. In particular, A is the generator of a semigroup of contractions $\{T(t)\}_{t\geq 0}$ on X and the compactness of the semigroup is consequence of the compactness of the inclusion $D(A) \subset X$ (notice that, since $N = 2$, $\mathbf{W}^{1,4}(\Omega) \subset \mathbf{W}^{1,4/3}(\Omega) \subset \mathbf{C}(\overline{\Omega})$ with compact imbedding) and some regularity results for nonsymmetric systems. A study of the eigenvalues of A can be found, e.g., in [65]

Concerning the rest of the terms of the equation in (P_2), we define $g(\mathbf{u}) = -(1+i\alpha)\left|\mathbf{u}\right|^2 \mathbf{u}$ with $D(g) = \mathbf{L}^{12}(\Omega)$. By using the characterization of the semi inner-bracket $[,]$ for the spaces $L^p(\Omega)$ (see, e.g., [20]), it is easy to see that $\mathbf{B} = -\mathbf{g}$ is an accretive operator on X, which is dominated by A, i.e.,

$$D_X(A) \subset D_X(B) \quad \text{and} \quad |Bu| \leq k\left|A^0 u\right| + \sigma(|u|)$$

for any $u \in D_X(A)$, some $k < 1$ and some continuous function $\sigma : \mathbb{R} \rightarrow \mathbb{R}$.

Here and in what follows, $|\cdot|$ denotes the norm in the space X (in contrast to the norm in space C which will be denoted by $\|\cdot\|$ if there is no ambiguity, when handling two spaces X and Y the corresponding norms will be indicated), $\left|A^0 u\right| := \inf\{|\xi| : \xi \in Au\}$ for $u \in D_X(A)$. In particular, the operator $A+B$ is also an accretive operator on X.

In order to calculate the Fréchet differential of Nemitsky operator $g(\mathbf{u})$, it is useful to start analyzing the Gateaux derivative of the complex function $\mathbf{h}(\mathbf{z}) := \|\mathbf{z}\|^2 \mathbf{z}$ in the direction of an arbitrary vector $\mathbf{v}$ of $\mathbb{C}$

$$\lim_{\substack{\beta\in\mathbb{R} \\ |\beta|\to 0}} \frac{\mathbf{h}(\mathbf{z}_0 + \beta\mathbf{v}) - \mathbf{h}(\mathbf{z}_0)}{|\beta|} = \mathbf{z}_0^2\overline{\mathbf{v}} + 2\left\|\mathbf{z}_0\right\|^2 \mathbf{v}.$$

Then, we identify the Fréchet differential of operator $g(\mathbf{u})$ as

$$D\mathbf{B}(\mathbf{y})\mathbf{v} = (1+i\alpha)[\mathbf{y}^2\overline{\mathbf{v}} + 2\left\|\mathbf{y}\right\|^2 \mathbf{v}]. \tag{17.59}$$

Since we have $\|D\mathbf{B}(\mathbf{y})\| \leq c\|\mathbf{y}\|^2$, by the results on the Fréchet differentiability of Nemitsky operators (see [7, Theorem 2.6] (with $p = 4$)) we get that, if we take $Y = \mathbf{L}^{4/3}(\Omega)$, then there exists $\delta^B > 0$ such that $\mathbf{B}$ is Fréchet differentiable as function from $B_{\delta^B}(w) = \{z \in D(\mathbf{B}); |w - z| < \delta^B\}$ into Y, and that the Fréchet derivative is locally Lipschitz continuous.

The nonlocal term is defined by

$$F(\mathbf{u}_t) = (1 + i\theta)\mathbf{u}(t) + \mu e^{i\xi}[m_1\mathbf{u}(t) + m_2\langle\mathbf{u}(t)\rangle + e^{i(\omega+\theta)\tau}(m_3\mathbf{u}(t-\tau) + m_4\langle\mathbf{u}(t-\tau)\rangle)],$$

is locally Lipschitz continuous and its Fréchet derivative is given by

$$\mathrm{D}F(\widehat{\mathbf{y}})\mathbf{v}(t) = -(1 + i\theta)\mathbf{v}(t) - \mu e^{i\xi}[m_1\mathbf{v}(t) + m_2\langle\mathbf{v}(t)\rangle - e^{i(\omega+\theta)\tau}(m_3\mathbf{v}(t-\tau) - m_4\langle\mathbf{v}(t-\tau)\rangle)].$$

In consequence, the operator $y \to Ay + \mathrm{D}B(w)y - \mathrm{D}F(\widehat{w})\left(e^{\omega^*\cdot}y\right)$ belongs to $\mathcal{A}(\omega^* : Y)$, for some $\omega^* \in \mathbb{C}$ with $\mathrm{Re}\,\omega^* = \gamma^* < 0$. This means that the operator $y \to Ay + \mathrm{D}B(w)y - \mathrm{D}F(\widehat{w})\left(e^{\omega^*\cdot}y\right) + \omega^* y$ is accretive in $Y = \mathbf{L}^{4/3}(\Omega)$. We recall (see [7]) that this differentiability of B does not hold if we take $X = Y = \mathbf{L}^2(\Omega)$.

We also recall that in [30] the existence (and uniqueness) of a mild solution of problem (P_2) was obtained through a pseudolinearization argument near a stationary solution $\widehat{w}$.

Theorem 8 ([30]). *Assume (H_1)–(H_7). Then, there exist $\alpha > 0$, $\beta > 0$ and $M \geq 1$ such that if $u_0 \in B_\beta^X(\widehat{w})$, $u_0(s) \in D_X(B)$ for any $s \in [-\tau, 0]$ then the solution $u(\cdot : u_0)$ of (17.59) exists on $[-\tau, +\infty)$ and*

$$|u(t : u_0) - w| \leq Me^{-\alpha t}\|u_0 - \widehat{w}\|, \textit{ for any } t > 0.$$

Moreover, there exist $\alpha^* > 0$, $\beta^* \in (0, \beta]$ and $M^* \geq 1$ such that if $u_0 \in B_{\beta^*}^{X\cap Y}(\widehat{w})$, $u_0(s) \in D_X(B) \cap D_Y(B)$ for any $s \in [-\tau, 0]$ then, for any $t > 0$,

$$|u(t : u_0) - w|_X + |u(t : u_0) - w|_Y \leq M^* e^{-\alpha^* t}(\|u_0 - \widehat{w}\|_X + \|u_0 - \widehat{w}\|_Y).$$

We can get better a priori estimates on the sup norm of the solution $\mathbf{u}$ if we assume more regular initial data in such a way that $u_0 \in B_{\beta^*}^{X\cap Y}(\widehat{w})$, $u_0(s) \in D(A) \cap D_X(B) \cap D_Y(B)$ for any $s \in [-\tau, 0]$. Indeed, the solution can be found (after technical arguments) as a fixed point for the application $f \to Q_1(Q_2(f))$, with $w = Q_2 f$ (for $f \in W^{1,1}(0, T : X)$, for any arbitrary $T > 0$) being the solution of the problem

$$\begin{cases} \dfrac{dw}{dt}(t) + Aw(t) + B(w(t)) = f(t) \quad \text{in } X, \\ w(0) = w_0, \end{cases}$$

and Q_1 a suitable operator (see [66, Theorem 5.3.1]). Since X is a reflexive Banach space, we know (see, e.g., [20, Lemma 7.8]) that $w_0 \in D(A) \cap D_X(B)$ implies that $w(t) \in D(A) \cap D_X(B)$ for a.e. $t \in (0, T)$ and that

$$\|Aw(t)\|_X \leq C(\|Aw_0\|_X + \|B(w_0)\|_X, \|f\|_{W^{1,1}(0,T:X)}).$$

Thus, by the Sobolev imbedding theorems we know that

$$\|w(t)\|_{\mathbf{C}(\overline{\Omega})} \leq M$$

for a.e. $t \in (0, T)$ with $M = M(\|Aw_0\|_X + \|B(w_0)\|_X, \|f\|_{W^{1,1}(0,T:X)})$. In particular, this property remains true for the fixed point of $Q_1(Q_2(f))$ (see [66, Theorem 5.3.1]) and thus

$$\|\mathbf{u}(t)\|_{\mathbf{C}(\overline{\Omega})} \leq M^*$$

for a suitable $M^* = M*(\|Au_0\|_{C([-\tau,0];X)} + \|B(w_0)\|_{C([-\tau,0];X)}, F)$. In consequence, without any loss of generality, we can replace function $\mathbf{g}$ by the truncated one $\mathbf{g}_{M^*}(\mathbf{u})$:

$$\mathbf{g}_{M^*}(\mathbf{u}) = \begin{cases} -(1 + i\alpha)\,|\mathbf{u}|^2\,\mathbf{u} & \text{if } |\mathbf{u}| \leq M^*, \\ -2(1 + i\alpha)\,(2M^*)^2\,\mathbf{u} & \text{if } |\mathbf{u}| \geq M^*, \end{cases}$$

and with $\mathbf{g}_{M^*}(\mathbf{u})$ a C^k-smooth function generating an accretive operator $\mathbf{B}_{M^*} = -\mathbf{g}_{M^*}$ on X dominated by A as before. This proves that, at least for regular initial data, **u** coincides with the solution of

$$\begin{cases} \dfrac{du}{dt}(t) + Au(t) = L(\mu, u_t(.)) + g_{M^*}(u_t(.)) & \text{in } X, \\ u(s) = u_0(s), & s \in [-\tau, 0]. \end{cases}$$

Thanks to this argument we can verify now the assumption (H_5) since by the results of Ambrosetti and Prodi (see [7, Sec. 1.3, Chap. 1]) we know that the Nemitsky operator associated to g_{M^*} has kth-continuous Fréchet derivatives on any $\mathbf{L}^p(\Omega)$, $p > 1$.

Remark 19. By introducing the representation operator $\mathbf{P} : \mathbb{R}^2 \to \mathbb{C}$, $\mathbf{P}(\rho,\phi) = \rho e^{i\phi}$ it is clear that the quasilinear operator $A\mathbf{P}(\mathbf{q})$ obtained from the operator $A\mathbf{u} = -(1+i\beta)\Delta\mathbf{u}$ satisfies also condition $A \in \mathcal{A}(\omega)$ (since $\mathbf{P}$ is merely a change of variables). We point out that

$$A\mathbf{P}(\mathbf{q}) = -(1+i\beta)[\Delta\rho - \rho\left|\nabla\phi\right|^2 + i(2\nabla\rho\cdot\nabla\phi + \rho\Delta\phi)]e^{i\phi}.$$

Then, the formal linearization of the operator $\mathbf{E}(\mathbf{q}) := A\mathbf{P}(\mathbf{q})$ at $\mathbf{q}^*(x,y) := \mathbf{y} \equiv \rho_0$ becomes

$$D\mathbf{E}(\mathbf{q}^*)(\rho e^{i\phi}) = -(1+i\beta)[\Delta\rho + i\rho_0\Delta\phi]e^{i\phi}.$$

Notice that the linearization of $\mathbf{C}(\mathbf{q})^{-1}A\mathbf{P}(\mathbf{q})$ needs a slight modification of the above linear expression. Nevertheless by applying the representation operator $\mathbf{P}$, after the linearization used in the abstract theorem, we get a curious result relating two nonlinear problems which are closed (in some sense) in the same spirit as the *pseudo-linearization principle* obtained in [30].

17.4.3.3. *Some comments on the associated transversality assumption*

Concerning problem (P_2), we give an outline of the study of eigenvalues and its implications on the associated transversality condition. The eigenvalue equation can be obtained by a linearization argument involving the Fréchet derivative of the nonlinear part, as in the preceding section.

As usual, the linear structure of the equation leads to the search of nontrivial solutions $\mathbf{z}(x)$ of the form $\mathbf{A_k}w^j_{\mathbf{k}}(x)$, with $j = 1, 2,$

where $w_{\mathbf{k}}^j(x)$ are the eigenfunctions for the usual Laplacian operator Δ with periodic boundary conditions on $\Omega = (0, L_1) \times (0, L_2)$. The eigenvalues of this problem are given by $\lambda_0^0 = 0$, $\lambda_{\mathbf{k}}^0 = 4\pi\left(\frac{k_1^2}{L_1^2} + \frac{k_2^2}{L_2^2}\right)$; $k_1, k_2 \in \mathbb{N}$ with the associate eigenfunctions

$$
\begin{aligned}
w_0 &= \frac{1}{\sqrt{|\Omega|}}, w_{\mathbf{k}}^1 = \sqrt{\frac{2}{|\Omega|}} \cos 2\pi\mathbf{kx}, w_{\mathbf{k}}^2 \\
&= \sqrt{\frac{2}{|\Omega|}} \sin 2\pi\mathbf{kx}, \text{ with } |\Omega| = L_1 L_2,
\end{aligned}
$$

where we have written $\mathbf{kx} := \left(\frac{k_1}{L_1}x_1 + \frac{k_2}{L_2}x_2\right)$. This study can be found in [65]. We introduce the notation $\lambda_{\mathbf{k}} = a_{\mathbf{k}} + ib_{\mathbf{k}}$ for the real and imaginary parts of the eigenvalues of the problem, and taking into account Fréchet derivative of the nonlinear part (17.59), the eigenvalue equations for the problem (P_2) are

$$
\begin{cases}
(a_{\mathbf{k}} + ib_{\mathbf{k}})[v_r + iv_i] - (1 + i\beta)(-\lambda_{\mathbf{k}})[v_r + iv_i] \\
\quad = (1 + i\theta)[v_r + iv_i] - (1 + i\alpha)[3\rho_0^2 v_r + i\rho_0^2 v_i] \\
\qquad + \mu e^{i\xi}\left[m_1 + m_2\delta_{0\mathbf{k}} + e^{-a\tau + i(\omega+\theta-b)\tau}(m_3 + m_4\delta_{0\mathbf{k}})\right][v_r + iv_i],
\end{cases}
$$

where v_r and v_i are the real and imaginary parts of the linearization $\mathbf{v}$, and $\delta_{0\mathbf{k}}$ denotes the Kronecker delta function. We arrive at

$$
\begin{cases}
a_{\mathbf{k}}v_r - b_{\mathbf{k}}v_i = -\lambda_{\mathbf{k}}^0 v_r + \beta\lambda_{\mathbf{k}}^0 v_i + \left(\left[1 - 3\rho_0^2\right]v_r + \left[\alpha\rho_0^2 - \theta\right]v_i\right) \\
\quad +\mu(m_1 + m_2\delta_{0k})\left[v_r\cos\xi - v_i\sin\xi\right] + \{\mu e^{-a_{\mathbf{k}}\tau}(m_3 + m_4\delta_{0k}) \\
\quad \times\left[\cos(\xi + (\omega + \theta - b_{\mathbf{k}})\tau)v_r - \sin(\xi + (\omega + \theta - b_{\mathbf{k}})\tau)v_i\right]\}, \\
b_{\mathbf{k}}v_r + a_{\mathbf{k}}v_i = -\beta\lambda_{\mathbf{k}}^0 v_r + \lambda_{\mathbf{k}}^0 v_i + (v_i + \theta v_r) - \left[\rho_0^2 v_i - 3\alpha\rho_0^2 v_r\right] \\
\quad + \mu(m_1 + m_2\delta_{0k})\left[v_r\sin\xi + v_i\cos\xi\right] + \{\mu e^{-a_{\mathbf{k}}\tau}(m_3 + m_4\delta_{0k}) \\
\quad \times\left[\sin(\xi + (\omega + \theta - b_{\mathbf{k}})\tau)v_r + \cos(\xi + (\omega + \theta - b_{\mathbf{k}})\tau)v_i\right]\}.
\end{cases}
$$

To show the procedure, without loss of generality, we consider the case

$$
m_3 + m_4\delta_{0\mathbf{k}} = 0. \tag{17.60}
$$

This represents a special, and important, choice of the combination of instantaneous and delayed terms in the global feedback, none of them necessarily zero. The equations for the eigenvalues become

$$\begin{cases} a_{\mathbf{k}}v_r - b_{\mathbf{k}}v_i = -\lambda^0_{\mathbf{k}}v_r + \beta\lambda^0_{\mathbf{k}}v_i + \left(\left[1-3\rho_0^2\right]v_r + \left[\alpha\rho_0^2-\theta\right]v_i\right) \\ \quad +\mu(m_1+m_2\delta_{0k})\cos\xi v_r - \mu(m_1+m_2\delta_{0k})\sin\xi v_i, \\ b_{\mathbf{k}}v_r + a_{\mathbf{k}}v_i = -\beta\lambda^0_{\mathbf{k}}v_r + \lambda^0_{\mathbf{k}}v_i + (v_i+\theta v_r) - \left[\rho_0^2 v_i - 3\alpha\rho_0^2 v_r\right] \\ \quad +\mu(m_1+m_2\delta_{0k})\sin\xi v_r + \mu(m_1+m_2\delta_{0k})\cos\xi v_i. \end{cases}$$

If we call

$$C_1\left(\mu,m_1,m_2,\xi,\lambda^0_{\mathbf{k}}\right) = 1-\lambda^0_{\mathbf{k}} - \mu(m_1+m_2\delta_{0k})\cos\xi,$$
$$C_2\left(\mu,m_1,m_2,\xi,\lambda^0_{\mathbf{k}}\right) = 1+\lambda^0_{\mathbf{k}} + \mu(m_1+m_2\delta_{0k})\cos\xi,$$
$$D\left(\beta,\mu,m_1,m_2,\xi,\lambda^0_{\mathbf{k}}\right) = -\beta\lambda^0_{\mathbf{k}} + \mu(m_1+m_2\delta_{0k})\sin\xi,$$

we obtain

$$\begin{cases} \left(a_{\mathbf{k}} - \left[C_1-3\rho_0^2\right]\right)v_r - \left(b_{\mathbf{k}} + \left[\alpha\rho_0^2-\theta-D\right]\right)v_i = 0, \\ \left(b_{\mathbf{k}} - \left[-3\alpha\rho_0^2+\theta+D\right]\right)v_r + \left(a_{\mathbf{k}} - \left[C_2-\rho_0^2\right]\right)v_i = 0. \end{cases}$$

The compatibility of this system implies

$$\det\begin{pmatrix} a_{\mathbf{k}} - \left[C_1-3\rho_0^2\right] & -b_{\mathbf{k}} - \left[\alpha\rho_0^2-\theta-D\right] \\ b_{\mathbf{k}} - \left[-3\alpha\rho_0^2+\theta+D\right] & a_{\mathbf{k}} - \left[C_2-\rho_0^2\right] \end{pmatrix} = 0,$$

that is

$$\begin{cases} \left(a_{\mathbf{k}} - \left[C_1-3\rho_0^2\right]\right)\left(a_{\mathbf{k}} - \left[C_2-\rho_0^2\right]\right) \\ \quad = \left(b_{\mathbf{k}} - \left[-3\alpha\rho_0^2+\theta+D\right]\right)\left(b_{\mathbf{k}} + \left[\alpha\rho_0^2-\theta-D\right]\right). \end{cases} \tag{17.61}$$

This expression is of the same type as (17.57) and, similarly, there is no general analytic solution for $a_{\mathbf{k}}$ and $b_{\mathbf{k}}$. Thus, Eq. (17.61) must also be solved numerically for a given set of parameters, to find the numerical values of the eigenvalues as in Eq. (17.57). One of the relevant parameter spaces of the representation is the one of (τ,μ) because they are the parameters of the perturbation.

Although the explicit analytical representation of the functions $a_{\mathbf{k}}$ and $b_{\mathbf{k}}$ is not possible, we can still say something analytic in the study of the transversality, already proved by the above numerical computation. From the equation (17.61), it is possible to find the implicit derivative

$$\left[\frac{d}{d\tau}a_{\mathbf{k}}\right]_{a_{\mathbf{k}}=0}.$$

The analytic computation are rather involved. We show how to proceed in a simpler, and still very important example

$$m_1+m_2\delta_{0\mathbf{k}} = 0, \tag{17.62}$$

where a remark similar as the one made for the expression (17.60) remains valid, in this case for the local part of the perturbation. For the case (17.62), we have

$$\begin{aligned} C_1\left(\mu, m_1,m_2, \xi, \lambda^0_{\mathbf{k}}\right) &= 1 - \lambda^0_{\mathbf{k}}, \\ C_2\left(\mu, m_1,m_2, \xi, \lambda^0_{\mathbf{k}}\right) &= 1 + \lambda^0_{\mathbf{k}}, \\ D\left(\beta, \mu, m_1,m_2, \xi, \lambda^0_{\mathbf{k}}\right) &= -\beta\lambda^0_{\mathbf{k}}. \end{aligned}$$

If we expand Eq. (17.61) for this case,

$$\begin{cases} a^2_{\mathbf{k}} - 2\left[1-2\rho_0^2\right] a_{\mathbf{k}} + \left(\left[1-\lambda^0_{\mathbf{k}} - 3\rho_0^2\right]\left[1+\lambda^0_{\mathbf{k}} - \rho_0^2\right]\right) \\ \quad = -b^2_{\mathbf{k}} + 2\left[-\beta\lambda^0_{\mathbf{k}} + \alpha\rho_0^2 + \theta\right] b_{\mathbf{k}} \\ \qquad + \left(\left[-\beta\lambda^0_{\mathbf{k}} + 3\alpha\rho_0^2 + \theta\right]\left[+\beta\lambda^0_{\mathbf{k}} + \alpha\rho_0^2 - \theta\right]\right), \end{cases}$$

and differentiate implicitly

$$\begin{cases} 2a_{\mathbf{k}}\dfrac{d}{d\tau}a_{\mathbf{k}} - 2\left[1-2\rho_0^2\right]\dfrac{d}{d\tau}a_{\mathbf{k}} - a_{\mathbf{k}}\dfrac{d}{d\tau}\left(2\left[1-2\rho_0^2\right]\right) \\ \quad +\dfrac{d}{d\tau}\left(1-\left(\lambda^0_{\mathbf{k}}\right)^2 - 2\left[2+\lambda^0_{\mathbf{k}}\right]\rho_0^2 + 3\rho_0^4\right) \\ \quad = -2b_{\mathbf{k}}\dfrac{d}{d\tau}b_{\mathbf{k}} + 2\left[-\beta\lambda^0_{\mathbf{k}} + \alpha\rho_0^2 + \theta\right]\dfrac{d}{d\tau}b_{\mathbf{k}} \\ \qquad -b_{\mathbf{k}}\dfrac{d}{d\tau}\left(2\left[-\beta\lambda^0_{\mathbf{k}} + \alpha\rho_0^2 + \theta\right]\right) \\ \quad +\dfrac{d}{d\tau}\left(\left[-\beta\lambda^0_{\mathbf{k}} + 3\alpha\rho_0^2 + \theta\right]\left[+\beta\lambda^0_{\mathbf{k}} + \alpha\rho_0^2 - \theta\right]\right). \end{cases}$$

The derivative of the real part $a_{\mathbf{k}}$ in the value $a_{\mathbf{k}} = 0$ can be written as

$$\begin{cases} \left[-2(1-2\rho_0^2)\frac{d}{d\tau}a_{\mathbf{k}}\right]_{a_{\mathbf{k}}=0} \\ \quad = \left[-\frac{d}{d\tau}\left(1-(\lambda_{\mathbf{k}}^0)^2 - 2\left[2+\lambda_{\mathbf{k}}^0\right]\rho_0^2 + 3\rho_0^4\right)\right]_{a_{\mathbf{k}}=0} \\ \quad +2\left[-b_{\mathbf{k}}\frac{d}{d\tau}b_{\mathbf{k}} + \left[-\beta\lambda_{\mathbf{k}}^0 + \alpha\rho_0^2 + \theta\right]\frac{d}{d\tau}b_{\mathbf{k}} \right. \\ \quad \left. -b_{\mathbf{k}}\frac{d}{d\tau}\left(\left[-\beta\lambda_{\mathbf{k}}^0 + \alpha\rho_0^2 + \theta\right]\right)\right]_{a_{\mathbf{k}}=0} \\ \quad + \left[\frac{d}{d\tau}\left(\left[-\beta\lambda_{\mathbf{k}}^0 + 3\alpha\rho_0^2 + \theta\right]\left[+\beta\lambda_{\mathbf{k}}^0 + \alpha\rho_0^2 - \theta\right]\right)\right]_{a_{\mathbf{k}}=0}. \end{cases}$$

The coefficient of the derivative of $a_{\mathbf{k}}$,

$$-2(1-2\rho_0^2) = -2\left[1-2(1+\mu\cos\xi)\right] = 2(1+2\mu\cos\xi),$$

does not vanish either for stability reasons as can be seen, e.g., in [30] and references therein.

Acknowledgments

The research of AC and JID was partially supported by the project ref. MTM2017-85449-P of the DGISPI (Spain). Moreover, AC acknowledges support also from the research group MMNL from UPM and JID was member of the Research Group MOMAT (Ref. 910480) supported by UCM.

References

[1] A. Aftalion and F. Pacella, Uniqueness and nondegeneracy for some nonlinear elliptic problems in a ball, *J. Differential Equations* **195**(2) (2003) 380–397.

[2] A. Aftalion and F. Pacella, Morse index and uniqueness for positive radial solutions of p-Laplace equations, *Trans. Amer. Math. Soc.* **356**(11) (2004) 4255–4272.

[3] S. Agarwal and D. Bahuguna, Existence and uniqueness of strong solutions to nonlinear nonlocal functional differential equations, *Electron. J. Differential Equations* **52** (2004) 1–9.

[4] S. Antontsev, J. P. Dias and M. Figueira, Complex Ginzburg–Landau equation with absorption: existence, uniqueness and localization properties, *J. Math. Fluid, Mech.* **16** (2014) 211–223.

[5] H. Amann, Dynamic theory of quasilinear parabolic equations: II Reaction–diffusion systems, *Diff. Int. Equ.* **3** (1990) 13–75.

[6] I. S. Aranson and L. Kramer, The world of the complex Ginzburg–Landau equation, *Rev. Mod. Phys.* **74** (2002) 99–143.

[7] A. Ambrosetti and G. Prodi, *A Primer of Nonlinear Analysis* (Cambridge University Press, Cambridge, 1993).

[8] S. N. Antontsev, J. I. Díaz and S. Shmarev, *Energy Methods for Free Boundary Problems: Applications to Non-linear PDEs and Fluid Mechanics*, Progress in Nonlinear Differential Equations and Their Applications, Vol. 48 (Bikhäuser, Boston, 2002).

[9] F. M. Atay, *Complex Time-Delay Systems* (Springer, Berlin, 2010).

[10] P. Auscher, L. Barthélemy, Ph. Bénilan and E. M. Ouhabaz, Absence de la L^∞-contractivité pour les semi-groupes associés aux opérateurs elliptiques complexes sous forme divergence, *Potential Anal.* **12** (2000) 169–189.

[11] D. Battogtokh, A. Preusser and A. S. Mikhailov, Controlling turbulence in the complex Ginzburg–Landau equation II. Two-dimensional systems, *Physica D* **106**(3–4) (1997) 327–362.

[12] D. Battogtokh and A. S. Mikhailov, Controlling Turbulence in the complex Ginzburg–Landau equation, *Physica D* **90** (1996) 84–95.

[13] P. Bégout and J. I. Díaz, On a nonlinear Schrödinger equation with a localizing effect, *C. R. Acad. Sci. Paris* **342** Sér. I (2006) 459–463.

[14] P. Bégout and J. I. Díaz, Localizing estimates of the support of solutions of some nonlinear Schrödinger equations: The Stationary Case, *Ann. Inst. Henri Poincaré (C) Non Linear Anal.* **29** (2012) 35–58.

[15] P. Bégout and J. I. Díaz, Existence of weak solutions to some stationary Schrödinger equations with singular nonlinearity, *Rev. R. Acad. Cien. Serie A. Mat. (RACSAM)* **109**(1) (2015) 43–63.

[16] P. Bégout and J. I. Díaz, A sharper energy method for the localization of the support to some stationary Schrödinger equations with a singular nonlinearity, *Discrete Continuous Dynam. Syst.* **34**(9) (2014) 3371–3382.

[17] P. Bégout and J. I. Díaz, Self-similar solutions with compactly supported profile of some nonlinear Schrödinger equations, *Electron. J. Differential Equations* **90** (2014) 1–15.

[18] P. Bégout and J. I. Díaz, Finite time extinction for the strongly damped nonlinear Schrödinger equation in bounded domains, *J. Differential Equations* (2019); doi: 10.1016/j.jde.2019.10.016.

[19] R. Bellman and R. Kalaba, *Quasilinearization and Nonlinear Boundary Value Problems* (Elsevier, New York, 1965).
[20] Ph. Benilan, M. G. Crandall and A. Pazy, *Nonlinear Evolution Equations in Banach Spaces*, in preparation.
[21] R. Bermejo and J. A. Infante, A multigrid algorithm for the p-Laplacian, *SIAM J. Sci. Comput.* **21** (2000) 1774–1789.
[22] C. Beta and A. S. Mikhailov, Controlling spatiotemporal chaos in oscillatory reaction–diffusion systems by time-delay autosynchronization, *Physica D* **199** (2004) 173–184.
[23] C. Beta, M. Bertram, A. S. Mikhailov, H. H. Rotermund and G. Ertl, Controlling turbulence in a surface chemical reaction by time-delay autosynchronization, *Phys. Rev. E* **67** (2003) 046224.
[24] F. Bethuel, H. Brezis, and F. Hélein, *Ginzburg–Landau Vortices* (Birkhäuser, Boston, 1994).
[25] M. E. Bleich and J. E. S. Socolar, Controlling spatiotemporal dynamics with time-delay feedback, *Phys. Rev. E* **54** (1996) 17–20.
[26] H. Brezis, *Opérateurs Maximaux Monotones* (North-Holland, Amsterdam, 1973).
[27] S. Carl and V. Lakshmikantham, Generalized quasilinearizarion and semilinear parabolic problems, *Nonlinear Anal.* **48** (2002) 947–960.
[28] A. C. Casal, J. I. Díaz, J. F. Padial and L. Tello, On the stabilization of uniform oscillations for the complex Ginzburg–Landau equation by means of a global delayed mechanism. *Actas XVIII CEDYA/VIII CMA* (Servicio de Publicaciones de la Univ. de Tarragona, 2003), in *CD-Rom Actas.*
[29] A. C. Casal and J. I. Díaz, On the principle of pseudo–linearized stability: application to some delayed parabolic equations, *Nonlinear Anal.* **63** (2005) 997–1007.
[30] A. C. Casal and J. I. Díaz, On the complex Ginzburg–Landau equation with a delayed feedback, *Math. Models Methods Appl. Sci.* **16**(1) (2006) 1–17.
[31] A. C. Casal, J. I. Díaz and M. Stich, On some delayed nonlinear parabolic equations modeling CO oxidation, *Dynam. Continuous, Discrete Impulsive Systems, Series A* **13b** (Supplementary Volume) (2006) 413–426.
[32] A. C. Casal, J. I. Díaz and M. Stich, Control of turbulence in oscillatory reaction–diffusion systems through a combination of global and local feedback, *Phys. Rev. E* **76** (2007) 036209.
[33] A. C. Casal, J. I. Díaz, M. Stich and J. M. Vegas, Hopf bifurcation and bifurcation from constant oscillations to a torus path for delayed

complex Ginzburg–Landau equations, in *Modern Mathematical Tools and Techniques in Capturing Complexity*, L. Pardo, N. Balakrishnan, M. A. Gil eds. (Springer, Berlin, 2011), pp. 57–76.

[34] A. C. Casal and J. I. Díaz, On the principle of pseudo-linearized stability: aplication to some delayed nonlinear parabolic equations, eds. *Proc. 4th World Cong. Nonlinear Analysts* (Orlando, FL, June 30–July 7, 2004).

[35] E. N. Chukwu, *Differential Models and Neutral Systems for Controlling the Wealth of Nations* (World Scientific, Singapore, 2001).

[36] M. C. Cross and P. C. Hohenberg, Pattern formation outside of equilibrium. *Rev. Mod. Phys.* **65** (1993) 851–1112.

[37] J. I. Díaz and G. Hetzer, A functional quasilinear reaction–diffusion equation arising in climatology, in *Équations aux dérivées partielles et applications. Articles dédiés à J.-L. Lions*, Elsevier, Paris, (1998) pp. 461–480.

[38] J. I. Díaz, J. F. Padial, J. I. Tello and L. Tello, Delayed complex Ginzburg–Landau equations Complex Ginzburg–Landau equations with a delayed nonlocal perturbation, *Electron. J. Differential Equations* **2020** (2020) 1–9.

[39] Th. Erneux, *Applied Delay Differential Equations* (Springer, Berlin, 2009).

[40] V. L. Ginzburg and L. D. Landau, On the Theory of superconductivity. *Zh. Eksp. Teor. Fiz.* **20** (1950) 1064–1082.

[41] J. Ginibre and G. Velo, The Cauchy problem in local spaces for the complex Ginzburg–Landau equation. I. Compactness methods, *Physica D* **95** (1996) 191–228.

[42] J. Ginibre and G. Velo, The Cauchy problem in local spaces for the complex Ginzburg–Landau equation. II. Contraction methods, *Comm. Math. Phys.* **187** (1997) 45–79.

[43] B. Guo and R. Yuan, The time-periodic solution to a 2D generalized Ginzburg–Landau equation, *J. Math. An. Appl.* **266** (2002) 186–199.

[44] J. K. Hale, *Theory of Functional Differential Equations* (Springer, New York, 1977).

[45] J. Hernández, F. Mancebo and J. M. Vega de Prada, On the linearization of some singular nonlinear elliptic problems and applications, *Ann. Inst. Henri Poincaré Non Linear Anal.* **19** (2002) 777–813.

[46] M. Ipsen, L. Kramer, and P. G. Sørensen, Amplitude equations for description of chemical reaction–diffusion systems. *Phys. Rep.* **337** (2000) 193–235.

[47] M. Kim, M. Bertram, M. Pollmann, A. von Oertzen, A. S. Mikhailov, H. H. Rotermund, and G. Ertl, Controlling chemical turbulence by global delayed feedback: pattern formation in catalytic CO oxidation reaction on Pt(110). *Science* **292** (2001) 1357–1360.

[48] Y. Kuramoto, *Chemical Oscillations, Waves, and Turbulence* (Springer, Berlin, 1984).

[49] V. Lakshmikantham and A. S. Vatsala *Generalized Quasilinearization for Nonlinear Problems* (Kluwer, Dordrecht, 1995).

[50] L. P. Lévy, Ginzburg–Landau theory, in *Magnetism and Superconductivity*, Texts and Monographs in Physics (Springer, Berlin, 2000).

[51] V. A. Liskevich and M. A. Perelmuter, Analyticity of sub-Markovian semi-groups, *Proc. Amer. Math. Soc.* **123**(4) (1995) 1097–1104.

[52] Wolfram Research, Inc. *Mathematica* Edition 5.2. Champaign, IL (2005).

[53] F. Mertens, R. Imbihl and A. S. Mikhailov, Turbulence and standing waves in oscillatory chemical reactions with global coupling, *Chem. Phys.* **101** (1994) 9903–9908.

[54] A. S. Mikhailov and K. Showalter, Control of waves, patterns and turbulence in chemical systems. *Phys. Rep.* **425** (2006) 79–194.

[55] A. Newell and J. Whitehead, Finite bandwidth, finite amplitude convection, *J. Fluid Mech.* **38** (1969) 279–403.

[56] N. Okazawa and T. Yokota, Monotonicity method applied to the complex Ginzburg–landau and related equations, *J. Math. Anal. Appl.* **267** (2002) 247–263.

[57] M. E. Parrot, Linearized stability and irreducibility for a functional differential equation, *SIAM J. Math. Anal.* **23** (1992) 649–661.

[58] K. Pyragas, Continuous control of chaos by self–controlling feedback. *Phys. Lett. A* **170** (1992) 421–428.

[59] W. M. Ruess, Linearized Stability for nonlinear evolution equations, *J. Evolution Equations* **3**(2) (2003) 361–373 (Dedicated to Philippe Bénilan).

[60] E. Schöll and H. G. Schuster (eds.), *Handbook of Chaos Control* (Wiley-VCH, Weinheim, 2007).

[61] M. Stich, Target patterns and pacemakers in reaction–diffusion systems, Ph.D. thesis, Technische Universität Berlin (2003).

[62] M. Stich, A. C. Casal, and J. I. Díaz, Control of turbulence in oscillatory reaction–diffusion systems through a combination of global and local feedback. *Phys. Rev. E* **76** (2007) 036209.

[63] M. Stich and C. Beta, Control of pattern formation by time–delay feedback with global and local contributions, *Physica D* **239** (2010) 1681–1691.

[64] P. Takáč, Invariant 2-tori in the time-dependent Ginzburg–Landau equation, *Nonlinearity* **5** (1992) 289–321.
[65] R. Temam, *Infinite-Dimensional Dynamical Systems in Mechanics and Physics* (Springer, New York 1988).
[66] I. I. Vrabie, *Compactness Methods for Nonlinear Evolutions*, 2nd edn. (Pitman, Longman, Essex, 1995).
[67] J. Wu, *Theory and Applications of Partial Functional Differential Equations* (Springer, New York, 1996).

https://doi.org/10.1142/9789811221903_0018

Chapter 18

Nonlinear Analysis and Control of Quasi-passive Dynamic Walker Based on Entrainment Effect

Longchuan Li*,‡, Isao T. Tokuda*,§,
Fumihiko Asano†,¶, and Cong Yan†,‖

*Department of Mechanical Engineering, Ritsumeikan University, 1-1-1 Nojihigashi, Kusatsu, Shiga 525-8577, Japan
†School of Information Science, Japan Advanced Institute of Science and Technology, 1-1 Asahidai, Nomi, Ishikawa 923-1292, Japan
‡llt2144@fc.ritsumei.ac.jp
§isao@fc.ritsumei.ac.jp
¶fasano@jaist.ac.jp
‖s1810218@jaist.ac.jp

This chapter overviews basics and applications of passive dynamics walking as a typical example of nonlinear dynamics in the field of robotics. The passive dynamic walker, proposed by Tad McGeer three decades ago, provides one of the most efficient gaits among all other types of locomotions in the sense that its actuation power comes solely from the gravity. With an appropriate initial condition, the walking cycles are sustained on a gentle downhill and the dynamics converges to a limit-cycle because of the phase reset by the ground collision, in which gravitational potential energy and energy loss cancel each other. Despite its highly efficient walking, versatility of the limit-cycle walking is rather limited since it shows only one steady-state gait pattern with fixed physical parameters. To resolve this issue, a control mechanism is introduced, which positively utilizes the intrinsic walking property, while preserving its efficiency. The control is based on the entrainment mechanism that makes the walking frequency the same as that of the wobbling mass attached to the walker. Using combined rimless wheels as a simplistic realization of the limit-cycle walker, we demonstrate both numerically and experimentally that the well-known *Arnold* tongues are observed as a typical entrainment property of the limit-cycle walker in two-dimensional parameter space spanned by the wobbling forcing and frequency.

18.1. A Brief Overview of Passive Dynamic Walking

18.1.1. *The most natural gait of walking robots*

It has been just three decades since Tad McGeer proposed the concept of passive dynamic walking [1]. Still today his walker is more human-like than any other state-of-the-art techniques of legged locomotion robots, see [2–4]. Through ingeniously designed configuration, McGeer's biped walker can take steps passively without any active control on a gentle downhill, with an appropriate initial condition. With a simplified point-feet compass-like model, the detailed passive dynamics have been further investigated [5]. As shown in Fig. 18.1(a), the masses, which are indicated by the circles, are distributed on the hip and the center of legs. Consequently, the walker shows a pendulum-like behavior during the leg-swing phase. Namely, the kinetic energy is converted to a gravitational potential energy until its center of mass (COM) reaches the highest point in one walking cycle, so that the walker can step forward by inversely transforming the potential energy into the kinetic energy. Although a part of the kinetic energy is consumed due to the foot-landing, the gravitational

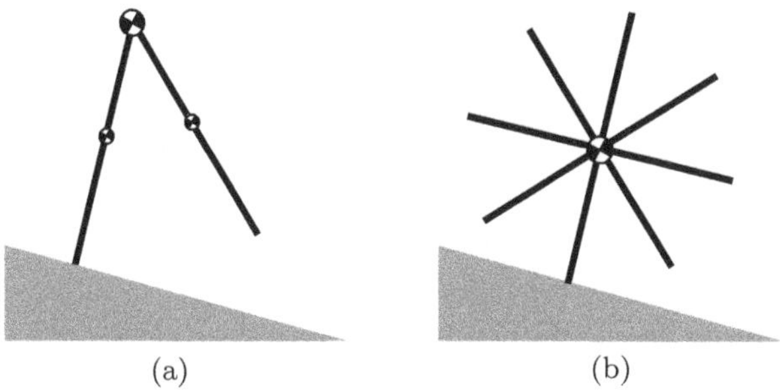

Fig. 18.1. Passive dynamic walking models: (a) compass-like biped walker; (b) rimless wheel.

potential energy provided by the downhill enables the next walking cycle. Consequently, the walking dynamics converges to a limit-cycle, where energy gain and loss balance. Inspired by this phenomenon, various sorts of limit-cycle walkers [6] have been designed and widely applied to robot locomotion control, towards achieving efficient and stable gait, see [7–10]. The dynamics of the biped walker can be highly complicated, if additional degrees of freedom (DOF), e.g., knee and ankle joints, are supplied. Fortunately, its mathematical model can be approximated by a simple rimless wheel model (Fig. 18.2(b)), with the assumption that the ratio between its hip and leg mass is large enough. Since the rimless wheel model gives rise to a limit-cycle walking, in which gravitational energy is consumed by the wheel-landing, such a 1-DOF system can sufficiently capture essence of the passive dynamic walking [11]. Benefited from this fact, several researches have been studying the rimless wheel model to unveil fundamental principles of the passive walker, see [12–14].

The aim of the present chapter is to overview the basics and applications of the passive dynamics walking as a typical example of nonlinear dynamics in the field of robotics. To easily illustrate crucial properties of the passive dynamic walking as well as its increasing the versatility, we introduce "Combined Rimless Wheels", which is composed of two identical rimless wheels connected by a beam without incorporating joint friction, as shown in Fig. 18.2. The advantage of the combined rimless wheels is that a control mechanism can be implemented by attaching a wobbling mass to the beam. By controlling the wobbling mass to move up and down periodically, the passive dynamic walker can be entrained to the periodic forcing of the wobbling mass. This provides an indirect control of the walking frequency of the limit-cycle system. We demonstrate both numerically and experimentally that the well-known *Arnold* tongues are observed as an entrainment property of the limit-cycle walker in two-dimensional parameter space spanned by the wobbling amplitude and frequency. Such technique is idealistic for efficiently controlling the limit-cycle walker without suppressing its inherent dynamic properties.

(a)

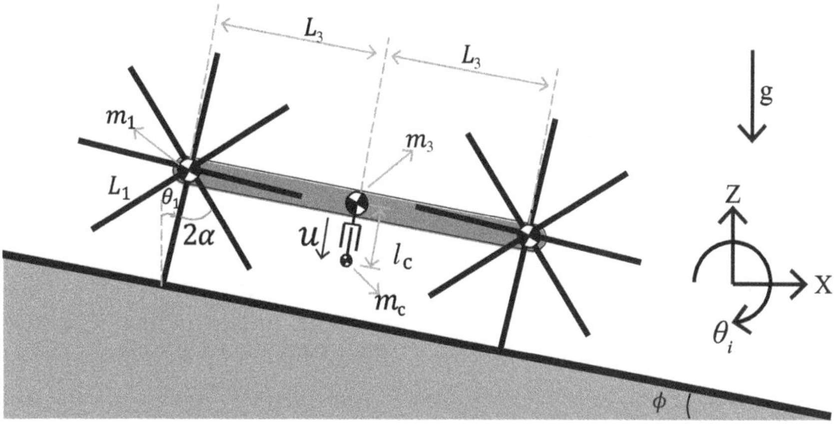

(b)

Fig. 18.2. Combined rimless wheels with wobbling mass. (a) Experimental machine for combined rimless wheels with wobbling mass and (b) mathematical model for combined rimless wheels with wobbling mass.

18.1.2. *Equation of motion*

Here we introduce equations of motion for the combined rimless wheels (details can be found in Ref. [15]). The rear and fore rimless wheels are mechanically synchronized with each other and rotate passively. The wobbling mass is attached to the beam and controlled by the input u [N]. Here, u is set to 0 in the passive dynamic walking, where it enables tracking control as will be introduced in the next section. Similar to Fig. 18.1, the circles in Fig. 18.2(b) indicate the distribution of masses, where the weights of wheel, beam and wobbling mass are given by m_1, m_3 and m_c, respectively. The system can therefore be sufficiently described by the generalized coordinate vector $\boldsymbol{q} = \begin{bmatrix}\theta_1 \; l_c\end{bmatrix}^{\mathrm{T}}$, where θ_1 is the angular position of the support legs with respect to vertical direction and l_c represents trajectory of the wobbling mass. The combined rimless wheels obey the following dynamics:

$$\ddot{\theta}_1 = \frac{N_{\theta_1}}{D_{\theta_1}}, \quad \ddot{l}_c = \frac{N_{l_c}}{D_{l_c}}. \tag{18.1}$$

The numerators and denominators are given by

$$\begin{aligned} N_{\theta_1} &= (2m_w + m_c)g\sin\theta_1 - m_c g\sin(\theta_1 - 2\phi) \\ &\quad + m_c L_1 \dot{\theta}_1^2 \sin(2(\theta_1 - \phi)) - 2\sin(\theta_1 - \phi)u, \end{aligned} \tag{18.2}$$

$$D_{\theta_1} = 2L_1(m_w + m_c\cos^2(\theta_1 - \phi)), \tag{18.3}$$

$$N_{l_c} = (m_w + m_c)(u + m_c(g\cos\theta_1 - L_1\dot{\theta}_1^2)\cos(\theta_1 - \phi)), \tag{18.4}$$

$$D_{l_c} = m_c(m_w + m_c\cos^2(\theta_1 - \phi)), \tag{18.5}$$

where $m_w := 2m_1 + m_3$ [kg]. As indicated in Eqs. (18.2) and (18.3), the wheels dynamics are affected by the input u, which controls trajectory of the wobbling motion as a function of l_c. In addition, the weight of mass should be assigned to the beam in passive dynamic walking condition, i.e., $m_3 \to m_3 + m_c$, $m_c \to 0$, since it is fixed on the beam.

18.1.3. *Collision equations*

The fore feet of the combined rimless wheels land on the ground when $\theta = \alpha + \phi$ due to its geometry. Consequently, the system dynamics is discretized by this event. We assume that the feet tips always contact with the ground without slipping. Moreover, we assume that the rear feet of both rimless wheels leave the ground immediately after the landing of the fore feet in accordance with the inelastic collision model. Then, the state transition is specified as follows:

$$\dot{\theta}_1^+ = \frac{m_w \cos(2\alpha) + m_c \cos^2\alpha}{m_w + m_c \cos^2\alpha}\dot{\theta}_1^-, \tag{18.6}$$

$$\dot{l}_c^+ = \dot{l}_c^- + \frac{(m_w + m_c)L_1 \sin(2\alpha)\cos\alpha}{m_w + m_c \cos^2\alpha}\dot{l}_c^-, \tag{18.7}$$

where the superscripts of "−" and "+" denote immediately before and immediately after the collision. Since $\cos(2\alpha) < 0$, the wheels rotation is decelerated by the ground collision according to Eq. (18.6). The angular position and length of the wobbling mass are reset as:

$$\theta_1{}^+ = \theta_1{}^- - 2\alpha, \quad l_c{}^+ = l_c{}^-. \tag{18.8}$$

18.1.4. *Typical gait of passive dynamic walking*

Passive dynamic gait can be generated by setting $l_c = 0$ and $\dot{l}_c = 0$ (i.e., fixing the wobbling mass m_c on the beam m_3). The numerical simulations were conducted using the parameters listed in Table 18.1 and the results are shown in Fig. 18.3. Angular position of the support leg θ_1 increases monotonically until its foot-landing. To realize this walking step, kinetic energy of the support leg θ_1 should be high enough to raise its COM to the highest point, that is called "potential barrier". Otherwise, the gait stops and the walker falls down. The walking frequency f_w [Hz] is calculated as the reciprocal of the duration of one walking cycle. After the system converges to a limit-cycle walking, the walking frequency f_w [Hz] also converges to natural frequency f_n of this walker. On the phase space, Poincaré

Table 18.1. Physical parameters used for simulations.

Parameter	Value	Unit	Physical meaning
m_1	1.0	kg	Wheel mass
m_3	1.0	kg	Beam mass
m_c	0.05	kg	Wobbling mass
L_1	0.15	m	Leg length
L_3	0.20	m	Beam length
g	9.81	m/s^2	Gravitational acceleration

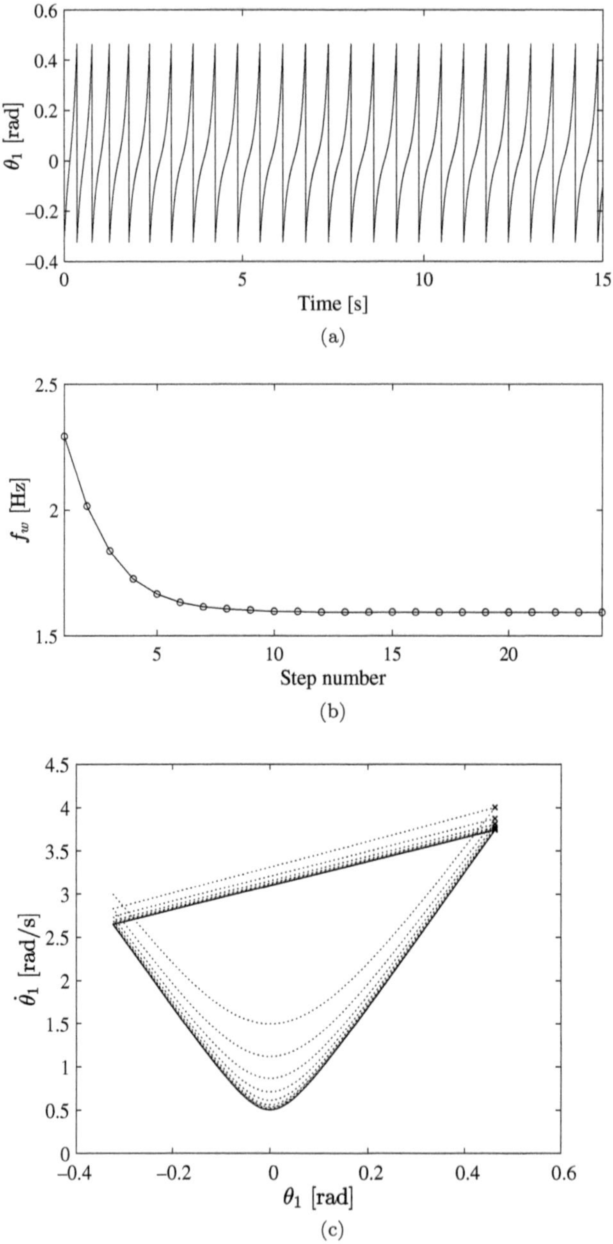

Fig. 18.3. Numerical simulation of the passive dynamic walking. (a) Time trace of angular position $\theta_1(t)$; (b) walking frequency f_w as a function of walking steps; and (c) trajectory of the walker in phase-space.

section can be defined as the timing at which the feet land on the ground. As the trajectory gradually converges to the limit-cycle, it crosses the Poincaré section at the same point. Such point is Lyapunov stable, if the system converges to the same point even when small perturbations are applied to the walker.

18.2. Entrained Limit-Cycle Walking

18.2.1. *Entrainment-based control*

The passive dynamic walker provides one of the most efficient gaits among all other types of locomotions in the sense that its actuation power comes solely from the gravity. Its versatility, on the other hand, is rather limited, due to its property that the passive walker has only one steady-state gait pattern with fixed physical parameters. To address this issue, a control mechanism should be designed in such a way that the intrinsic walking property is positively utilized, while guaranteeing its efficiency. To positively utilize the pendulum-like dynamics of the walker, we have proposed an entrainment-based control, inspired by the load carrying mechanism of quadruped mammals' dynamic walking, see [16–18], that widens applicability of the passive dynamic walker. We keep the joints between the rimless wheels and the beam passive, while we actively control the wobbling mass to track a periodic waveform with a frequency close to the natural frequency f_n of the walker. This machine, therefore, becomes a quasi-passive dynamic walker [19], since the main body remains passive and is expected to be entrained to the active wobbling motion. To enable the wobbling mass to follow a desired waveform, a tracking control method is applied as follows:

$$\ddot{l}_c = A(\theta_1)u + B(\theta_1, \dot{\theta}_1), \tag{18.9}$$

where

$$A(\theta_1) = \frac{m_w + m_c}{m_c(m_w + m_c\cos^2(\theta_1 - \phi))}, \tag{18.10}$$

$$B(\theta_1, \dot{\theta}_1) = \frac{(m_w + m_c)(g\cos\theta_1 - L_1\dot{\theta}_1^2)\cos(\theta_1 - \phi)}{m_w + m_c\cos^2(\theta_1 - \phi)}. \tag{18.11}$$

Control input u to achieve $l_c \to l_d$ is therefore defined as:

$$u = A(\theta_1)^{-1}(v - B(\theta_1, \dot{\theta}_1)), \tag{18.12}$$

$$v = \ddot{l}_d(t) + K_{\mathrm{D}}(\dot{l}_d(t) - \dot{l}_c) + K_{\mathrm{P}}(l_d(t) - l_c), \tag{18.13}$$

where l_d [m] represents the desired wobbling trajectory and K_{D} [s^{-1}] and K_{P} [s^{-2}] are the PD control gains. Here, we first set l_d to a sinusoidal waveform F_{sin}:

$$F_{\mathrm{sin}}(t) = A_m \sin(2\pi f_c t), \tag{18.14}$$

where A_m [m] and f_c [Hz] are the desired wobbling amplitude and frequency, respectively.

18.2.2. *Typical gait of entrained limit-cycle walking*

As observed in Fig. 18.3, natural frequency of the walker was estimated to be about 1.61 [Hz]. Accordingly, the wobbling frequency f_c was set to 1.7 [Hz] so as to easily induce an entrainment effect. Numerical simulation was conducted to entrain the limit-cycle walker, using the parameter values listed in Tables 18.1 and 18.2, and the results are shown in Fig. 18.4. Although the wobbling mass tracks the target sine waveform fairly well, collisions between the legs and the ground induces occasional perturbations to the wobbling motion as observed by comparing the support leg angular position θ_1 and the wobbling velocity $\dot{l}_c$. Despite such perturbations, the walking frequency gradually converges to the wobbling frequency, indicating the occurrence of entrainment.

The entrainment-based method can successfully control the walking frequency, as far as the wobbling frequency is set to be around

Table 18.2. Control parameters used for simulations.

Parameter	Value	Unit	Physical meaning
K_{D}	60	s^{-2}	Derivative gain
K_{P}	900	s^{-1}	Proportional gain
A_m	0.03	m	Wobbling amplitude

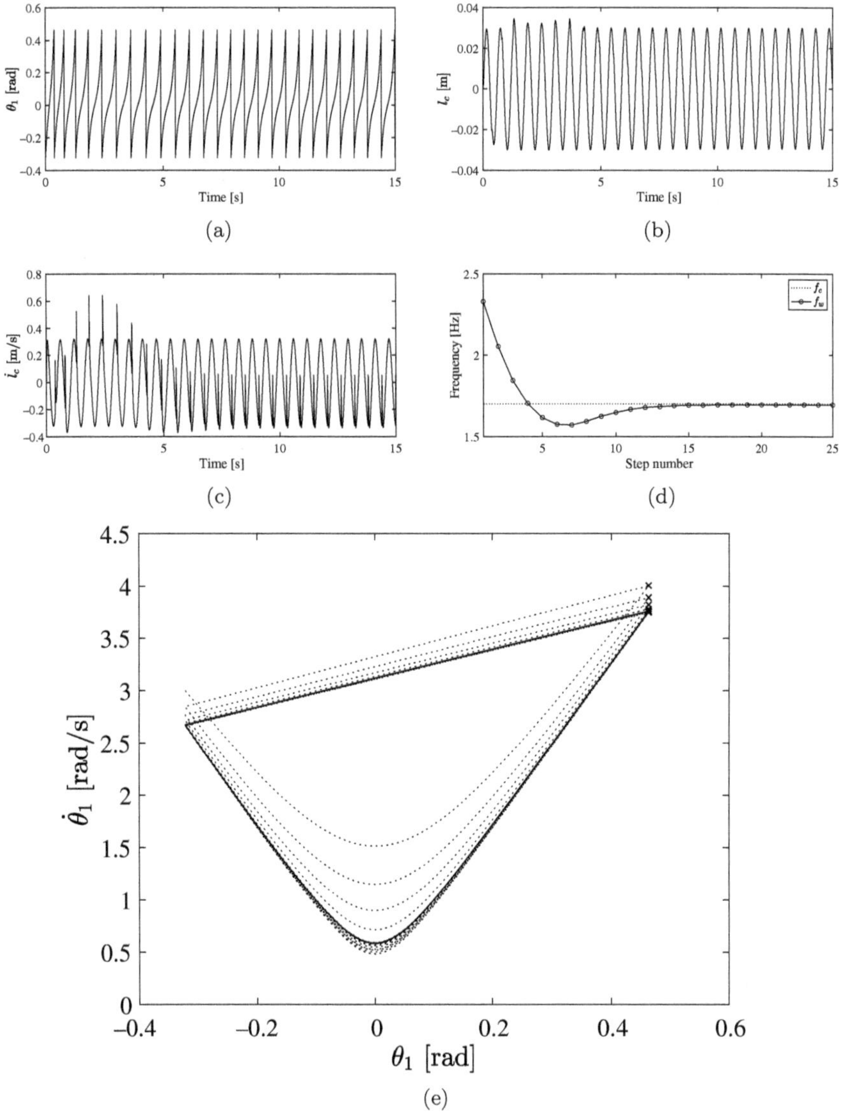

Fig. 18.4. Numerical simulation of limit-cycle walking entrained to wobbling mass with a frequency of $f_c = 1.7$ [Hz]. (a) Angular position $\theta_1(t)$; (b) wobbling mass position $l_c(t)$; (c) wobbling mass velocity $\dot{l}_c(t)$; (d) walking frequency f_w; and (e) trajectory of the walker in phase-space.

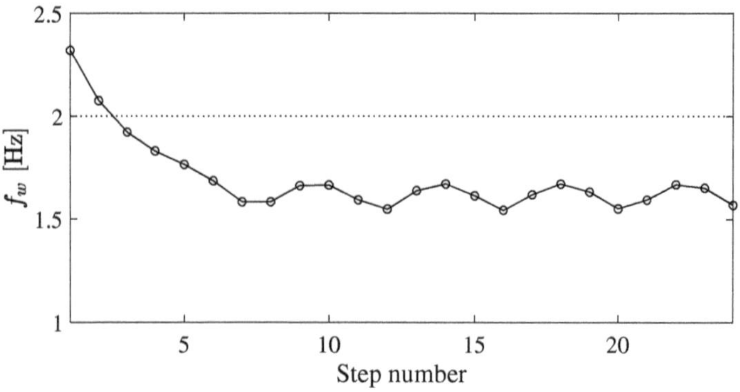

(a)

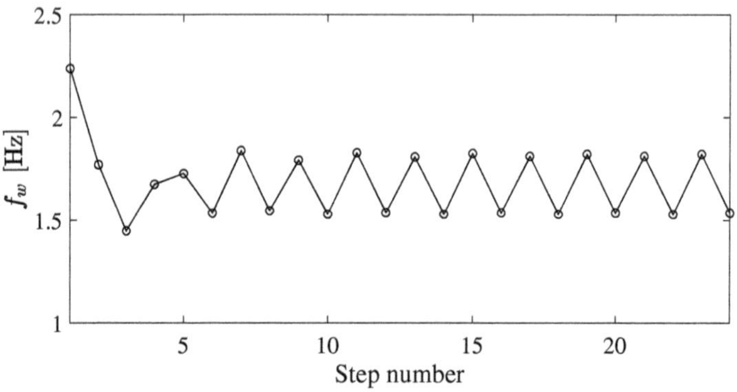

(b)

Fig. 18.5. Walking frequency f_w as a function of walking steps. (a) Wobbling frequency of $f_c = 2$ [Hz]; and (b) wobbling frequency of $f_c = 5$ [Hz].

natural frequency of the walker. Its entrainability, however, becomes limited as the mismatch between natural frequency of the walker and the wobbling frequency grows large. Although the walking frequency is affected by the wobbling motion due to their coupling effect, it does not converge to the wobbling frequency as shown in Fig. 18.5(a). Figure 18.5(b) shows that when the wobbling frequency is about three times as high as the natural frequency ($f_c = 5$ [Hz]), the dynamics becomes quite complicated. The walker generates period-2 gait and is entrained to the wobbling mass with a frequency ratio of 1 : 3, i.e., $\bar{f_w} : f_c = 1 : 3$ ($\bar{f_w}$: averaged walking frequency).

18.2.3. *Evaluation of entrainability*

The advantage of our entrainment-based control method is that it positively utilizes an intrinsic property of the quasi-passive walker. It, however, has an obvious limitation. Namely, a desired walking frequency cannot be achieved, if the wobbling frequency is out of the entrainment range [20]. To evaluate the general performance of this control method as well as to observe basic tendency of the effect of the wobbling waveform, the *Arnold* tongue [21] of 1 : 1 entrainment is drawn for various trajectories of the wobbling mass. Here, four waveforms are considered for $l_d(t)$:

- $F_{\sin}(t) = A_m \sin(2\pi f_c t)$,
- $F_{\sin 7}(t) = A_{m7} \sin^7(2\pi f_c t)$,
- $F_{\sin 11}(t) = A_{m11} \sin^{11}(2\pi f_c t)$,
- $F_{\sin 19}(t) = A_{m19} \sin^{19}(2\pi f_c t)$,

with a normalized forcing power:

$$\langle F_{\sin}^2 \rangle = \langle F_{\sin 7}^2 \rangle = \langle F_{\sin 11}^2 \rangle = \langle F_{\sin 19}^2 \rangle, \tag{18.15}$$

where $\langle \cdot \rangle$ indicates averaging over one wobbling cycle. As shown in Fig. 18.6, sharpness of the desired wobbling trajectories differs from one waveform to another.

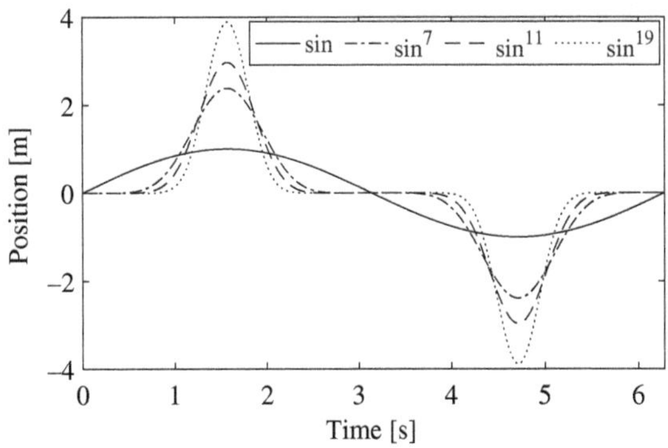

Fig. 18.6. Four types of desired wobbling trajectories.

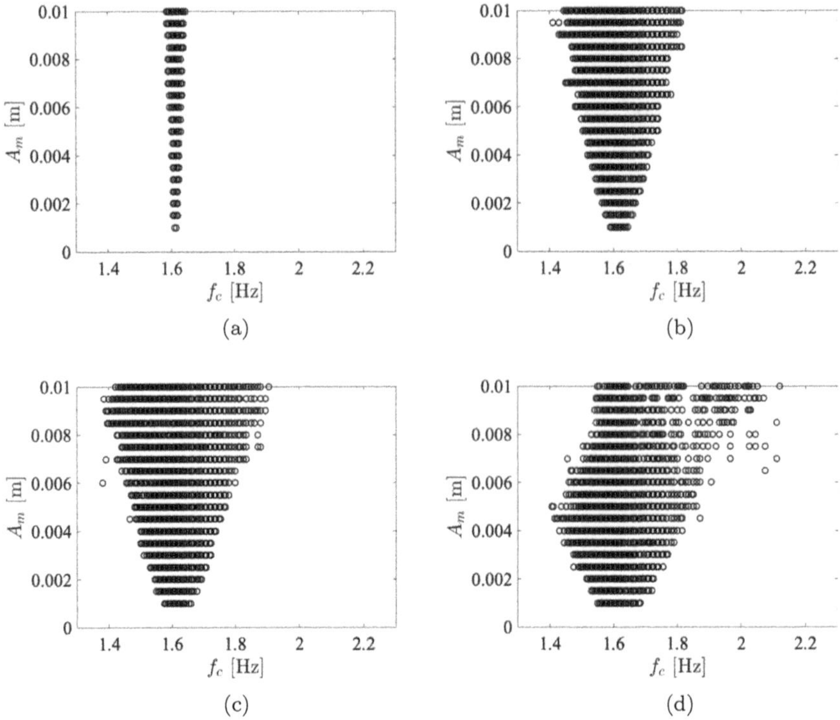

Fig. 18.7. *Arnold* tongues obtained from numerical simulations. (a) Desired trajectory of $F_{\text{sin}}(t)$; (b) desired trajectory of $F_{\text{sin}\,7}(t)$; (c) desired trajectory of $F_{\text{sin}\,11}(t)$; and (d) desired trajectory of $F_{\text{sin}\,19}(t)$.

In two-dimensional parameter space spanned by the wobbling amplitude and frequency, the *Arnold* tongues, which represent $1:1$ entrainment, are drawn for the four waveforms in Fig. 18.7. The simple sinusoidal waveform F_{sin} results in a typical triangle shape of the *Arnold* tongue. The entrainment range can be enlarged by increasing the sharpness of the desired wobbling trajectory from F_{sin} to $F_{\text{sin}\,7}, F_{\text{sin}\,11}, F_{\text{sin}\,19}$. However, the entrainment becomes rather unstable as the desired wobbling trajectory are set to be too sharp. The entrainment rate, calculated as the ratio between the entrainment range and the whole parameter space ($A_m \in [0.0005, 0.01]$, $f_c \in [1.3, 2.3]$), is shown in Fig. 18.8. Among the four waveforms, $F_{\text{sin}\,11}$ is found to be optimal to entrain the limit-cycle walker with the

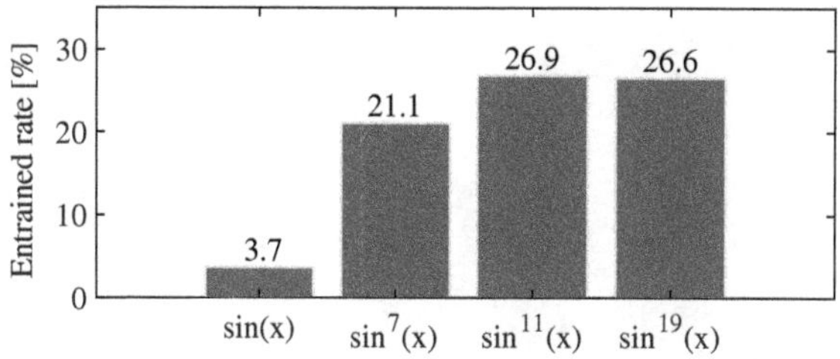

Fig. 18.8. Comparison of entrainment ranges [%] for four wobbling trajectories.

Table 18.3. Parameters used for experiment.

Parameter	Value	Unit	Physical meaning
m_1	0.455	kg	Wheel mass
m_3	2.4	kg	Beam mass
m_c	0.186	kg	Wobbling mass
L_1	0.1	m	Leg length
L_3	0.27	m	Beam length
g	9.81	m/s^2	Gravitational acceleration

largest entrainment range. A systematic method for optimizing the desired wobbling trajectory, which takes into account phase response property of the walker, has been introduced also in [22].

In order to examine practical applicability of the control method in real-world, an experimental machine was designed as shown in Fig. 18.2(a). Synchronization between the fore and rear feet were realized by adding a rigid connection rod between them, where the wobbling motion was achieved by a piston-crank mechanism. Parameter setting of this machine is listed in Table 18.3. The experimental results using a sinusoidal waveform $F_{\sin}$ as the desired wobbling trajectory are shown in Fig. 18.9. Again, a clear *Arnold* tongue structure is discernible. Namely, as the wobbling amplitude is increased, the entrainment range becomes larger. A too large amplitude, however, induces instability in the walking. The experimental system is therefore highly consistent with the numerical simulations.

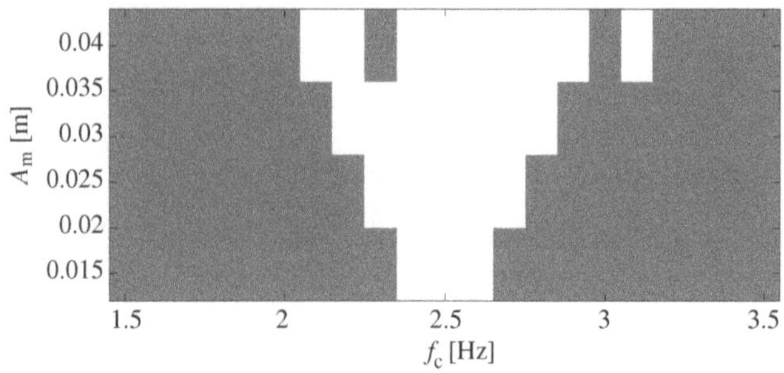

Fig. 18.9. *Arnold* tongues obtained from experiment. The white region corresponds to 1 : 1 entrainment, while gray region indicates no entrainment.

18.3. Summary

In this Chapter, basics and applications of the quasi-passive dynamics walking are briefly reviewed. If the walking slope is steep enough, the kinetic energy overcomes the potential barrier to raise walker's COM to the highest point to sustain the passive walking. The limit-cycle walking is achieved at the point, where gravitational potential energy and energy loss due to foot-landing balance. As one of the key features of the limit-cycle oscillator, we focus on its entrainment capability to periodic external forcing. We then introduce an entrainment-based control method and demonstrate that the walking frequency can be controlled by the wobbling mass. Up-and-down movement of the wobbling mass entrains the limit-cycle walker efficiently by making good use of the intrinsic dynamic property of the walker. Numerical simulations as well as experimental study validate the effectiveness of this control method. Basic entrainment property of the walking system has been characterized by the *Arnold* tongues.

References

[1] T. McGeer, Passive dynamic walking, *Int. J. Robotics Res.* **9**(2) (1990) 62–82.
[2] L. Amy, Make robot motions natural, *Nature* **565** (2019) 422–424.

[3] ATLAS, Boston dynamics: Humanoid robot ATLAS, Available: https://www.bostondynamics.com/atlas/.

[4] G. Bledt, M. J. Powell, B. Katz, J. D. Carlo, P. M. Wensing, and S. Kim Mit cheetah 3: Design and control of a robust, dynamic quadruped robot, *Proc. IEEE/RSJ Int. Conf. Intelligent Robots and Systems* (2018), pp. 2245–2252.

[5] A. Goswami, B. Thuilot, and B. Espiau, Compass-like biped robot part I: Stability and bifurcation of passive gaits, *Res. Rep. INRIA* (1996).

[6] D. G. E. Hobbelen and M. Wisse, Limit cycle walking, in *Humanoid Robots, Human-like Machines* (IntechOpen, 2007).

[7] A. Goswami, B. Espiau and A. Keramane, Limit cycles in a passive compass gait biped and passivity-mimicking control laws, *Autonom. Robots* **4**(3) (1997) 273–286.

[8] M. W. Spong, Passivity-based control of the compass gait biped, *The IFAC 1999 World Congr.*, 1999, pp. 19–23.

[9] F. Asano, Z. Luo and M. Yamakita, A novel gait generation for biped walking robots based on mechanical energy constraint, *IEEE Trans. Robotics Automat.* **20**(3) (2004) 565–573.

[10] S. H. Collins, M. Wisse, A. Ruina and R. Tedrake, Efficient bipedal robots based on passive-dynamic Walkers, *Science* **307** (2005) 1082–1085.

[11] J. P. Laumond, M. Benallegue, J. Carpentier and A. Berthoz, The yoyo-man, *Int. J. Robotics Res.* **36**(13–14) (2017) 1508–1520.

[12] M. J. Coleman, A. Chatterjeel and A. Ruina, Motions of a rimless spoked wheel: a simple three-dimensional system with impacts, *Dynam. Stability Syst.* **12**(3) (1997) 139–159.

[13] F. Asano and Z. Luo, Asymptotically stable biped gait generation based on stability principle of rimless wheel, *Robotica* **27**(6), (2009) 949–958.

[14] W. Jia, J. Yang, L. Bi, Q. Zhang, Y. Sun, H. Pu and S. Ma, Modelling and analysis of the passive planar rimless wheel mechanism in universal domain, *IEEE/RSJ Int. Conf. Intelligent Robots and Systems* (2017), pp. 4969–4975.

[15] F. Asano and I. Tokuda, Indirectly controlled limit-cycle walking of combined rimless wheel based on entrainment to active wobbling mass, *Multibody Syst. Dynam.* **34**(2) (2015) 191–210.

[16] I. S. Young, R. M. Alexander, A. J. Woakes, P. J. Butler and L. Anderson, The synchronization of ventilation and locomotion in horses (Equus Caballus), *J. Experimental Biol.* **166**(1) (1992) 19–31.

[17] L. C. Rome, L. Flynn, E. M. Goldman and T. D. Yoo, Generating electricity while walking with loads, *Science* **309**(5741) (2005) 1725–1728.

[18] L. C. Rome, L. Flynn and T. D. Yoo, Biomechanics: Rubber bands reduce the cost of carrying loads, *Nature* **444**(7122) (2006) 1023–1024.

[19] L. Li, I. Tokuda and F. Asano, Energy-efficient locomotion generation and theoretical analysis of a quasi-passive dynamic walker, *IEEE Robotics Automat. Lett.* **5**(3) (2020) 4305–4312.

[20] L. Li, F. Asano and I. Tokuda, Nonlinear analysis of an indirectly controlled sliding locomotion robot, *Proc. IEEE/RSJ Int. Conf. Intelligent Robots and Systems*, 2018, pp. 8221–8226.

[21] A. S. Pikovsky, M. G. Rosenblum and J. Kurths, *Synchronization: A Universal Concept in Nonlinear Science* (Cambridge University Press, Cambridge, 2001).

[22] L. Li, I. Tokuda and F. Asano, Optimal input waveform for an indirectly controlled limit-cycle walker, *Proc. IEEE/RSJ Int. Conf. Intell. Robots and Systems* (2018) pp. 7454–7459.

https://doi.org/10.1142/9789811221903_0019

Chapter 19

Physics-Inspired Swarm Optimization: The General Algorithmic Search

Sergio Hernández*,‡, Guillem Duran*,§ and José M. Amigó†,¶

*Fragile Technology, Murcia, Spain
†Centro de Investigación Operativa,
Universidad Miguel Hernández, Elche, Spain
‡sergio@fragile.tech
§guillem@fragile.tech
¶jm.amigo@umh.es

In this chapter, we present a particle swarm algorithm for optimization called Global Algorithmic Search (GAS). Specifically, GAS is a stochastic, single-objective method that evolves a swarm of agents in search of a global extremum. In general, swarm optimization algorithms are inspired in natural processes, be they biological, physical, chemical or social. GAS belongs to the algorithms based on physical processes or, in mathematical terminology, dynamical systems. For numerical simulations, we use a sample of 31 test functions, all of which are considered hard for the sake of global optimization, with dimensions ranging between 2 and 45. The results show that GAS outperforms three popular algorithms, namely, Basin Hopping, Cuckoo Search, and Differential Evolution, especially in concurrent optimization, i.e., when several runs with different initial settings are executed and the first best wins. Python codes of all algorithms and complementary information are available online.

19.1. Introduction

Optimization is a basic task in many areas of science, economy, and technology, not least because many questions can be (re)formulated (usually in more elegant ways) as optimization problems. A well-known example is the equations of motion of a mechanical system, which amount to the minimization of its action, the so-called Principle of Least Action. Needless to remind, mechanical systems are the forerunners of modern dynamical systems; in other words, the latter are a mathematical generalization of the former. Concurrently, some optimization algorithms have been inspired by the traditional sciences, e.g., biology (Genetic Algorithms [9], Differential Evolution [16], Cuckoo Search [21]), physics (Simulated Annealing [13], Charged System Search [11]), chemistry (Basin Hopping [19]), social (Altruism Algorithm [18]) and more. Unlike conventional, calculus-based optimization methods, those algorithms rely on swarm (or multiagent) interaction and information sharing. Nowadays optimization remains a vigorous research field in which novel techniques are developed to deal with old and new challenges such as those that arise in control theory, operations research and complex systems, in particular, dynamical systems composed of many interacting parts [3]. As the above few examples show, the development of swarm methods belongs certainly to this effort.

This chapter deals with unconstrained function optimization, i.e., the search for global maxima or minima of a single real-valued function (the objective or cost function) without constrains. General Algorithmic Search (GAS), the global optimization method presented here, is a probabilistic, single-objective technique that evolves a population of agents (the swarm) in search of a global extremum. Therefore, GAS belongs to the family of Particle Swarm Optimization (PSO) algorithms, whose common denominator is the search of a point (e.g., an optimal solution) on an n-dimensional surface (e.g., defined by a cost function) via a swarm of candidate solutions (the "particles"). The evolution of the agents in GAS is loosely inspired in the Simulated Annealing algorithm (see Sec. 19.1.2). Its performance is supported by numerical evidence and benchmarking against some of the most popular heuristics currently used. For the present work we have chosen three of them: Basin Hopping, Cuckoo Search and Differential Evolution.

In turn, PSO algorithms belong to the overarching field of Swarm Intelligence (SI), which studies how self-organization and global "intelligent" behavior emerge in decentralized systems, natural or artificial, from the local interactions among their constituents. In particular, SI deals with the modeling of swarming systems in the biological realm, as well as many-particle systems with complex dynamics in the physical realm and engineering [2, 6]. Prototypical examples are insect colonies [5] and cellular robotic systems [1]. Let us mention in passing that algorithms based on ant colonies are applied to optimize paths in networks, especially in combinatorial optimization (travelling salesman problem, vehicle routing), urban transportation and location of communication infrastructures. In modern parlance, the approach of SI to optimization could be called crowd search, very much in the sense that it is used in social sciences, where the agents of the crowd share information and cooperate with a common goal. The interested reader is referred to the recent surveys [15, 23].

Next, we elaborate a bit on PSO, before focusing on the specifics of GAS.

19.1.1. *Particle swarm optimization*

Particle swarm optimization [12] was originally meant to model such social behaviors as bird flocking and fish schooling, although it was soon adapted to optimization [6]. As Kennedy and Eberhart asked in their seminal paper [12]: "Why is social behavior so ubiquitous in the animal kingdom? Because it optimizes. What is a good way to solve engineering optimization problems? Modeling social behavior." Conventional, calculus-based optimization algorithms start with a single, randomly chosen point in the search landscape (defined by the objective function), and use the derivative of the objective function to "move" it downhill towards a lower position (without loss of generality). Instead, PSO draws on a population of candidate solutions (particles, agents, walkers, scouts, etc.) that survey the search landscape in parallel, interacting with each other and updating their current positions according to their own position and the best global position found so far by the other particles. Since the search is global and parallelized, the chances of the particle swarm rapidly converging to a global extremum are high, although it goes without saying that success is not guaranteed. On the positive side, PSO makes few assumptions, so it can be applied to a broad variety of optimization problems.

PSO algorithms have the following basic structure. The initial position and velocity of the particles are generally drawn from a uniform distribution. Of course, the more irregular the search landscape, the higher the density of particles should be. From then on and until the stopping criterion is met (i.e., a satisfactory solution has been found or the maximum number of iterations has been reached), positions and velocities are updated according to some simple rules. These rules take into account the best known position of the particle (its lowest position so far) as well as the best known global position of the particles, and involve model parameters and random numbers. The model parameters determine the performance and efficiency of the algorithm, and they are in general empirically fine tuned. The role of the random perturbations is to introduce noise into the search, thus preventing the particles from getting stuck in suboptimal regions. At the end of each search iteration, both the best known position of each particle and the best global position are updated.

From a mathematical point of view, PSO algorithms have not been shown to converge to a/the global optimal solution (global extremum of the objective function), which classifies them as heuristic methods. On the other hand, given an optimization problem, metaheuristics is a procedure for selecting a heuristic method (e.g., within the PSO algorithms) that can provide a good enough solution for that problem; see e.g., [17] for a review of metaheuristics.

Regarding dynamical systems, PSO has been applied, in particular, to the detection of nonlinear dynamical and the identification of their parameters. See, for instance, [10, 14, 24].

19.1.2. *General algorithmic search*

General Algorithmic Search (GAS) is a thermodynamics-inspired optimization algorithm that uses a swarm of particles (which we will call walkers) to mimic the behavior of an ionized gas when diffusing to fill a volume (the search domain), where an electric potential, representing the function values to be maximized, is present, attracting the particles towards the sought global optimum. A further ingredient is the "flow", which is a proxy of the contribution of each walker to the swarm entropy [25].

GAS combines an evolutionary approach, where particles can jump to another position to maximize the entropy of the gas (in particular, the cross-entropy of the gas distribution and the potential over the gas cloud volume) with a simulated annealing-like schema. The temperature of the gas varies over time, producing an algorithm that is well suited for nonlinear systems. GAS can be applied to functions that are discontinuous and defined over complex, possibly nonconvex, domains.

The rest of this chapter is organized as follows. In Sec. 19.2, we describe the GAS algorithm via a flowchart and comment some of its features. The benchmark of GAS against Differential Evolution (DE), Cuckoo Search (CS), and Basin Hopping (BH) is detailed in Sec. 19.3. The test-bed consists of 15 functions defined on two-dimensional intervals (see the Appendix) together with 16 functions defined on higher-dimensional intervals, all of which are considered especially hard for the sake of global optimization. The higher dimensional functions comprise the Lennard–Jones potentials of m particles for $3 \leq m \leq 10$, which depend on the resulting $\frac{1}{2}m(m-1)$ inter-particle distances (up to 45 variables) and the so-called d-dimensional Rastrigin functions for $3 \leq d \leq 10$. The results of the benchmark are discussed in Sec. 19.4. They show that GAS outperforms BH clearly, and CS and DE marginally in the case of only one run. More importantly, the computational edge increases significantly with the number of runs. We also take a closer look at the optimization of Lennard–Jones potentials because BH was specially developed to cope with these functions. Here again the performance of GAS scales nicely with the number of runs. The main messages of this chapter are summarized in the conclusion (Sec. 19.5).

Finally, all the materials referred to in this chapter are available online at the URL [20]. In particular, Python codes of the above algorithms as used in the computations reported here can be found in the file Solvers.ipynb.

19.2. The GAS Algorithm

Given a continuous map $f : \mathbb{R}^d \to \mathbb{R}$, a typical (single-objective) optimization problem has the form (without restriction)

$$\text{minimize}\, f(\mathbf{x})$$

subject, in general, to some constraints on $\mathbf{x} = (x^{(1)}, \ldots, x^{(d)})$. Thus, our objective is to find the position $\mathbf{x}^*$ of a global minimum of f on a finite *search domain* $\mathcal{D} \subset \mathbb{R}^d$, i.e.,

$$\mathbf{x}^* = \arg\min_{\mathbf{x}\in\mathcal{D}} f(\mathbf{x}),$$

where $\mathcal{D}$ might have a nontrivial topology due to hypothetical constraints among the variables $x^{(n)}$, $1 \leq n \leq d$. As a general topological requisite, we assume that $\mathcal{D}$ is compact (i.e., closed and bounded), which guarantees the existence of $\mathbf{x}^*$. In applications, $\mathcal{D}$ is usually an interval.

Roughly speaking, the GAS algorithm evolves a swarm of agents in pseudo-time according to some rules. These agents will be called hereafter *walkers* and they are characterized by both external and internal variables. Specifically, the state of a walker comprises the current space coordinates as well as the current value of its so-called *flow*, which is a proxy for its individual contribution to the swarm entropy. Following the basic idea of Tabu Search [7, 8], local minima found during the search will be flagged as "tabu" to avoid confinement in suboptimal regions. As a new ingredient, the diffusion of the walkers in the search domain is enhanced by a process called *cloning* in which the state of a given walker can be replaced by the state of another one.

Thus, the state of a walker i, $1 \leq i \leq N$, is given by its position $\mathbf{x}_i = (x_i^{(1)}, \ldots, x_i^{(d)}) \in \mathcal{D}$ and its flow $F_i > 0$, to be defined in Eq. (19.2) below. For notational convenience, we denote the walker i just by $\mathbf{x}_i$. Given a swarm of N walkers $\mathbf{x}_i$, set

$$f_{\min} = \min_{1\leq i\leq N} f(\mathbf{x}_i), \quad f_{\max} = \max_{1\leq i\leq N} f(\mathbf{x}_i),$$

and

$$\phi_i = \frac{f(\mathbf{x}_i) - f_{\min}}{f_{\max} - f_{\min}} \tag{19.1}$$

so that $0 \le \phi_i \le 1$ for $i = 1, \dots, N$, i.e., ϕ_i is a scaling of $f(\mathbf{x}_i)$ that mimic a physically plausible potential defined over the walker positions. For the distance between walkers, we use the Euclidean distance in $\mathbb{R}^d$:

$$\text{dist}(\mathbf{x}_i, \mathbf{x}_j)^2 = \sum_{n=1}^{d} ((x_i^{(n)})^2 - (x_j^{(n)})^2).$$

The following flowchart describes how GAS searches for a global minimum $\mathbf{x}^*$ of f; some steps will be commented.

Step 0. Initialization

0.1 The initial positions of the walkers, $\mathbf{x}_i$ $(1 \le i \le N)$, are chosen randomly in the search domain $\mathcal{D}$.

0.2 Compute

$$\mathbf{x}_{\min} = \arg \min_{1 \le i \le N} \phi(\mathbf{x}_i).$$

In the rare event of multiplicity, choose one of the minima randomly.

0.3 Use $\mathbf{x}_{\min}$ and the algorithm L-BFGS-B (see [4]) to find a local minimum $\mathbf{x}^*$.

0.4 Fill out the *tabu memory list* with $\mathbf{x}^*$, i.e., set

$$\mathbf{t}_r = \mathbf{x}^*, \quad 1 \le r \le N.$$

0.5 Set BEST$= \mathbf{x}^*$.

Comment. The search loop starts at this point.

Step 1. Walkers Flow & Cloning

1.1 For each walker $\mathbf{x}_i$ choose randomly another walker $\mathbf{x}_j$, $j \ne i$, and compute $d_{i,j}^2 = \text{dist}(\mathbf{x}_i, \mathbf{x}_j)^2$.

1.2 For each walker $\mathbf{x}_i$ choose randomly one tabu memory $\mathbf{t}_r$ and compute $\delta_{i,r}^2 = \text{dist}(\mathbf{x}_i, \mathbf{t}_r)^2$. If $\mathbf{t}_r = \mathbf{x}_i$, set $\delta_{i,r}^2 = 1$.

1.3 Compute the flow of $\mathbf{x}_i$,

$$F_i = (\phi_i + 1)^2 \cdot d_{i,j}^2 \cdot \delta_{i,r}^2, \tag{19.2}$$

for $i = 1, \dots, N$.

1.4 For each walker $\mathbf{x}_i$ choose randomly another one $\mathbf{x}_k$, $k \neq i$, and compute the *probability of* $\mathbf{x}_i$ *cloning* $\mathbf{x}_k$:

$$P_{i,k} = \begin{cases} 0 & \text{if } F_k > F_i, \\ \min\{1, (F_i - F_k)/F_i\} & \text{if } F_k \leq F_i. \end{cases} \tag{19.3}$$

Note that there is only one k for each $i = 1, \ldots, N$.

1.5 Take a random number $\rho \in [0, 1]$. If $\rho < P_{i,k}$ then the walker $\mathbf{x}_i$ copies the state of $\mathbf{x}_k$, i.e., $\mathbf{x}_i \leftarrow \mathbf{x}_k$ and $F_i \leftarrow F_k$.

Comments.

- To force the swarm to evolve as a cloud of gas that maximizes its entropy, we compare the walkers flows (proxis to their contributions to the swarm entropy) to decide whether to clone in Step 1.4. By increasing the flow, as the value function decreases, and the distance to other walkers and taboo states increases, we obtain the desired effect of entropy growth.
- Steps 1.3–1.5 promote the mobility of the walkers in such a way that walkers can jump to distant, lower points of the landscape.

Step 2. Local searches

2.1 Find the "center of mass" of the walkers as follows:

$$\mathbf{x}_{cm} = \sum_{i=1}^{N} \phi_i \mathbf{x}_i.$$

2.2 Compute

$$\mathbf{x}_{\min} = \arg \min_{1 \leq i \leq N} \phi(\mathbf{x}_i)$$

(in the first loop, $\mathbf{x}_{\min} = \mathbf{x}^*$). In the rare event of multiplicity, choose one of the minima randomly.

2.3 Use the algorithm L-BFGS-B with $\mathbf{x}_{cm}$ and the current swarm of walkers $\mathbf{x}_i$ to find a local minimum $\mathbf{t}$.

2.4 Use the algorithm L-BFGS-B with $\mathbf{x}_{\min}$ and the current swarm of walkers $\mathbf{x}_i$ to find another local minimum $\mathbf{t}'$.

2.5 Add $\mathbf{t}$ to the tabu memory list by overwriting a randomly chosen tabu memory $\mathbf{t}_i$, and execute then the external routine Memory Flow & Cloning (see below).

2.6 Add $\mathbf{t}'$ to the tabu memory list by overwriting a randomly chosen tabu memory $\mathbf{t}_j$, and execute then the external routine Memory Flow & Cloning (see below).

2.7 Compute the minimum tabu memory,

$$\mathbf{x}^* = \min\{\mathbf{t}_r : 1 \le r \le N\},$$

and update BEST$\leftarrow \mathbf{x}^*$.

Comment. Steps 2.5 and 2.6 do not change the number of tabu memories. These steps allow to find a global minimum which is close to a local one.

Halt criterion The usual criteria to exit a search loop include the stability of the BEST, a maximum number of loops or function reads, a maximum execution time, or possibly a mix of some of them. As stability criterion we suggest to compare the current BEST with its average over a certain number of the preceding loops; exit then the search loop at this point if the absolute value of the difference is smaller than the precision sought. In any case, output

$$\mathbf{x}^* = \text{BEST}$$

if the halt criterion is met; otherwise, continue.

Step 3. Position update of the walkers and loop end

3.1 Define the "jump" of $\mathbf{x}_i$ as

$$\Delta_i = 10^{-(5-4\phi_i)} \in [10^{-5}, 10^{-1}].$$

3.2 If $L^{(n)}$ is the length of the nth dimension of $\mathcal{D}$ $(1 \le n \le d)$, then update $\mathbf{x}_i$ as follows:

$$x_i^{(n)} \leftarrow x_i^{(n)} + L^{(n)}\xi^{(n)},$$

where each $\xi^{(n)}$ is a random variable drawn from the normal distribution $\mathcal{N}(0, \Delta_i)$.

3.3 If the resulting $\mathbf{x}_i$ does not belong to the search domain $\mathcal{D}$, repeat Step 3.2 with Δ_i replaced by $\Delta_i/2$ until $\mathbf{x}_i \in \mathcal{D}$.

3.4 Go to Step 1.

Comments.

- By using a greater perturbation on the walkers with a higher value of f, we simulate the heating of the less promising areas as a form of continuous simulated annealing. This analogy to simulated annealing was more explicit in earlier versions of GAS.
- The scope of Steps 3.1–3.2 is that walkers with a small ϕ_i move slower than walkers with a large ϕ_i, thus favoring the accumulation of walkers on the lower zones of the landscape.
- In the case of searching for a global *maximum* of f (i.e., a global minimum of $-f$), the jump of $\mathbf{x}_i$ in Step 3.1 is defined by

$$\Delta_i = 10^{-(1+4\phi_i)}.$$

External routine: Memory Flow & Cloning

R.1 For each tabu memory $\mathbf{t}_r$ choose randomly another tabu memory $\mathbf{t}_s$, $s \neq r$, and compute $\tilde{d}^2_{r,s} = \mathrm{dist}(\mathbf{t}_r, \mathbf{t}_s)^2$.

R.2 Compute the flow of $\mathbf{t}_r$,

$$\tilde{F}_r = (\tilde{\phi}_r + 1)^2 \cdot \tilde{d}^2_{r,s}, \tag{19.4}$$

for $r = 1, \ldots, N$, where

$$\tilde{\phi}_r = \frac{f(\mathbf{t}_r) - \tilde{f}_{\min}}{\tilde{f}_{\max} - \tilde{f}_{\min}}, \tag{19.5}$$

and

$$\tilde{f}_{\min} = \min_{1 \le r \le N} f(\mathbf{t}_r), \quad \tilde{f}_{\max} = \max_{1 \le r \le N} f(\mathbf{t}_r).$$

R.3 Compute the probability $\tilde{P}_{r,s}$ of cloning $\mathbf{t}_s$. To compute $\tilde{P}_{r,s}$ use formula (19.3) with F_i, F_k replaced by $\tilde{F}_r, \tilde{F}_s$, respectively.

R.4 Take a random number $\rho \in [0, 1]$. If $\rho < \tilde{P}_{r,s}$ then $\mathbf{t}_r$ copies the state of $\mathbf{t}_s$, i.e., $\mathbf{t}_r \leftarrow \mathbf{t}_s$ and $\tilde{F}_r \leftarrow \tilde{F}_s$.

Comment. Steps R.2–R.4 are meant to prevent static tabu memories via cloning.

19.3. Benchmark

In this section, the performance of the algorithm presented in Sec. 19.2 is compared against Basin Hopping (BH), Cuckoo Search (CS), and Differential Evolution (DE). The reason for selecting BH and DE is that they are standard algorithms of the Python libraries, what is a good indication of their widespread use. As for CS, it was shown in [22] to be potentially more efficient that Swarm Particle Optimization [12] and Genetic Algorithms. For the sake of this benchmark, we boosted even more the efficiency of CS by adding to its standard implementation a local search every 100 search loops. As mentioned in Sec. 19.1, the interested reader will find the Python codes used on the website [20].

The set of test functions is composed of 15 functions of two variables and 16 functions of more than two variables. Specifically, the two-dimensional test functions,

$$f_n(x_1, x_2), \quad 1 \leq n \leq 15, \tag{19.6}$$

are as follows: *Ackley* ($n = 1$), *Beale* ($n = 2$), *Booth* ($n = 3$), *Easom* ($n = 4$), *Eggholder* ($n = 5$), *Goldstein–Price* ($n = 6$), *Levy N. 13* ($n = 7$), *Matyas* ($n = 8$), *MacCormick* ($n = 9$), *Rastrigin 2D* ($n = 10$), *Rosenbrock* ($n = 11$), *Schaffer N. 2* ($n = 12$), *Schaffer N. 4* ($n = 13$), *Sphere* ($n = 14$), and *Three-hump-camel* ($n = 15$). See the Appendix for definitions, search domains, and global minima.

For the higher dimensional test functions we use the following:

(i) The eight *Lennard–Jones potentials* of m particles, $3 \leq m \leq 10$, at positions $\mathbf{r}_1, \ldots, \mathbf{r}_m \in \mathbb{R}^3$ in reduced units (i.e., such that both the depth of the pair potential well and the finite distance at which the pair potential vanishes are set equal to 1),

$$L_m(\mathbf{r}_1, \ldots, \mathbf{r}_m) = 4 \sum_{1 \leq i < j \leq m} \left(\left(\frac{1}{r_{i,j}} \right)^{12} - \left(\frac{1}{r_{i,j}} \right)^{6} \right), \tag{19.7}$$

where $r_{i,j} = \mathrm{dist}(\mathbf{r}_i, \mathbf{r}_j) = r_{j,i}$. The number of terms on the right hand side of (19.7) is $\frac{1}{2}m(m-1)$, i.e., the number of independent variables of L_m is $\frac{1}{2}m(m-1)$). The global minima of L_m for

$2 \leq m \leq 110$ can be found in Table I of Ref. [19]. Here $\mathcal{D} = [-1.1, 1.1]^{3m}$.

(ii) The eight d-dimensional *Rastrigin functions*,

$$R_d(\mathbf{x}) = 10d + \sum_{n=1}^{d} \left(x_n^2 - 10\cos 2\pi x_n\right), \quad 3 \leq d \leq 10,$$

with $\mathcal{D} = [-5.12, 5.12]^d$, $\mathbf{x}^* = (0, \ldots, 0)$, and $R_d(\mathbf{x}^*) = 0$. Needless to say, $f_{10}(x_1, x_2) = R_2(x_1, x_2)$.

All the above test functions belong to a collection of functions generally used for benchmarking in global optimization because of their challenging geometrical properties; visit the website [20] for graphical information. Furthermore, the Lennard–Jones potentials were included because BH was originally designed to cope with these functions and shown in [19] to improve the efficiency of genetic algorithms, the most successful global optimization method for Lennard–Jones clusters till then.

As for the computational specifics, the calculations were done on an Intel i7 with 8 cores, at 3.8 GHz, 16 GB of RAM, running Ubuntu 16.4. The halt criterion was an error (absolute value of the difference between the current BEST and the known global minimum) $\leq 10^{-6}$.

19.4. Results

Figure 19.1 summarizes the benchmark of GAS against BH, CS, and DE using the 31 multi-dimensional functions listed above as test functions; see the inlets and caption for the color and line style codification.

First of all, panel (a) shows the percentage of global minima found with each algorithm against the number of function reads. In each case, the code was executed only once to generate the plots. In view of panel (a), we conclude that GAS outperforms BH clearly and DE marginally, with CS performing somewhere in-between.

More importantly, this favorable performance of GAS improves significantly in the case of concurrent optimization. By these we mean that each algorithm is run several times, each time with different

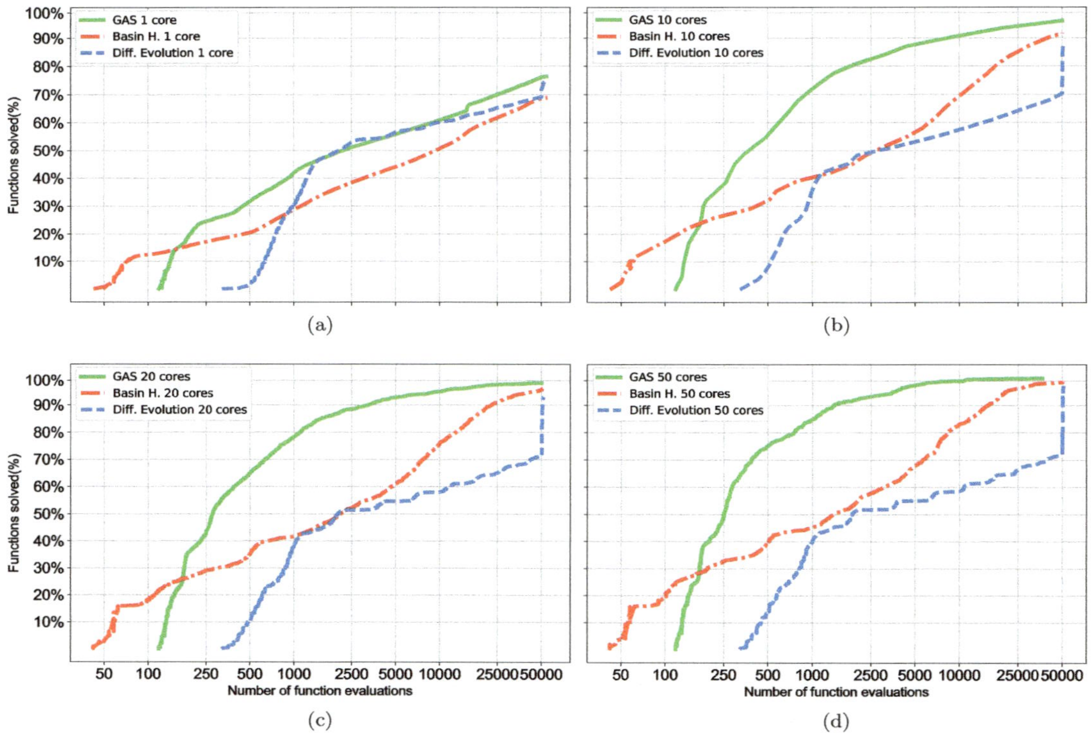

Fig. 19.1. To compare the performances of GAS (continuous thick line), BH (discontinuous segment-dot thick line), CS (continuous thin line with transversal segments) and DE (discontinuous segment-segment thick line) we plot the percentage of optimized test functions against the number of function reads when executing the algorithm one time (panel a). The same information is shown for concurrent optimization with 10 runs (panel b), 20 runs (panel c) and 50 runs (panel d).

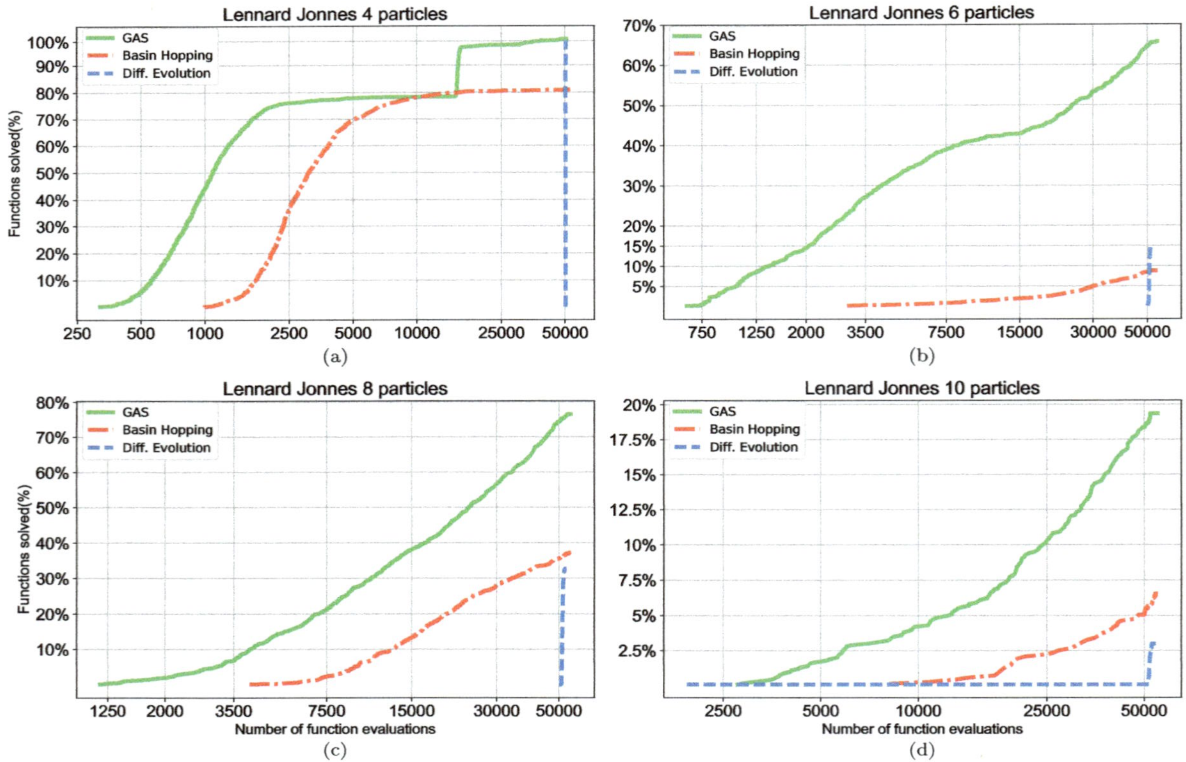

Fig. 19.2. Line style codification as in Fig. 19.1. Percentage of test functions optimized in one single run against the number of function reads for the Lennard–Jones potentials of 4 particles (panel a), 6 particles (panel b), 8 particles (panel c), and 10 particles (panel d).

initial settings (e.g., walker positions and random generator's seeds in the case of GAS), till a given number of function reads has been achieved. Panels (b)–(d) plot the highest percentage of minima found with $T = 10, 20, 50$ runs, respectively, vs the number of function reads. Alternatively, one can think that each algorithm is run simultaneously with T different initial settings. We see at panels (b)–(d) that the GAS performance scales nicely with the number of runs, while the CS and DE performances also improve but to a lesser extent than the GAS performance does, and the BH performance hardly changes.

As a complement to the above results, Figure 19.2 shows the performances of GAS, BH, CS and DE when optimizing the Lennard–Jones potentials (19.7) of $m = 4$, 6, 8, and 10 particles at randomly chosen positions (panels (a)–(d), respectively) with a single run. For the reason mentioned in Sec. 19.3, we expected BH to score highest in this particular benchmark but this was not the case. For $m = 4$, the best performer is CS, followed by GAS and BH. For $m = 6$, GAS overtakes CS while BH remains third; the performance gap between GAS and the other algorithms widens. Finally, for $m = 8, 10$ BH overtakes CS but it still trails GAS at a considerable distance; in both cases, the distance increases with the number of reads. DE fails badly in all cases. Let us finally remark that GAS is an unbiased algorithm, meaning that it does not exploit any specific insight into this particular optimization problem nor into any other problem, for that matter.

19.5. Conclusion

In this chapter, we presented the General Algorithmic Search (GAS), a physics-inspired swarm algorithm for the search of global extrema (Sec. 19.2). Among its features, we highlight the use of walkers with an internal label (called flow), randomization, and cloning (i.e., the replacement of a walker or a tabu memory by a copy of another one). Numerical simulations with a representative sample of test functions (Sec. 19.3) show that GAS compares favorably to Basin Hopping, Cuckoo Search, and Differential Evolution, the performance gap increasing with the number of runs in a concurrent optimization (Sec. 19.4). We may conclude, therefore, that GAS is a powerful tool for global optimization. The Python codes of the algorithms used in the present work as well as additional information are posted on the website [20].

Appendix: 2D Test Functions

Search domains ($\mathcal{D} \subset \mathbb{R}^2$), global minima ($\mathbf{x}^* \in \mathcal{D}$), and minimum values of the 2D test functions (19.6).

(1) *Ackley function:*

$$\begin{aligned} f_1(x_1, x_2) &= -20\exp(-0.2\sqrt{0.5(x_1^2 + x_2^2)}) \\ &\quad - \exp(0.5(\cos 2\pi x_1 + \cos 2\pi x_2)) + e + 20 \end{aligned}$$

with $\mathcal{D} = [-5, 5] \times [-5, 5]$, $\mathbf{x}^* = (0, 0)$, and $f_1(\mathbf{x}^*) = 0$.

(2) *Beale function:*

$$\begin{aligned} f_2(x_1, x_2) &= (1.5 - x_1 + x_1 x_2)^2 + (2.25 - x_1 + x_1 x_2^2)^2 \\ &\quad + (2.625 - x_1 + x_1 x_2^3)^2 \end{aligned}$$

with $\mathcal{D} = [-4.5, 4.5] \times [-4.5, 4.5]$, $\mathbf{x}^* = (3, 0.5)$, and $f_2(\mathbf{x}^*) = 0$.

(3) *Booth function:*

$$f_3(x_1, x_2) = (x_1 + 2x_2 - 7)^2 + (2x_1 + x_2 - 5)^2$$

with $\mathcal{D} = [-10, 10] \times [-10, 10]$, $\mathbf{x}^* = (1, 3)$, and $f_3(\mathbf{x}^*) = 0$.

(4) *Easom function:*

$$f_4(x_1, x_2) = -\frac{\cos x_1 \cos x_2}{\exp\left((x_1 - \pi)^2 + (x_2 - \pi)^2\right)}$$

with $\mathcal{D} = [-100, 100] \times [-100, 100]$, $\mathbf{x}^* = (\pi, \pi)$, and $f_4(\mathbf{x}^*) = -1$.

(5) *Eggholder function*:

$$f_5(x_1, x_2) = -(x_2 + 47)\sin\sqrt{\left|\frac{x_1}{2} + x_2 + 47\right|} - x_1 \sin\sqrt{|x_1 - (x_2 + 47)|}$$

with $\mathcal{D} = [-512, 512] \times [-512, 512]$, $\mathbf{x}^* = (512, 404.2319)$, and $f_5(\mathbf{x}^*) = -959.6407$.

(6) *Goldstein–Price function*:

$$\begin{aligned} f_6(x_1, x_2) = {} & (1 + (x_1 + x_2 + 1)^2(19 - 14x_1 + 3x_1^2 \\ & - 14x_2 + 6x_1x_2 + 3x_2^2)) \\ & \times (30 + (2x_1 - 3x_2)^2(18 - 32x_1 + 12x_1^2 + 48x_2 \\ & - 36x_1x_2 + 27x_2^2)) \end{aligned}$$

with $\mathcal{D} = [-2, 2] \times [-2, 2]$, $\mathbf{x}^* = (0, -1)$, and $f_6(\mathbf{x}^*) = 3$.

(7) *Levy function N. 13*:

$$f_7(x_1, x_2) = \sin^2 3\pi x_1 + (x_1 - 1)^2(1 + \sin^2 3\pi x_2) + (x_2 - 1)^2(1 + \sin^2 2\pi x_2)$$

with $\mathcal{D} = [-10, 10] \times [-10, 10]$, $\mathbf{x}^* = (1, 1)$, and $f_7(\mathbf{x}^*) = 0$.

(8) *Matyas function*:

$$f_8(x_1, x_2) = 0.26(x_1^2 + x_2^2) - 0.48x_1x_2$$

with $\mathcal{D} = [-10, 10] \times [-10, 10]$, $\mathbf{x}^* = (0, 0)$, and $f_8(\mathbf{x}^*) = 0$.

(9) *McCormick function*:

$$f_9(x_1, x_2) = \sin(x_1 + x_2) + (x_1 - x_2)^2 - 1.5x_1 + 2.5x_2 + 1$$

with $\mathcal{D} = [-1.5, 4] \times [-3, 4]$, $\mathbf{x}^* = (-0.54719, -1.54719)$, and $f_9(\mathbf{x}^*) = -1.9133$.

(10) *Rastrigin 2D function*:

$$f_{10}(x_1, x_2) = 20 + x_1^2 - 10\cos 2\pi x_1 + x_2^2 - 10\cos 2\pi x_2$$

with $\mathcal{D} = [-5.12, 5.12] \times [-5.12, 5.12]$, $\mathbf{x}^* = (0, 0)$, and $f_{10}(\mathbf{x}^*) = 0$.

(11) *Rosenbrock 2D function*:

$$f_{11}(x_1, x_2) = 100(x_2 - x_1^2)^2 + (x_1 - 1)^2$$

with $\mathcal{D} = \mathbb{R} \times \mathbb{R}$, $\mathbf{x}^* = (1, 1)$, and $f_{11}(\mathbf{x}^*) = 0$.

(12) *Schaffer function N. 2*:

$$f_{12}(x_1, x_2) = 0.5 + \frac{\sin^2(x_1^2 - x_2^2) - 0.5}{(1 + 0.001(x_1^2 + x_2^2))^2}$$

with $\mathcal{D} = [-100, 100] \times [-100, 100]$, $\mathbf{x}^* = (0, 0)$, and $f_{12}(\mathbf{x}^*) = 0$.

(13) *Schaffer function N. 4*:

$$f_{13}(x_1, x_2) = 0.5 + \frac{\cos^2\left(\sin\left|x_1^2 - x_2^2\right|\right) - 0.5}{(1 + 0.001(x_1^2 + x_2^2))^2}$$

with $\mathcal{D} = [-100, 100] \times [-100, 100]$, $\mathbf{x}^* = (0, 1.25313)$, and $f_{13}(\mathbf{x}^*) = 0.292579$.

(14) *Sphere function*:

$$f_{14}(x_1, x_2) = x_1^2 + x_2^2$$

with $\mathcal{D} = \mathbb{R} \times \mathbb{R}$, $\mathbf{x}^* = (0, 0)$, and $f_{14}(\mathbf{x}^*) = 0$.

(15) *Three-hump-camel function*:

$$f_{15}(x_1, x_2) = 2x_1^2 - 1.05x_1^4 + \frac{x_1^6}{6} + x_1x_2 + x_2^2$$

with $\mathcal{D} = [-5, 5] \times [-5, 5]$, $\mathbf{x}^* = (0, 0)$, and $f_{15}(\mathbf{x}^*) = 0$.

Acknowledgments

The authors were financially supported by the Spanish Ministry of Science and Innovation, grant PID2019-108654GB-I00.

References

[1] G. Beni and J. Wang, Swarm intelligence in cellular robotic systems, in *Proc. NATO Advanced Workshop on Robots and Biological Systems*, Tuscany, Italy, June 26–30 (1989).

[2] E. Bonabeau, G. Theraulaz and M. Dorigo, *Swarm Intelligence: From Natural to Artificial Systems* (Oxford University Press, Oxford, 1999).

[3] R. Bouffanais, *Design and Control of Swarm Dynamics* (Springer, Singapore, 2016).

[4] R.H. Byrd, P. Lu and J. Nocedal, A limited memory algorithm for bound constrained optimization, *SIAM J. Sci. Statist. Comput.* **16** (1995) 1190–1208.
[5] M. Dorigo and T. Stützle, *Ant Colony Optimization* (A Bradford Book, 2004).
[6] R. C. Eberhart, Y. Shi and J. Kennedy, *Swarm Intelligence* (Academic Press, San Diego, 2001).
[7] F. Glover, Tabu Search — Part 1, *ORSA J. Comput.* **1** (1989) 190–206.
[8] F. Glover, Tabu Search — Part 2, *ORSA J. Comput.* **2** (1990) 4–32.
[9] J. H. Holland, *Adaptation in Natural and Artificial Systems* (University of Michigan Press, Ann Arbor, 1975).
[10] X. Hu and R.C. Eberhart. Adaptive particle swarm optimization: detection and response to dynamic systems, in *Proc. Congr. Evolutionary Computation (CEC'02)*, Cat. No.02TH8600, Vol. 2, Honolulu (2002), pp. 1666–1670.
[11] A. Kaveh and S. Talatahari, A novel heuristic optimization method: Charged system search, *Acta Mechanica* **213** (2010) 267–289.
[12] J. Kennedy and R. Eberhart, Particle swarm optimization, in *Proc. IEEE Int. Conf. Neural Networks*, (1995) pp. 1942–1948.
[13] S. Kirkpatrick, C.D. Gelatt, M.P. Vecchi, M.P, Optimization by simulated annealing, *Science* **220**(4598) (1983) 671–680.
[14] H. Matsushita and T. Saito, Application of particle swarm optimization to parameter search in dynamical systems, *Nonlinear Theory Its Appl.* **2** (2011) 458–471.
[15] A. Slowik (ed.), *Swarm Intelligence Algorithms: A Tutorial* (CRC Press, Boca Raton, 2020).
[16] R. Storn and K. Price, Differential evolution — a simple and efficient heuristic for global optimization over continuous spaces, *J. Global Optim.* **11** (1997) 341–359.
[17] S. Voss, Meta-heuristics: the state of the art, in *Local Search for Planning and Scheduling*, ed. A. Nareyek, Lecture Notes in Artificial Intelligence, Vol. 2148 (Springer, Berlin, 2001), pp. 1–23.
[18] M. Waibel, D. Floreano and L. Keller, A quantitative test of Hamilton's rule for the evolution of altruism, *PLoS Biol.* **9** (2011) e1000615.
[19] D. J. Wales and J. P. K. Doye, Global optimization by basin-hopping and the lowest energy structures of Lennard Jones clusters containing up to 110 atoms, *J. Phys. Chem. A* **101** (1997) 5111–5116.
[20] https://github.com/HCsoft-RD/Optimization.
[21] X.-S. Yang and S. Deb. Cuckoo search via Lévy flights, in *World Congress on Nature & Biologically Inspired Computing (NaBIC 2009)* (IEEE Publications, 2009) pp. 210–214.

[22] X.-S. Yang and S. Deb, Engineering optimization by cuckoo search, *Int. J. Math. Modelling Num. Optim.* **1**(4) (2010) 330–343.

[23] X.-S. Yang, ed., *Nature-Inspired Computation and Swarm Intelligence: Algorithms, Theory and Applications* (Academic Press, San Diego, 2020).

[24] M. Ye, *Parameter Identification of Dynamical Systems Based on Improved Particle Swarm Optimization.* in eds. D.S. Huang, K. Li and G.W. Irwin, *Intelligent Control and Automation (ICIC 2006)*, *Lecture Notes in Control and Information Sciences*, Vol. 344 (Springer, Berlin, 2006) pp. 351–360.

[25] S. Hernandez and G. Duran, *Fractal AI: A fragile theory of intelligence.* arXiv:1803.05049 [CS] (2018).

Lightning Source UK Ltd.
Milton Keynes UK
UKHW021319290721
387966UK00001B/64